CHARACTERIZATION OF CERAMICS

CERAMICS AND GLASS: SCIENCE AND TECHNOLOGY

A Series of Monographs

Edited by

JOHN B. WACHTMAN, JR.
Inorganic Materials Division
National Bureau of Standards
Washington, D.C.

Volume 1
Radome Engineering Handbook: Design and Principles
edited by J. D. Walton, Jr.

Volume 2
Physics of Electronic Ceramics *(in two parts)*
edited by L. L. Hench and D. B. Dove

Volume 3
Characterization of Ceramics
edited by L. L. Hench and R. W. Gould

IN PREPARATION
Electrical Conductivity of Ceramics
edited by N. M. Tallan

Ceramic Processing
edited by P. J. Jorgensen

Mechanical Properties of Ceramics
edited by J. B. Wachtman, Jr.

CHARACTERIZATION OF CERAMICS

edited by L. L. Hench and R. W. Gould
Department of Metallurgical and Materials Engineering
College of Engineering
University of Florida
Gainesville, Florida

MARCEL DEKKER, INC. New York 1971

MARCEL DEKKER, INC.
95 Madison Avenue, New York, New York 10016

LIBRARY OF CONGRESS CATALOG CARD NUMBER: 74-160516

ISBN NO.: 0-8247-1302-8

PRINTED IN THE UNITED STATES OF AMERICA

CERAMICS AND GLASS: SCIENCE AND TECHNOLOGY

Series Editor:

J. B. WACHTMAN, JR.
Chief, Inorganic Materials Division,
National Bureau of Standards

FOREWORD

Ceramics and glass are ancient materials which continue to be used in the old ways but which have also been refined and developed by modern science and technology into many new and useful types of material. Ceramics and glass can be manufactured with unique chemical, thermal, electrical, magnetic, optical, mechanical, and mass transport properties. Their applications are much too long to list here in detail, but some examples will serve to illustrate the great diversity and importance of these applications. Special chemical and thermal properties are exploited in furnaces and crucibles which are essential to the refining, alloying, and casting of metals and to other types of high temperature processing. Special electrical properties form the basis for applications as insulators, dielectrics, piezoelectric transducers, and an ever-increasing family of semiconductor devices. Magnetic ceramics include cores for computer memories and high-coercive-force permanent magnets for electric motors. Special optical properties of some ceramics make them suitable for laser hosts, modulators, infrared windows, etc. The high elastic moduli and hardness of certain ceramics make them suitable for special applications requiring mechanical stability such as gyroscope mounts and wear surfaces while the extremely low thermal expansion achievable is appropriate to telescope mirrors and to applications involving thermal shock. The high intrinsic strength of many ceramics has never been fully exploited because of their brittleness and accompanying sensitivity to flaws, but great improvements in the strength of bulk commercial ceramics and the technique of surface compression combine to offer much promise in this area while the recent development of high-strength, high-temperature ceramic fibers opens new possibilities for composite materials. The high speed of sound in certain ceramics has led to their successful use in armor and the low damping characteristic of many ceramics makes them useful as delay lines. The great range of rates of mass transport achievable in ceramics makes them useful as protective coatings at the extreme of low diffusion and makes them promising as fuel cell components at the other extreme of high diffusion rates. The thermal stability of certain nuclear ceramics has led to their adoption as fuel elements for most power reactors.

Additional examples could be quoted to illustrate proven uses of ceramics in modern technology and the promise of other uses to come, but the continuing improvement of traditional ceramics with large volume use should also be noted. Glass, electrical, and sanitary whitewares, structural clay products, enamels, and chinaware are still benefiting from science and technology through improved properties and reliability.

Ceramic materials are of great interest to scientists quite apart from their many practical applications. Certain phenomena occurring in ceramics lie at the frontier of science and stimulate scientific work of the highest sophistication and quality. These phenomena include the detailed behavior of superconducting polycrystalline oxides which are semiconductors at room temperature. The voltage and time-dependent behavior of semiconducting glasses is not well understood, and the question of the structure of glasses and its relation to their properties remains the basis of an active and challenging field of science. The discovery of the widespread occurrence of phase separation in glass and the use of controlled nucleation and controlled crystallization in glass to produce crystalline ceramics has raised stimulating new scientific questions as well as enriching technology. The range of behavior from glass to crystalline ceramics is so rarely continuous that the term "ceramic" is today frequently used to include the whole range of these materials. The chemistry of crystalline ceramics continues to provide scientific challenges as well as the discovery of useful new materials. Layer structures, stacking-block structures, magnetic structures, the extent and character of solid solution, the type and interaction of point defects, and the structure of nonstoichiometric solids all need to be systematized and quantitatively understood. The chemistry of surfaces, including grain boundaries, is receiving increasing study. Mechanical properties have been extensively studied, but the details of mechanical behavior in specific materials offer many opportunities for research and development. The list of scientific challenges could be extended; ceramic materials are the seat of many phenomona worthy of study.

Some of the above applications are appropriate to a specific type of engineering and some of the scientific problems can be viewed simply as part of the solid state physics and chemistry. It is, however, usually advantageous to view both applications of ceramics and relevant aspects of science together from a materials science viewpoint in which the underlying chemistry, the kinetics of processing to produce a material which is often in a metastable state, and the physics of properties are all taken into account. Understanding the properties of ceramics depends upon understanding how microstructure and small changes of composition affect these properties. Experimental knowledge of the factors controlling properties depends critically upon measurement techniques, particularly upon techniques for measuring properties at high temperature and upon characterization techniques. The series "Ceramics and Glass: Science and Technology" is intended to present a survey of current knowledge of processing characterization, and physical properties of ceramics taking into account the underlying chemistry; it is made up of two parts: (1) textbooks, monographs, and handbooks and (2) selected conference proceedings. The first component in the series is made up of volumes each of which gives a complete treatment of its subject area rather than just a report of progress. Sufficient knowledge of the basic physics, chemistry, and mathematics is assumed to treat each subject on the level of a first-year graduate or advanced undergraduate course in materials science. This level is deliberately chosen to make the textbooks, monographs, and handbooks broadly useful to scientists and engineers in industry and in research and development laboratories as well as to students. The second component of the series is made up of conference proceedings selected for their subject, quality, and timeliness. The growth of technical society meetings in size and diversity of subject matter has caused increasing emphasis to be placed on small, separate topical conferences. Such conferences typically cover the selected topic with a series of invited papers plus a limited number of contributed papers. When successful, the proceedings of such a conference give a state-of-the-art presentation useful for several years until the new information can be assimilated in an organized manner into textbooks and handbooks. Good conferences and their proceedings have an important intermediate place between individual research papers and textbooks. The series "Ceramics and Glass: Science and Technology" accordingly includes selected conference proceedings but recognizes and maintains the distinction between these and the textbook, monograph, and handbook series component.

The present volume, "Characterization of Ceramics" edited by L. L. Hench and R. W. Gould, surveys methods for measuring the chemical and structural composition of ceramics. These methods provide the experimental foundation on which the whole technology of ceramic processing ultimately rests. Processing depends not only on bulk composition but also on chemical additives present only in trace amounts; accordingly, methods for accurate chemical analysis of both major components and trace components are summarized. The process of sintering depends on the character of the powder used as starting material. Powder characterization is one of the most difficult experimental areas of materials science. A wide variety of complex techniques exist, so that a careful choice must be made to obtain adequate characterization for the purpose at hand without requiring an excessively expensive and lengthy effort. The present book systematically surveys these techniques and the information which they can provide. Many properties of a finished ceramic depend on microstructure so that techniques for characterization of microstructure are important both to the technology of ceramic production and to the scientific interpretation of physical property studies. Accordingly, the techniques for qualitative and quantitative determinations of microstructure are surveyed. Finally, the character of the surface of a ceramic is quite important to such properties as mechanical strength and adhesion of coatings, so that modern techniques for surface characterization are surveyed. The present book thus provides a single-source reference to the whole spectrum of characterization techniques relevant to ceramic processing and use. Many of these techniques are also relevant to characterization of inorganic materials in general including minerals, ores, aggregates, and metals. It is hoped that this book will be broadly useful to workers concerned with these materials as well as to those concerned with ceramics.

John B. Wachtman, Jr.
Chief, Inorganic Materials Division
National Bureau of Standards

PREFACE

The ceramist has been concerned with characterization of his products for many centuries. By gaining a subjective sense of the reproducibility of his raw materials, he in turn achieved a measure of control over the uniformity of his final product. However, the advent of mass production made quantitative descriptions of the chemical and mineralogical compositions of raw materials essential for rapid production schedules to be maintained. As specialized technical ceramics appeared on the production line, careful control of properties of the final product became paramount if costs were to be minimized and sales insured. However, identifying the source of property variations within the complex processing sequence was often a frustrating detective game, and may still be for many products. In addition, new customer requirements demand frequent product property alterations which in turn necessitate time-consuming and expensive "guess and try research."

It is in response to the dual motivations of maintaining production controls and producing new products that the characterization of ceramics has become an important element in nearly every ceramic organization's total activity. However, the consideration of characterization as a separate discipline requiring specialized training and programs is a recent concept. Such a development is a result of two major driving forces. First, physical ceramics research has established that the properties of a final product may be controlled by a variety of independent factors. These factors comprise the five major sections of this book. The second driving force toward characterization sophistication has been the rapid development of a large number of high resolution analytical techniques appropriate for ceramic materials. A separate literature for each technique has appeared within the last decade or so. Consequently, the design of an appropriate ceramics characterization program has become an ever more difficult job. This volume attempts to present in a single source a review of each of the major instrumental analytical techniques available for ceramic characterization. A discussion of limitations encountered in applying the methods to ceramics is also presented.

The book is designed for the ceramics or materials engineer who is concerned with designing a characterization program, evaluating characterization results, or attempting to understand processing-property relations. It should also be an especially timely textbook for a materials science curriculum. The level of treatment of the topics should be appropriate for a senior-level course. Sufficient references have been provided so that each topic can be readily pursued into the specialized literature as required by the course or problem at hand.

L. L. Hench
R. W. Gould

CONTRIBUTORS TO THIS VOLUME

S. R. BATES, Department of Metallurgical and Materials Engineering, College of Engineering, University of Florida, Gainesville, Florida

L. BERRIN, Bell Telephone Laboratories, Inc., Allentown, Pennsylvania

JOHN H. BOUGHTON, 602 Marsh Road, Wilmington, Delaware

R. T. DeHOFF, Department of Metallurgical and Materials Engineering, College of Engineering, University of Florida, Gainesville, Florida

V. D. FRÉCHETTE, State University of New York, College of Ceramics at Alfred University, Alfred, New York

S. W. FREIMAN,* IIT Research Institute, Chicago, Illinois

R. W. GOULD, Department of Metallurgical and Materials Engineering, College of Engineering, University of Florida, Gainesville, Florida

L. L. HENCH, Department of Metallurgical and Materials Engineering, College of Engineering, University of Florida, Gainesville, Florida

WILLIAM J. JAMES, Departments of Chemistry and Ceramic Engineering, Graduate Center for Materials Research, University of Missouri-Rolla, Rolla, Missouri

DAVID R. LANKARD, Columbus Laboratories, Battelle Memorial Institute, Columbus, Ohio

GORDON LEWIS, Departments of Chemistry and Ceramic Engineering, Graduate Center for Materials Research, University of Missouri-Rolla, Rolla, Missouri

R. W. NEWMAN, Florida Institute of Technology, Melbourne, Florida

DALE E. NIESZ, Columbus Laboratories, Battelle Memorial Institute, Columbus, Ohio

*Present address: Naval Research Laboratory, Washington, D. C.

PAT RANKIN, Department of Chemistry, University of Florida, Gainesville, Florida

R. C. SUNDAHL, Bell Telephone Laboratories, Inc., Allentown, Pennsylvania

CHARLES F. TUFTS, The Bayside Research Center of General Telephone and Electronics Laboratories, Inc., Bayside, New York

ROBERT K. WARE, Owens-Illinois, Toledo, Ohio

J. D. WINEFORDNER, Department of Chemistry, University of Florida, Gainesville, Florida

CONTENTS

PART 3 - SIZE, SHAPE, STRAIN, AND SURFACE OF POWDERS

CHARACTERIZATION OF CERAMICS

Chapter 1

INTRODUCTION TO THE CHARACTERIZATION OF CERAMICS

L. L. Hench

University of Florida
Gainesville, Florida

Characterization has been defined by the Materials Advisory Board of the National Research Council[1] as:

> "Characterization describes those features of the composition and structure (including defects) of a material that are significant for a particular preparation, study of properties, or use, and suffice for the reproduction of the material."

In order to characterize a ceramic material, therefore, it is necessary to evaluate its composition and structure sufficiently that the material and its properties can be reproduced. As discussed by both the Materials Advisory Board's panels on Characterization[1] and Ceramic Processing[2], quantitative characterization is the very foundation upon which the science of ceramics must rest. It is the objective of this book to provide a synthesis of the stages of quantitative characterization that are possible, a review of the experimental techniques available for characterization, and a source of literature from which to pursue detailed analytical methods.

There are various stages of characterization of ceramics which correspond to the sequence of processing steps that a ceramic product undergoes. Figure 1 illustrates a general flow chart of ceramic processing and the output resulting from each processing step. In some cases a final product is achieved after a few processing steps. In other cases, certain processing steps may be skipped. Some ceramic ware may require every stage of refinement indicated.

In every case, the properties and end use behavior of the ceramic

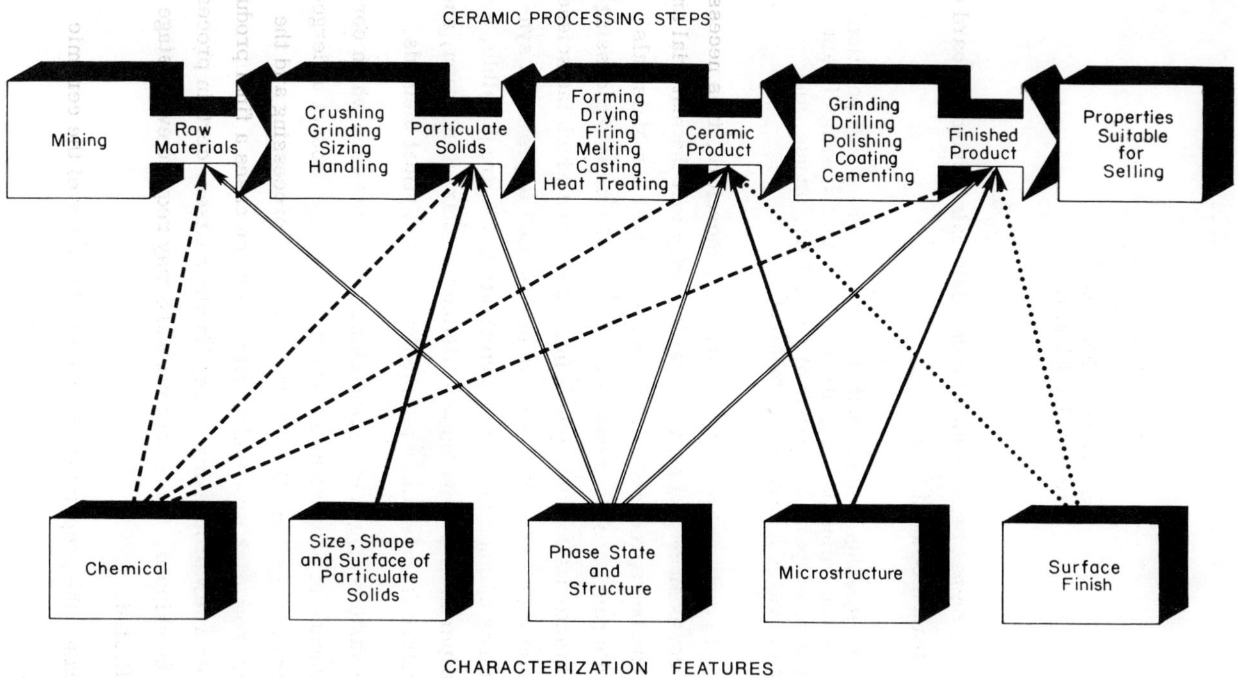

Fig. 1. The relationship between characterization features and products of the ceramics processing sequence.

product will be dependent on each of the processing steps used in its manufacture. Therefore, a number of alternative opportunities for characterization are present throughout the processing sequence.

Figure 1 illustrates five major classes of characterization. The arrows relate the characterization features to the appropriate product in the ceramic processing sequence. It is important to emphasize two points here. First, each characterization feature indicated can be quantitatively evaluated. Second, characterization is directed towards the output of the processing steps and not to the processing method itself. Of course, the same analytical techniques may often be used to investigate processing parameters but it is not materials characterization. This is an important distinction since different processing steps can lead to similar properties. Therefore, in order to establish the physical origin of properties it is necessary to determine the composition and structure of the product and not just the specifics of the process used in the product's manufacture.

With many characterization paths possible, establishing which characterization areas to evaluate is perhaps the most critical, and most difficult, step in a characterization program. For a commercial product, the single criteria for establishing a characterization program is: select those methods which provide the property control required for application and sales at least cost. Characterization costs are additive to production costs, and, therefore, must be economically justified in terms of increased sales, less waste, etc.

Commercial characterization methods may often be different from the characterization steps required to understand the properties of new products obtained from a research and development effort. The economics of the R & D characterization are relatively unimportant as long as improvement in product properties, reduction in production costs, or new products warrant the expense eventually. However, the necessity for long range economic justification must be recognized as part of the criteria for selecting R & D characterization methods.

Because of economic factors, it is usually prudent that the least expensive characterization methods be examined first. For example, if the combination of a four element semiquantitative chemical analysis and a quantitative petrographic phase examination is sufficient to characterize the end-use behavior of a refractory, then more elaborate tests are of little merit. However, if the tests are insufficient to predict behavior in a new application of the refractory, the characterization plan must be modified. The analytical method necessary to predict the more specialized behavior of the refractory may be considerably more costly and must be considered in establishing prices and sales projections.

Any characterization measurement that is not necessary to predict processing response or end-use behavior should be avoided.

Data taking alone is not valid characterization regardless of how sophisticated the technique may be — correlation of the data with properties and behavior is essential for characterization to be meaningful. Therefore, development of a valid characterization program should first be considered as a research effort even when a production item is involved. The research results should establish that each characterization method serves a particular function in predicting the properties of the material. If this is accomplished, and the properties selected are correctly related to sales or cost, a characterization plan will be optimized.

The economic aspects of both production characterization and R & D characterization provide much of the interest in instrumental analytical techniques. Instrumental methods can often provide rapid, reproducible, and relatively automated means of characterization. Consequently, instrumental methods may be applied to on-line characterization or require only semiskilled operators, reducing the analytical costs.

Another important feature of instrumental techniques is that they often can provide new capabilities for characterization. Limits of compositional detection or degree of resolution of a structural parameter may often be extended by employing new instrumental methods. Several chapters in this volume, notably those on x-ray and electron microscopy techniques, emphasize this point.

For the above reasons this volume will concentrate primarily upon instrumental methods of characterization. The methods can be subdivided into five areas shown in Fig. 1, and are presented in this manner in the book. The ordering of the chapters in this way closely approximates the order in which characterization steps will be performed during the processing of ceramic products.

The reader who is interested in additional discussions of the nature and philosophy of materials characterization, a commentary on industrial needs, or a variety of case histories is referred to the Materials Advisory Board reports[1,2] and the study of E. R. Stover[3].

REFERENCES

(1) "Characterization of Materials," The Materials Advisory Board Publication MAB-229-M, National Academy of Sciences-National Academy of Engineering, Washington, D.C., March, 1967.

(2) "Ceramic Processing," Materials Advisory Board Division of Engineering National Research Council, Publication 1576, National Academy of Sciences, Washington, D.C., 1968.

(3) E. R. Stover, "A Critical Survey on Characterization of Particulate Ceramic Raw Materials," Technical Report AFML-TR-67-56, May, 1967.

PART 1

Chemical Analysis

Chapter 2

GENERAL ANALYTICAL CHEMISTRY

Pat Rankin

Department of Chemistry
University of Florida
Gainesville, Florida

I. INTRODUCTION

Analytical chemistry encompasses an ever broadening spectrum of tools and techniques. Even to chemists the mass of varied instrumental techniques is sometimes overwhelming. As a subdivision of chemistry it overlaps inorganic, organic, biological and physical chemistry, physics, electronics, materials and other branches of engineering.

Physics has provided us with principles of spectra upon which we base many modern instrumental techniques. Theories concerning physical properties such as molecular vibrations, conditions of excitation, and transition possibilities help explain absorption and emmission spectra. It is not surprising that in recent years, over 50% of the experimental work published in the journal, Analytic Chemistry, was done by people other than analytical chemists! Recent technological innovation and advancement have increasingly required the use of analytics, thus drawing it into the forefront of research in many fields. This fact has erased former boundaries of the discipline, and, unfortunately, complete categorical breakdown of techniques is nearly impossible due to their variety. Classification often depends upon the point of view. For example, chemists concerned with trace analysis might divide methods into gas, solution and solid sample types, while others may be concerned with the division by mechanical, electrical, thermal, or nuclear properties. Circuitry, instrumental makeup, and speed of analysis are other possible bases for classification. From a materials standpoint, we are interested in becoming familiar with a broad range of techniques so as to gain a simple orientation in the field of analytical chemistry. In this chapter, we will discuss the nature and advantages of several general analytic catagories and a few specific techniques.

The basic purpose of quantitative analysis is to find the relative amount of a constituent in a sample by finding a numerical value of some property of the sample.

There are, however, two major classes of determination methods. The first involves the use of chemical techniques.

These methods rely directly upon chemical reactions. The two main groups in this class are gravimetric* and volumetric† analysis. Almost all remaining methods fall under the second major class called physicochemical methods. They measure mass-dependent properties and not the mass itself. In some of these methods, chemical reactions play a part. Physical methods are those in which reactions play no direct part. Almost all physicochemical and physical methods are instrumental and will be the major concern of the chapter. This class primarily relies upon relations between energy and mass for its analytical methods. For example, in spectrophotometry, the amount of radiant energy absorbed is a function of the mass of the absorbing species (6,11)

*Separation of object from other substituents in a sample and weighing as a compound of definite composition.

†Finding the amount of reagent required to react with a certain constituent thus yielding its weight.

The remainder of this chapter concerns the five major methods used in analytical chemistry. Generally the divisions are based upon the types of physical or chemical properties involved. The categories include: separation, optical, electrochemical, thermal, and nuclear radiation methods. Procedures for choosing appropriate analytical techniques are discussed at the end of the chapter.

II. SEPARATION METHODS

Analysts are always hopeful that direct analysis of a sample can be made, but pretreatment is usually necessary. This happens most often when there are one or more substances present in the sample which might interfere with the determination in question. Today separation is probably the most useful and versatile form of pretreatment. Although chemical pretreatment is usually preferable because of its speed, physical separation techniques are usually necessary when the sample consists of a mixture of substances having only slightly differing physical and chemical properties. Although most separation techniques prepare a sample for quantitative determination, some serve as analytical techniques themselves.

Similar to analytical chemistry as a whole, classification of separation techniques is difficult. Methods of classification overlap and cannot be tabled satisfactorily, therefore, several methods will be listed. Two general types of heterogeneous separations are: (1) batch and (2) continuous. They refer to the way a sample is analyzed and represent one type of classification. Another classification type divides into categories the forces used to separate components. These are mechanical, physical, and chemical. These groups overlap (all separation requires some mechanical operation) but the major function involved in a separation generally decides the appropriate category for the method. Several methods familiar to most students are difficult to classify because the force of separation is very complex (see Table I). Table II lists physical separation processes involving heterogeneous equilibria [1, 12].

The two primary goals in separation are recovery and purity. It is important to recover as close to 100% of the material separated at the greatest possible purity so later analysis can be as accurate as possible. Efficiency and capacity are two other important aspects when considering the goals of

TABLE I

Classification of Typical Separation Processes on the Basis of Principal Separative Force[a]

Mechanical	Solid phases; sieving; ultrafiltration	
Physical	Miscible phases	Magnetic and centrifugal methods Thermal diffusion Electromigration
	Miscible phases separated by permeable or semi-permeable membrane	Dialysis; ultrafiltration; sieving Gaseous diffusion
	Immiscible phases	Bulk equilibria, liquid-liquid extraction; electrodeposition Surface equilibria, adsorption
Chemical	Absorption (CO_2 in NaOH) Precipitation ($BaSO_4$) Electrodeposition, i.e., ($Cd^{+2} \rightleftharpoons Cd^{\circ}$)	

[a]From Ref. (12).

separation. As a sample becomes more difficult to separate, the amount of work necessary for a certain degree of separation accuracy and purity becomes increasingly important. The capacity of a technique is the size sample the technique will separate along with the rate at which the sample is separated. If a large sample must be separated and analyzed in gas chromotography (sample capacity < 0.1 ml) a number of runs might have to be made, combining results. When choosing a separation technique for a particular purpose the mentioned factors play an important role[(3)].

There are many methods which can be considered separatory, but we shall touch upon the highlights of some of the major areas: mechanical, diffusion, electrophoresis, distillation, liquid-liquid extraction, precipitation and crystallization, ion exchange, and chromotography.

A. Mechanical

Three general groups divide this method:

(1). Exclusion by size. Filtration by means of a selectively porous material, where the concern is liquid-solid separation or liquid purification before analysis.

(2). Exclusion by size, mass, or density. Sedimentation, centrifugation, impaction, and impingement are useful here, (the latter two separate

TABLE II

Typical Separative Processes Involving Heterogeneous Equilibria[a]

	Gas	Liquid	Solid
Bulk phases			
Liquid	Absorption Condensation Distillation Gas-liquid chromatography	Electrodeposition (metals soluble in Hg)[c] Solvent extraction	Crystallization[d] Ion exchange (resin) Ion exclusion (?) Precipitation[c,d] Solvent extraction Zone refining[d] (fractional melting and freezing) Enfleurage[d]
Solid	Sublimation		
Interfaces (sorption)			
Liquid	Foam chromatography	Emulsion chromatography	
Solid	Adsorption (Hypersorption) (Gas-solid chromatography)	Adsorption Electrodeposition (fractional monolayers on "inert" solid electrodes Ion exchange on minerals (?) Zone refining (?)	

[a]Table taken from Kolthoff (12).

[b]Most forms of "chromatography" have been omitted because they are primarily operational terms.

[c]Considered by some to be "chemical"

[d]True equilibrium with entire bulk phase may be difficult to establish.

solids from gases and liquids). Flotation, decantation, screening, and seiving are familiar terms which fall under this category.

(3). Exclusion by other physical properties. This group includes magnetic, sonic, thermal, and electrical separations.

B. Diffusion

Separations that are difficult or even impossible by conventional techniques are possible here. The process is one in which mixture separations are possible through varying experimental conditions and different diffusion rates in various molecular types. The main methods are:

(1). Thermal diffusion. A temperature gradient is established and molecular types move toward or away from the heat. (Physical state may be gaseous, solid, or liquid.)

(2). Barrier diffusion. A liquid or gas phase flows through an irreversible barrier with pore dimensions smaller than the average mean free path of the molecules in the mixture.

C. Electrophoresis

When charged colloid particles and large ions under an electric field migrate to an appropriate cathode or anode, electrophoresis has occurred. Separation and analysis is based upon the migration velocity differences of various substances.

D. Distillation

This method is excellent for identifying the components of a relatively complex multicomponent liquid sample. It is based upon the differences in the boiling point of various substances. Fractional and sample distillations are the two general distillation categories.

E. Liquid; Liquid Extraction

If a liquid substance soluble in two immiscible liquids is combined with the solvents, a definite proportion of solute concentration in one solvent to the other solvent develops. When removal of the more concentrated portion occurs, and fresh solvent is returned, solute leaves the now more concentrated solvent to reestablish the definite proportion. Extractions are continued in this manner until the needed accuracy is reached. The two immiscible solvents are usually organic and aqueous combinations.

F. Precipitation and Crystallization

These methods, for the separation and purification of compounds are fairly self explanatory. Precipitation is the solid deposition from solution caused by a chemical reagent, heat, etc. Crystallization is the formation of crystals during cooling from a fusion state. The main property that makes these techniques effective in separation is that crystals of similar ions or molecules grow by addition.

G. Ion Exchange

This method deals with the exchange of ionized substances to establish quantitative separations. Like chromatography, a liquid sample percolates through a granular solid contained in a glass tube, but separation is based upon principles different from chromatography. Two general procedures exist: cation and anion exchange. The process involves the exchange of ions of like sign between a solid and a solution, where the solid is the ion exchanger. This ion exchange is usually a synthetic resin.

H. Chromatography

Covering a number of related techniques, chromatography separates similar substances by differential migration. This migration takes place in a narrow zone of porous media, being produced by electrical potential or by flow of liquid or gas. This flow is a driving force on the sample and acts selectively on the solutes so they will migrate at different rates. Thus as the solvent flow or potential gradiant drives particles uniformly, different migration rates occur due to resistive forces. These resistive forces can be properties such as viscosity, permeability, and adsorption, which are the most common and selective. Chromatography is divided into two general categories: Those methods with a liquid mobile phase and those with a solid mobile phase. Principal types are column, thin layer, paper, and gas chromatography.

Column and paper chromatography provide very good separation. With the help of dyes and developing agents, fairly good quantitative results can be established. The relative volume retention of an unknown substance provides strong qualitative information in chromatography. Chromatographic peaks in gas chromatography provide quantitative results.

III. OPTICAL METHODS

Optical methods of analysis comprise a large division of analytical chemistry. Primarily instrumental in nature, these methods are concerned with the emission and absorption of radiant energy by matter (see Fig. 1).

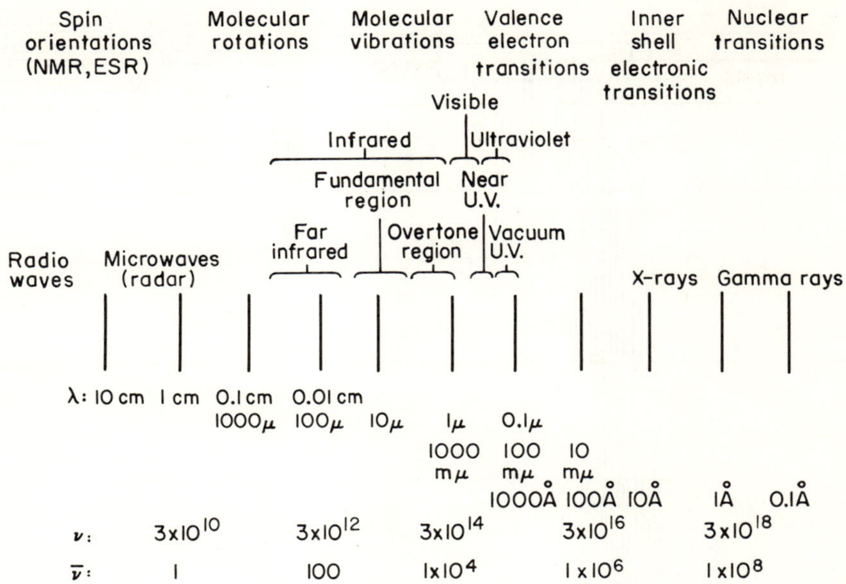

Fig. 1. The electromagnetic spectrum. After Reilley and Sawyer (17).

To become familiar with the nature of this very important category, it is necessary to recall some of the physical energy principles of the atom and molecule.

Atomic energy levels (Fig. 2) designate the levels between which electronic transition are permitted. The undisturbed (normal) atom orbitals are occupied as needed, according to the quantum rules. When sufficient energy is provided, an electron in an outer level (least held) can easily be excited to the next highest level. When this electron returns to its former energy site (which it has strong tendency to do), a quantum of radiation, or photon is emitted. The spacing of the energy levels designates the discrete amount of energy the photon will have.

The process of an outer electron being raised to one energy level, then returning is resonance radiation. If enough energy is supplied, some of the inner as well as outer electrons can be raised to a higher orbital level. Upon returning to the former levels the electrons pause at intermediate levels almost like a ball bouncing down steps. Thus an atom may be excited to varying degrees, yielding as many as several thousand discrete and reproducible wavelengths. This emission on an atomic level is the basis of emission

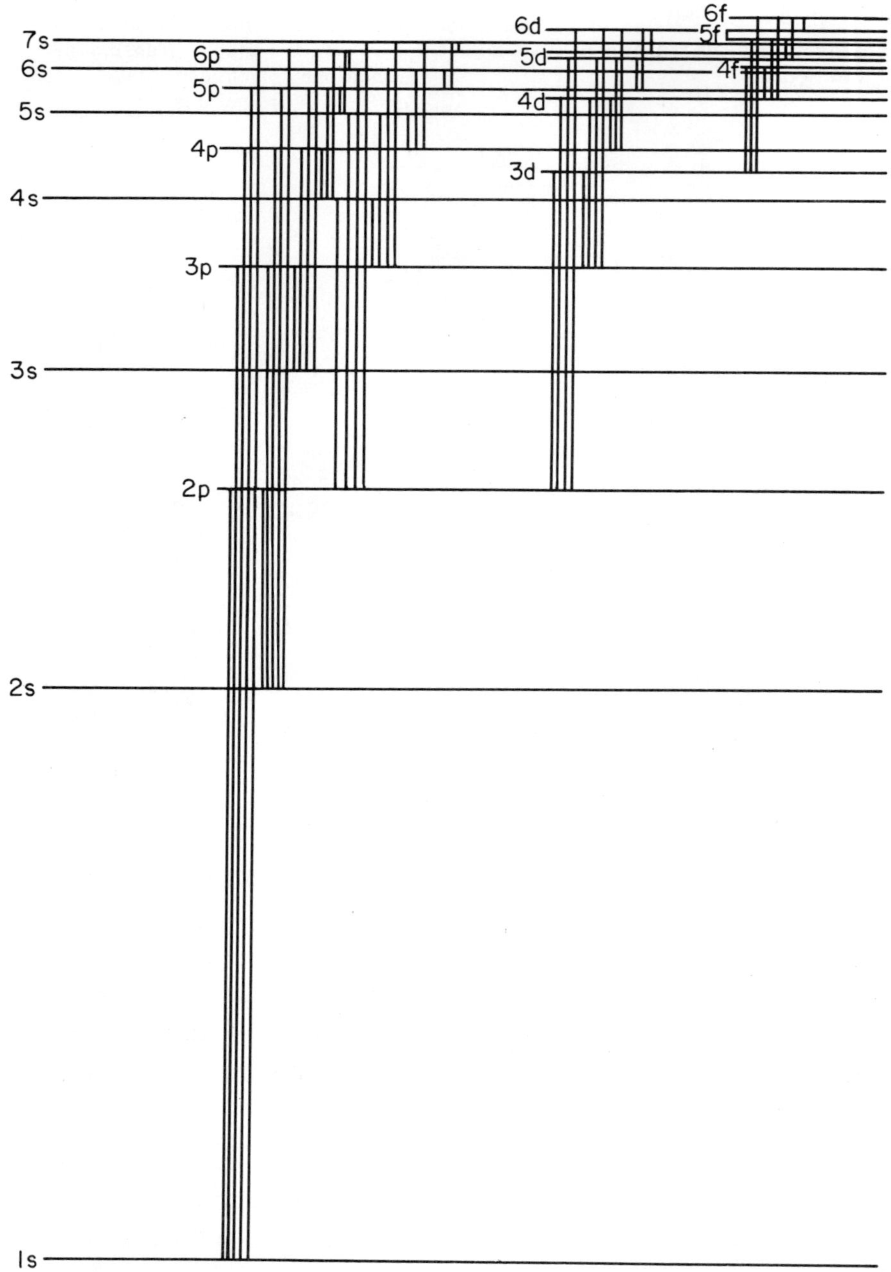

Fig. 2. Schematic representation of atomic energy levels. After Ewing (10).

spectroscopy. When certain atoms are bombarded by a beam of fast electrons or energetic x rays, emission of x rays occur. Here an inner electron in the atom is excited. A higher level electron then drops into its place. Thus the photons emitted will have a shorter wavelength and a much greater frequency than optical emmission spectra.

On a molecular scale (Fig. 3), there exist major energy levels called electronic levels. They are each divided into a number of vibrational levels which correspond to variations of the whole molecules' energy in relation to interatomic vibrations. The vibrational levels are each further subdivided into rotational levels which correlate with the rotation of atoms within the molecule. When dealing with the molecular levels, absorption of radiation passing through the levels becomes important, because energy transitions are not usually observed as emission spectra. Electronic level transitions are recorded in the visible and ultraviolet regions. Between the vibrational levels, transitions occur in the near infrared while the far infrared and microwave regions show rotational level transitions.

Various substances differ in the details of their energy-level patterns, yielding varied absorption spectra throughout the visible, ultraviolet, infrared, and microwave regions. These differences provide the basis for analytic absorption techniques within each range of wavelengths. Variations of principles along with additional optical properties of matter help comprise a great number of other optical methods. A few examples are given[10,2,3].

A. Dispersion

The variation of the refractive index of a substance with wavelength is its dispersion. The dispersion of a material throughout the spectrum is closely related to the degree to which radiation is absorbed. In high absorbance regions, the index rises abruptly with increasing wavelength while it decreases with increasing wavelength in areas of high transparency[10].

B. Polarimetry

The ability of some crystals and solutions to rotate the plane of polarized light passing through them is called optical activity. The optical activity varies with wavelength and this is known as rotary dispersion. The list of optical methods is very large and we will discuss the applicability of only the most important ones[10].

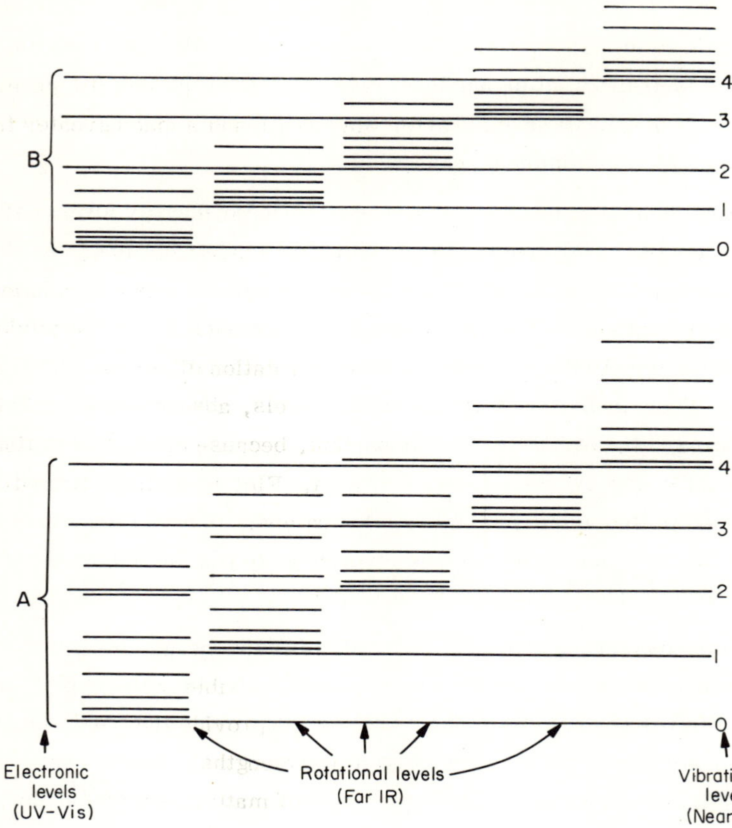

Fig. 3(a). Schematic molecular energy levels. After Ewing (10).

C. Absorption in the Visible and Ultraviolet Regions

This method is nondestructive, rapid and measurements are fairly simple for the novice. Complete separation is not usually required before analysis. The structural information it yields is highly specific and it carries a high sensitivity (detects to 10^{-8} moles/liter). Samples used should be homogeneous. Note that many substances do not absorb in the accessible region.

Useful application lies in quantitative determination of many organic and inorganic substances. Qualitatively it can give clues in determining organic structures[1,10,2,3,7].

D. Absorption in the Infrared Region

This is a nondestructive, fairly rapid method applicable to all phases. Most substances have an infrared spectrum which usually yields extensive

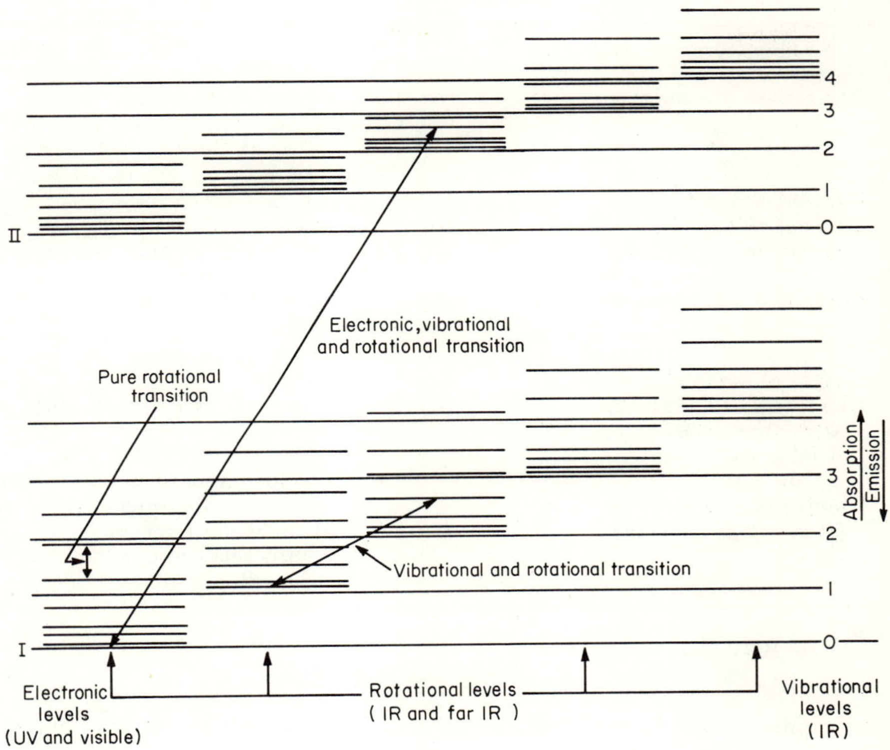

Fig. 3(b). Vibrational and rotational energy transitions. After Reilley and Sawyer (17).

information. Instruments are comparatively easily operated. Mixtures are difficult to work with because of the large number of absorption bands created. Homonuclear diatomic molecules and single atoms will not absorb in the infrared. All samples must be homogeneous.

This method is good for identification of complex molecules. Information about the number of atomic groups and their linkage in an unknown molecule can be derived. Quantitative analysis is not very good for mixtures[1,2,3,7,10].

E. Spectropolarimetry

The new instrumental methods available are superior to previous methods in that they yield more information in less time from a smaller sample of optically active material. The method is good for the determination of absolute configurations and stereochemistry of molecules[10].

TABLE III

A Summary of Electrometric Methods of Analysis[a]

Quantity measured	Name or description of the method
Resistance (conductance)	(A) Conductometric titration (B) Concentrations measured by resistance after calibration with known mixtures
Current-voltage	Polarography
Current vs concentration of component of interest	(A) Amperometric titration (B) Amperometry at small constant applied potential
Coulombs (current x time)	(A) Coulometry at constant electrode potential (B) Coulometric titration
Weight of separated phase	Electrogravimetry
Electromotive force vs concentration of component of interest or Electromotive force vs time (at constant current)	(A) At zero current drawn from the electrodes (1) Potentiometric titration (2) Measurement of ionic activities or concentrations; e.g., pH (B) At small constant current (1) Potentiometry at constant current (2) Chronopotentiometry

[a]From Ref. (1).

F. X Ray Fluorescence

This method involves the excitation of samples by high energy x rays producing fluorescent radiation whose wavelength and intensity provide a basis for quantitative measurements[10].

IV. ELECTROCHEMICAL METHODS

Electrochemical methods are characterized by variations in potential and current which produce an electrical response dependent upon the specific nature of the solution. The field is governed by a number of fundamental principles based on the various phenomena occurring within an electrochemical cell. Specific expansion of these principles provide for the great variety of methods available today. Advanced electronic equipment can now handle low voltages and currents and increase method versatility and mechanization in instrumentation (see Table III).

The cell concentration of a component produces an electrical response of three general types:

(1). Effect upon an electrode in solution

(2). Effect upon an applied voltage

(3). Changing chemical state due to applied current

The electrical response may depend upon factors such as the magnitude and nature of the solution composition; temperature; properties of the electrical circuit, electrode type, size, and spacing; stirring conditions; the mechanical features of the instrumentation; and current flow direction.

Electrical interactions with matter may occur in one of the following specific ways.

(1). Electron exchange may occur by discharge, ionization, or redox of the chemical species.

(2). Charged particles are attracted or repulsed and their movement will cause electrical conduction because of electron or ion movement.

(3). Optical and electrical polarization may occur because of alignment of molecules with permanent dipole moments as the field is imposed.

Electroanalytic techniques can be divided into several general areas: potentiometry, voltammetry, amperometry, conductimetry, coulometry, and chronopotentiometry. There are many modifications of these methods and sometimes the difference between them are difficult to see[9,3,4,10].

A. Potentiometry

One of the simplest methods, potentiometry, uses a zero current cell of nonpolarized electrodes to determine relative potential. The well-known Nernst equation establishes a relationship between electrode potential and the concentration of species in solution producing that potential; thus quantitative determinations are possible. Potentiometry also measures the changes in the electromotive force of an electrolytic cell caused by the addition of a titrant. Probably the most valuable application is the determination of solution pH. When no suitable indicator is available, titrations can often be determined by potentiometry.

In the 0.1 to 1 M concentration range errors will not usually exceed 0.1 to 0.3%, depending on how good the titrimetric procedure is. The sensitivity of pure potentiometry is limited when working with dilute solutions of titrants. Selectivity is moderately good, although preliminary separations may be necessary when analyzing mixtures.

B. Voltammetry and Polargraphy

These methods differ from potentiometry in that we now have appreciable current flowing through the cell. Also we are dealing with one or more polarized electrodes. Analysis is based upon production and interpretation of current-voltage curves. Polarography consists of a specific type of current-voltage curves obtained when a solution of an oxidizable or reducible substance is electrolyzed in a cell consisting of a mercury dropping electrode. This method provides both a qualitative and quantitative analysis, as every substance possesses a characteristic curve and concentration can be determined from this curve.

C. Amperometry

In our classification this method is concerned with the current readings in an electrolytic system at a constant applied potential. The current depends on the kinetics of the reaction at the electrode and upon the rate of mass transfer to the electrode. Quick reversible reactions limit the electrolytic current through the mass transfer processes of diffusion, migration and convection. For quiet ionic solutions and nonelectrolytes in supporting electrolytes, the diffusion limits the current.

D. Conductivity

This general method deals with those electrical properties of solutions not depending on the specific nature of electrode reactions. Electrical conductivity measures the ability of a solution to carry current. Ions in solution (electrolytes) migrate to appropriately charged electrodes in an electric field. Ohms law is obeyed. When a constant electromotive force exceeding the decomposition voltage of the electrolyte is applied, the current flow between electrodes varies inversely with the resistance of the electrolytic solution. Conductance is the reciprocal of this resistance. This conductance depends upon the sum of all the equivalent conductances of the ions in solution. Strong electrolytes in solution generally show a linear increase in conductance with increasing concentration, up to 10 to 20 % by weight. Any greater concentration results in the conductance passing through a maximum. Conductance measurements are applied in equibalance point determinations of titrations involving acid-base neutralization and precipitation reactions. They are seldom used in redox reactions.

E. Coulometry

These methods of analysis measure the exact quantity of electricity passing through a solution during an electrochemical reaction. Generally two things can happen in coulometry.

(1). Primary coulometric analysis: oxidation or reduction at one of the electrodes occurs.

(2). Secondary coulometric analysis: a substance can react quantitatively with a single product of electrolysis.

All coulometric analysis must conform to only one comprehensive reaction for which the electrode reaction proceeds with 100% current efficiency. The electron is the primary standard and titrant and reactant is generated in various amounts. The amount of reactant formed during current flow is related directly to the net charge transfer and can be calculated from Faraday's law when the current and time are known, i.e., $Q=it$. Coulometry is applied in the determinations of volatile or gaseous substances such as the halogens and unstable chemicals such as Cr^{+2} and Cu^{+}. Accuracy is excellent in the microgram region. Literally hundreds of procedures have been developed for the qualitative determination of numerous elements and ions.

F. Chronopotentiometry

One polarized electrode, a constant current with variable voltage and time characterize this method of analysis. It involves limited electrolysis and diffusion control of solution species. There are several conditions necessary for chronopotentiometry:

(1). The electroactive species must move by linear diffusion only.

(2). A constant electrolysis current must always be greater than the diffusion current of the analytic species.

(3). Transition time is the measured variable (the time it takes for the working electrode to become polarized).

(4). The bulk concentration in solution is proportional to the square root of the transition time.

Chronopotentiometry is useful in the quantitative determination of solutions. Within seconds several ionic species can be measured with moderate accuracy. With high current densities, catalytic or kinetic effects at the electrode can be determined, while in polarography they cannot[1,3,4,9,10].

V. THERMAL METHODS

All of these methods have their origins in the principles of chemical thermodynamics. In general, most thermal analyses can be considered as first law and/or second law methods, depending on the type of measurement relation involved. In the first law method, the quantity of heat liberated or absorbed in a chemical reaction is proportional to the amount of reaction taking place. The second law method involves thermochemical relationships and a short definition is impractical here. A fairly comprehensive list of thermal techniques include: cryoscopy, ebulliometry, calorimetry, thermometric titrations, differential thermal analysis and thermogravemetric analysis[16]. For more details on the last two subjects, the reader is referred to Chapter 9.

A. Cryoscopy

This method of analysis deals with the determination of freezing points and freezing point depression of liquids containing dissolved substances. We shall expand this topic to include melting points as both points are thermodynamically the same. In general, cryoscopy includes temperature determinations from liquid-solid equilibria, freezing point determinations of liquids, and melting points of solids. Some of the primary analytical applications of cryoscopy are listed below:

(1). Using the change in freezing and melting points to determine sample purity.

(2). Determining the molecular weights of unknown solutes.

(3). Identifying samples by the mixed melting point method.

(4). Studying binary solutions to determine solute, solid solution and phase system ionization.

(5). Determining solid phases in equilibrium with liquids[16].

B. Ebulliometry

This is the science of measuring and using boiling points. The boiling point of a liquid is defined as the temperature at which its vapor pressure equals the system pressure, under liquid-vapor thermal equilibrium. Thermodynamics assigns a significance to boiling points and they can be used to identify materials. Other analytical uses include determination of molecular weights, azeotrope study, liquid and vapor state reaction following, or molecular association study[16].

C. Calorimetry

With refined electrical and temperature measurements this field is useful to analytical chemistry. It can be used to distinguish molecules from one another by their reaction energies. Quantitative evaluation of the purity of a substance by high precision energy measurements is also possible[16].

D. Thermometric-Enthalpy Titrations

Although heats of reaction determinations have been conducted for many years, the thermometric titration is relatively new to analytical chemistry. The solution in question is titrated with a suitable reagent and the changes in temperature during the titration are noted. The method is conducted under effectively adiabatic conditions. The information learned from this method is extensive. Provided the kinetics are fast, thermometric titration curves may supply information on all the thermodynamic functions of a reaction[16].

E. Differential Thermal Analysis (DTA)

Generally, differential thermal analysis, similar to thermogravimetric analysis, is concerned with the detection of physical or chemical changes which may occur when matter is heated. Specifically, DTA studies the change in the absorption or liberation of heat by a sample measured as a function of the temperature. This change is compared to a reference substance that has been submitted to the same conditions. The method is especially suited for studying the structural changes within a solid at elevated temperatures. The method is mostly qualitative but some quantitative work has been done in obtaining calorimetric data. The method is used extensively in mineralogical investigations and standard tests. Most of the literature concerning DTA deals with minerals, ceramics, and clays. DTA can detect impurities present to the extent of 1 to 2% [16,1,10,2].

F. Thermogravimetric Analysis (TGA)

This technique studies the weight changes of a system as the temperature is raised a certain rate. Dynamic thermogravimetry subjects a sample to a continuous increase in temperature, usually linear with time. The second type, isothermal thermogravimetry subjects a sample to a constant temperature for a certain length of time and weight changes are noted. The instrument used is an automatic thermobalance. This method is often used in determining drying ranges of precipitates. Information gained from TGA is often enhanced by the application of DTA, as the two techniques are complementary. When both methods are utilized, fairly complete thermoanalytical data are obtained. When x-ray diffraction, optical and chemical analysis is found, the combination of this information yields a good quantitative estimation of solid-state reactions[1,10].

VI. NUCLEAR RADIATION AND OTHER METHODS

Radioactivity is one of the best analytical tools available today. There are several methods which convert unknowns into radioactive substances and apply radioactive tracers to analysis. This section will discuss activation analysis and tracer methods, plus several extremely important techniques which do not easily lend themselves to classification in our scheme. These methods include nuclear magnetic resonance, electron spin resonance, and mass spectrometry.

A. Activation Analysis

This method is a popular nuclear method applicable to both qualitative and quantitative analysis. The elements to be analyzed in the sample are bombarded with certain nuclear particles and thus made radioactive. This radioactive sample can then be identified and quantitatively measured. In general the quantity of radioisotope produced in the sample is proportional to the quantity of its precursor isotope. Thus we are provided with a measure of the element giving rise to the isotope. This method has a very low detection limit, and there is no danger of sample contamination. The accuracy is limited and the method is somewhat slow. A neutron source is used for the bombardment[2,3].

B. Radioactive Tracer Methods

A sample substance is "labeled" by substituting a radioactive tracer for a fraction of the atoms of an element in the sample. By following the path of the labeled compound, diffusion processes, reaction mechanisms, and analytical information can be found. Isotope dilution and radiometric analysis are two main tracer methods. Almost all elements can be determined by one of these methods, as isotopic tracers are available for most elements. Both types can have extreme sensitivity[3].

C. Nuclear Magnetic Resonance (NMR)

In brief, NMR is a form of spectroscopy in which radio-frequency radiation is absorbed in a magnetic field due to the magnetic properties of nuclei. The nuclei of certain atoms behave as nonrotating particles, (those of even atomic weight and number) while others rotate. NMR is concerned with the rotating nucleus of the proton (H^+). Because of this rotation the nuclei have magnetic properties, or nuclear magnetic momentum. Therefore, we are concerned with interaction between the magnetic momenta of protons and an applied external magnetic field. The magnetic property of the proton is affected by its relation to other protons in a molecule and thus the magnetic field interaction varies for the various protons in a molecule. The differences in field interaction form the basis for the qualitative analysis of many organic structures. The NMR is also useful in measuring reaction velocities [1,5,4, 10].

D. Electron Spin Resonance (ESR)

Known also as electron paramagnetic resonance, ESR takes advantage of the fact that atoms, molecules and ions having an odd number of electrons exhibit characteristic magnetic properties (see Chapter 20). These properties can arise from unpaired electrons spinning or orbiting about the nucleus. When a magnetic field is applied a magnetic interaction takes place and the results are observed and serve as the basis for this method. ESR and MNR are closely related in principle[1, 10].

E. Mass Spectrometry

This method is used exhaustively in quantitative and qualitative analysis of gases, liquids, and solids. Ions are separated according to their mass by the repetitive action of an electric and magnetic field. Initially an ionization chamber beams electrons so as to remove an electron from the unknown atoms or molecules. The leftover positive ions are accelerated by a potential gradient into a magnetic field. The field applies right angle lines of force. All of the particles in the field move in a circular motion. Equal massed particles follow trajectories with the same radius. A collector then collects the particles of similar mass. As the accelerating potential gradient is varied within given limits, successive ion groups of different masses are taken by the collector and a mass spectrum is obtained. The position of the peaks in the spectrum indicate ion mass while the height indicates the number of ions detected[1,7,3,10]. An alternative form of this method is to use a "time of flight" spectrometer. The length of part traveled by the ions down the flight tube is proportional to the mass of the ions. Coupled with a Knudsen cell this is a standard method for characterizing the vaporization properties of ceramics.

VII. FACTORS GOVERNING THE SELECTION OF AN ANALYTICAL METHOD

A number of factors are listed below which should be considered when deciding which methods would be most suitable for a particular analysis.

(1). May the sample be destroyed in the process?

(2). Is it a trace or bulk analysis?

(3). How fast are the results needed?

(4). What information are you seeking (% impurities; presence or absence of something; qualitative; quantitative)?

(5). What is the nature of the sample (phase state, ionic, alloy, etc.)?

(6). Which instruments are available?

(7). What are possible interferences with the method?

After these particular conditions of the analysis have been established, it will often prove necessary to use more than one technique.

Table IV compares the applicability of some of the more popular techniques[10]. Table V shows a comparison of analytical sensitivities of several of these techniques[3]. Some additional information concerning the techniques best suited for various sample physical states are given in Table VI. In general, accuracies of 0.1% can be obtained with x-ray fluorescence, several electroanalytical techniques and certain spectrophotometric techniques. Most of the other methods give results accurate to 1-5%.

TABLE IV

The Comparative Applicability of Various Analytical Procedures [a]

Type of Sample	Procedure	Application
Alloys, ores	Spectrography	General; rapid
	Electrodeposition	General; slower; less-expensive apparatus
	Colorimetry	More specific; especially for lesser constituents
	Activation	Specific; less convenient except in special cases
	X-ray absorption	Where sought element and impurities vary widely in atomic weight
	X-ray fluorescence	General; rapid
Traces of metal ions	Colorimetry Nephelometry Fluorimetry Polarography	These are of comparable sensitivity and accuracy; highly specific
Gaseous mixtures	Gravimetric	Especially for carbon dioxide or water
	Volumetric (Orsat, etc.)	Mixtures; to determine several constituents
	Manometric (Warburg)	Evolution or uptake; small samples
	Thermal conductivity Velocity of sound	For binary mixtures, where applicable
	Infrared absorption	Routine assay for single component
	Mass spectra	General; expensive apparatus
Mixtures (complete separation not required)	Infrared spectra Raman spectra	Especially for organic compounds
	X-ray diffraction	Crystalline solids
	Isotope dilution	Analysis for single component
	Mass spectra	For simple volatile compounds
	NMR	For liquids
Mixtures (separation procedures)	Ion exchange	For ionic materials
	Countercurrent distribution Partition chromatography	Must be partially soluble in each of two immiscible liquids
	Adsorption chromatography	Chiefly for organic compounds
	Electrodeposition	For cations

[a]From G. W. Ewing (10).

TABLE V

Absolute Detection Limits in Nanograms[a]

Element	Absorption spectro-photometry	UV-visible fluores-ence	Atomic absorption	Flame spectro-photometry	Neutron activation	Spark source mass spectroscopy	Emission spectrography		
							DC arc	Copper spark	Graphite spark
Ar						0.03			
Ag	5		1	3	0.01	0.2	8	200	0.5
Al	0.5	0.5	100	4	1	0.02	30	20	3
As	10		100	100	0.1	0.06	600	500	100
Au	5		5	50	0.05	0.2	100	40	
B	50	40		5		0.01	30	10	0.3
Ba	100		30	3	5	0.2	4	10	
Be	8	1	10	20		0.008	5	0.2	
Bi	600		8	100	50	0.2	4		5
Br	10				0.5	0.1			
C						0.01			
Ca	100	3	1	0.2	100	0.03	5	20	
Cd	3	2000	0.3	20	5	0.3	100	100	20
Ce	40			200	10	0.1	100	40	
Cl	100				1	0.04			
Co	3	20	2	8	0.5	0.05	100	40	5
Cr	7		2	0.5	100	0.05	80	20	1
Cs			3	0.4	50	0.1	10,000	300	
Cu	2	30	0.3	1	0.1	0.08	20	100	0.5
Dy			300	4	0.0001	0.5	50	40	
Er			300	8	0.1	0.5	50	30	
Eu			100	1	0.0005	0.2	30	2	
F		1			100	0.02		10	
Fe		20	3	3	5000	0.05	30	80	3
Ga		1	50	2	0.5	0.09	100	60	
Gd			10,000	10	1	0.5	50	20	

Ge	20	2000	200	500	0.5	0.2	100		
H						0.0008			
He						0.003			
Hf		100		4000	100	0.4	300	50	
Hg	5		30	100	1	0.6	100	500	10
Ho			300	4	0.01	0.1	50	20	
I	10			500	0.5	0.1			
In	50	40	20	1	0.005	0.1	100	80	10
Ir	6			5000	0.01	0.3	300	500	
K	200		0.3	0.01	5	0.03	100	20	
Kr						0.1			
La	4000			20	0.1	0.1	10	5	
Li	3	200	0.2	0.01		0.006	8	0.2	
Lu				6	0.005	0.1	100	100	
Mg	2	0.2	0.1	0.8	50	0.03	8	10	
Mn	5		0.5	0.5	0.005	0.05	10	3	0.3
Mo	10		10	20	10	0.3	100	5	
N	10					0.01			
Na	8000		0.4	0.01	0.5	0.02	10	30	
Ne						0.02			
Nb	5	100	5000	200	0.5	0.08	300	20	
Nd	4000		5000	10	10	0.4	80	30	
Ni	4		2	5	5	0.07	80	1	1
Np								200	
O		2				0.01			
Os	50			500	5	0.4			
P	100			40	50	0.03	1000	2000	10
Pa								200	
Pb	6		4	0.5	1000	0.3	10	30	
Pd	10		30	5	0.05	0.3	200	50	
Pr	4000		15,000	50	0.05	0.1	80	30	
Pt	10		20	60	5	0.5	300	2	
Pu								200	
b			2	0.1	5	0.1	30	20	

Table V (Continued)

Element	Absorption spectro-photometry	UV-visible fluores-cence	Atomic absorption	Flame spectro-photometry	Neutron activation	Spark source mass spectroscopy	Emission spectrography DC arc	Copper spark	Graphite spark
Re	200		1000	100	0.05	0.2	300	200	
Rh	50		20	10	0.05	0.09	200		
Ru	100	1000	20	10	1	0.03	300		
S	100	200		2000	500	0.03			
Sb	4		10	30	0.5	0.2	100	500	10
Sc	8	100	200	3	1	0.04	30	0.8	
Se	200	2	100	600	500	0.1	10,000		
Si	100	80	6000	700	5	0.03	30	10	
Sm	1000	100	1000	10	0.05	0.5	50	40	
Sn	60	100	100	30	50	0.3	10		
Sr	10		2	0.2	0.5	0.09	100	30	
Ta	40			1000	5	0.2	3000	100	10
Tb	1000	3000	6000	10	5	0.1	300	100	
Te	400		20	100	5	0.3	4000	300	
Th	50	20		8000	5	0.2	5000	40	
Ti	10		100	20	5	0.05	20	10	3
Tl	20	30	5	3		0.2	30	200	50
Tm			300	8	1	0.1	50	10	
U	300	0.1		200	0.5	0.2	4000	200	
V	10	2500	100	10	0.1	0.04	80	8	1
W	30	1000	250	200	0.1	0.5	600	80	10
Xe						0.4			
Y	1000	20	1000	10		0.07	3	0.8	
Yb			100	2	0.1	0.5	50	5	
Zn	100	1000	0.3	80	10	0.1	300	200	10
Zr	50	20		500	100	0.1	100	20	3

[a] From Trace Analysis, Physical Methods (3).

TABLE VI

Phase States of Sample, Listing Methods Appropriate to These States

Gas	Liquid	Solid
Gas chromotography Mass spectrometry	Gas chromotography Liquid chromotography Electroanalysis Absorption spectro-photometry Flame photometry Atomic absorption photometry Fluorescence Mass spectrometry	Emission spectrography X-ray fluorescence Neutron activation Spark source Mass spectrometry

AMOUNT OF UNKNOWN TO BE ANALYZED

Trace	Concentrations > 0.1%
Emission spectrography Activation analysis Mass spectrometry Gas chromatography Absorption spectrophotometry	X-ray fluorescence Flame photometry Differential analysis

VIII. SUMMARY

It is not possible to summarize so broad a field as analytical chemistry in a single chapter. This chapter has outlined the spectrum of commonly used analytical techniques providing the reader with a broad general survey of the field. A separate bibliography included at the end of the chapter lists a variety of general references from each major analytical technique.

REFERENCES

(1). H. H. Willard, L. L. Merritt, and J. A. Dean, Instructional Methods of Analysis, Van Nostrand, Princeton, New Jersey, 1958.

(2). J. Krugers and A.I.M. Keulemans (eds.), Practical Instrumental Analysis, Elsevier, London, 1965.

(3). G. H. Morrison (ed.), Trace Analysis, Physical Methods, Wiley (Interscience), New York, 1965.

(4). Paul Delahay, Instrumental Analysis, MacMillan, New York, 1957.

(5). W. G. Beel (ed.), Physical Methods in Chemical Analysis, Vol. III, Academic, New York, 1956.

(6). R. A. Day, Jr. and A. L. Underwood, Quantitative Analysis, Prentice-Hall, Englewood Cliffs, N. J., 1967.

(7). W. Wagner, C. H. Hull, and G. E. Mackle, Advanced Analytical Chemistry, Reinhold, New York, 1956.

(8). E. J. Bair, Introduction to Chemical Instrumentation, McGraw-Hill, New York, 1962.

(9). H. A. Strobel, Chemical Instrumentation, Addison-Wesley, Reading, Mass., 1960.

(10). G. W. Ewing, Instrumental Methods of Chemical Analysis, McGraw-Hill, New York, 1960.

(11). I. M. Kolthoff, P. J. Elving, and E. B. Sandell, Treatise on Analytical Chemistry, Part I: Vol. 1, Wiley (Interscience), New York, 1959.

(12). I. M. Kolthoff, P. J. Elving, and E. B. Sandell, Treatise on Analytical Chemistry, Vol. 2, Wiley (Interscience), New York, 1961.

(13). I. M. Kolthoff, P. J. Elving, and E. B. Sandell, Treatise on Analytical Chemistry, Vol. 3, Wiley (Interscience), New York, 1961.

(14). I. M. Kolthoff, P. J. Elving, and E. G. Sandell, Treatise on Analytical Chemistry, Vol. 4, Wiley (Interscience), New York, 1961.

(15). I. M. Kolthoff, P. J. Elving, and E. G. Sandell, Treatise on Analytical Chemistry, Vol. 6, Wiley (Interscience), New York, 1965.

(16). I. M. Kolthoff, P. J. Elving, and E. G. Sandell, Treatise on Analytical Chemistry, Vol. 8, Wiley (Interscience), New York, 1968.

(17). C. N. Reilley and D. T. Sawyer, Experiments for Instrumental Methods, McGraw-Hill, New York, 1961.

BIBLIOGRAPHY

A. Optical Methods

Light Scattering

W. Heller, Record Chem. Progress, Kresge-Hooker Sci. Lit., 20, 209 (1959).

Colorimetry

F. D. Snell and C. T. Snell, Colormetric Methods of Analysis (6 Volumes), Van Nostrand, Princeton, N. J., 1963.

Ultraviolet and Visible

C. N. R. Rao, Ultraviolet and Visible spectroscopy," Chemical Applications, Butterworths, Eng., (1961).

Fluorimetry

M. Curie, Fluorescence and Phosphorescence, Humamn and Cie, Paris, (1946).

X-rays

E. P. Bertin, Principles and Practice of X-ray Spectrometric Analysis Plenum Press, 1970.

Electron Probe Analysis

V. E. Cosslett, "Techniques in Electron Probe Microanalysis," in E. R. Lippincott and M. Margoshes, Eds., Proceeding of the Xth Colloquim Spectroscopium Internationale, Spartan Books, Wash., (1963) pp. 357-381.

Microwave Spectrophotometry

C. H. Townes and A. L. Schawlow, "Microwave Spectroscopy," McGraw-Hill, New York, 1955, pp. 427-555.

Emission Spectroscopy

American Society for Testing and Materials, Committee E-2 on Emission Spectroscopy, "Methods for Spectrochemical Analysis," 3rd Ed., ASTM, Phil., (1964).

Flame Photometry

R. Mavrodineanu, Appl. Spectroscopy, 10, 51 (1956).
R. Mavrodineanu, Appl. Spectroscopy, 10, 137, (1956).
R. Mavrodineanu, Appl. Spectroscopy, 13, 132, 149 (1959).
R. Mavrodineanu, Appl. Spectroscopy, 14, 17 (1960).

Infrared Spectroscopy

H. M. Hershenson, Infrared Absorption Spectra, Academic, New York, (1959).

Raman Spectroscopy

W. T. Elwell and J. A. F. Gidley, Atomic-Absorption Spectrophotometry, Macmillan (Pergamon), New York, 1961.

Electron Diffraction

C. E. Hall, Introduction to Electron Microscopy, 1st Ed., McGraw-Hill, New York, 1953.

Refractometry and Dispersometry

S. S. Batsanov, Refractometry and Chemical Structure, Consultant Bureau, New York, (1961).

Elving Kalthoff, Treatise on Analytical Chemistry, Vol. 6, Chapt. 70, 1965.

Polarimetry

C. Djerassi, Optical Rotatory Dispersion, McGraw-Hill, New York, 1960.

Chemical Microscopy

A. E. J. Vickers, Modern Methods of Microscopy, Butterworths, London, 1956.

B. Electrochemical Analysis

H. A. Strobel, Chemical Instrumentation, Addison-Wesley, Reading, Mass., 1960.

J. J. Lingane, Electroanalytical Chemistry, 2nd Ed., Wiley (Interscience), New York, 1958.

G. Kortuim and J. O'M. Bockris, Textbook of Electrochemistry, Elsevier, Houston, Tex., 1951.

H. H. Willard, L. L. Merritt, and J. A. Dean. Instrumental Methods of Analysis, Van Nostrand, Princeton, N. J., 1958.

C. Thermal Analysis

Natl. Bur. Std. Tech. Note 270-1, Selected Values of Chemical Thermodynamic Properties, U. S. Government Printing Office, Washington, D. C., 1965.

I. M. Kolthoff, R. J. Elving, and E. B. Sandell, Treatise on Analytical Chemistry, Vol. 8, Wiley (Interscience), New York, 1968.

D. Nuclear and Miscellaneous Analysis

G. H. Morrison, (ed.), Trace Analysis, Physical Methods, Wiley (Interscience), New York, 1965, pp. 374-376 (Ref. listed of 68).

R. T. Overman and H. M. Clark, Radioisotope Techniques, McGraw-Hill, New York, 1959.

Symposium on Nucelonics, Anal. Chem. 21, 318-368 (1949).

R. C. Kock, Activation Analysis Handbook, Academic, New York, 1960.

D. W. Stewart, Mass Spectrometry, in Technique of Organic Chemistry, (A. Weissberger, ed.), 3rd Ed., Vol., Part IV, Wiley (Interscience), New York, 1960.

G. P. Barnard, Modern Mass Spectrometry, The Institute of Physics, London, 1953.

J. D. Roberts, Nuclear Magnetic Resonance, McGraw-Hill, New York, 1959.

J. A. Pople, W. G. Schneider, and H. J. Bernstein, High-Resolution Nuclear Magnetic Resonance, McGraw-Hill, New York, 1959.

E. Separation Techniques

G. D. Dickey and C. L. Bryden, Theory and Practice of Filtration, Reinhold, New York, 1946.

A. Henley and G. L. Schuettlei, Electrophoresis Bibliography, 2nd Ed., Biochemical Instruments Division, American Instrument Co., Silver Springs, Md., 1955.

H. G. Cassidy, Fundamentals of Chamatography, Wiley (Interscience), New York, 1957.

I. M. Kolthoff and P. J. Elving, Treatise on Analytical Chemistry, Part I, Vol. 2 and 3, 1961.

Chapter 3

X-RAY SPECTROSCOPY

R. W. Gould

Department of Metallurgical and Materials Engineering
University of Florida
Gainesville, Florida

I. INTRODUCTION

We have witnessed in the past decade a phenomenal growth in the technology associated with the mineral, ceramic, and metals industries. This growth has prompted and required new thinking concerning old problems. Some technological areas are finding it difficult indeed to keep pace with the expanding frontier of knowledge. The growth of the mineral industries has been paralleled by an increased demand for more rapid and precise methods of chemical analysis. Science has answered the challenge by providing the analyst with many new and exciting analytical methods. Unfortunately, however, the apparent complexities of these new techniques have discouraged many industries from adopting them for plant use. The minerals industries have been particularly guilty of this attitude and are only now realizing what many industries, i.e., chemicals, metals, and electronics, to name a few, realized 8 to 10 years ago. That is, automation must extend down to the laboratory level in order to provide rapid and precise quality control. Many industries, forseeing the vast potential of instrumental analysis, have already installed on-stream continuous analyzers which detect changes in the composition of the flow stream and automatically relay this information so that corrective changes can be made.

The purpose of this chapter is to provide the minerals and ceramics industries with a brief but comprehensive look at one of the most important instrumental methods of analysis, x-ray emission spectrography, or more simply x-ray fluorescence analysis. This method has many advantages over other methods of analysis of minerals, a fact which we hope will become more apparent throughout the pages of this chapter.

II. PRINCIPLES OF X-RAY SPECTROCHEMICAL ANALYSIS

We wish to emphasize that the contents of this chapter are directed to those individuals in the minerals industries who wish to obtain a background in the fundamentals of x-ray spectrographic analysis. There is no intention to compete with the several excellent text books that are currently available on the subject[1, 2, 17] rather, it is hoped that the reader will become familiar with the advantages, problems, limitations, and horizons of the x-ray spectrochemical procedure.

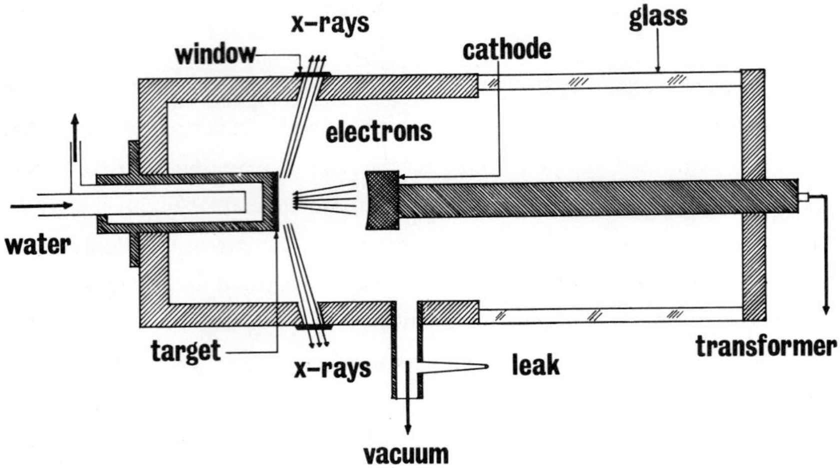

Fig. 1. Gas filled x-ray tube.

III. ORIGIN AND PROPERTIES OF X RAYS

X rays are electromagnetic radiation not unlike visible light, ultraviolet, or infrared. The credit for their discovery goes to Prof. Wilhelm Röntgen who, in 1895, while experimenting with evacuated cathode ray tubes noticed that some sort of penetrating invisible radiation was being emitted. Röntgen showed that this radiation would penetrate thick objects, cast shadows, blacken photographic emulsion, and cause a barium platinocyanide phosphor to glow brightly. Early experimentation with this new radiation used gas tubes such as the one shown in Fig. 1 in which a very small amount of continuously admitted gas is ionized.

The production of x rays generally requires a source of electrons, a high accelerating voltage, and a metal target. In present day tubes the electrons are obtained from a hot tungsten filament (cathode). These electrons are focused by means of a biased cathode cup onto a metal target (anode). A large potential difference between the anode and the cathode (10 kV - 100 kV) serves to accelerate the electrons to the required energy. A schematic drawing of a modern tube (sealed tube) is shown in Fig. 2. In a tube such as this the important features are: (1) purity of target material, generally W or Cr; (2) accelerating voltage (variable from 10 to 100 kV); (3) filament current (variable from 0 to 100 mA); (4) anode to window distance; (5) window material (generally thin Be metal); and (6) removing heat from target (water cooling).

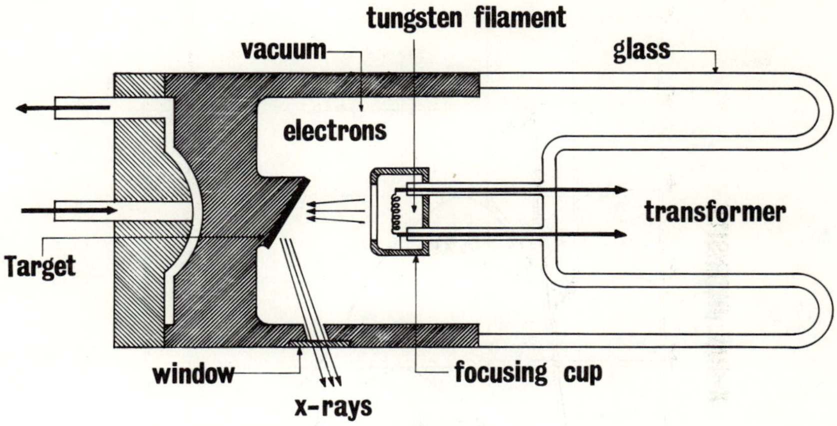

Fig. 2. Modern sealed x-ray tube.

Modern x-ray tubes because they are sealed* often develop contamination on the anode. This contamination which frequently can be troublesome, comes from the cathode and consists of W from the filament and Fe and Ni from the filament supports. This point will be discussed in greater detail in a subsequent paragraph.

IV. EXCITATION OF X RAYS WITH ELECTRONS

X rays and heat are produced whenever electrons of sufficient energy hit the target or anode of an x-ray tube. The resulting x-radiation may be classified into two categories, i.e., continuous radiation and characteristic radiation. Since the origin of these two types of x-radiation differ they will be discussed separately.

V. CONTINUOUS RADIATION

Electrons produced by the tungsten filament are rapidly drawn or accelerated toward the anode (target) by the high potential difference existing between the anode and cathode (10 - 50 kV). This acceleration imparts a large kinetic energy to the electrons, which, of course, must be given up upon impact on the target. Most of this energy (99%) is dissipated as heat with less than 1% going into x rays. When the x-ray spectrum of an x-ray tube is spread out (by a single crystal used as a grating) and examined, a picture

*Several companies (e.g., Jarell Ash) produce a demountable high vacuum filament tube in which the anode may be changed. These tubes are used predominantly for x-ray diffraction studies.

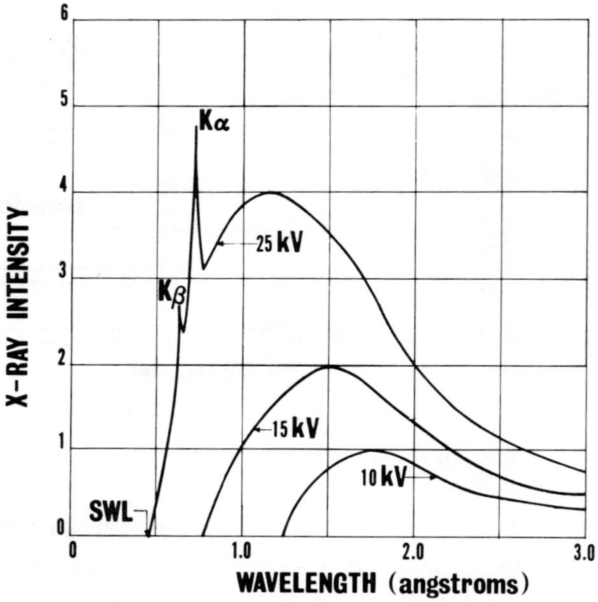

Fig. 3. Spectrum of Mo target x-ray tube.

like Fig. 3 results. In Fig. 3 the tube target is Mo and the different curves are the result of using different accelerating voltages.

Several important questions may be asked about this figure. (1) Why does the spectrum terminate abruptly at the short wavelength end and not at the longer wavelengths? (2) Why does this abrupt cut-off (called the short wavelength limit or SWL) at the short wavelengths vary with kilovoltage? (3) Why are the sharp peaks found on the 25 kV curve and not on the 20 kV curve? (4) What relationship do the smooth maxima in the continuous radiation curve have to the SWL?

These questions can be answered if we look at the mechanisms by which an electron loses its energy. When a charge particle is rapidly decelerated electromagnetic radiation is produced. This is the situation at the target of an x-ray tube. All electrons do not lose their entire kinetic energy in a single impact. Those that do, however, give rise to a wavelength corresponding to the SWL. Most of the electrons are deviated many times by target atoms before all their energy is lost. These electrons give rise to the rest of the continuous spectrum. The SWL is independent of target material and is given by a simple relationship:

$$\lambda \text{ SWL } (\breve{A}) = \frac{12{,}400}{\text{Tube potential in volts}}$$

Thus a tube operated at 50,000 V (50 kV) has a SWL of approximately 0.248 Å.

The low intensity of the continuous radiation curve at the longer wavelength end results from the inability of a long wavelength (3 Å) to penetrate the window of the x-ray tube, or, for that matter, the air between the tube and the detection system.

Note that each continuous curve has a smooth maximum. This maximum corresponds to the most frequent energy loss suffered by electrons when they hit the target. The position of this maximum has been found empirically to be a function of λ_{SWL} according to the following relationship:

$$\lambda_{max} \simeq \frac{3}{2} \lambda_{SWL}$$

The very sharp maxima (approximately 0.001 A wide) that appear on the 25 kV curve of Fig. 3 are caused by characteristic radiation. Their origin and the answer to question 3 asked in the second paragraph of this section will be discussed in the next section.

VI. CHARACTERISTIC RADIATION

Let us still assume that we are discussing the process of x-ray production by high energy electrons. Consider the situation shown in Fig. 4(a) where we see a target atom of our Mo tube drawn schematically. The two innermost electrons occupy the K electron shell and are held in this energy level by the strong electrostatic attraction of the nucleus. If we allow a sufficiently energetic beam of electrons (Fig. 4(b)) to hit the Mo target atom we can cause one (or both) of the K shell electrons to be completely ejected. The electron beam energy needed to produce the effect is approximately 20,000 V. In other words a tube potential of 5000 V cannot produce an electron vacancy in the K shell of Mo. However, look at the L shell with its 3

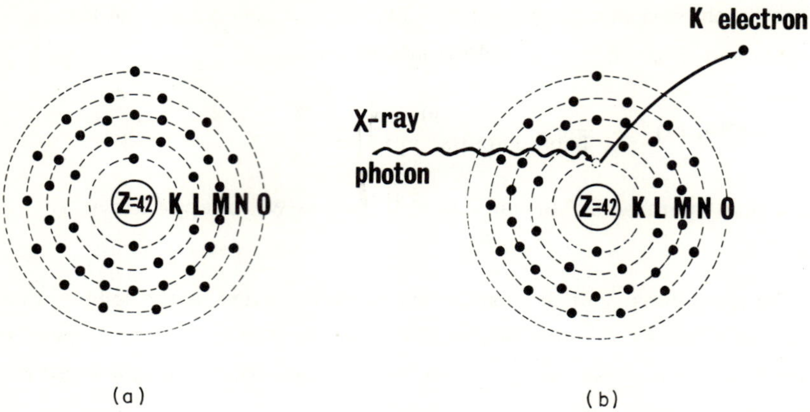

Fig. 4. (a). Schematic of Mo atom; and (b). Interaction of electrons with Mo target atom.

subshells L_I, L_{II}, and L_{III}. It is intuitively obvious that it is not as difficult (doesn't require as much energy) to remove an electron from the L_I shell as it is to remove it from the K shell. This is due to the weaker attraction of the nucleus caused, in part, by the shielding of the K shell electrons. Specifically, it requires an electron beam having an energy of 2870 V to eject an L_I energy electron from Mo.

Vacancies in inner electron shells are not stable and, in fact, they are filled instantly by electrons from the outer shells. Consider one electron hole in the K shell. If we could slow down the process of filling this vacancy we would see that most of the time this hole is filled by an electron from the L_{III} subshell. The next most frequency contribution of electrons is the L_{II} subshell. These electron jumps from one shell to another are called transitions. When they occur (i.e., when a charged particle is accelerated from one shell to another) radiation is produced. The wavelength of the radiation is directly related to the energy difference between the shells in which the transition is taking place. The greater the energy difference in the transition the shorter is the resulting radiation wavelength. In such a way an atom is said to have an x-ray spectrum consisting of many wavelengths. Some of these wavelengths are too long to be detected by present day instruments for this and other reasons x-ray spectra are generally very simple (few lines) when compared to optical emission spectra. Note that it is possible to produce L radiation without producing K radiation but the reverse can not be true. This is easy to understand if you consider the transition to form a cascade, i.e., when a K shell hole is filled, the filling electron leaves a vacant site which in turn must be filled. Figure 5 shows a so-called energy level diagram for U. Here we can see in addition to K and L spectra the M series of lines. Spectral lines vary in intensity within one series and between the various series. The intensity of a line is directly related to the frequency of the electron transition involved in the production of that line. The relative intensity of lines within the K and L spectra (with the strongest line in each series being given the intensity of 100) are shown in the following tabulation. According to this the $K\beta_1$ line of Mo would be only ~1/5 as strong as the MoK α ($MoK\alpha_1 + MoK\alpha_2$).

Relative Spectral Line Intensities

K Spectrum		L Spectrum	
$K\alpha_1$	100	$L\alpha_1$	- 100
$K\alpha_2$	50	$L\beta_1$	- 50
$K\beta_1$	21	$L\beta_2$	- 20
$K\beta_2$	3	$L\alpha_2$	- 11
		$L\gamma_1$	- 10

VII. PRODUCTION OF X RAYS BY HIGH ENERGY X RAYS (X-RAY FLUORESCENCE)

It is more convenient to produce fluorescent x rays with a high energy x-ray beam than with electrons. The energy conditions for spectral excitation are the same as those outlined in the paragraphs above with the exception that the exciting electrons are replaced by a continuous spectrum of x rays. These x rays are normally produced by a tungsten target x-ray tube operated at a rather high kilovoltage (up to 100 kV). As we shall see later, Cr or Ti target tubes are preferred over W when low Z elements are being excited.

Consider for a moment the spectral output of a W target x-ray tube used in a commercial spectrograph. Assume that the tube is operating at a potential of 50 kV. The partial spectrum, as shown in Fig. 6 consists mainly of continuous or "white" radiation. Note that there are no K series lines present (excitation potential for WK α is approximately 70 kV). A few of the

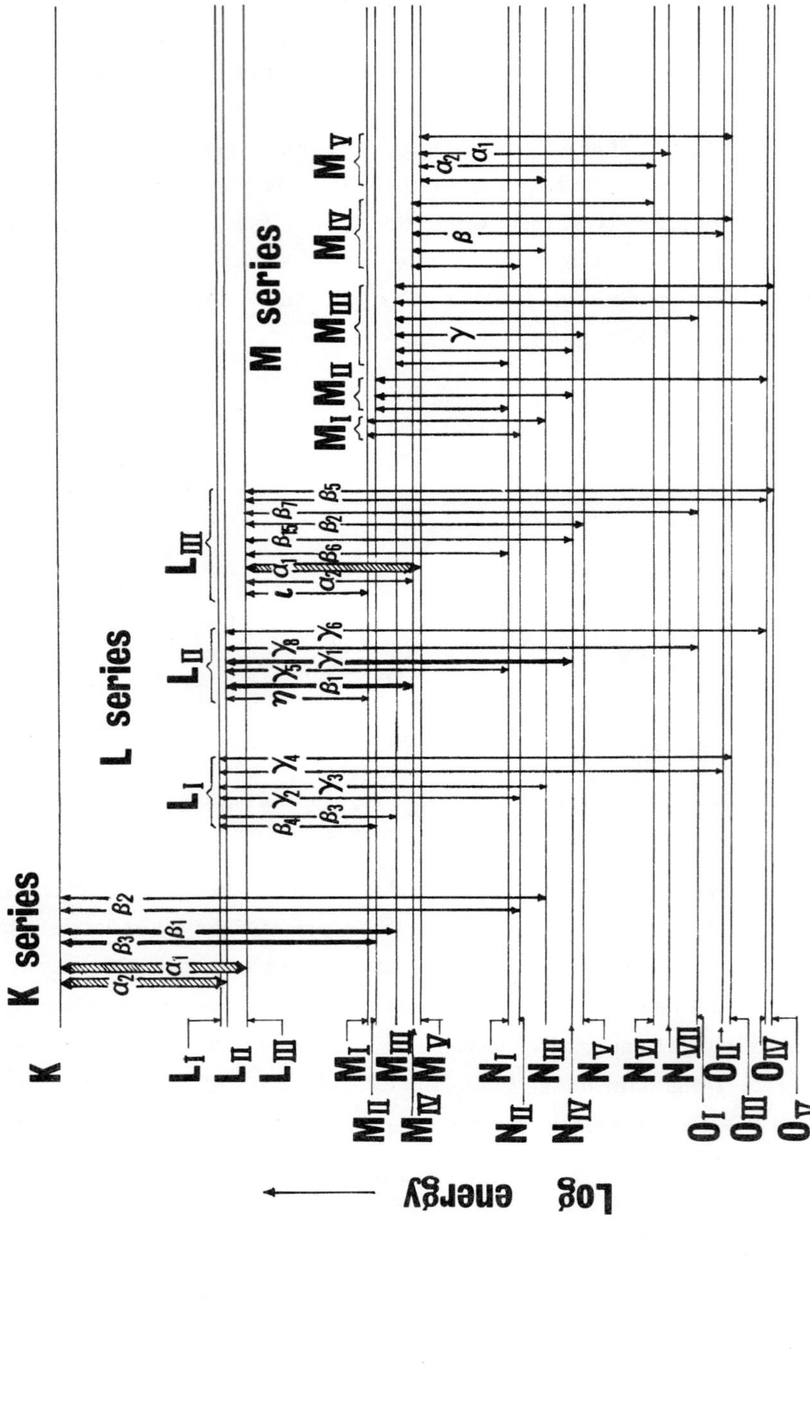

Fig. 5. Energy level diagram for uranium.

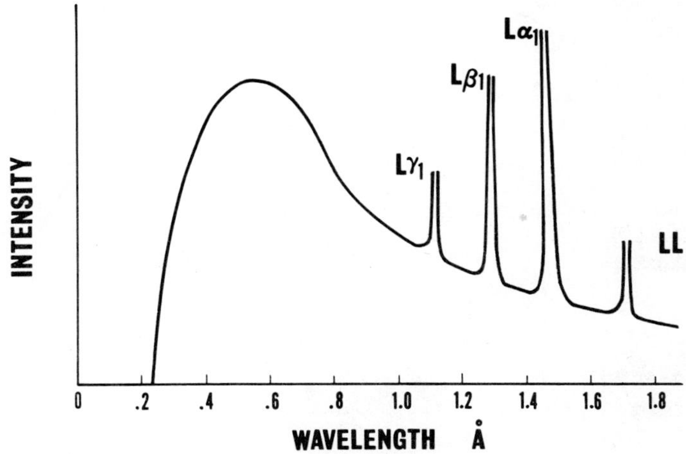

Fig. 6. Spectrum of W target spectrograph tube.

stronger L lines are shown for comparison with the SWL (short wave length limit) and the maximum in the white radiation curve. The intensity of the continuous spectrum is proportional to the atomic number of the target material (Z), tube current (i), and voltage (V) through the following equation.

$$\text{Intensity of continuous spectrum} = iZV^m$$

where m is approximately 2. This equation clearly shows why W (Z = 74) is often chosen as a target material in x-ray spectrography tubes and why continuous spectrum intensity is so sensitive to changes in excitation voltage.

VIII. EXCITATION OF FLUORESCENT X-RAYS FROM A SAMPLE

Figure 7 is a schematic drawing of a simple x-ray spectrograph having a W target x-ray tube. Let us assume that the sample is an alloy of the three metals, aluminum, zinc, and silver. The question one might ask is, "How much energy will it require to excite the K spectra from each of these elements?" or phrased in another way, "How much energy is required to remove a K shell electron from each of these atoms?" One could assume as Moseley (Moseley's law) did that it would require more energy to excite the K spectra of the heavy element Ag than it would for Al. Following this reasoning, it would appear that the electrons in the K shell of Al atom were not as tightly bound to their nucleus as the K-shell electrons in Ag. Now the energy needed to perform this operation is generally given in Tables of

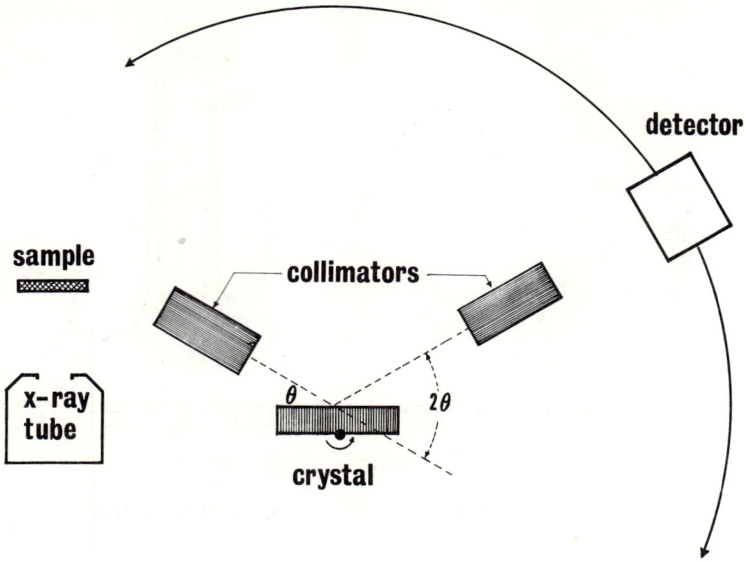

Fig. 7. Schematic drawing of x-ray spectrograph.

Critical Absorption Energies in units of electron volts, i.e., how many volts must be applied to the electron stream in an x-ray tube so that the electrons become sufficiently energetic to eject K shell electrons from the target of the tube. However, you will recall that now we are not discussing x-ray production by electrons but rather x-ray production (x-ray fluorescence) by other x-rays of high energy. As it turns out, the same tables may be used because the maximum energy of an x-ray beam coming from an x-ray tube operating at 50,000 V is the same as the energy in a stream of electrons whose accelerating potential is 50,000 V. Thus for our alloy of Al, Zn, and Ag we can determine from tables (for example, such as those in Ref. 3, page 2670) that our W tube must operate at a minimum of 1559 V to excite AlKα, 9657 V to excite ZnKα, and 25,509 V to excite AgKα. This minimum operating voltage for excitation may also be found by using the simple expression:

$$V_{min} \cong \frac{12,350}{\lambda_{K\ \text{Absorption Edge}}}$$

The intensity of a characteristic line is strongly dependent on the so-called over-voltage $(V - V_m)$ where V is the tube voltage and V_m was defined in the last paragraph, and we have:

$$I_{characteristic} \propto i(V - V_m)^k$$

where i is the tube current and k varies from 1.5 to 2.

IX. OTHER METHODS OF FLUORESCENT X-RAY EXCITATION

While high energy x rays are the most practical method for exciting fluorescent x-ray spectra, other methods may be used. We have already mentioned the use of electrons and this method is the heart of the electron probe microanalyzer which uses a highly focused beam ($\sim 1\mu$ diameter) of electrons to excite fluorescent x rays from a small selected area of a sample. The use of electrons as x-ray exciters demands that the sample be placed in the high vacuum (10^{-5} mm Hg) which is required for the production of the electron beam. This is a very serious limitation when liquid samples are being investigated.

The use of radioactive isotopes to excite characteristic x rays is gaining some acceptance particularly where portability and lower cost are important factors. These units use low intensity radioactive sources (approximately 1 mCi) which eliminates the need for expensive stabilized high voltage equipment. Several manufacturers have marketed such units which use a high energy resolution solid state detector (diode) and multichannel pulse height analyzer (to be discussed in a later chapter).

X. X-RAY ABSORPTION

All matter will absorb x rays to a greater or lesser extent. The fraction of monochromatic (single wavelength) incident x-ray beam lost as a result of the absorption process is dependent upon the physical state and chemical composition of the absorbing medium as well as the wavelength of the x rays. Intuitively one expects "hard," high energy (short wavelength) x rays to penetrate a substance easier than soft (long wavelength) x rays. Also intuition would predict that for a fixed wavelength, a substance composed of element Z in the gaseous state would absorb less of the incident beam than the same weight of substance in the solid state, i.e., iodine and iodine vapor being good examples.

While the behavior of x rays does follow these simple "laws," there are some other complications which we will call "absorption discontinuities." Consider the experimental arrangement given in Fig. 8 below. On the left we have a hypothetical constant intensity (I_0) source wherein we can produce any single wavelength that we choose. The sample, for simplicity sake, is chosen to be a thin piece of pure element, say lead (Pb). On the right of the figure is a radiation detector which can detect all wavelengths with equal efficiency (as we will see later, a scintillation type radiation counter would

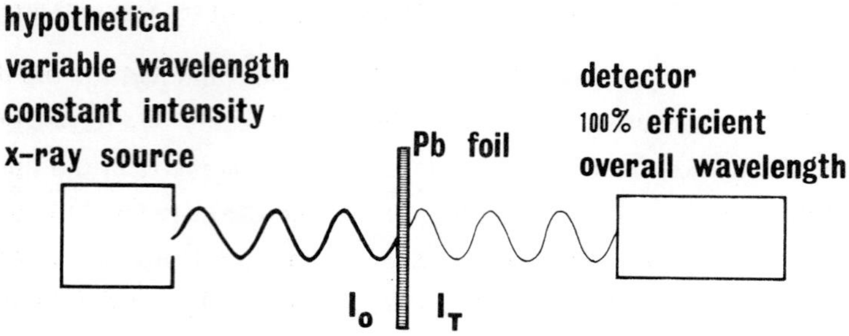

Fig. 8. Hypothetical x-ray absorption measuring device.

very nearly fit the description). The experiment goes as follows: Vary the wavelength of the x rays while measuring the intensity (I_T) of the beam which passes through the absorber. If the absorption followed the simple "laws" of the preceding paragraph, a curve such as Fig. 9 might result. However, as we have said, this model for the behavior of x rays when passing through a material is too simple and for the real case, a curve such as Fig. 10 would result. Let us examine Fig. 10 as we decrease the wavelength, i.e., go from right to left in the figure. We see that the transmitted intensity I_T appears to follow the simple "law" for awhile then suddenly I_T takes a large drop at a λ of 0.95 Å. This pattern is repeated three more times in the figure. What is the reason for these "jumps" in I_T? The answer

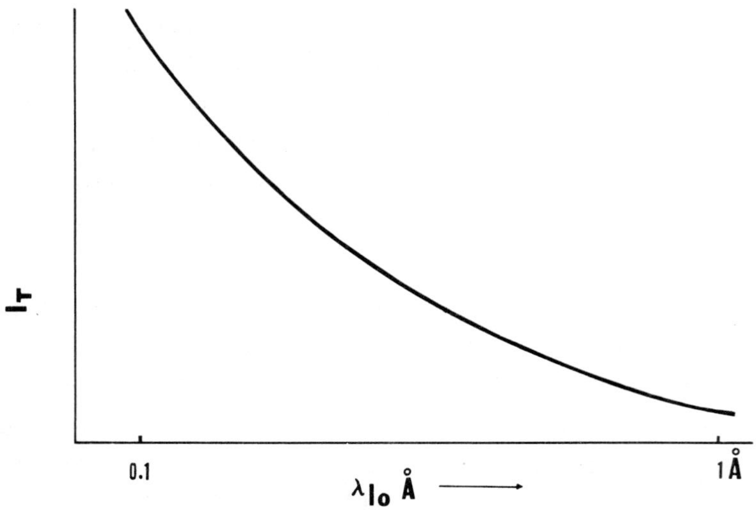

Fig. 9. Simple absorption of energy as a function of wavelength.

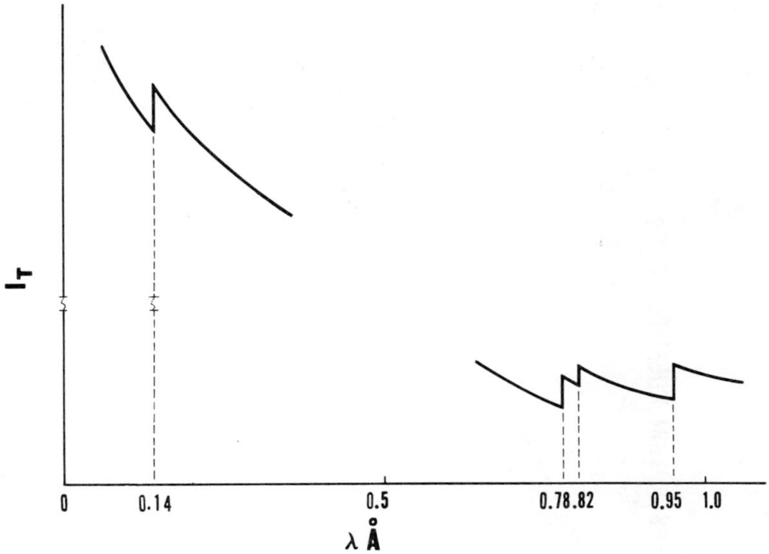

Fig. 10. Intensity transmitted by a sheet of Pb as a function of λ.

to this question is intimately related to the production of characteristic x rays. When the λ of the incident beam (I_o) becomes short enough (has enough energy) to eject an electron from a shell, it (I_o) will be strongly absorbed (resonance) at that particular wavelength so that I_T will decrease abruptly. We have seen that the energy necessary to eject an electron (i.e., the energy with which an electron is bound to the nucleus) is a very discrete quantity. Therefore, these "jumps" or absorption edges as they are called are very sharp. In the figure the three closely spaced edges correspond to ejection of electrons from the three L subshells, i.e., L_I, L_{II}, and L_{III}. The single edge at short wavelength is the K absorption edge. Note that the curve I_T versus λ must be drawn broken between the three L edges and the K edge because I_T is many orders of magnitude higher for wavelengths near the absorption edge than those near the three L edges.

XI. ABSORPTION EQUATION AND THE MASS ABSORPTION COEFFICIENT

It can be shown easily the I_o and I_T are related through a very simple equation, namely:

$$I_T = I_o e^{-\mu/\rho \, \rho t}$$

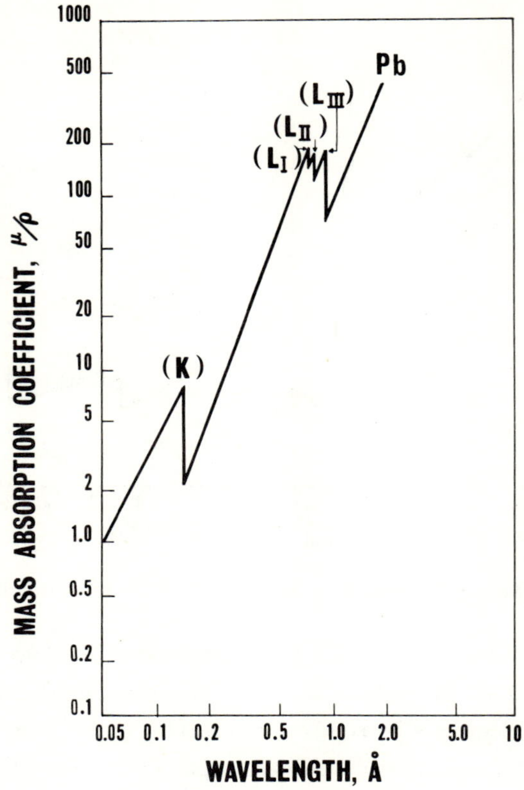

Fig. 11. The mass abosrption coefficient (μ/ρ) for Pb as a function of λ.

where μ is the linear absorption coefficient, ρ is the density, t is the sample thickness, and μ/ρ is called the mass absorption coefficient. By measuring I_T, I_o, and the sample thickness t, we can obtain values for μ/ρ as a function of the incident wavelength λ. This is shown in Fig. 11 where as before the sample is a thin sheet of Pb.

According to the equation $I_T = I_o e^{-\mu/\rho \, \rho t}$, we can see that if μ/ρ jumps up abruptly then I_T must decrease abruptly. In this way we can see that the absorption jumps in Fig. 11 must correspond exactly to the jumps in Fig. 10. Tables of μ/ρ versus λ are available[4 – 6] for most elements covering a large range of available wavelengths. Remember that the radiation must be monochromatic when using these tables but the sample need not be a pure element. The mass absorption coefficient of any compound may be determined if the weight fractions of various elements that make up the compound are known. Thus, for example, the μ/ρ of $CaCo_3$ when using $AlK\alpha$ radiation is given by:

$$\left[\mu/\rho\right]^{CaCO_3}_{AlK\alpha} = (\mu/\rho)^{Ca}_{AlK\alpha} \text{ w/o Ca in } CaCO_3$$

$$+ \quad (\mu/\rho)^{C}_{AlK\alpha} \text{ w/o C in } CaCO_3$$

$$+ \quad (\mu/\rho)^{O_2}_{AlK\alpha} \text{ w/o } O_2 \text{ in } CaCO_3$$

XII. THE MEANING OF THE MASS ABSORPTION COEFFICIENT

It is not the intent of this book to present the theoretical foundations of x-ray spectroscopy. In the case of the mass absorption coefficient (μ/ρ), however, a little more detail will be helpful particularly for the discussion of background intensity which will appear in a subsequent paragraph. Liebhafsky[2] has given a detailed view of the mass absorption coefficient which is presented in a summary fashion in Fig. 12 below. This figure clearly shows the fate of a monochromatic x-ray beam when it impinges upon a substance. We can see in the figure that two general processes are occurring. Both of these processes involve the loss of energy from the incident x-ray beam. One of these processes, designated τ/ρ, involves the production of characteristic x rays and the emission of photoelectrons while the other process, designated σ/ρ, is related to two types of x-ray scattering. The first type of scattering is called unmodified scattering, and in this

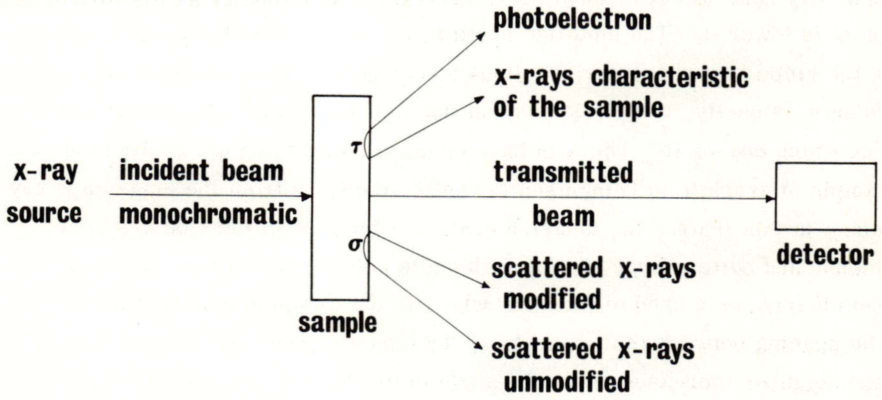

Fig. 12. The mass absorption coefficient (μ/ρ) as outlined by Liebhafsky[2].

case there is no change of wavelength when the incident beam I_o is scattered from the substance. In the second type of scattering the wavelength is changed and the scattered wavelength is slightly longer than the incident x-ray wavelength. This type of scattering is called Compton modified or simply modified scattering. The mass absorption coefficient (μ/ρ) is considered to be the sum of τ/ρ and σ/ρ and the relative importance of these two terms will vary with the position of the absorbing element on the periodic chart. Elements as light as boron, carbon, and lithium are particularly good scatters of x rays. Carbon, for example, scatters x rays to such a degree that the σ/ρ term dominates the τ/ρ term for certain incident wavelengths. At the same time it is well to point out that the fraction of the scattered x rays that are modified (that is, changed in wavelength) increases as the atomic number decreases. Thus, for a heavy element such as lead, the percentage of scattered radiation that has changed wavelength is quite low, whereas for an element such as lithium the amount of scattering is high and most of the x rays which are scattered are changed in wavelength. We have already discussed τ/ρ in previous chapters and have shown that it is related to the production of characteristic x rays. It is associated with the absorption edge of a material and, in general, it may be said that for most of the element on the periodic chart, τ/ρ is the most important part of μ/ρ.

What happens to the scattered radiation and where does the scattered radiation go? A large portion of it finds its way into the background, i.e., the region between the analytical lines. The unmodified scattering appears as a very general background which increases in intensity as the diffraction angle is lowered. The modified scattering, on the other hand, also appears in the vicinity of characteristic lines and often has the appearance of a low to medium intensity, very wide peak on the high wavelength side of the spectral line which causes it. This can be seen quite clearly in Fig. 13 where a sample of graphite is being used to scatter the lines from the tungsten x-ray tube. In this figure, the tungsten scattered lines from the tube are very intense and quite sharp. On the high angle side of these lines, however, one can clearly see a band of intensity which is the Compton modified line. The spacing between the parent line, the tungsten line, and its Compton modified neighbor increases as θ, the angle of diffraction, increases. These Compton lines can often interfere with spectral intensity measurements.

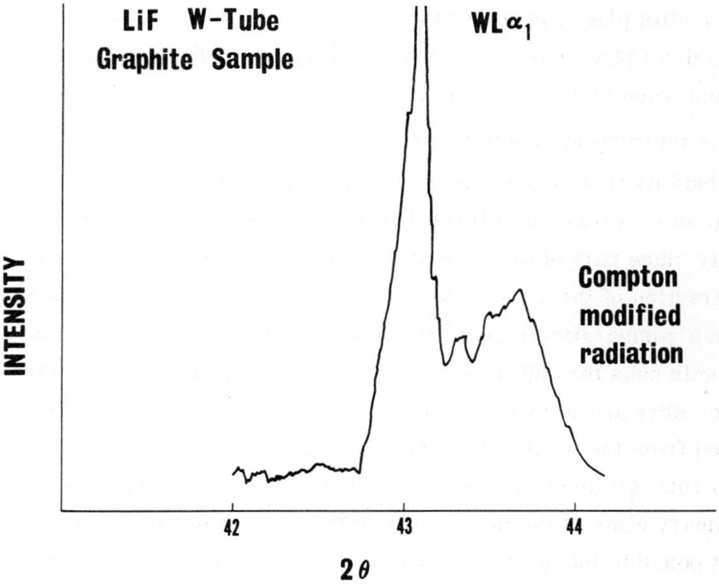

Fig. 13. Scattering of x rays by graphite.

XIII. MEASUREMENT OF X-RAY SPECTRA

Geometry and Basic Principles of Detection and Measurement of X-Ray Spectra

Figure 7 shows a schematic view illustrating the basic elements of an x-ray spectrometer. In this particular illustration, the sample is placed above the x-ray tube, in the so-called inverted optics position. The high energy x-ray tube A is placed so that its window is within a few centimeters of the sample S. The fluorescent radiation produced at the sample S passes out of the tube-sample chamber through a set of slits called soller slits. These slits consist of parallel plates stacked to form a bundle having a small spacing between the plates. They serve to direct the fluorescent x-ray beam onto the analyzing crystal at position C. The responsibility of the analyzing crystal is to spread the fluorescent x-ray spectrum over a large angular region so that the movable detector at position D can record the spectrum. In order for Bragg's law to be satisfied, the tilt angle of the crystal (θ) must be exactly one half that of the detector (2θ) at all times. This relationship is maintained by an appropriate gearing arrangement. In addition to the entrance collimator slit, which we have discussed, there is usually another set of soller slits placed in front of the detector. These slits serve to reduce vertical divergence from rays diffracted by the crystal and are extremely important when high resolution is desired.

It is interesting to note that each of the steps that the x-ray beam takes as it winds its way through the spectrograph results in a loss of intensity. Passage of an x-ray beam through a set of soller slits obviously reduces the intensity since part of the area of the slit is blocked by solid metal. Even the diffraction of the x-ray beam by the crystal is rather inefficient and results in a comparatively high loss of x-ray intensity. This diffracted beam must again pass through a set of soller slits which are usually finer than the entrance slits and in this step another large fraction of the x-ray intensity is removed from the beam. Finally, this beam passes into a detector which is seldom 100% efficient in detecting x rays. It is clear, therefore, that one of the primary aims of the spectrograph design is to produce at the sample the highest possible intensity so that collimation, etc. will not reduce the intensity to such a level that measurements take an undue amount of time. This problem is particularly troublesome for the light elements where the efficiency of x-ray production is very low, that is, where the quantity of photons produced is much lower than for the medium or heavy elements. It should also be emphasized that for light elements, the spectrograph must be provided with a vacuum x-ray path or alternatively the entire x-ray path must be filled with a low μ/ρ gas such as helium.

XIV. CRYSTALS AND X-RAY DIFFRACTION THEORY

A. X-Ray Diffraction

The diffraction of x rays by a single crystal can be easily understood if one considers the interaction of the x-ray beam with the individual electrons surrounding each atom. Let us consider for a moment the effect that an incoming x-ray beam has upon a single electron. When the electron is struck by an x-ray photon, it begins to oscillate with the same frequency as the incoming radiation. Since the electron is now an oscillating charged particle it becomes a source of electromagnetic radiation. This point source of electromagnetic radiation now radiates a wavelength of the same magnitude as that of the incoming electromagnetic radiation providing there is no loss of energy in the original photon-electron impact.

In general, there will be very little energy loss in this impact if the electrons are tightly bound as is the case for elements where $Z > 20$. The electrons in the single crystal will scatter x rays in this fashion. Now in certain directions these scattered x rays will be in phase and thus reinforce each other. In other directions they will be out of phase and cancel each other. In the direction in which the scattered x rays are in phase we get a high intensity of scattered x rays. The angular position at which this occurs can be obtained from Bragg's law according to Fig. 14 below. Here we see an edge of a single crystal showing two parallel planes of atoms. The atoms on these planes are in a dense regular array and we may look upon these planes as planes containing a high density of electrons. The final result, $n\lambda = 2d \sin\theta$, given in Fig. 14, is known as Bragg's law which relates the angle of incidence of the x-ray beam (θ) to the angle of emergence of the x-ray beam (θ). It is clear from this figure that the angle of incidence equals the angle of diffraction. In the small figure to the upper right we see the arrangement of the single crystal with respect to the two angles, θ and 2θ. Recall that the detector is always at an angle of 2θ and the crystal is at an angle of θ. These angles are measured from the direction of the incoming x-ray beam. In

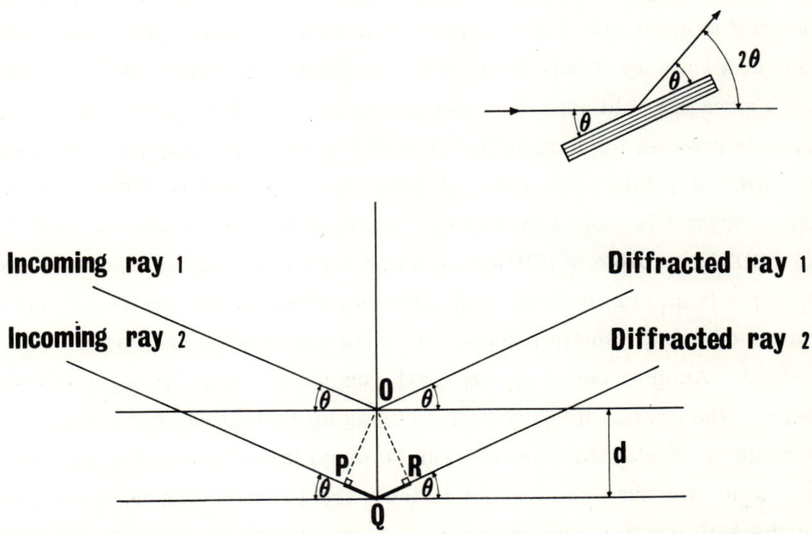

$\overline{PQR}$ must equal $n\lambda$ for diffraction to occur (where n is an integer)

$$\overline{PQR} = \overline{PQ} + \overline{QR} = d\cos(90-\theta) + d\cos(90-\theta) = 2d\sin\theta$$

$$n\lambda = 2d\sin\theta$$

Fig. 14. Diffraction of x rays and Braggs law.

Bragg's law, n stands for the order of diffraction, λ is the wavelength of the incoming x-ray beam, d is the spacing between the set of planes, and θ is the angle of diffraction. If you divide both sides of Bragg's law by 2d, we obtain:

$$\sin\theta = \frac{n\lambda}{2d}$$

Since sin θ is limited to lie between the values of 0 and 1, a crystal with a fixed d spacing is only capable of diffracting a limited range of wavelength. For example, a lithium fluoride single crystal which has a 2d spacing perpendicular to the (200) planes of 4.026 Å can only diffract wavelengths up to 4.026 Å. However, for this to occur, the detector would have to be making an angle of $2\theta \simeq 180°$ which, in practice, is impossible to obtain because of the geometry of the spectrometer. Most spectrometers are capable of 2θ angles of 150-160° at a maximum, and consequently, a crystal can never be used to its fullest extent. Fortunately, there are many crystals available today and almost the entire x-ray wavelength region can be covered. One might reasonably ask why not use one crystal for all wavelengths, that is, why not pick a large d spacing crystal and use this crystal for all wavelengths you want to analyze? Unfortunately, this practice would place many short λ lines at a very low θ angle where the background is very high. In addition, this approach would not take advantage of the high diffracting power some crystals possess for certain wavelength x rays. For example both quartz and EDDT crystals are capable of diffracting phosphorus $K\alpha$ x rays but the quartz crystal is more efficient by a factor of nearly 2. On the other hand, the EDDT is capable of diffracting Al$K\alpha$ x rays as well as many of the other elements from aluminum through calcium, whereas the quartz crystal is limited to a much smaller region since its maximum d spacing is only 6.686 Å. Another factor governing the choice of a crystal is the fluorescence of the crystal itself when it is being hit by high energy x rays. A good example of crystal fluorescence can be found in the analysis of phosphorus. One could, for example, use an ADP crystal to diffract phosphorus x rays but the ADP crystal (ammonium dihydrogen phosphate) contains phosphorus and would be excited by the continuous tungsten x rays which come in through the entrance collimator. Thus the crystal itself would have a very high background of phosphorus even when the sample you were analyzing contained no phosphorus at all.

B. Crystal Selection

Many factors govern the choice of an analyzing crystal. Some of these factors have been discussed in the preceding paragraphs, and a more complete list is given in Table I. Perhaps the single most important feature of a crystal is the interatomic or d spacing since this governs what range of wavelengths can be diffracted. Figure 15 shows the range of wavelengths that can be diffracted by the commonly used analyzing crystals. If we limit our attention to the elements between Z = 13 and Z = 92, it is clear that for most applications LiF and EDDT would be sufficient.

TABLE I

Factors which Govern the Selection of the Analyzing Crystal

Interatomic spacing
Diffracting power
Rocking curve breadth (resolution)
Chemical composition
Chemical stability
Stability under x-rays
Size available
Cost

C. Curved Crystal Optics

The preceding paragraphs have discussed some of the aspects of flat crystal optics, namely the choice of the flat crystal. Before discussing other parts of the x-ray crystal optics system, we will take a brief look at what is called "focusing" curved crystal optics. Analytically useful x rays have too short a wavelength to be focused like light, i.e., through a lens. Nor can they be focused like electrons by an electromagnetic lens since x rays carry no charge. However, there is a way that x rays may be "concentrated" into a small region. This is accomplished using nothing more than simple Bragg diffraction with the exception that the crystal is curved and ground in a special way. Figure 16 illustrates the manner in which the crystal must be curved and ground. In order to satisfy Braggs Law the following conditions must be met: The crystal is curved [either plastically (LiF, NaCl, etc.) or elastically (quartz, topaz)] to a radius R and ground to a radius of R/2. This second step puts the crystal surface on the so-called

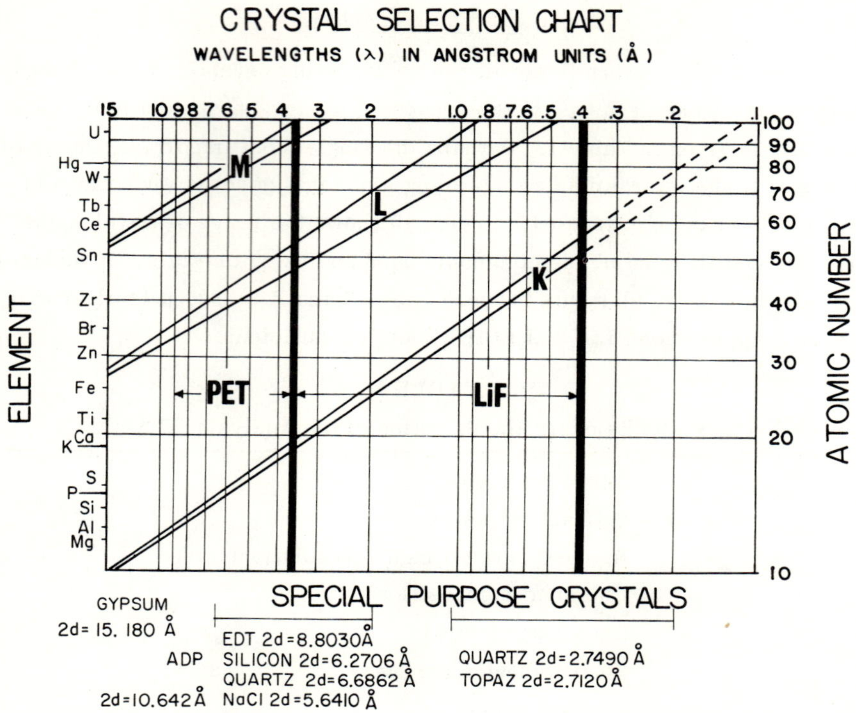

Fig. 15. Range of wavelengths diffracted by commonly used analyzing crystals.

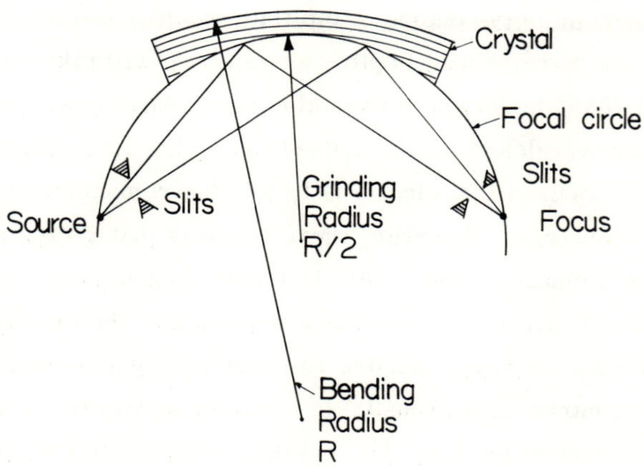

Fig. 16. Curved and ground single crystal.

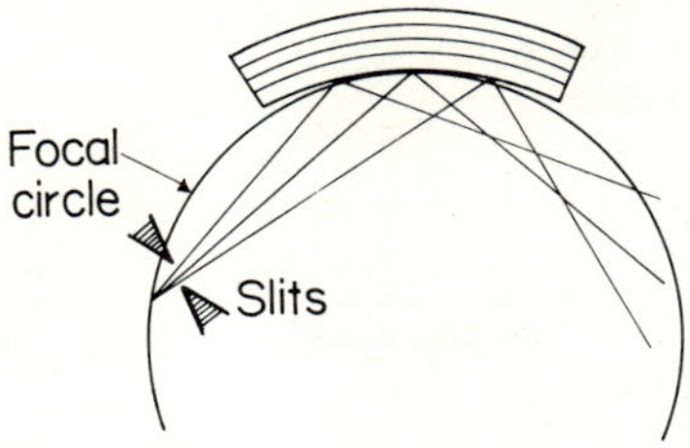

Fig. 17. A crystal which has been curved but not ground.

"focusing" circle whose radius is R/2. Note that the source (a line-shaped x-ray beam perpendicular to the drawing at S) of x rays (which must be divergent) and the detector slit are also constrained to lie on this "focusing" circle. If the crystal is simply curved but not ground (Fig. 17), it can be seen that many of the divergent rays coming from the line-shaped source S will not be such that the angle of incidence and the angle of diffraction θ are equal except at the center of the crystal (point of tangency with focusing circle)*. If R is large, this later condition may not be very critical. The details of curved crystal optics are beyond the scope of this book but several excellent treatments of this subject exist[1-2], and should be read by those interested in x-ray spectroscopy having this type of x-ray optical system. The x-ray spectroscopist always prefers a system that will produce well separated analytical lines (high resolution), a high peak to background ratio, and a high overall total line intensity. In general, these conditions are difficult to obtain simultaneously, but a very acceptable compromise is obtained from present day equipment. In our treatment of crystals we will confine our attention to flat crystal optics and briefly discuss the problems of resolution in the paragraphs below.

XV. RESOLUTION OF X-RAY SPECTROGRAPHS

A. Crystals

In general, achieving a high degree of spatial resolution of analytical lines is accomplished at the expense of total line intensity. Often, however,

*The reader can easily verify these previous statements concerning curved crystals by constructing a detailed drawing of the curved crystal optical system.

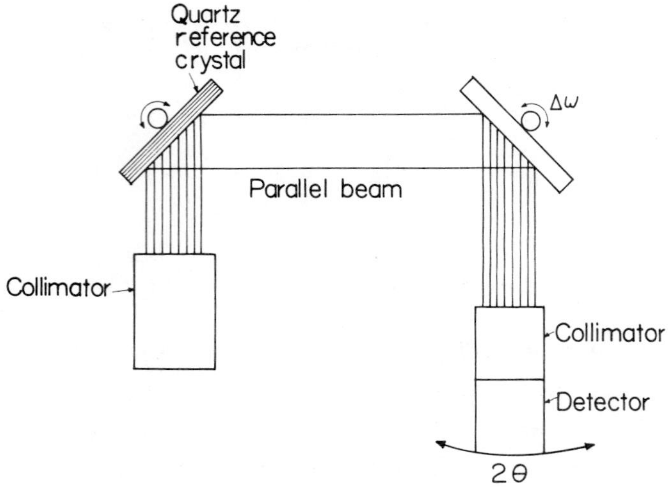

Fig. 18. Double crystal spectrometer.

the peak to background ratio is significantly improved, and this after all is the limiting factor when low concentrations are being measured. The resolution is dependent upon the type and perfection of the crystal and the characteristics (length and plate spacing of the collimators).

In order to discuss the perfection and diffracting power of the analytical crystal we must define a few crystal parameters. These terms are associated with the so-called "rocking curve" of the crystal. This curve is obtained by "rocking" (slowly moving the crystal through the Bragg angle) the test crystal in a highly parallel beam of monochromatic x rays. Figure 18 above illustrates a double crystal spectrometer used to obtain "rocking curve" data. The rocking curves look very much like a Gaussian or bell-shaped curve. The parameters of interest are labeled in Fig. 19. These parameters are W, the angular width of the peak at half maximum intensity, P, a ratio of quanta diffracted (precisely at the Bragg angle) to quanta incident, and R, the area under the curve which is related to the total reflecting power of the crystal. Table II lists these parameters for several crystals in common use. Birks shows that the wavelength resolution r_λ is related to W through the relation:

$$r_\lambda = 2dW \cos \theta$$

Table II clearly illustrates that no one crystal has all the desired properties. In general, most crystals used in x-ray spectrographs today have small enough W values for adequate resolution. A more limiting factor, generally, is the diffracting power R and the availability of the correct 2d spacing. This latter fact is particularly troublesome when the very light elements are being analyzed, i.e., sodium to boron.

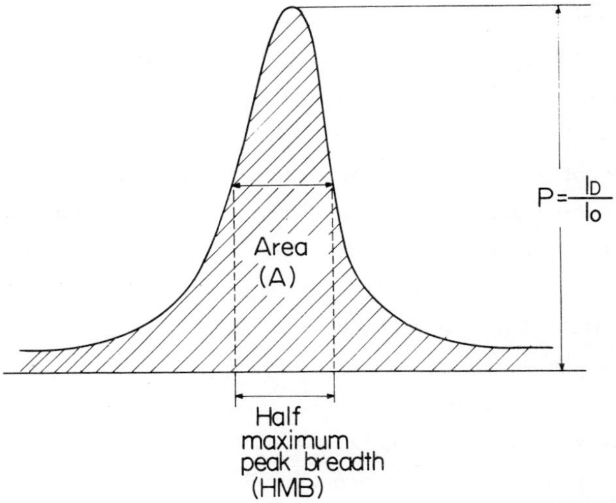

Fig. 19. Crystal parameter measured from rocking curve.

TABLE II

Crystal Parameters for Some Common Crystals Used in X-Ray Spectroscopy

Crystal	2d (Å)	Condition	W	P	R
Topaz	2.7	Abraded and etched	1'45"	6.5%	4×10^{-5}
LiF	4.02	Abraded and etched	1'50"	51.6%	4.5×10^{-4}
Quartz	8.46	Abraded and etched	10"	33.0%	2.7×10^{-5}
EDDT	8.8	Abraded and etched	55"	47.5%	1.7×10^{-4}
Graphite[a]	6.68	Pyrolytic	~0.30°		

[a] Graphite monochromators are now in use having a scattering power twice that of LiF.(15)

B. Collimators in X-Ray Spectroscopy

A collimator, as the latin root implies, is a device used to produce a parallel beam. The most frequently used type of collimator consists of a series of equally separated thin, flat, parallel plates. The "resolving power" of a collimator increases as (1) the length increases and (2) the plate spacing decreases.

Figure 20 shows a more detailed view of a collimator in action. Note that only those rays which are within the angular range $\theta \pm \Delta\theta$ will be able to enter a detector placed behind the collimator. $\Delta\theta$ is determined by the design of the collimator. As $\Delta\theta$ increases, the collimator can pass more energy with a resulting loss of resolution. How much resolution is necessary? It might, at first, appear foolish to answer this with the statement, "Only as much as you need," but let us examine the problem. Any collimator reduces the peak intensity, but it also reduces the background intensity at the same time. Unless you are trying to separate two very closely spaced lines, use only enough collimation to maximize the peak to background ratio. A high resolution flat crystal system using a good LiF crystal and two fine spacing 4 inch long collimators will produce a line having a width at half-maximum intensity of 0.05°. This is sufficient to resolve $CuK\alpha_1$ from $CuK\alpha_2$. It is very instructive to calculate the intensity loss ($I_{transmitted}/I_{incident}$) in such a system. This ratio will be in the neighborhood of 10^{-5}. The aesthetic value of a separated $K\alpha_1$-$K\alpha_2$ seems hardly worth the trouble.

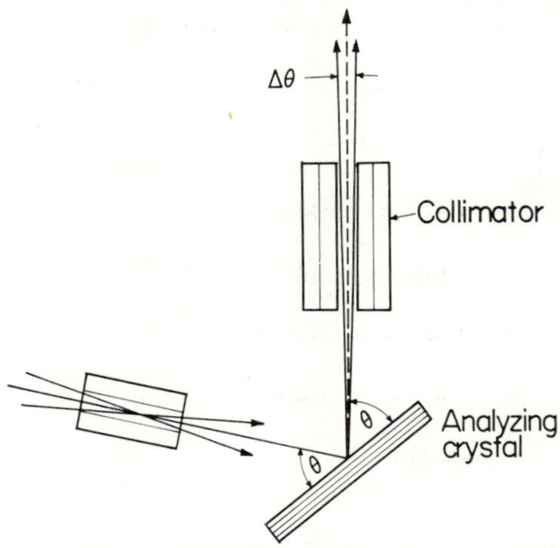

Fig. 20. Detailed view of collimator action.

XVI. RADIATION DETECTORS

A short twenty years ago this section of the chapter would have been very short. Indeed we would have needed to discuss only two methods of radiation detection, mainly, photographic film, and Geiger counters. In present day instruments, these methods of measuring x-ray intensity have been almost completely replaced by the more sophisticated and far more useful proportional "type" detectors. This group of detectors includes the scintillation counter, the sealed proportional counter, the gas flow proportional counter, and most recently, the recently emerging total solid state detectors.

All of the detectors in this list share at least one common important feature, namely energy discrimination. This means that the detector can "tell" what energy radiation is incident upon it. Now energy is, of course, related to the frequency and therefore to λ, the wavelength of the incident beam. Thus, proportional type detectors when used with appropriate electronics (pulse height selectors, etc.) have the ability to separate out and measure that part of the incident beam intensity corresponding to a single band of energies ($E \pm \Delta E$) which implies, of course, a wavelength band of $\lambda \pm \Delta\lambda$.

A. Choice of Detector

Just as with crystals, there is not one detector that can be used with all wavelengths. Therefore, a choice of detector must be made when one uses the x-ray spectrograph. Fortunately, this choice is not too difficult. Figure 21 shows the efficiency of several detectors to various wavelength radiations. This figure shows that for $\lambda \leq 2.5$ Å, a scintillation counter is nearly 100% efficient. For longer wavelength x-rays (i. e., from elements Z = 12 to Z = 20) a thin window flow type detector is preferable. Counting efficiency is not the only criterion for choosing a detector. Several other factors such as energy selectivity, noise level, and dead time must be considered. Before briefly discussing each type of detector in more detail, we will turn our attention again to the energy discrimination ability of these proportional type detectors.

B. Energy Discrimination of Radiation Detectors

Perhaps the easiest way to visualize the energy discrimination ability of a detector is to examine the nature of the process by which a photon is

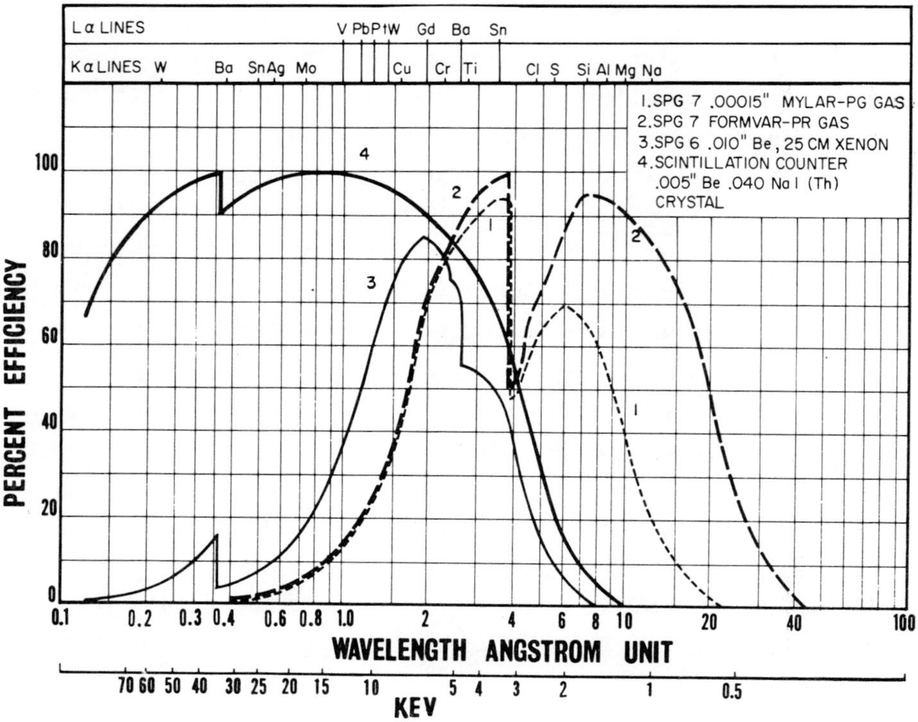

Fig. 21. Radiation detector efficiency as a function of λ.

detected. In all types of detectors, including film, energy must be absorbed in some way in order to be detected. The different types of detectors only differ in the manner in which this absorption takes place.

Consider the sealed gas proportional counter tube shown in Fig. 22. The energy absorption process takes the form of ionization (formation of a positive ion and a free electron)of the gas that fills the tube. Assume that the filling gas in the tube is krypton whose ionization potential is 13.93 V. Let us further assume that two equally intense, monochromatic x-ray beams are simultaneously incident upon the tube window, say MoKα and NiKα. (For the moment let us also neglect the fact that the detector is approximately seven times more sensitive to NiKα radiation than to MoKα radiation.) One only has to look in the critical x-ray excitation energy tables to see that MoKα possesses a photon energy of 19,996 V while NiKα has an energy of 8329 V. How many ionization events will occur when a photon of MoKα passes

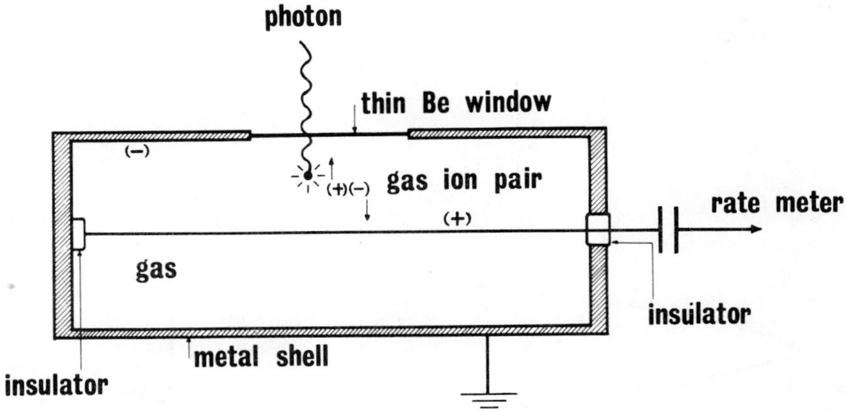

Fig. 22. Sealed gas proportional detector.

into the counter tube? If the process of ionization were 100% efficient, the answer would be 19,996 V/13.93 V or about 1435 events. Experience has shown that this process is closer to 50% efficient so that 717 is a better figure. Using the same procedure for NiKα, we find 300 ionizations will be produced. One might reasonably ask the question, "Does each MoKα photon produce exactly 717 ion pairs and each NiKα exactly 300 ion pairs?" Nature being what it is, the answer would have to be, "No." Rather for MoKα we do find that the most probable number of ion pairs will be 717, but that around this number there will be a spread. The statistical variation in the number of ion pairs formed will be $\sigma = N^{1/2}$ (i.e., one standard deviation).

$$1\ \sigma_{MoK\alpha} = 27; \quad 1\ \sigma_{NiK\alpha} = 17$$

Thus if we plot the number of ionization events for MoKα and NiKα, we will not get sharp lines but rather a distribution of events about the average (Fig. 23). Thus the process occurring in the detector tube may be summarized as follows:

Photon → N ± ΔN simultaneous ionization event

Output pulse whose amplitude (1-2 mV) is proportional to the number of ionizations

Thus (neglecting detector efficiency), each entering photon produces a distribution of pulses whose amplitudes are centered about a mean value which

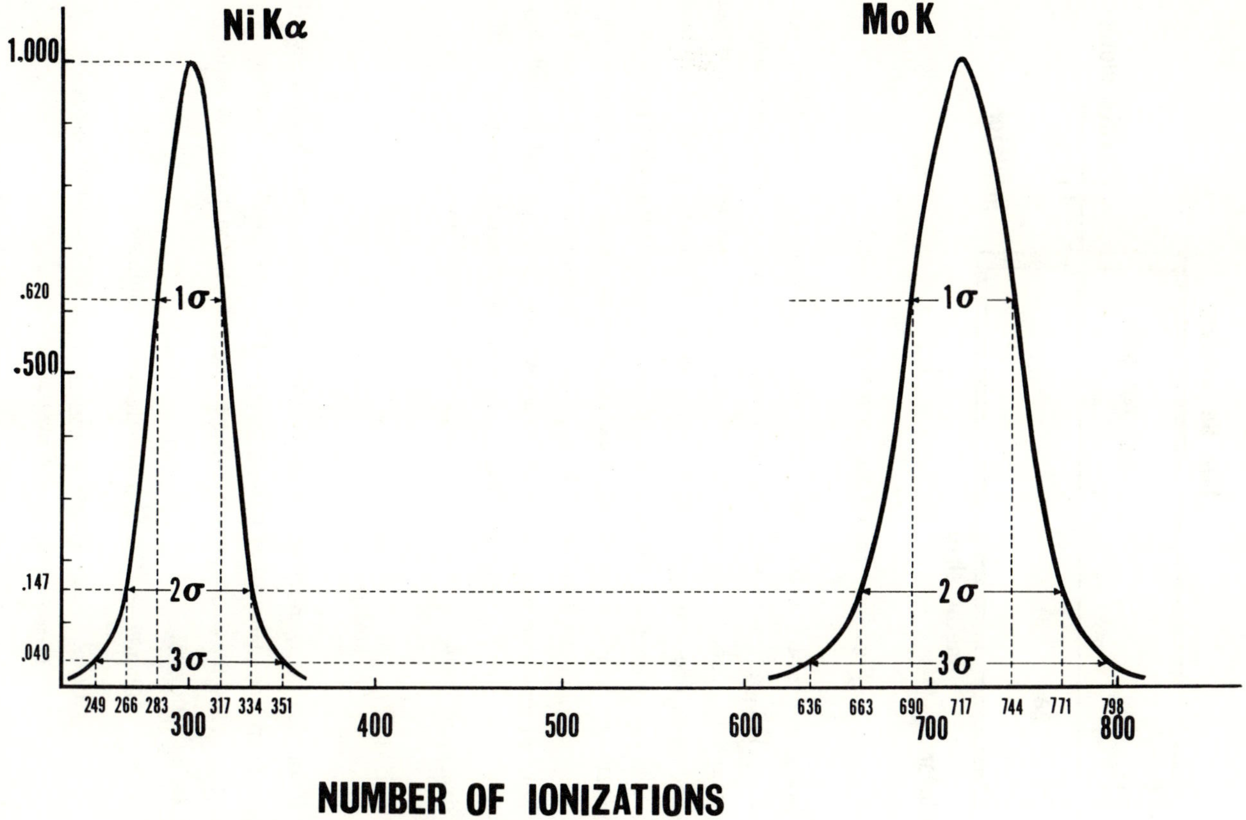

Fig. 23. Width of pulse amplitude distribution curves and the number of ionization event.

depends on the energy of the photons. Pulses coming from the detector are amplified in a linear manner so as not to alter the relationship between photon energy and pulse amplitude. These amplified pulses (now in the volt range) are fed into a pulse amplitude analyzer and discrimator which we will discuss in the next paragraph.

C. Pulse Amplitude Analysis and Discrimination

We know that the pulses caused by MoKα photons have amplitudes (in volts) larger than the NiKα pulses. In actual fact these pulses are generally amplified to lie in the voltage range of approximately 0 to 100 V. If we could sample a small piece of the voltage range at a time we could easily locate the voltages (amplitudes) of our MoKα and NiKα pulses. In Fig. 24 this narrow sampling voltage (1-2 V) is appropriately called a window. The general practice is to move this window along the voltage scale without changing the window width. The "window sill" is called the base line voltage. This voltage can generally be varied from 0 to 100 volts of 0 to 10 V depending on the equipment. In some equipment the base line is "scanned" manually

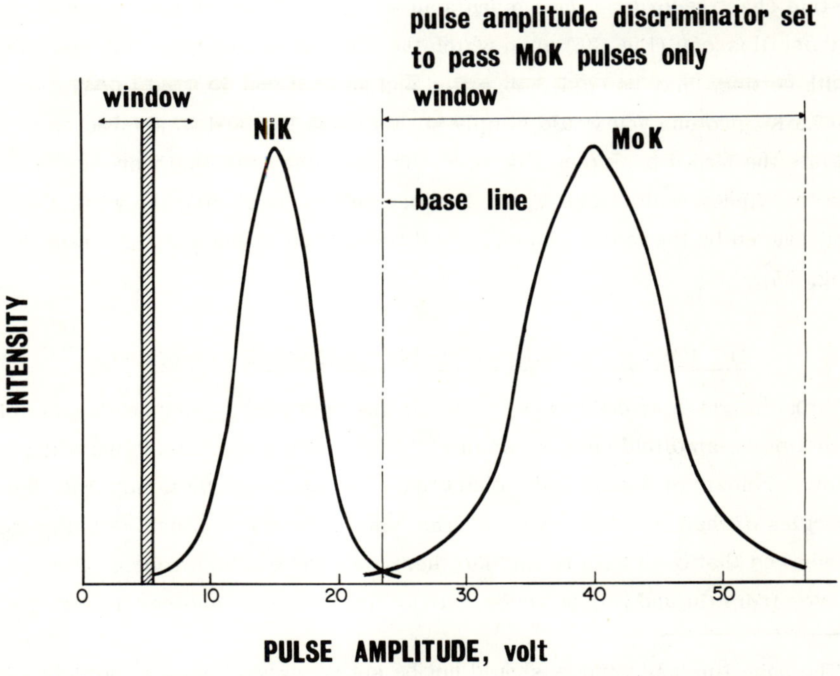

Fig. 24. Pulse amplitude (volts) for MoKα and NiKα x rays.

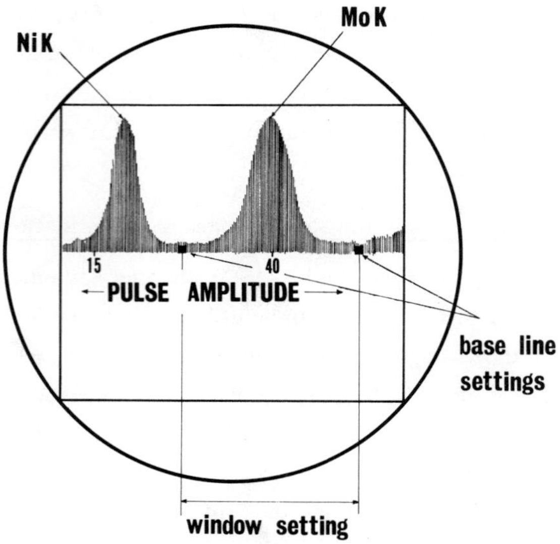

Fig. 25. Cathode ray oscilloscope display of pulse amplitude distortion.

while in others it may be automatically scanned so that the results appear on a strip chart recorder. One manufacturer displays the pulses on a cathode ray oscilloscope (Fig. 25) upon which the discriminating base line and window width settings may be easily observed and set. If you desired to count only the MoKα photons you would simply set the base line and window so as to isolate the MoKα peak (see Fig. 24). Thus, if the base line were set at 25 V, and the window width at 30 V,* only those pulses (excluding the weak NiKα tail) caused by the MoKα photons could be counted in the scaling circuitry (Fig. 25).

D. Energy Resolution of Pulse Amplitude Discriminator

One might reasonably ask, "What is the best energy resolution possible with a pulse amplitude discriminator?" Or perhaps said in another way, "How far apart on the periodic chart must two elements be to separate the energies of their Kα lines with a pulse amplitude discriminator?" Figure 24 showed that even with Ni and Mo there was some interference. PAD curves from Cu and Zn, however, would blend into one another almost

* The base line and window should not be set so tightly that a minor shift in peak position will cause a large change in output intensity.

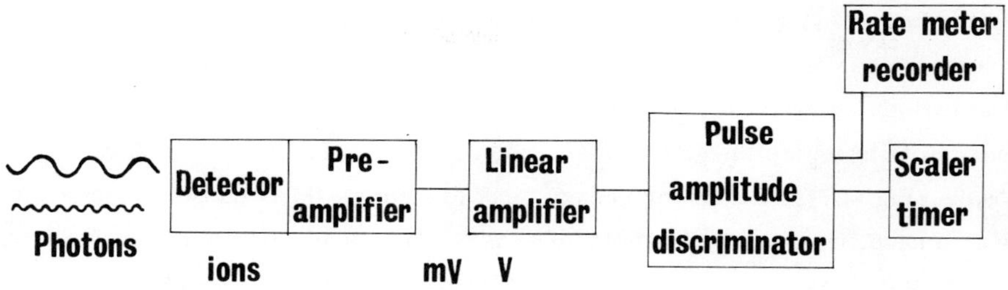

Fig. 26. Schematic pulse counting electronics.

imperceptably, thus preventing visual separation.* At present, the best energy dispersion in gas type detectors is approximately 2-3 keV. However, in the past several years, a new type of solid state diode detector [7] has been introduced. The manufacturers [8-9] state that the energy resolution is better than 1 keV, (~250 V). The next few years should see a rapid development of total solid state detectors and increased use of high energy non-dispersive spectrographs.

E. Summary of Pulse Amplitude Discrimination

Pulse amplitude discrimination may be used to eliminate undesired radiations that are simultaneously incident upon the detector. This is best exemplified by a so-called order interference when a 1st order line is being interfered with by a higher order one of a heavier atomic species. For example, second order $CaK\beta_5$ will be diffracted at approximately the same angle as 1st order $PK\alpha$, but the energy of second order $CaK\beta_5$ is proportional to 3.074 Å, which, of course, is much more energetic than 1st order $PK\alpha$. High energy cosmic noise, fluorescence, and scatter from the analyzing crystal and slits, etc. can also be eliminated successfully by PAD. This results in a lower background and a higher peak to background ratio.

A process which may occur with the use of proportional counters and PAD is called the escape peak phonomenon. It is due to the x-ray fluorescence of the gas in the detector tube by the incident radiation. This x-ray fluorescence of the detector gas requires energy which must be taken from the incident

* Mathematical separation is still possible (see Birks[1]) and provides the basis for a unique method of analysis called nondispersive (no crystal) x-ray spectroscopy.

x-ray beam. In general, these "gas" x-rays escape from the detector because the gas is transparent to its own x-rays. Thus, the energy of many of the incident quanta have been reduced, but their remaining energy is still sufficient to ionize the gas atoms. The result is a PAD curve with two peaks (Fig. 27). The lower energy peak (escape peak) is due to those quanta which have first caused the detector gas to fluoresce. If conditions are just right, this peak may be much larger than the higher energy peak.

XVII. DETECTORS - BRIEF DESCRIPTIONS OF TYPES

A. Gas Proportional Detector

We have already seen this type of detector in Figure 22. A modification called the flow-gas proportional counter is used for low atomic number ($Z < 20$) analysis. It has the same electronic properties as a sealed gas detector, but as the name implies, a gas mixture is slowly flowed through the detector. This allows the detector to be constructed with a very thin window which is needed for the longer wavelength x-rays coming from elements such as aluminum.

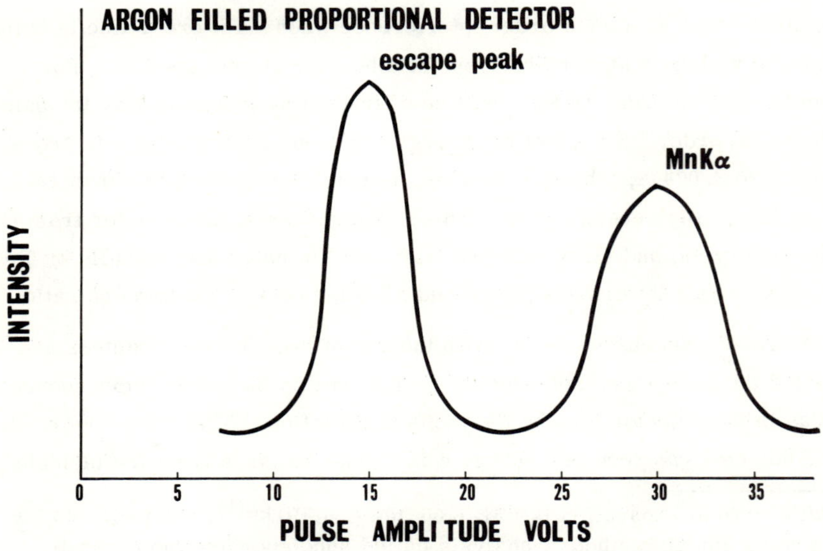

Fig. 27. Escape peak in gas-filled detector.

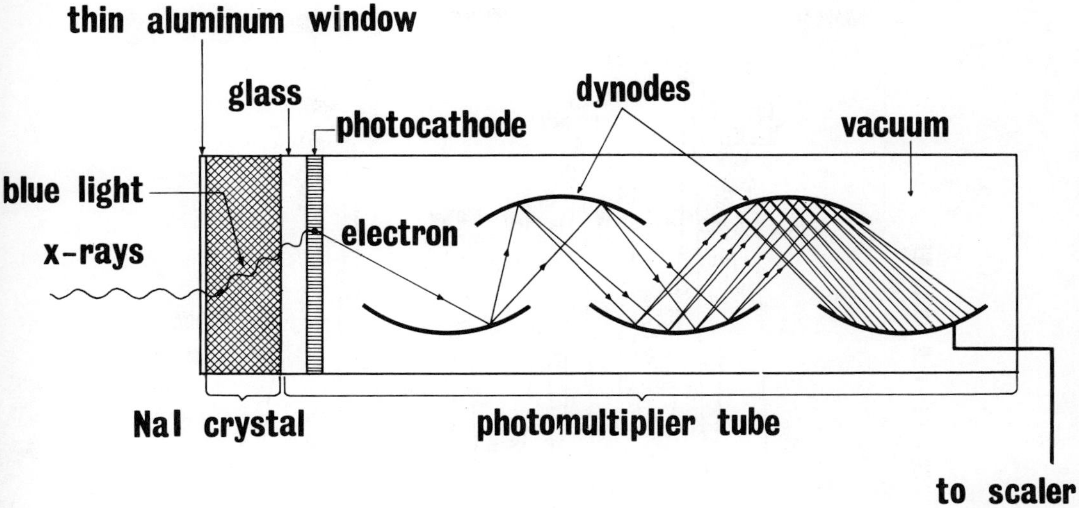

Fig. 28. Schematic drawing of scintillation detector.

B. Scintillation Detector

Figure 28 shows the construction of the highly sensitive scintillation detector. In analogy to the gas detector, the x-ray photons are absorbed in the (light shielded) thallium doped NaI crystal by a process which produces a blue light whose intensity is proportional to the energy of the incident beam. This blue light strikes a photocathode material which emits electrons in proportion to the intensity of the incident light. These electrons are amplified by the photomultiplier tube. Thus, large pulses are produced which are proportional to the energy of the incident x-ray beam. In the past, these detectors have been noisier (when no x-rays are incident) than the gas-filled counterparts. Recent solid state models, however, are very quiet and have somewhat improved energy resolution. This detector is recommended for general purpose x-ray spectroscopy of heavier elements due to its high sensitivity.

C. Solid State Detectors

Several companies have developed a new compact nondispersive x-ray spectroscopy system which utilizes various high energy gamma ray sources. In addition, the energy spectra is detected with a lithium drifted Si diode which must be maintained at cryogenic temperatures along with the transistorized

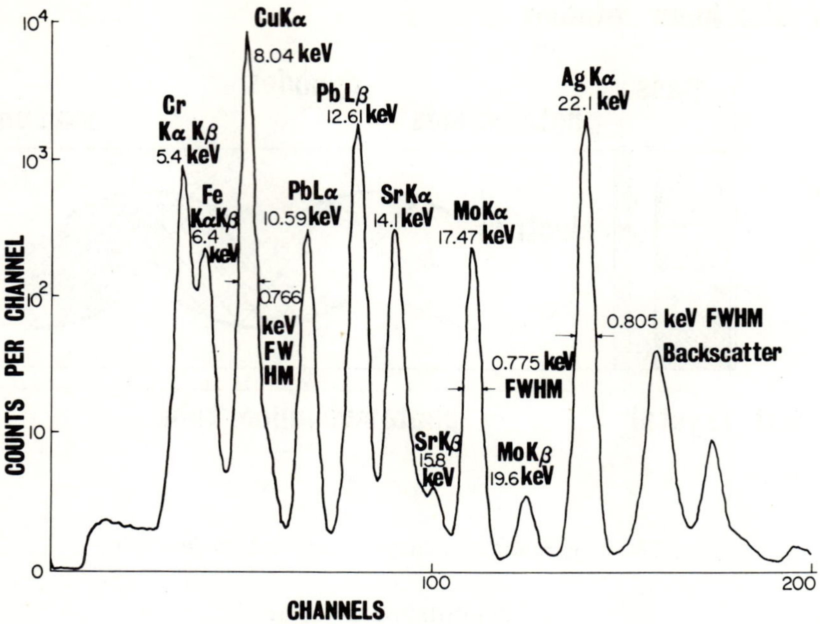

Fig. 29. Energy spectrum resolved by solid state diode detector.

preamplifier. The energy spectra is displayed on a multichannel (400-1000 channels) pulse height analyzer and the systems are now reported to have resolutions of about 180 eV and lower (Fig. 29).

D. Counting Rate Limitations of Detectors and Electronics

Modern x-ray spectrographs have been designed to produce high intensity analytical lines. In the earlier Geiger-Müller counter days very high intensities were unnecessary because the Geiger counter began to "choke up," i.e., lose linearity, when the incident x-ray beam was stronger than several hundred counts per second. All detectors, however, have a time period when they are "dead," i.e., when they are preparing themselves to count. For the older Geiger-Müller detectors this was of the order of 50 to 200μ sec, whereas, for the gas proportional and scintillation detectors it is approximately 0.2μ sec. This last figure would imply that 50,000 counts could arrive in one second and all of them would be counted!

With the exception perhaps of the most recent solid state models, most electronics used for counting pulses cannot handle count rates as high as

50,000 cps.* In fact, electronic "dead" times of the order of 1 μsec (10,000 cps maximum count rate) are very common.

Regardless of any manufacturers' claims, it is always very wise to check the linearity of every detector and its associated electronics and determine the dead time. A simple method for doing this has been outlined by Cohen.[10]

E. Spectral Interference and Background Intensity

We have already discussed one form of spectral interference, namely Compton modified scattering. There are, of course, many other contributors of unwanted radiation and noise present in a spectrograph system. These are listed in Table III.

TABLE III

Common Spectral Interference and Contributors to Background Intensity

Type	Remarks
Compton scattering	Common in low Z matrices
Tube line scatter	Strong spectra from target of excitation tube. Intensity of these lines increases as Z of matrix decreases.
Higher order lines from Elements of greater Z	Can generally be eliminated by PAD
Tube contamination lines	Tube contamination should be checked periodically by scattering its radiation from lucite (Cr, Ni, Fe, and Cu)
Analyzing crystal and component fluorescence and scatter	Can generally be eliminated by PAD
Detector and electronic circuit noise	Check detector and electronics with x-rays off regularly. Use lower amplifier gain.

* One symptom of lost counts is a premature flattened working curve.

F. Interelement and Matrix Effects in X-Ray Spectroscopy

In principle, x-ray spectroscopy consists, as we know, of producing characteristic x rays from elements in the sample using a high energy x-ray source as the exciter. These characteristic x rays are produced from a finite thickness of the sample. Much of the intensity that is measured has had to pass through the sample on its way to the detector. Several processes may occur during their passage. We know that the characteristic x rays produced from a layer beneath the surface will be weakened by absorption. But they may also be augmented by a process known as enhancement. At any rate it is rarely true in a multielement sample that the intensity of a characteristic line will be linearly related to weight percent. Let us briefly discuss absorption and enhancement and their effect upon expected x-ray intensity.

G. Absorption

As was pointed out in an earlier section, all matter absorbs x rays to a greater or less extent depending upon (1) the wavelength of the x rays, and (2) the mass absorption coefficient μ/ρ of the materials for that particular wavelength. Several extreme examples of the effect of the matrix composition upon emergent fluorescent intensity can be seen in Fig. 30 (a-c). In curve 30a a moderately heavy element such as Mo is being measured in a matrix consisting mainly of Al_2O_3. Here the μ/ρ for the matrix for MoKα is very low (<5) when no Mo is present, but begins to rise as more Mo is added. Thus, two opposing processes occur as the concentration of Mo increases: (1) the fluorescent MoKα x-ray intensity increases. (2) it becomes progressively more difficult for the MoKα fluorescent radiation to get out. The net result is a sharp initial rise in the curve followed by a gradual flattening out as we approach nearly pure Mo (self absorption). In curve b the reverse situation is illustrated. Here the matrix consists of relatively heavy elements such as W $[(\mu/\rho)^{W}_{MoK\alpha} = 102$, compared to $(\mu/\rho)^{Mo}_{MoK\alpha} = 18]$. The initial slope in curve b is very low because the MoKα is so heavily absorbed. Eventually the MoKα intensity begins to rise as the composition of the matrix again approaches pure Mo. Finally, curve c depicts the case where the matrix (μ/ρ) does not change appreciably with the addition of the element being measured. Mo added to a $TiCl_2$ matrix would be a good example. The most severe case of absorption occurs when the sample contains an

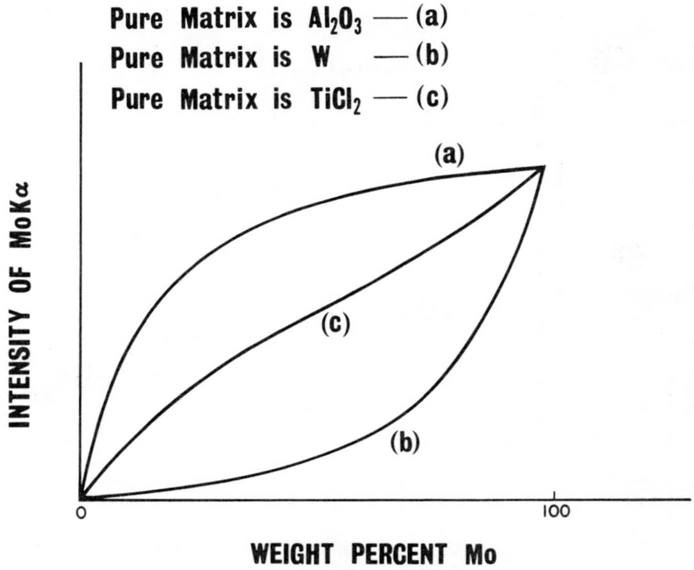

Fig. 30. Effect of matrix composition upon emergent fluorescent intensity of MoKα radiation: (a). matrix is Al_2O_3, (b). matrix is W, and (c). matrix is $TiCl_2$.

element whose absorption edge is at a wavelength slightly greater than the wavelength you are trying to measure. Two elements which interact in this manner are Fe and Cr. Cr has its Kα absorption edge at 2.07 Å while FeKα radiation has a wavelength of 1.93 Å. Thus the $(\mu/\rho)^{Cr}_{FeK\alpha} = 445$, which is a very large mass absorption coefficient.

H. Enhancement

Just as the name implies, it is possible with certain element combinations to obtain an increase in the expected fluorescent x-ray intensity of a certain element in a sample. In this situation one element in the sample has strong emission line with a wavelength slightly shorter than the absorption edge of the element being measured. We again can use Fe and Cr as a good example of this effect. As we saw, Cr strongly absorbs FeKα radiation but the end result of this absorption is the emission of CrKα x rays. Thus in a sample containing both Cr and Fe, the measured CrKα intensity would be the sum of CrKα excited by the tube radiation and CrKα excited by the strong FeKα (enhancement).

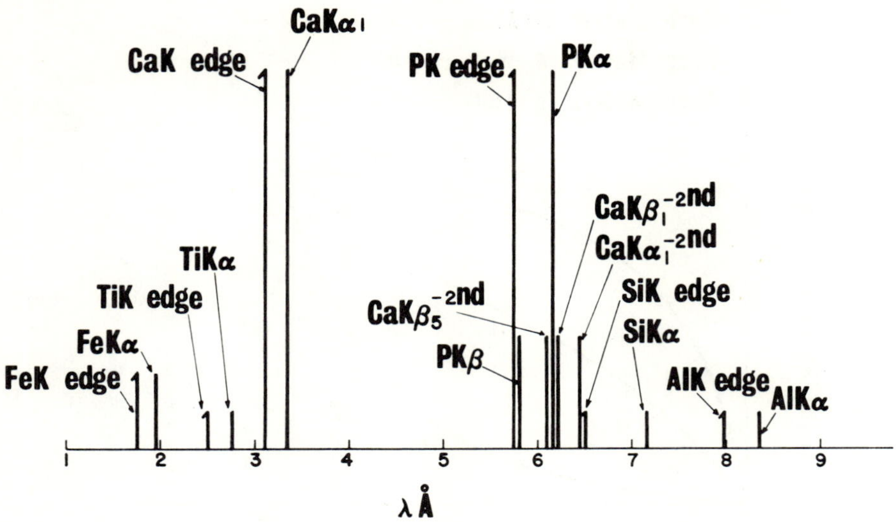

Fig. 31. Emission lines and absorption edges of the major elements in Florida phosphate rock.

I. Prediction of Interelement and Spectral Interference Effects

As a first step in setting up a quantitative analysis by x-ray spectroscopy, it is a good idea to locate on a diagram the strong emission lines and the absorption edges for all elements in the sample. Then it will be relatively easy to qualitatively evaluate what interelement effects and spectral interferences are present. Bertin[11] has treated the problem of spectral line interference in great detail and this paper should be consulted. As an added precaution one should also plot the weaker lines for the major components in the sample. An example of this procedure is illustrated in Fig. 31 for the analysis of Florida phosphate rock. Table IV lists a typical wet chemical analysis of Florida phosphate neglecting fluorides, TiO_2, etc. Note that there are two important major constituents, Ca and P,* and several minor elements such as Fe, Si, and Al. In Fig. 31 the emission lines and absorption edges of the major components are indicated by larger symbols than those used for the minor components. Note that there is a clear case of spectral interference between two 2nd order CaKβ lines and the PKα 1st order line preventing the direct use of this strong PKα line. (This problem can be completely eliminated using either pulse amplitude selection or a crystal such

* O_2, F, and C are also present in moderate quantities.

TABLE IV

Typical Dry Basis Analysis of Florida Phosphate Rock

P_2O_5	35.24%
CaO	50.33
Fe_2O_3	1.11
Al_2O_3	0.83
Acid, insoluble	2.74

as Ge which will not diffract second order lines.) Other than this effect there are no serious interelement effects in this system that will influence the measurement of phosphorus except for large changes in matrix composition.

J. Preparation of Samples for X-Ray Spectroscopy

The x-ray spectrochemical method may be applied to a sample in almost any state. For this reason there are a great many sample preparation methods and the reader is referred to references 12 and 16. These references outline many sample preparation procedures and also provide a valuable catalog of small parts and devices frequently used by x-ray spectroscopists in preparing samples for analysis.

K. Sources of Error in X-Ray Spectrographic Analysis

Neglecting the ever-present problem of obtaining a representative 10-20 g sample from a 10 lb core section or worse yet from a 10 ton stockpile, we still are left with several important sources of error in x-ray spectrographic analysis. Some of these are listed in Table V. A very brief discussion of some of the topics listed in this table is given in the following paragraphs.

TABLE V

Sources of Error in X-Ray Spectrographic Analysis

1.	Improper chemical analysis of standards
2.	Nonreproducible sample preparation procedure
3.	Drift in detector electronics
4.	Drift in x-ray tube intensity
5.	Resetting the goniometer
6.	Contamination of the optical path
7.	Gradual misalignment of spectrograph geometry
8.	Precision of mixing standards
9.	Vacuum fluctuation or He gas flow rate changes
10.	Exceeding count rate linearity of detectors or electronics

L. Sample Preparation Error

The tendency of most beginners in x-ray spectroscopy is to try and short cut the all important sample preparatory procedures. This is not unnatural since it generally stands out as the most difficult and time consuming step in the entire process. Unfortunately, there is no panacea in sample preparation except care, attention to details, and a thorough knowledge of the relationships between the physical state of the sample and the emergent x-ray intensity. In order to evaluate any sample preparation procedure one generally becomes involved with statistical techniques. This will be briefly discussed in a subsequent section of this paper.

M. Drift in Electronic Circuits

Electronic components have become more stable since the introduction of the transistor. However, at the same time, x-ray spectrographic electronics are becoming increasingly more complex. Electronic drift becomes particularly serious when using pulse amplitude discrimination since a small shift in the PAD curve can severely alter the counting rate. It is often difficult to determine the source of the drift in an x-ray spectrograph (see next section). Stable x-ray sources (radioactive isotopes) may be used to advantage to isolate drift problems.

N. X-Ray Intensity Drift

The intensity of an x-ray tube often changes slowly throughout the day. This effect may be related to the stability of the high voltage supply and the milliamp stabilizer, but it can also be due to changes in the cooling water supply (pressure, flow rate, etc.) resulting in irregular cooling of the x-ray tube header.

O. Contamination in the Optical Path

Loose powdered samples, liquid samples, and a careless operator often lead to a gradual buildup of contamination in the spectrograph. Inverted optics spectrographs (sample above the x-ray tube) should be periodically checked for contamination which may have fallen on the tube window. A single small particle of brass scratched from the overhanging sample holder housing can produce a high intensity source of Cu and Zn radiation. Spilled solutions will dry on the tube window and produce a nearly invisible source of contamination. This type of contamination can be monitored by periodically obtaining a spectrum scattered from a pure scattering sample (such as lucite). The lucite is placed in a sample position and the spectrograph is run at high intensity using the most sensitive detector and collimator system. Note that the procedure will also scatter any impurities in the x-ray tube target and allow that part of the system to be checked at the same time.

Particulate matter may also find its way into other parts of the spectrograph, such as between the blades of a collimator, on the crystal surface, etc. It is a good idea to periodically check and clean these areas of the instrument.

P. Exceeding the Count Rate Linearity on Detectors or Electronics

Several of these effects have been discussed previously where it was mentioned that the actual dead times of the detectors, and electronics should be measured by the users of the equipment. This is particularly true for users who routinely obtain and utilize high counting rates. Another serious effect is the shifting of PAD curve with counting rate changes. This effect has been discussed by Birks[1] and can lead to erroneous intensity results.

Q. Use of Standards in X-Ray Spectroscopy

The measured x-ray intensity from an element in a multielement sample can be related theoretically to the quantity of that element present. At the present time, however, the accuracy of this method is limited by the inherent small errors in available mass absorption coefficients. Thus the x-ray spectroscopists seeking highly precise results must in general turn to reference standards of some sort. In this section we will discuss some of the various types of standards that are currently being used (see Table VI).

TABLE VI

Types of Standards Used in X-Ray Spectroscopy

1. Chemically analyzed comparison standards
2. Synthetic standards
3. Internal standards

R. Comparison Standards

The best comparison standards should be nearly identical with the unknown in every way. This similarity must go beyond a simple chemical equality, i. e., the elements in the standard should be present in the same chemical compounds as in the unknown, and the physical states should also be identical. In mineral samples it is nearly impossible to satisfy these requirements unless a set of standards is obtained (by careful chemical analysis) from the same ore body from which the unknown samples will be taken. Even this procedure will not always yield satisfactory results if the matrix of the ore body is subject to large variations in composition. In the metals and ceramics industries, however, it is often very easy to prepare and use sets of chemically analyzed comparison standards because the overall composition of alloy melts are subject to rather tight controls.

S. Synthetic Standards

For reasons mentioned above, it is almost impossible to produce synthetic comparison standards for mineral systems if the unknown samples are to be examined directly. On the other hand, if one is willing to reduce the unknown samples to a common matrix composition by solution or fusion techniques, then synthetic standards become very useful. Some pitfalls are present in solution and fusion techniques, of which the x-ray spectroscopist should be aware. Notwithstanding the usual chemical problems encountered when trying to dissolve or fuse a stubborn mineral sample there is the greater problem of diluting the element of interest below a reasonable intensity level.* In addition, fusion techniques can often lead to a loss of a volatile constituent and a subsequent error in the analysis. If these problems are not serious in your particular system, then you can obtain from the solution or fusion technique the ultimate in precision and accuracy available with the x-ray spectrograph.

T. Internal Standards

Another method which has long been used in optical emission spectroscopy is the method of internal standards. This method is very effective for controlling interelement effects in the sample. Here one chooses an element to add to the sample whose excited x-radiation is acted upon by other elements in the sample in much the same manner as the radiation from the element of interest. Naturally there should not be any interelement or interference effects between the internal standard and the element being analyzed.

A set of comparison or synthetic standards is still needed and to each of these standards a fixed amount of the internal standard is added. The ratio of the analyte intensity to the internal standard intensity is plotted against the percent analyte in the sample. This method requires the additional step of weighing, and the added problem of proper mixing of the internal standard with the samples.

Every sample has its own "built-in" internal standard in the form of the scattered tube radiation. For instance, when using a W target tube the $WL\alpha_1$ line is scattered by the sample and its intensity will react to the chemical composition of the sample in much the same way as if W were in the sample itself.† Unfortunately this method has many limitations but may be used successfully in some instances.

*This is particularly true for the light elements for which a vacuum or He path is required.

†It will not, of course, be enhanced by any element in the sample.

U. Statistics in X-Ray Spectroscopy

No book on x-ray spectroscopy, or any other analytical method, would be complete without a section on statistical methods. Unfortunately, such a treatment is beyond the scope of this short chapter. However, it should be pointed out that statistics and statistical methods are a very important part of x-ray spectrochemical analysis, and as such should not be avoided. The statistical method can be a labor saving device when designing an experiment, finding a hidden source of error, or evaluating alternate procedures. Even the simple act of counting involves the statistical concept which states that a sufficiently large number of counts must be taken for the answer to be meaningful. Chapter 10 of Liebhafsky et al.[2] provides a good introduction into the basic statistical methods that are needed in x-ray spectrochemistry.

V. Suggestions for Further Study in X-Ray Spectrochemical Analysis

Throughout this brief presentation we have tried to point out the important features of x-ray spectrochemical analysis. In doing so, we have had to leave out many facts which will be important to the serious user of x-ray spectroscopy. For those who intend to utilize x-ray spectrochemical analysis, we suggest further reading. A list of available literature is given in Table VII.

TABLE VII

Some Suggestions for Further Study in X-Ray Spectroscopy

BOOKS

Birks, X-ray Spectrochemical Analysis [1]

Liebhafsky et al., X-ray Absorption and Emission in Analytical Chemistry [2]

Bertin, Principles and Practice of X-ray Spectrometric Analysis [16]

PROCEEDINGS

Advances in X-ray Analysis (Denver Research Conferences), Plenum, New York, Vol. 1-14.

JOURNALS

Norelco Reporter [13]
Analytical Chemistry
Applied Spectroscopy
ARL Spectrographers New Letter [14]
Spectrochemica Acta
X-Ray Spectrometry

REFERENCES

(1). L. S. Birks, X-ray Spectrochemical Analysis, Wiley (Interscience), New York, 1959.

(2). H. A. Liebhafsky, H. G. Pfeiffer, E. H. Winslow, and P. D. Zemany, X-ray Absorption and Emission in Analytical Chemistry, Wiley, New York, 1960.

(3). Handbook of Chemistry and Physics (40th Ed.) Chemical Rubber, Cleveland, Ohio, 1959.

(4). Norelco Reporter, May - June, 1962.

(5). International Tables for X-ray Crystallography, Vol. III, Kynoch, Birmingham, England, 1959, p. 175.

(6). L. S. Birks, Electron Probe Microanalysis, Wiley (Interscience), New York, 1963.

(7). H. R. Bowman, E. K. Hyde, S. G. Thompson, and R. C. Jared, Science, 151 [3710] (Feb. 4, 1966) 562.

(8). Technical Measurement Corp., 441 Washington Ave., North Haven, Conn.

(9). Ortec Inc., 102 Midland Rd, Oak Ridge, Tenn.

(10). J. B. Cohen, Diffraction Methods in Materials Science, Macmillan Series in Materials Science, Macmillan, New York (1966).

(11). E. P. Bertin and R. J. Longobucco, "Spectral Line Interference in X-ray Fluorescence Spectrometry," Norelco Reporter, 9 [3] (1962) 64.

(12). E. P. Bertin and R. J. Longobucco, Norelco Reporter, 6, 104.

(13). Norelco Reporter, Philips Electronic Instruments, 750 South Fulton Ave., Mount Vernon, N. Y.

(14). Applied Research Labs., Glendale, Calif.

(15). R. W. Gould and S. R. Bates, "Application of the Graphite Monochromator to Light Element, X-Ray Spectroscopy," Applied Spectroscopy, Vol. 22 5, Part 1, Oct. 1968.

(16). E. P. Bertin, Principles and Practice of X-ray Spectrochemical Analysis, Plenum, New York, 1970.

Chapter 4

ATOMIC ABSORPTION FLAME SPECTROMETRY

J. D. Winefordner

Department of Chemistry
University of Florida
Gainesville, Florida

I. INTRODUCTION

Fraunhofer [1] early in the nineteenth century observed many dark lines in the spectrum of the sun, and Kirchhoff [2] in 1860 used atomic absorption spectra to determine the presence of certain elements in the solar atmosphere. Kirchhoff and Bunsen [3,5] proved that the positions of these emission lines were identical to certain atomic emission lines. Therefore, the qualitative composition of stellar atmospheres could be determined by observing the atomic absorption spectra superimposed on the continua emitted from the central regions of stellar bodies. Since then, the approximate quantitative composition of stellar atmospheres has been determined by measurement of the total absorption of spectral lines for which the oscillator strengths of the transitions are accurately known.

Atomic absorption spectrometry — except for the measurement of mercury vapor in the atmosphere — was virtually unknown as an analytical tool until 1955 when Walsh [6] and Alkemade and Milatz [7] simultaneously published papers describing the application of atomic absorption flame spectrometry to chemical analysis. Walsh discussed the theoretical factors of importance in an atomic absorption analysis and also discussed some of the possible advantages of atomic absorption over atomic emission flame spectrometry. Because of the emphasis on analytical measurements, Walsh [6,8] is generally recognized as the discoverer of atomic absorption spectrometry.

The rather slow development of atomic absorption spectrometry as an analytical tool is somewhat surprising in view of the vast amount of research done in the area of astrophysics on the qualitative and quantitative composition of the solar and stellar atmospheres, especially the works regarding the hyperfine structure of atomic spectral lines and the studies concerning resonance radiation from atoms. The great success of flame emission spectrometry undoubtedly obscured some of the fundamental studies performed concerning atomic absorption spectrometry.

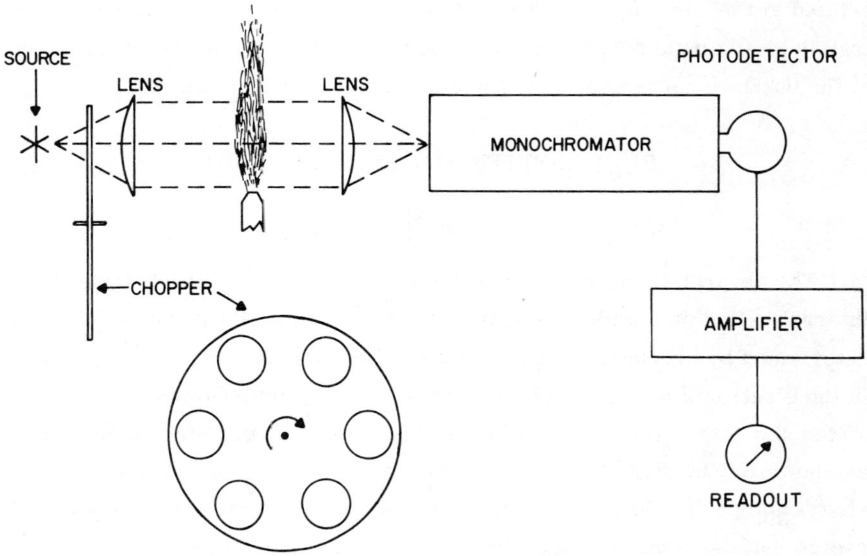

Fig. 1. Schematic diagram of an atomic absorption flame spectrometer.

This chapter is concerned with theoretical and experimental aspects of atomic absorption flame spectrometry as well as uses of this sensitive, selective method for trace metal analysis. Only articles published prior to December 31, 1966, are reviewed in this chapter.

II. CONCEPT OF ATOMIC ABSORPTION FLAME SPECTROMETRY

In atomic absorption flame spectrometry, a sample solution is introduced into the flame to produce an atomic vapor of the element in concern. The absorption of radiation from a suitable source is measured using standard absorption techniques (see Fig. 1 for a schematic diagram of an atomic absorption flame spectrometer). To minimize measurement of radiation emitted from the flame gases and from the atom being measured by atomic absorption, the source radiation is chopped and a tuned ac amplifier is used to amplify the photodetector signal. Therefore, since emission of radiation from the flame gases is predominately constant in intensity, such radiation striking the photodetector will produce a dc current which is not amplified by the tuned ac amplifier; of course, any intensity fluctuations within the frequency response bandwidth of the amplifier-readout system will produce a noise signal. The absorption of radiation, specifically the absorbance, is directly related to the concentration of the atomic absorbers. The unknown concentration of a sample solution can thus be found by comparing the absorbance of the unknown concentration to the absorbance of a standard (s).

III. THEORETICAL CONSIDERATIONS

A. Atomic Absorption Lines

The absorption of radiation of atoms occurs over a narrow frequency interval called an atomic absorption line [9,10]. The frequency interval is determined by broadening paramenters. The absorption of radiation results in the electronic activation of an atom from the ground (sometimes metastable states above the ground) electronic state, l, to an upper electronic state, u, as shown in Fig. 2. The excited atom remains in the upper level, u, for a short time period, e.g., 10^{-8} sec*, and then undergoes in most cases a nonradiational deactivation via collisions with flame gas molecules, e.g., H_2O, CO_2, O_2, N_2, CO, etc. If the excited atom undergoes a radiational deactivation, the emitted radiation, which is isotropic, is called atomic fluorescence†. Since relatively few atoms undergo radiational deactivation, e.g., one out of a hundred, and since fluorescence is isotropic†† whereas the radiation collected in atomic absorption measurements is only over a small solid angle, atomic fluorescence usually causes negligible interference in atomic absorption spectrometry except perhaps at the highest sample concentrations, e.g., at solution concentrations exceeding 100 ppm.

The intensity of radiation transmitted through an atomic vapor of absorbers, I_λ, varies with wavelength as shown in Fig. 1 and is related to the incident intensity I_λ^o at the same wavelength λ and to the product of the atomic absorption path length l, in cm, times the atomic absorption coefficient k_λ, in cm^{-1}. The expression relating I_λ to the above factors is

$$I_\lambda = I_\lambda^o \exp(-k_\lambda l) = I_\upsilon^o \, 10^{-0.43 k_\lambda l} \tag{1}$$

---

* The reciprocal of this time is the transition probability, i.e., number of transitions per second.

† Emitted in all directions.

†† This is basis of new analytical flame method called atomic fluorescence flame spectrometry (9, 10, 11, 12).

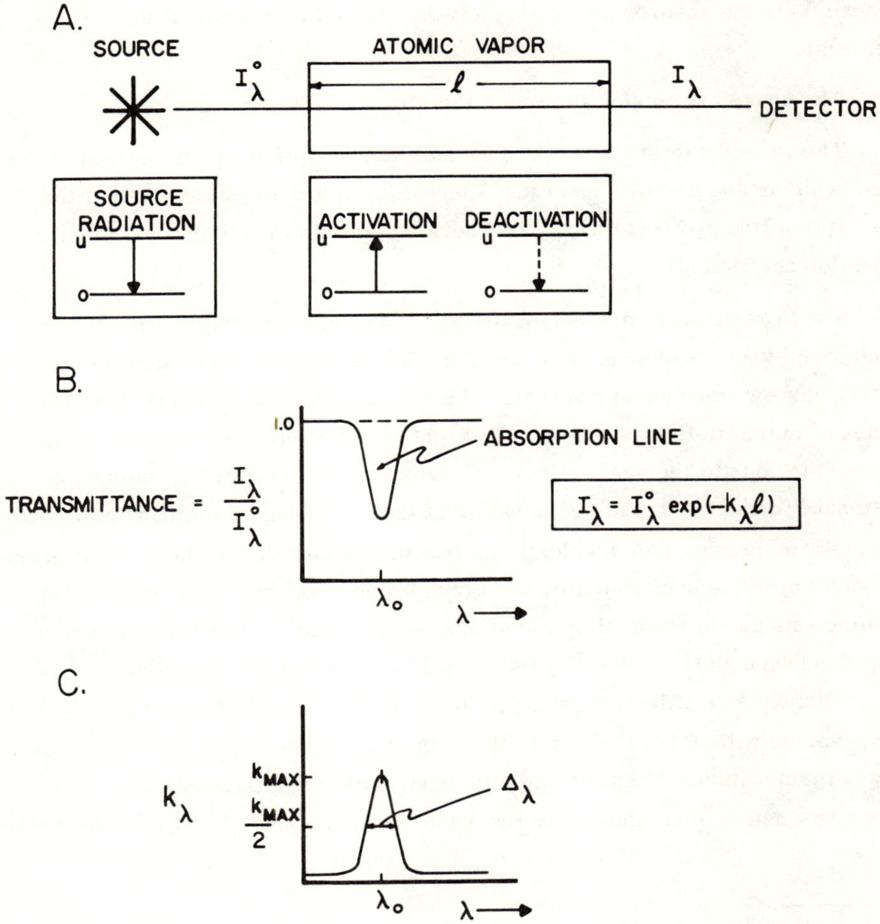

Fig. 2. Schematic representation of absorption of radiation by an atomic vapor (A), of the variation of intensity of radiation passing through an atomic vapor as a function of wavelength (13) and of the variation in the atomic absorption coefficient k_λ as a function of wavelength (λ).

The exponent in the right-hand expression is more commonly called the absorbance (or the optical density*), i.e.,

$$A_\lambda = 0.43k_\lambda L = \log_{10}\frac{I^0_\lambda}{I_\lambda} \qquad (2)$$

---

* Sometimes called the extinction.

where A_λ is the absorbance at wavelength λ for an infinitesimal wavelength interval.

The above expression in simply the absorption law*.

The wavelength interval $\Delta\lambda$ of an atomic absorption line is defined as the half-width of the plot of k_λ versus λ (see Fig. 1) and so is the width of the absorption line profile at $k_{max}/2$, where k_{max} is the maximum atomic absorption coefficient.

The broadening of a spectral line[11, 12] results in the absorption of radiation by atoms over a finite wavelength interval $\Delta\lambda$ rather than an infinitely narrow wavelength interval. The major causes of spectral line broadening of atoms in flames and in hollow-cathode discharge lamps are: Doppler broadening due to thermal agitation of the absorbing or emitting atoms; collisional broadening due to perturbation of the absorbing or emitting atoms by foreign gas atoms or molecules;† natural broadening due to the finite lifetime of an atom in the excited state; and Stark broadening due to the splitting of an atomic line due to the application of an electric field††. For flame temperatures between 1500° - 3000°K, the Doppler half-width for most lines[12] of most elements is of the order of 0.006 - 0.09 Å. The Doppler half-width $\Delta\lambda_D$ varies with $T^{\frac{1}{2}}$, where T is the temperature, in °K, of the atomic vapor, i.e., temperature of flame or atomic vapor within a hollow cathode. For most resonance lines (generally resonance lines are used in atomic absorption

---

* Commonly called the Beer law.

† Collisions by foreign species cannot only broaden the spectral line but also can cause shift in the peak wavelength λ_o, and asymmetry of the spectral line shape. In the discussion in this chapter, however, shift and asymmetry effects will be assumed negligible.

Resonance broadening is generally negligible in atomic absorption studies and need not be considered. Resonance broadening is similar to collisional broadening but is due to collisions between the same type of absorbing or emitting atoms.

†† Stark broadening is negligible for atoms in flames and is also generally not significant for atoms in hollow-cathode discharge tubes. It is important only in arcs or sparks of high charge density.

Zeeman broadening is a result of splitting of atomic lines into hyperfine components due to the application of a magnetic field. This effect is negligible unless an extremely strong external magnetic field is applied, which is not done in analytical studies.

flame spectrometry) of most atoms, the collisional half-widths [12] are between 0.01 and 0.1 Å for analytical flames at 1 atm pressure, i.e., for C_2H_2/air, H_2O_2, etc., flames. The collisional half-widths vary with $1/T^{1/2}$. The natural half-width of most resonance lines [12] is about 10^{-4} Å, which is a negligible contribution to the total width of a spectral line which is primarily determined by Doppler and collisional broadening*.

It should be stressed that self-absorption of emission lines often occurs in sources of extended depth, e.g., hollow-cathode discharge lamps or metal vapor arc lamps, particularly if the sources are run at high currents where the atomic vapor concentration in the source is sufficiently high that the line center is appreciably absorbed. Self-absorption [11] is the absorption of radiation by an atom of the same kind emitting the radiation; the absorption is much greater in the line center than in the line wings and results in a broad flat-topped spectral line. If an emission line is completely absorbed in the line center due to absorption of the emitted light from the hot central plasma by cool atoms in the cooler atomic vapor surrounding the hot plasma, this is called self-reversal. Self-reversed lines cannot be used in atomic absorption flame spectrometric studies† because essentially no radiation is present in the region over which the major absorption of radiation occurs.

B. Atomic Absorption Coefficient

The atomic absorption coefficient as a function of frequency...k_λ, in cm^{-1} can be expressed by the famous Voigt profile expression which accounts for line broadening due to a Gaussian effect (Doppler broadening) and due to a Lorentzian effect (collisional and natural broadening) [9,10,13]. The general Voigt profile expression will not be given here but instead the simplified expression of

$$k_\lambda = \frac{k_o a \rho}{\pi} \tag{3}$$

where "a" is the damping constant, which is defined by $(\ln 2)^{1/2}$ times the ratio of collisional half-width to Doppler half-width, is a complex function of the types of broadening (fortunately tables of ρ values are available; in such

---

* Doppler and collisional half-widths add quadratically to give the total half-width of the line.

† Osram and Philips metal vapor discharge lamps when run at currents specified by the manufacturer are often self-reversed.

tables ρ is given for a number of "a" values* and relative wavelengths), and k^o is the atomic absorption coefficient at wavelength λ_o for the case of pure Doppler broadening and is given by

$$k_o = \frac{2(\ln 2)^{\frac{1}{2}} X \lambda_o^2 N_o f}{\pi^{\frac{1}{2}} c \Delta\lambda_D} \tag{4}$$

where λ_o is the wavelength, in cm, at the line center, X is a constant given by $\pi e^2/mc$, where e and m are the charge and mass of the electron and c is the speed of light, N_o is the concentration, in atoms/cm^3, in the ground state in the hot gases, $\Delta\lambda_D$ is the Doppler half-width, in cm, and f is the absorption oscillator strength of the transition. Therefore, as long as "a" and ρ remain constant as N_o varies for a given absorbing atom, then k_λ will be directly proportional to atomic concentration N_o.

C. Integrated Atomic Absorption Coefficient

As discussed previously, k_λ will be dependent on the type and wavelength of the absorption transition and on the physical conditions of the system, e.g., hot gas temperature and pressures and types of all flame gas products and the absorbing atom in concern. From the Einstein theory of radiation, it can be shown that the integrated atomic absorption coefficient, i.e., the area under a plot of k_λ versus λ, for a given absorption line is given by

$$\int_{\text{line}} k_\lambda \, d\lambda \cong \frac{\pi^{\frac{1}{2}} k_o \Delta\lambda_D}{2(\ln 2)^{\frac{1}{2}}} \tag{5}$$

as long as the number of atoms (per cm^3 of flame gases) in the excited electronic state, u, is negligible compared to the number of atoms (per cm^3 of flame gases) in the ground, l, electronic state. This is true for most atoms in flames of the type used in atomic absorption flame spectrometry. It should be stressed that the integrated atomic absorption coefficient is constant for any given atomic concentration N_o, no matter what physical processes are

---

* Tables of "a" values are given by C. Young, see Tables for Calculating the Voigt Profile, University of Michigan, 1965 and J. Quant. Spectrosc. Radiat. Transfer, 5, 549 (1965).

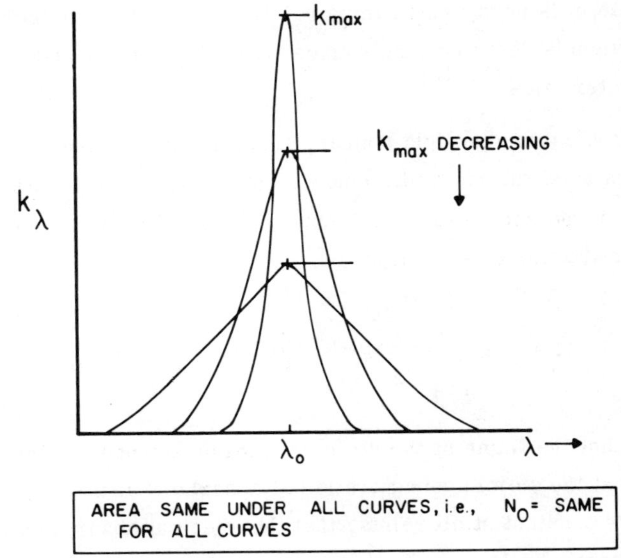

Fig. 3. Schematic representation of influence of broadening of an atomic absorption line for a given atom at a constant atomic concentration.

responsible for formation of the line. In Fig. 3, several spectral line profiles for a certain atomic concentration N_0 of some absorbing atom are given. It should be noted that areas under each curve are identical, whereas the peak absorption coefficient varies greatly due to broadening processes.

If the integrated atomic absorption coefficient could be measured simply and inexpensively, this parameter would certainly be most ideal for atomic absorption flame spectrometric studies. Unfortunately for such measurements to be done simply and accurately a very intense continuum source, a very high resolution monochromator, and a double beam system should be used. Such a system would be extremely expensive and bulky and so is not practical for routine atomic absorption measurements. (See Refs. [9, 10] .)

D. Measured Absorption Signal

In atomic absorption spectroscopy, it is difficult to measure either A_λ or $\int k_\lambda d\lambda$ because of insufficient resolution of the monochromator. Therefore, one of two systems are used: either a continuum source-monochromator system in which the monochromator spectral bandwidth is much greater than the absorption line width or a narrow line source-monochromator in which the source linewidth is narrow compared to the absorption line width (the narrow line source system is used in all commercial instruments). In both systems, the measured signals depend on the source type and intensity and the absorption line characteristics (14-16).

When using a continuum light source, i.e., a source-monochromator system having a spectral bandwidth much greater than the absorption line half-width, the irradiance (ergs sec^{-1} cm^{-2}) absorbed from the source by the atomic vapor in the flame is given by

$$I_A^C = \frac{\Omega_A}{4\pi} I_{oc} \int_o^{\infty} [1 - \exp(-k_\lambda l)] d\lambda \tag{6}$$

This assumes that the flame at the region of measurement is in thermal equilibrium, that the atomic concentration of absorbers at the region of measurement is constant at all points, that the spectral absorption line is broadened primarily by Lorentzian and Doppler broadening, and that only one isolated absorption line is being measured. The symbols are: $\frac{\Omega_A}{4\pi}$ is the functional solid angle of radiation collected from the source by the entrance optics, I_{oc} is the spectral irradiance of the continuum source at λ_o, the peak absorption wavelength (units of ergs sec^{-1} cm^{-2} $\mathring{A}^{-1}$), k_λ is the atomic absorption coefficient, in cm^{-1}, and ℓ is the absorption path length, in cm. The integral in Eq. (6) is given for low optical densities (low atomic concentrations of absorbers) by

$$\int_o^{\infty} [1 - \exp(-k_\lambda l)] d\lambda = l \int_o^{\infty} k_\lambda d\lambda = \frac{l k^o \pi^{1/2} \Delta\lambda_D}{2 (\ln 2)^{1/2}} \tag{7}$$

where all terms were defined previously. The integral for high optical densities is given by

$$\int_o^{\infty} [1 - \exp(-k_\lambda l)] \, d\lambda = \left(\frac{\pi^{1/2} k_o l a \Delta\lambda_D^2}{\ln 2} \right)^{1/2} \tag{8}$$

where all terms were defined in a previous section. Therefore, the absorbed irradiance I_A for low and high optical densities is given by

$$I_A^C \text{ (low)} = \frac{\Omega_A \pi^{1/2} I_{oc} k_o l \Delta\lambda_D}{8\pi (\ln 2)^{1/2}} \tag{9}$$

$$I_A^C \text{ (high)} = \frac{\Omega_A I_{oc}}{4\pi} \left(\frac{\pi^{1/2} k_o l a \Delta\lambda_D^2}{\ln 2} \right)^{1/2} \qquad (10)$$

which indicates that the absorbed irradiance is linear with atomic concentration $N_o^{1/2}$ [see Eq. (4)] at low optical densities but proportional to $N_o^{1/2}$ at high optical densities. Generally the fraction of radiation absorbed is measured, i.e.,

$$\alpha = \frac{I_A}{I_\lambda d\lambda} \qquad (11)$$

where $\int I_\lambda d\lambda$ is the energy transmitted per unit time per unit cross-sectional area of flame gases when blank solution is aspirated into the flame gases. For the case of a continuum source-monochromator:

$$\int_s I_\lambda d\lambda = \frac{\Omega_A}{4\pi} I_{oc} s \qquad (12)$$

where s is the spectral bandwidth of the monochromator. Thus α for low and high concentrations is given by

$$\alpha_{low}^C = \frac{\pi^{1/2} k_o l \Delta\lambda_D}{2(\ln 2)^{1/2} s} \qquad (13)$$

$$\alpha_{high}^C = \frac{1}{s} \left(\frac{\pi^{1/2} k_o l a \Delta\lambda_D^2}{\ln 2} \right)^{1/2} \qquad (14)$$

The fraction of radiation absorbed is related to the absorbance A by

$$A = -\log(1 - \alpha) \qquad (15)$$

Therefore the absorbance at low and high concentrations is given by the relatively complex expressions of Eqs. (16) and (17):

$$A_{low}^C = \log \left[1 - \frac{\pi^{1/2} k_o l \Delta\lambda_D}{2(\ln 2)^{1/2} s} \right]^{-1} \qquad (16)$$

$$A_{high}^C = \log \left[1 - \left(\frac{\pi^{1/2} k_o l \Delta\lambda_D^2}{\ln 2} \right)^{1/2} \right]^{-1} \qquad (17)$$

When using a narrow line source (the most common system in atomic absorption spectrometry), the irradiance absorbed from the source by the atomic vapor is given by

$$I_A^L = \frac{\Omega_A}{4\pi} \left[1 - \exp(-\bar{k}l)\right] \int_{\Delta\lambda_S} I_\lambda \, d\lambda \tag{18}$$

where $\Omega_A/4\pi$ is the fractional solid angle of radiation collected from the source by the entrance optics, the term in brackets represents the fraction of radiation absorbed by the atomic vapor, and the integral term is the integrated source intensity for the spectral line having a half-width, $\Delta\lambda_S$. The term $\bar{k}$ is the average atomic absorption coefficient over the source line width and is given by

$$\bar{k} = k_o \delta \tag{19}$$

where δ is a correction factor accounting for the finite width of the source line compared to the absorption line and also accounting for the profiles of both lines. The integral in Eq. (18) is given by

$$\int_{\Delta\lambda_S} I_\lambda \, d\lambda = \frac{\Omega_A}{4\pi} \tag{20}$$

where I_L is the integrated source irradiance, in ergs sec^{-1} cm^{-2}.

The term in square brackets in Eq. (18) is given by:

$$\left[1 - \exp(-kl)\right] = \bar{k}l \tag{21}$$

at low atomic absorber concentrations and:

$$\left[1 - \exp(-\bar{k}l)\right] = 1 \tag{22}$$

at high atomic absorber concentrations. Therefore the values of I_A, α, and A for low and high concentrations of atomic absorbers is given by combining the preceeding equations with the use of the preceding definitions:

$$I_{A\,(low)}^{L} = \frac{\Omega_A k_o \delta \, l \, I_L}{4\pi} \tag{23}$$

$$I_{A\,(high)}^{L} = \frac{\Omega_A I_L}{4\pi} \tag{24}$$

$$\alpha_{(low)}^{L} = k_o \delta l \tag{25}$$

$$\alpha^{L}_{(\text{high})} = 1 \qquad (26)$$

$$A^{L}_{(\text{Low or high})} = k_o \delta l \qquad (27)$$

Therefore, when using a line source, the expressions with absorbance are useful for higher concentration levels since α approaches unity and I_A approaches I_L, i.e., complete absorption of the source line. However, both α and A are quite useful at low α values, i.e., α's $\lesssim 0.05$, which corresponds to low atomic absorber concentrations (dilute atomic gas). Since phototubes give a linear output with radiant power striking the photocathode, a readout linear for α is directly obtainable whereas an output linear with absorbance requires a logarithmic conversion.

Therefore the so-called growth curves which are plots of the logarithm of the irradiance absorbed versus the logarithm of the atomic concentration (or linear functions related to these parameters) have slopes of 1.0 and 0.5 for the continuum source at low and high concentrations and slopes of 1.0 and 0.0 for the line source at low and high atomic concentrations. Actually, at very high concentrations, the continuum power curve has a slope approaching 0.0 since the absorption line width increases and is no longer small compared to the spectral bandwidth s.

E. Influence of Flame Temperature on Atomic Concentration

The atomic concentration N_u of atoms in state u of energy E_u is related to the atomic concentration N_o of atoms in state o ($E_o = 0$ if ground state) by the Boltzmann equation

$$N_o = N \frac{g_u}{g_o} \exp(-E_u/kT) \qquad (28)$$

For most atoms, $N_u/N_o << 1.00$, i.e., the number of excited to ground state atoms is very small (see Table I for some representative values of N_u/N_o). Therefore, the atomic concentration in state o, N_o, is not greatly affected by

TABLE 1

Fraction of Excited Atoms N_u/N_o as a Function of Wavelength and Temperature for Typical Resonance Lines

Resonance line wave-length, A	g_u/g_o	N_u/N_o			
		T = 2000°K	T = 3000°K	T = 4000°K	T = 5000°K
2000	2	3.1×10^{-16}	3.5×10^{-11}	6.8×10^{-8}	1.2×10^{-6}
4000	2	6.8×10^{-8}	1.4×10^{-5}	2.4×10^{-4}	1.6×10^{-4}
6000	2	1.5×10^{-5}	6.8×10^{-4}	5.0×10^{-3}	1.6×10^{-2}
8000	2	5.0×10^{-5}	6.6×10^{-4}	9.0×10^{-3}	2.7×10^{-2}

temperature changes if only electronic excitation is considered.* The total atomic concentration in all electronic states, N_t, is given by

$$N_t = N_o + N_u + \dots = \frac{N_o}{g_o}[g_o + g_u \exp(E_u/kT) + \dots]$$

$$N_t = \frac{N_o}{g_o} B(T) \tag{29}$$

where B(T) is the electronic partition function and is defined by the term in parenthesis. Because the terms with $\exp(-E_j/kT)$, where E_j is the energy of any electronic state, are generally much smaller than unity,

$$N_t \cong N_o \tag{30}$$

for many elements with simple atomic spectra. Therefore, the atomic concentration in all electronic states is essentially the same as in the ground electronic state.†

In addition to the influence of flame temperature on the degree of atomization and on the degree of electronic excitation, it should be mentioned that k_o varies linearly with $(1/T)^{\frac{1}{2}}$ and so k_o does vary slightly if the flame temperature is changed even if N_o remains the same. On the other hand, the integrated atomic absorption coefficient for a constant value of N_o is independent of flame temperature.

---

*See next section for influence of flame temperature on degree of atomization.

†The approximation given in Eq. (11) is valid for many elements in analytical flames and will be assumed for the rest of this discussion.

F. Variation of Atomic Concentration with Sample Solution Concentration

Winefordner and Vickers [18] give an expression relating the atomic concentration (in all electronic studies, N_t) to the sample solution concentration C (moles/liter) introduced into the flame gases. The derived expression is

$$N_t = \frac{10^{19} F \epsilon \beta C}{e_f Q}, \quad cm^3 \text{ atoms} \tag{31}$$

where F is the solution flow rate into the flame gases in cm^3/min, C is the sample concentration in moles/liter, Q is the flow rate in cm^3/sec of unburnt gases introduced into the flame at room temperature and 1 atm, e_f is the flame expansion factor (no units), ϵ is sample introduction efficiency (no units), and β is the efficiency of atomization (no units). The flame expansion factor is given by

$$e_f = \frac{n_T T}{n_{room} T_{room}} \tag{32}$$

where n_T and n_{room} are the number of moles of combustion products at the flame temperature T and room temperature T_{room}, respectively.

It is important to note that several parameters in Eq. (31) vary with flame temperature and composition, and those parameters are: ϵ, β, and e_f. The variation of e_f with T is approximately linear [see Eq. (32)] since $n_T/n_{room} T_{room}$ is approximately constant with change in flame temperature, i.e., $n_T/n_{room} T_{room} \cong 0.003$ for many analytical flames.

The variation of ϵ and β with T is much more complex. The efficiency of sample introduction, ϵ, is defined as the ratio of the moles of salt molecules produced in the flame gases per unit time to the moles of salt actually introduced into the flame per unit time and is, therefore, a complicated function of the type and geometry of the aspirator-burner used, e.g., total-consumption or premixed systems, of the flow rates and pressures of unburnt gases and solution into the flame, of the viscosity, surface tension, and density of the introduced solution, and of the rate of evaporation of the solvent and solute. All of the above factors influence the size of the droplets produced in a flame when as aspirator (nebulizer) is used and influence the rate or evaporation of the solution droplets to produce a salt mist. Winefordner, Mansfield, and Vickers [19] and later Parsons and Winefordner [20] measured approximate values of ϵ for several turbulent flames (H_2/O_2 and C_2H_2/O_2)

produced by total-consumption aspirator burners and found ϵ values of about 0.2 - 0.8 for most conditions used in analytical flame spectrometry. Bernstein[21] measured values of ϵ in premixed flames produced using chamber-type aspirator-burners and found values of about 0.05 - 0.1 for most conditions. The chamber-type aspirator-burners have a lower value of ϵ because most of the solution aspirated into the chamber is rejected (i.e., the number of droplets reaching the flame gases comprise only a small fraction of the solution volume introduced into the chamber — this ratio is commonly called the yield). However, the efficiency of conversion of the resultant aerosol mist to salt gas molecules is greater than that with total consumption nebulizer-burners.

The efficiency of atomization, β, is simply the ratio of the concentration of atoms in concern to the total concentration of all species containing that same atom, e.g., if $MgCl_2$ is aspirated into a flame, then β_{Mg} is defined[18] as

$$\beta_{Mg} = \frac{N_{Mg}}{N_{Mg} + N_{MgO} + N_{MgH} + N_{MgOH} + N_{MgCl} + N_{MgCl_2} + N_{Mg^+}} \tag{33}$$

where the N's correspond to the concentration of the species (in all electronic states) appearing as subscripts, i.e., particles per cm^3 of flame gases. If a flame is in thermal equilibrium (this assumption is approximately valid for many analytical flames), then the value of β can be estimated from the total pressure of all species, containing Mg, from the equilibrium constants of processes, such as $MgO \rightleftharpoons Mg + O$, $MgH \rightleftharpoons Mg + H$, etc., and from the partial pressures of flame gas products, e.g., O, H, OH, etc. If this is done, it is found that β is a very complex function of flame gas temperature and composition. De Galan and Winefordner[12] have given a method for measurement of β factors and have measured β factors for a number of salts of several elements introduced into C_2/H_2/air flames. It is evident from such studies that there is an optimum flame temperature and composition for which the β factor is maximal, and, therefore, an experimental design should be performed prior to an analysis by atomic absorption flame spectrometry in order to obtain optimum conditions of analyses.

It should be stressed that all atomic flame methods, i.e., atomic emission, atomic absorption, and atomic fluorescence flame spectrometry, depend directly on the concentration N_o of ground state atoms in the flames. Therefore, the change in any parameter causing a variation in N_o (e.g., flame temperature and composition, solution and gas flow rates, etc.) will cause the same type of interference (see discussion of interferences) in all three methods. De Galan and Winefordner [22] have shown that elements forming stable compounds in the flame gases, e.g., MgO, result in essentially the same interference in atomic absorption as in atomic emission flame spectrometry. In other words, a change in flame temperature, ΔT, results in essentially the same change in absorption and emission signals in atomic absorption and atomic emission flame spectrometry, respectively, because the most significant effect is the change in ground state concentration. This result is contrary to the opinions expressed in many books, articles, and manufacturer's literature on atomic absorption flame spectrometry.

IV. EXPERIMENTAL CONSIDERATIONS

In Fig. 1, a schematic diagram of a typical atomic absorption flame spectrometer is given. It should be noted that there is great similarity between an atomic absorption flame spectrometer and an atomic emission flame spectrometer (commonly called flame photometer). In fact, the atomic absorption flame spectrometer shown schematically in Fig. 1 can be converted to an atomic emission flame spectrometer * by simply moving the chopper from its present position to one between the flame and the monochromator (of course, the source is then turned off). Some commercial atomic absorption instruments have this facility.

A. Source of Excitation

If a monochromator of very high resolving power were available, then it would be more accurate analytically to use an intense continuum source and scan the atomic absorption line [6] (see Section III). However, as previously pointed out, such instrumentation would be extremely expensive.

---

* As has been pointed out by V. A. Fassel [23], flame emission spectrometry is an exceedingly useful tool and has not been supplanted by atomic absorption flame spectrometry for all elements. In fact flame emission spectrometry gives lower or comparative limits of detection compared to atomic absorption flame spectrometry for 43 elements and atomic absorption gave lower limits of detection for only 24 elements. It should be pointed out that Fassel used premixed fuel-rich acetylene-oxygen flames for these studies. Therefore, in many instances, atomic emission flame spectrometry will give more sensitive and better results and should be used. It is important that a flame spectrometer be useable for both absorption and emission studies or that the analyst have both types of instruments since these techniques are both complementary and competitive.

If a continuum source of radiation* and a medium resolution monochromator** which is not capable of resolving absorption lines in flames† is used and if measurements of intensity transmitted are performed only when the monochromator is set on the absorption line peak, then a loss in sensitivity (see Section III) occurs as the spectral bandwidth of the monochromator increases compared to the absorption line half-width[24]. This results because most of the radiation passing through the flame gases and out of the exit slit of the monochromator is not absorbed. Therefore, there is a loss in signal compared to the use of a narrow line source, i.e., $\alpha^{L}_{(low)} > \alpha^{C}_{(low)}$. However, even though signal is lost via radiation not absorbed but reaching the photodetector, the sensitivity (limit of detection) may not be changed appreciably if a very stable source of excitation is used. The sensitivity of analysis is dependent on the signal-to-noise ratio, and, therefore, if the noise is reduced, a smaller signal due to a smaller concentration can be detected. However, for elements having intense, narrow line sources, the continuum source gives poorer sensitivities and analytical curves. The continuum source is most useful for elements for which stable, intense, narrow-line sources are not available, e.g., the rare earths. Fassel et al.[25] have used a spectral continuum as the source of radiation and found that for 16 elements lower limits of detection were obtained than with line sources†† and that for another 16 elements comparable limits of detection were obtained using the spectral continuum or line sources. For eleven elements, line sources gave definitely superior results.

All commercial instruments and most studies so far performed have used narrow line sources of excitation. If the narrow line source has a line width considerably less than absorption line width of atoms in the flame gases, then the measured absorbance will be proportional to the peak atomic absorption coefficient (see Section III). The narrow line sources most commonly used in atomic absorption studies are the hollow-cathode discharge tubes (HCDT). The conventional HCDT's produce very sharp spectral lines of moderate intensity when operated at low currents.[26,27] The conventional HCDT consists

---

* Xenon arc discharge lamp or tungsten strip filament lamp.

** The 0.5-m Ebert grating monochromator with a minimum spectral bandwidth of about 0.2 Å (manufactured by Jarrell-Ash Co.).

† Absorption line half-widths of atoms[12] in flames is about 0.01 - 0.1 Å.

†† Line sources were not available for several of these elements.

simply of a glass cylinder with a quartz window sealed at one end via a graded-seal or some type of wax or cement and a tungsten-nickel or Kovar wire sealed in the opposite end to support the cathode. At the same end as the cathode, a second wire lead (the anode) is sealed into the cylinder. For elements that are easy to machine the cathode is made of the element itself. If the element is expensive, of low mechanical strength, or difficult to machine, then a cathode machined from an element with a simple spectrum, e.g., aluminum, is used, and a small quantity of the desired metal is placed inside the cathode as a foil liner. Most HCDT's contain one of the noble gases* at a pressure of a few mm. A high voltage** is applied between the electrodes to strike and maintain a gaseous discharge. Alternating current operation of HCDT's† result in higher peak currents, shorter tube life, poorer stability, etc., and so they are usually operated on a very stable dc power supply and the emitted radiation is modulated with a mechanical chopper. The performance of HCDT's as a function of fill gas, lamp current, etc. has been given by Slavin and Manning [27]. HCDT's should always be operated at the lowest possible currents to assure narrow emission lines.†† Conventional HCDT's for over half of the elements in the periodic table have been constructed and are commercially available.†††

Sullivan and Walsh [28] have described the construction of high-intensity HCDT's which give 100 times the intensity of conventional HCDT's and also narrower line widths. This is achieved by using two independent power supplies to produce discharges between two different sets of electrodes. One set of electrodes results in an arc discharge producing the atomic vapor which is excited by the second discharge. The first discharge producing the atomic vapor is run at low currents to minimize self-absorption broadening, and the second discharge is operated at high currents to excite an appreciable number of atoms to produce a high intensity.

---

* Usually argon.

** 800 V at variable output current of 5 to 100 mA.

† Since alternating current modulates the lamp, a mechanical chopper is not needed.

†† At high HCDT currents, the emission line broadens and a loss in sensitivity analogous to the continuous source and medium resolution monochromator results.

††† Manufacturers of HCDT's include: Westinghouse, Techtron, Perkin-Elmer, and Varian.

Metal vapor discharge lamps * can be used for a very few elemental determinations if care is taken to under-run the lamps, i.e., run at lower currents than specified by the manufacturers. Such lamps are available for the alkali metals, mercury, cadmium, zinc, thallium, indium, and gallium. Good conventional HCDT's are available for most of these elements except for some of the alkali elements, and so metal vapor discharge lamps are used to measure alkali metals.** Metal vapor discharge lamps are quite inexpensive and operate simply from 220 V ac with a ballast control.

Other line sources which have been used for atomic spectral studies include electrodeless discharge tubes operated in a microwave cavity at 2450 MHz and flames(29) doped with the element in concern. The former are inexpensive, extremely intense, and producers of narrow emission lines but suffer from instability. However, with technological advanced, electrodeless discharge tubes may eventually replace HCDT's for many analyses. The latter are useful only for elements which are appreciably excited in flames.† Also flames produce emission lines of about the same width as absorption lines resulting in a loss in sensitivity and flames do not produce a stable output over a long period of time.

B. Entrance Optics

Generally two lenses†† are used. The first lens is placed after the source and before the flame and is used to collimate the source radiation before it is passed through the flame. The second lens between the flame and monochromator entrance slit focuses the source radiation on the monochromator entrance slit.

C. Flame Cell and Burner Type

Both turbulent and premixed diffusion flames, e.g., H_2/O_2, C_2H_2/O_2, H_2/air and C_2H_2/air, C_2H_2/N_2O, respectively, have been used in atomic

---

* Osram or Phillips.

** The best method, however, is simply flame photometry where limits of detection near 0.00001 ppm result. Therefore, metal vapor discharge lamps are not recommended for any atomic absorption studies.

† For example, flames are poor sources for elements having resonance lines below 3000 Å, e.g., Zn, 2139 Å and Cd, 2288 Å.

†† Lenses are usually made of quartz.

absorption measurements. Turbulent flames are usually hotter and often of lower background than the premixed flames.* However, such flames are produced using a total-consumption aspirator-burner† which has small orifices for fuel and oxidant, and, therefore, such flames are small in size, i.e., l is about 1 cm. Such burners are direct injection resulting in an aspiration efficiency ϵ of 0.2 to 0.8, depending on conditions. Premixed flames usually have lower temperature, higher background, and lower efficiency of operation (ϵ <0.2 usually). However, such flames have been used almost exclusively for atomic absorption studies because long path lengths, e.g., $l \cong 10$ cm, can be achieved with ease.†† Also premixed flames have much less audible noise than turbulent flames. The premixed flames are usually produced with a slot-type burner[(30-33)]. The sample solution, however, is not introduced directly into a slot-type burner but rather is first dispersed in a chamber, and then the small droplets are allowed to enter the flame gases.

There are several means of increasing the effective path length of a turbulent flame or the residence time of atoms in the light path. A multiple-pass system[(34)] in which source light is reflected back and forth through the flame gases can be used. However, because of the small size of turbulent flames, only several (2 or 3) passes go through the strongly absorbing portion of the flame gases. In addition, there are light losses at each mirror, and so only two or three passes can be used anyway with conventional HCDT's which are not highly intense for most elements; of course, an extension of path length can also be accomplished by using three identical burners side-by-side and a multiple-pass system consisting of front surface mirrors. Such a system is more expensive, however, and also uses more sample solution. Several workers[(35-37)] have introduced the flame into a long (e.g., 50 cm) quartz or ceramic tube mounted between the HCDT source and the monochromator entrance slit. With such a device, limits of detection for some elements were lowered by 10 - 100 times. This system, however, is bulky, and the same tube length is not optimal for all elements. Also alignment of all components is quite critical.

---

* H_2 flames have lower background than C_2H_2 flames.

† For example, aspirator-burner of the Beckman DU flame attachment.

†† If l is increased much more than 10 cm, not much is gained because the solution droplets are no longer uniformly dispersed along the flame length.

For most atomic absorption studies, premixed flames and chamber-type slot-burners have been used. The most commonly used flame type has been C_2H_2/air[38]. Much of the work with turbulent flames has been done because much hotter flames, e.g., H_2/O_2[39,40] and C_2H_2/O_2[41,42], which reduce compound formation of the atom in concern with flame gas products,* can be used. For example the β factors for certain elements like Al, Si, and Hf are so small that the absorption by these elements is difficult to detect in the C_2H_2/air flame. However, the recent introduction of the C_2H_2/N_2O† flame[43], which has a maximum flame temperature of 2995°C as compared 3060°C for the C_2H_2/O_2 flame, has reduced the necessity of using turbulent flames for many elements.

D. Monochromator

In atomic absorption studies with line sources, the monochromator is used only to isolate the line in concern from other lines in the source and to reduce the background signal and noise from the flame gases and is not used to resolve the absorption line. Most monochromators in commercial atomic absorption instruments use gratings rather than prisms because greater resolution over the entire spectral range at lower cost can be obtained. The greater the dispersion of the monochromator the lesser the chance for radiational type interferences. For example, if the source emits two lines within the spectral bandwidth of the monochromator and at the wavelength setting of the monochromator and if only one of these lines is of the element in concern, e.g., Ni, and the other is of another metal, e.g., Fe, then the interfering line (Fe in this case) will reach the detector and this will reduce the sensitivity of the sample determination (Ni in this case). By reducing the spectral bandwidth of the monochromator, only the sample line can be isolated.

E. Photodetector

Multiplier phototubes†† are generally used in atomic absorption studies. When selecting a multiplier phototube, the following items should be considered: spectral range of response, gain, dark current, and cost. Since much of the

---

* Increase β.

† The C_2H_2/N_2O flame when used properly is not hazardous, but the user must follow directions specified by the manufacturer.

†† Commonly called photomultipliers.

atomic absorption studies are performed in the ultrviolet * region, a high gain, low dark current multiplier phototube, such as the RCA 1P28 or R106 or R135 † should be used. The RCA 1P28 is considerably less expensive, but the other two respond to wavelengths as low as 1650 Å. All three tubes are of the side-window type. If end-on multiplier phototubes are used, then the EMI 9558QB is recommended since it has very high gain, very low dark current, and is useful over the spectral range 1600-8500 Å.

F. Amplifier-Readout

Generally ac amplifiers are used to minimize recording of thermal emission from the flame gases [6,44]. However, if the elements in concern emit only weakly at the resonance wavelengths, e.g., Zn 2139 Å and Cd 2288 Å, then a dc amplifier can be used. In any event, the frequency response bandwidth of the amplifier-readout system, Δf, in hertz should be made as small as possible to minimize the recording of noise. The value of Δf can often be varied by a control on the amplifier itself or a resistance-capacitance network on the recorder input. It should be stressed that Δf should be made small, but values of Δf less than 1 Hz are of little use because of drift in the rate of aspiration and in the light source. Drift in the rate of aspiration can be minimized by using forced-feeding of solution, but this is inconvenient to do routinely. Drift in the light source can be minimized by using a double-beam [45] system in which the source around the flame and the source through the flame are alternately observed. A double-beam system does allow measurements to be taken faster than with a single-beam system if lamps are not premixed. In any event, drift of some type usually occurs in atomic absorption studies, and little is to be gained by making Δf smaller than 1 Hz by using signal averaging devices of the type used in NMR [46]. To improve the signal-to-noise ratio, it is best to set Δf at about 1 Hz and then take several readings of signal and background, alternately, of course. The readout is usually a meter, recorder, or digital voltmeter.

In Table II, a listing of some of the presently available commercial atomic absorption flame spectrometers is given. This listing is not complete but rather is given only to acquaint the reader with some typical instruments available at the time of writing of this chapter.

---

* 2100 - 4000 Å.

† Hamamatsu Television Co., Ltd., Hamamatsu City, Shizuoka Prefecture, Japan.

TABLE II

Manufacturers of Atomic Spectrometers[a, b, c]

Manufacturer and address	Model	Aspirator	Flame	Monochromator	Detector	Amplifier	Readout
Atomic Absorption and Electronics, Corp. 50 Broadway, New York City 10004 (sold by National Instrument Laboratories, Inc., 12300 Parklawn Drive, Rockville, Md.)	100	T.C.	H-O A-O H-A	Grating	PM	DC (S.B.)	M. or Rec.
Beckman Instruments, Inc. Fullerton, California	1300 1301 (accessories for DU, DU-2, and DB Spectrophotometers)	T.C. or Ch.	H-O, A-O H-A, A-A N-A	Prism	PM	DC or AC (S.B.)	M. or Rec.
Evans Electroselenium, Ltd. St. Andrews Works Halstead, Essex (sold by Fisher Scientific, Pittsburgh, Pennsylvania)	140	Ch.	A-A, A-N	Grating	PM	AC (S.B.)	M.
Hilger and Watts 98 St. Pancras Way Camden Road, London, N. W. 1 (sold in U. S. by Englis Equip. Co., Chicago 5, Illinois)	H909 (accessories for Uvispek Photoelectric Spectrophotometer)	Ch.	A-A	Prism	PM	DC (S.B.)	M.

Jarrell-Ash Co,, 590 Lincoln Street Waltham, Massachusetts	82-500 Series	T.C. or Ch.	H-O, A-O H-A A-A, H-A N-A, A-N	Grating	PM	AC (S.B.)	M. or Rec.
	82-600	T.C.	H-O, A-O H-A	Grating	PM	AC (S.B.)	M., Dig., or Rec.
	82-700	T.C.	H-O, A-O H-A	Grating	PM	AC (S.B.)	M., Dig., or Rec.
		Ch.	A-A, H-A N-A, A-N				
Optics S. P. A Via Calatafimi 7-9 Milano, Italy	AT-6	Ch.	A-A	Prism	PM	AC (D.B.)	M.
	6-1078 (attachment for single-beam F-4, and for double-beam 6-1100 and 6-1102 spectrophotometers)	Ch.	A-A	Grating	PM	DC and AC and S.B. and D.B.	M. or Rec.
Perkin-Elmer, Corp., Norwalk, Conn.	303	Ch.	A-A, N-A	Grating	PM	AC (S.B.)	M., Dig., or Rec.
	290	Ch.	A-A, N-A, A-N	Grating	PM	AC (S.B.)	M. or Rec.
	139-0421 (attachment for Model 139 spectrophotometer)	Ch.	A-A	Grating	PM	DC (S.B.)	M.

Manufacturer and address	Model	Aspirator	Flame	Mono-chromator	Detector	Amplifier	Readout
Techtron Pty., Ltd., Melbourne, Australia (sold in U. S. by Aztec Instruments, Inc. 2 Silverbrook Road Westport, Conn.)	AA-4 (the AA-4 is an improved version of the AA-3)	Ch. T.C.	A-A, A-N H-O, A-O H-A	Grating	PM	AC (S.B.)	M., Dig., or Rec.
	AA-100	Ch. T.C.	A-A H-O, A-O, H-A	Grating	PM	AC (S.B.)	M., Dig., or Rec.
Unicam Instruments, Ltd. York Street Cambridge, England (sold in U. S. by Wilkins Anderson, Co. 4525 W. Division Street Chicago 51, Illinois)	SP.900A	Ch.	A-A	Prism	PM	AC (S.B.)	M.

[a] Definition of symbols: T.C. = Total-consumption aspirator-burner; Ch. = chamber type aspirator-burner; H-O = Hydrogen Oxygen Flame; A-O = Acetylene-Oxygen Flame; H-A = Hydrogen-Air Flame; A-A = Acetylene-Air Flame; A-N = Acetylene-Nitrous Oxide Flame; N-A = Natural Gas-Air; PM = Multiplier Phototube; S.B. = Single Beam; D.B. = Double Beam; M. = meter; Rec. = Recorder; Dig. = Digital Readout.

[b] Westinghouse Electric Corp., Electronic Tube Division, Elmira, N. Y. has available a large assortment of hollow-cathode discharge tubes as well as Perkin-Elmer Corp., Beckman Instruments, Inc., and Techtron Pty., Ltd.

[c] Status of 1964-1965. No claim is made for completeness.

G. Optimization of Experimental Conditions

To obtain the best experimental conditions for measurement, i.e., conditions resulting in the greatest precision and the lowest detectable sample concentration, those experimental conditions giving the maximum signal-to-noise ratio, S/N, must be obtained. If sufficiently accurate expressions and values of spectral, flame composition, and instrumental parameters are available, then, in principle, optimization can be done completely from theory using the expressions developed by Winefordner and co-workers [14, 15, 18, 46-48]. Most workers prefer lengthy trial and error approach based on experimental measurements. It should be stressed, however, that conditions giving the maximum signal-to-noise ratio are desired because those conditions result in the highest precision and greatest sensitivity; of course, those same conditions may also give poor selectivity of measurement, e.g., if the monochromator slit width must be increased to give good sensitivity, this may result in interferences due to the detection of lines other than the one of interest. Therefore, optimum experimental conditions for precision and sensitivity must be chosen carefully.

H. Optimization Based on Theory

Winefordner and co-workers [14, 15, 18, 46-48] have derived expressions for the signal-to-noise ratio in atomic absorption flame spectrometry. The absorption signal S_A in atomic absorption measurements is the difference between the readout voltage E for the blank b and the sample s i.e.,

$$S_A = E_b - E_s \tag{34}$$

The readout voltage is determined by the photoanodic current i flowing out of the photodetector, the resistance R_L of the load resistor through which the current i flows, and the gain G of the amplifier readout system and so

$$S_A = GR_L(i_b - i_s) \tag{35}$$

The photoandoic current from a multiplier phototube is determined by the light power reaching the photocathode and is given by

$$i = \gamma K_o WI \tag{36}$$

where K_o is the optics factor, in units of cm, of the monochromator and is determined by the transmittance of the monochromator and the entrance optics, the monochromator slit height, and any magnification of the source image due to the entrance optics, W is the monochromator slit width, in cm, I is the irradiance of radiation from the source, and γ is the photoanodic sensitivity of the phototube, in amperes at photoanode to watts of radiant power at photocathode. Thus S_A is given by

$$S_A = GR_L \gamma K_o W I_A \tag{37}$$

where I_A is given by Eq. (9), (10), (23), and (24). Therefore, since S_A is directly proportional to I_A and since $GR_L \gamma K_o$ are independent of atomic concentration of absorbers, all conclusions regarding the relationship between I_A and k_o (and atomic concentration given in Section III also apply here).

Of course, associated with every signal S_A is a noise N, which is determined by phototube shot noise and which is due to random emission of electrons from a photoemission surface, by flame flicker, by source flicker, and by amplifier noise. The noise voltage N is related to the total root-mean-square (rms) photoanodic noise current $\overline{\Delta} i_T$ by

$$N = GR_L \overline{\Delta} i_T \tag{38}$$

The rms photoanodic noise current due to all source, i_T, is given by the following expression

$$\overline{\Delta} i_T = \left(\overline{\Delta} i_p^{\,2} + \overline{\Delta} i_f^{\,2} + \overline{\Delta} i_s^{\,2} + \overline{\Delta} i_s^{\,2}\right)^{\frac{1}{2}} \tag{39}$$

where $\overline{\Delta} i_p$, $\overline{\Delta} i_f$, $\overline{\Delta} i_s$, and $\overline{\Delta} i_a$ are, respectively, the rms noise currents due to the phototube shot effect, flame flicker, source flicker, and amplifier noise. Winefordner and co-workers (14, 15, 18, 46-48) have discussed the evaluation of the above noises. In many experimental systems in which a good electronics system and an intense line source of excitation are used, the noises $\overline{\Delta} i_p$, $\overline{\Delta} i_f$, and $\overline{\Delta} i_a$ will all be negligible compared to $\overline{\Delta} i_s$, and so for illustrative purposes for this specfic system, the noise expression will be given. Winefordner and Vickers (14) have shown that $\overline{\Delta} i_s$ is given by

$$\overline{\Delta} i_s^L = \xi (\Delta f)^{\frac{1}{2}} i_b = \xi (\Delta f)^{\frac{1}{2}} \gamma K_o W I_L , \tag{40}$$

for a line source and by

$$\overline{\Delta} L_s^C = \xi (\Delta f)^{\frac{1}{2}} \gamma K_o W s I_{oc} \tag{41}$$

for a continuum source, where Δf is the frequency response bandwidth of the amplifier readout in Hz, i.e. cps, and ϵ is the source flicker factor and is the ratio of the rms noise fluctuation signal due to the source to the signal due to the source intensity at a 1-Hz frequency response bandwidth. In terms of noise voltage,

$$N = GR_L \bar{\Delta} i_T \tag{42}$$

The signal-to-noise ratio S/N is, in general, given by

$$\frac{S}{N} = \frac{i_b - i_s}{\bar{\Delta} i_T} \tag{43}$$

which is independent of the amplifier-readout gain G and the value of the load resistor R_L. Now for the specific case given above where $\bar{\Delta} i_s$, the source flicker noise, is the major noise, S/N is given by

$$\left(\frac{S}{N}\right)^L = \frac{I_A}{\xi (\Delta f)^{1/2} I_L} = \frac{\alpha_A}{\xi (\Delta f)^{1/2}} \tag{44}$$

for the case of a line source and by

$$\left(\frac{S}{N}\right)^C = \frac{I_A}{(\Delta f)^{1/2} I_{oc}\, s} = \frac{\alpha_A}{\xi (\Delta f)^{1/2}} \tag{45}$$

which shows that S/N is independent of phototube gain and, therefore, of phototube voltage,* and so there is no need to optimize an atomic absorption instrument for phototube voltage or gain. Also S/N is independent of monochromator slit width.† Of course, as the slit width decreases, S/N will eventually decrease due to the phototube shot noise. Also at very large slit widths, $\bar{\Delta} i_f$ will become significant because $\bar{\Delta} i_f$ is due to flicker in flame background (band or continuum), which varies with the square of the slit width. Of course, S/N will increase as l increases, although there is a maximal length of l for most burner systems (anyway the longer l, the greater the contribution due to

* It is assumed, of course, that the phototube voltage is above the value where both S and N can be accurately measured with existing instrumentation and below the point of regenerative ionization within the phototube.

† Also independent of monochromator aperture, slit height, and transmittance.

flame flicker noise, $\bar{\Delta}i_f$). Also S/N will increase, if ξ can be decreased by using a more stable source. Finally, it should be pointed out that it is not practical to decrease Δf much below 1 Hz due to source and aspiration rate drift (if drift occurs, the above equations for S/N are not valid). It is interesting to note that the signal-to-noise ratio in atomic absorption flame spectrometry is not critically dependent on instrumental conditions.

Perhaps the best means of observing the dependence of S/N on some variable X is to plot curves of S/N versus X for various values of other variables, Y, Z, etc. The sensitivity of S/N to variation in variables X, Y, and Z can be determined as well as the interdependence between X, Y, and Z.

I. Optimization Based on Experiment

When optimizing experimental parameters it is important to use a proper statistical experimental design to determine the interrelationships between variables and to obtain the values of parameters giving the maximum signal-to-noise ratio. Experiments must be performed in such a manner that instrumental conditions resulting in the maximum S/N ratio can be determined in a minimum of steps and time. The reader is referred to any of several excellent books on the statistical design of experiments [47].

J. Limiting Detectable Sample Concentration

The limiting detectable concentration in atomic absorption flame spectrometry (or any spectrochemical method) is defined as that concentration resulting in a signal-to-noise ratio * of

$$\frac{S}{N} = \frac{3 \cdot 2^{\frac{1}{2}}}{n^{\frac{1}{2}}} \tag{46}$$

where n is the number of sample and blank measurements, the 3 represents 3 standard deviations or a 99.86% probability that the value of C_m actually belongs to the sample population, and the 2 is needed because the signal S is the difference between the blank and sample readings. Experimentally, the above definition is unambiguous. Theoretically, if an expression for S/N is

* It has been assumed in the derivation of Eq. (46), that the number of sample measurements, n, is the same as the number of blank measurements, n. For values of $n < 10$, the 3 in Eq. (46) should be replaced by the value of the student "t" and 2n-2 degrees of freedom and a 99.86% confidence level.

available, then the minimum detectable atomic concentration N_{min} can be calculated using the previous expressions. Once N_{min} is known, the theoretical minimum detectable sample concentration C_{min} can be found using Eq. (31). The value of being able to calculate C_{min} (and N_{min}) from theory is severalfold. First, such calculations can be used to decide whether atomic absorption flame spectrometry is even worth trying. Second, such calculations can be used to determine what instrument and experimental conditions give the best results. Third, such calculations are useful for explaining the influence of interferences on limits of detection and shapes of analytical curves. See Refs. [46] and [47].

K. Interferences in Atomic Absorption Flame Spectrometry

Interferences in flame spectrometry (41, 50) can be classified as blank interferences or radiation interferences. Blank interferences are primarily a result of two reasons: cross sensitivity due to inadequacy of monochromator or optics in general and background interferences due primarily to flame background and spectral overlap. If a modulation system with tuned ac amplifier is used, then blank interferences due to elements in the flame gases emitting light within the spectral bandwidth of the monochromator should be negligible; however, if nonabsorbed line radiation from the source itself is within the spectral bandwidth of the monochromator, this will cause a bending of analytical curves toward the abscissa as high sample concentrations and will in general give low absorbance readings. Blank interferences are generally not nearly as important in atomic absorption as in atomic emission flame spectrometry.

Radiation interferences do not act on the blank but rather act directly on the analytical line of interest. Radiation interferences can be classified as specific, in which the interferent * has a greater effect on some elements than on others, and nonspecific in which the interferent has the same effect on all elements.

Nonspecific radiation interferences include: sample transport interferences, solvent evaporation interferences, and flame shape interferences. The rate of sample transport (51) is critically dependent on the surface tension, viscosity, and vapor pressure of the solvent and aspirator type and geometry.

---

* The interferent is called the concomitant by many authors.

Changes in the solvent characteristics [51] greatly affects the sample transport rate F in Eq. (31), and the sample aspiration efficiency ϵ. The rate of sample evaporation is critically dependent on the type of solvent, the size of droplets, and the flame composition and temperature. All of these factors again influence ϵ greatly [19, 20]. The flame size [49], which is represented by e_f in Eq. (31), is dependent on the flame composition, solvent type, and burner type; the efficiency of sample solution aspiration, ϵ, is dependent upon the size (diameter) of the flames gases, l. Therefore, all nonspecific interferences must be considered when doing any kind of flame spectrometry. The analyst must carefully control flow rate and pressure of the aspirating gas and the temperature of the sample solution so that F and ϵ do not vary significantly during a series of measurements.

Specific radiation interferences can be subdivided as: solute vaporization interferences, dissociation interferences, ionization interferences, and excitation interferences. Solute vaporization, dissociation, and ionization interferences influence the number of atoms available for absorption, i.e., they decrease the efficiency of atomization, β. Solute vaporization interferences results when an element of interest, e.g., Ca, introduced into the flame as $CaCl_2$ in H_3PO_4, forms a compound of low volatility with an interferent, i.e., phosphorous in the aerosol droplet, resulting in a compound, a Ca-P-O complex, which may be thermodynamically stable or may decompose only slowly to atoms. In any event, solute vaporization interference reduces the atomic concentration of absorbers and results in a depressing effect. This type of interference as well as ionization and dissociation interferences are very sensitive to variation in flame temperature and composition, i.e., β varies greatly with flame temperature and composition. Dissociation interference occurs, e.g., when HCl influences the degree of dissociation of KCl in the flame gases and thus influences the atomic concentration of K.

Ionization interference occurs, for example, when KCl is sprayed into the flame gases; at low sample concentrations, K will be appreciably ionized in the flame gases — the degree of ionization increases as flame temperature increases or as atomic concentration of K decreases. Again, this is a depressant type of interference. If CsCl is added to the above KCl solution, the Cs atoms will also ionize increasing the electron concentration in the flame gases and therefore the excess of electrons will repress the ionization of K. This latter effect is an enhancement type of interference. The final type of specific radiation interference is called excitation interference and this interference is due to the change in number of excited atoms as the flame temperature changes. Since the fractional change in excited-to-ground state atomic concentration is generally small, excitation interference is negligible in atomic absorption flame spectrometry but very important in atomic emission methods. *

The above method of dividing interferences is more logical than most subdivisions commonly used. In most atomic absorption articles, the methods of subdividing interferences is extremely biased toward atomic absorption flame spectrometry whereas the above method is of general use for all types of flame spectrometry.

L. Quantitative Analysis

Most quantitative analyses [52] in atomic absorption flame spectrometry involve the use of an analytical curve in which the absorption signal † versus the sample concentration introduced into the flame gases is plotted. An analytical curve is plotted using standard solutions of the species of interest. The concentration range of the curve should bracket all unknown concentrations, and at least three measurements of each concentration should be made to assure a statistically valid result. The absorption signal of the unknown solution is then determined, and the unknown concentration is determined by using the analytical curve. Analytical curves are more useful if many (more than 10) samples are to be determined at one time. However, if only one or two samples are to be measured, an analytical curve requires far too much time, and therefore the standard addition method should be used. In this method, the absorption signal of the unknown is first measured. An accurately measured volume of the solution of unknown concentration and the modified

---

* Many authors call excitation interference the following: if the excitation energy of an emission line of an interferent is similar to the excitation energy of the sample atom, then there can result an exchange of energy. This does occur but this doesn't influence the equilibrium concentration of excited atoms if in fact thermal equilibrium does exist. The population ratio N_u/N_0 between an excited and ground state is dependent on only one variable parameter, the flame temperature. Thus the so-called interelement interferences in flame emission spectrometry based on the former explanation do not exist for gases in thermal equilibrium.

† The absorption signal is either absorbance A or fraction absorbed α.

unknown solution is measured. By simple algebra, the unknown concentration can be calculated from the above information.

M. Minimization of Interferences in Atomic Absorption Flame Spectrometry[49]

1. Changes in Flame Temperature and Composition

If ionization and solute volatilzation interferences are important, then a flame of lower or high temperature, respectively, could be used. Solute volatilization interferences can also usually be diminished by using a fuel-rich flame. Ionization interference can always be noted by the increased slope of the analytical curve at low sample concentrations, and solute volatilzation interferences can be noted by the decreased slope of the analytical curve at high sample concentrations. If a species is thermodynamically unstable but the dissociation process is kinetically slow, e.g., Ca-P-O mentioned previously, then an increased sensitivity results by taking measurements high in the flame to allow time for decomposition.

2. The Parametric Method

In this method a set of analytical curves are plotted. Each curve is for a different concentration of the interferent. Therefore, in an actual analysis, the solution of unknown concentration is measured to give the approximate concentration of the interferent and then measured for the species in concern. The concentration of the species in concern is then determined by using the proper analytical curve corresponding to the approximate interferent concentration.

3. The Simulation Method

In this method, all standard solutions used to prepare the analytical curve contain the same concentration of interferent as the unknown sample solution: this method is useful only if all unknowns have about the same concentration of the interferent.

4. The Method of Standard Addition

If the volume of standard solution added to the unknown solution is relatively small compared to the volume of unknown and if all absorption signals can be accurately measured, then this method is excellent and compensates for most types of interferences. This method is particularly useful for minimizing ionization interferences.

5. The Normalization Method

This method is useful for minimizing ionization in flames by raising the electron concentration of free electrons in the flame gases to a higher, nearly constant level. This is done by adding a salt having an easily ionizable element, called a normalizer e.g., CsCl.

6. The Saturation Method

This method is used if the concentration of the interferent varies greatly from sample to sample. Therefore, ionization is repressed in this case by adding an exceedingly large concentration of the normalizer to all unknowns.

7. The Buffer Method

This method consists in adding a third element which is not the interferent. The buffer element is called a releasing agent and effectively releases the analysis element from the interferent. For example, the effect of phosphorous on calcium can be suppressed by adding an element, such as strontium, which is also interferred with by the phosphorous, and thus the calcium is released by the action of the strontium.

8. The Dilution Method

This method can be used with interferences that decrease in relative magnitude with dilution. For example, the influence of phosphorous on calcium decreases with dilution.

Herrmann and Alkemade [49] discuss the above methods of minimizing interferences in detail and also discuss other more elaborate methods for minimizing interferences in flame emission spectrometry. Most of the methods discussed also apply directly to minimization of interferences in atomic absorption flame spectrometry.

N. Sensitivity of Analysis

The sensitivity of analysis is directly dependent on the signal-to-noise ratio discussed in a previous section. In Table III, some typical limits of detection for over 50 elements measured by atomic absorption flame spectrometry are given. For comparison purposes, some typical limits of detection for the same elements measured by flame emission spectrometry are also given in Table III.

TABLE III

Some Reported Limits of Detection (in ppm) in Atomic Absorption Flame Spectrometry and in Flame Emission Spectrometry[a]

Element	Emission	Absorption
Aluminum	0.01 (0.2)[b]	0.5 (0.1)[b]
Antimony	0.3 (20)[b]	0.2 (0.2)[b]
Arsenic	1.0 (50)[b]	0.04 (0.5)[b]
Barium	0.01 (0.03)[b]	1.0 (0.1)[b]
Beryllium	0.1 (1)[b]	0.05 (0.003)[b]
Bismuth	1.0 (40)[b]	0.2 (0.002)[b]
Boron	0.1 (30)[b]	250. (15)[d]
Bromine	100.	—
Cadmium	0.3 (6)[b]	0.0004 (1.001)[b]
Calcium	0.001	0.01
Carbon	10.	—
Cerium	1.0[c] (10)[b]	—
Cesium	0.0003 (0.005)[b]	0.05 (0.05)[b]
Chlorine	100.	—
Chromium	0.003 (0.1)[b]	0.001 (0.005)[b]
Cobalt	0.03 (1)[b]	0.003 (0.007)[b]
Copper	0.01 (0.1)[b]	0.005 (0.01)[b]
Dysprosium	0.1 (0.1)[b]	6.4 (0.2)[b]
Erbium	0.1 (0.3)[b]	3.0 (0.2)[b]
Europium	0.002 (0.003)[b]	2.5 (0.2)[b]
Fluorine	100.	—
Gadolinium	0.03[c] (2)[b]	— (4)[b]
Gallium	0.01 (0.7)[b]	1.0 (0.07)[b]
Germanium	3.0 (0.6)[b]	0.1 (2)[b]
Gold	0.1 (4)[b]	0.1 (0.1)[b]
Hafmium	80. (75)[b]	— (15)[b]
Holmium	0.1 (0.1)[b]	4.0 (0.3)[b]
Indium	0.003 (0.03)[b]	0.1 (0.05)[b]
Iodine	10.	—
Iridium	100. (100)[b]	— (4)[b]
Iron	0.03 (0.7)[b]	0.003 (0.01)[b]
Lanthanum	0.01 (1)[b]	10. (80)[b]
Lead	0.03 (3)[b]	0.01 (0.01)[b]
Lithium	0.00001 (0.000003)[b]	0.005 (0.005)[b]
Lutetium	0.1(0.2)[b]	— (50)[b]
Magnesium	0.003 (0.2)[b]	0.003 (0.0005)[b]
Manganese	0.001 (0.1)[b]	0.001 (0.005)[b]
Mercury	1.0 (40)[b]	0.01 (0.2)
Molybdenum	0.03 (0.03)[b]	0.2 (0.1)[b]
Neodymium	0.03 (1)[b]	— (2)[b]
Nickel	0.03 (0.6)[b]	0.003 (0.01)[b]
Niobium	1.0 (1)[b]	3.0 (20)[b]
Nitrogen	10.	—
Osmium	100. (10)[b]	—
Palladium	0.01 (1)[b]	0.1 (0.5)[b]
Phosphorous	1.0	100.

Platinum	3.0 (40)[b]	0.5 (0.5)[b]
Potassium	0.00001	0.005
Praseodymium	0.3[c] (2)[b]	— (6)[b]
Rhenium	0.3	25.
Rhodium	0.03 (0.3)[b]	0.1 (0.03)[b]
Rubidum	0.0001 (0.002)[b]	0.02 (0.005)[b]
Ruthenium	0.01 (0.3)[b]	0.3 (0.3)[b]
Samarium	0.1 (0.6)[b]	50. (5)[b]
Scandium	— (0.07)[b]	— (0.2)[b]
Selenium	—	0.02 (1)[b]
Silicon	3.0 (5)[b]	0.1 (0.2)[b]
Silver	0.01 (0.3)[b]	0.02 (0.01)[b]
Sodium	0.00001	0.005
Strontium	0.001 (0.004)[b]	0.02 (0.01)[b]
Sulfur	10.[c]	—
Tantalum	10. (18)[b]	— (6)[b]
Tellurium	2.0 (200)[b]	0.3 (0.3)[b]
Terbium	0.1[c] (1)[b]	— (2)[b]
Thallium	0.01	0.03
Thorium	1.0[d]	—
Thulium	0.1 (150)[b]	6.2
Tin	0.3 (4)[b]	0.02 (0.1)[b]
Titanium	0.03 (0.5)[b]	1.0 (0.1)[b]
Tungsten	1.0	250.
Uranium	1.0[d] (10)[b]	10000. (12)[b]
Vanadium	0.1 (0.3)[b]	0.5 (0.1)[b]
Ytterbium	0.01 (0.05)[b]	2.0 (0.04)[b]
Yttrium	0.01 (0.3)[b]	10. (0.3)[b]
Zinc	3.0 (50)[b]	0.0002 (0.002)[b]
Zirconium	10. (50)[b]	— (5)[b]

[a] Limits of detection for flame emission taken from P. T. Gilbert, Jr. Analysis Instrumentation - 1964, pp. 193-233, Plenum, New York, 1964. Limits of detection for atomic absorption taken from Slavin, W., Atomic Absorption Newsletter, No.

[b] Values in parenthesis taken from Fassel et al.(23) who used a premixed, fuel-rich, and oxyacetylene flame.

[c] Molecular band emission.

[d] Continuous emission.

Limits of detection can often be lowered even more by using organic solvents. Organic solvents often produce an increase in aspiration efficiency ϵ and an increase in atomization efficiency β. The reader is referred to the paper by Gibson, et al.(56) for a discussion of the influence of organic solvents on the atomic absorption signal.

O. Applications of Atomic Absorption Flame Spectrometry

In Table IV, a listing is given of elements which have been determined using atomic absorption flame spectrometry and of the areas and types of samples in which these measurements were made. This tabulation is a summary of a more complete compilation[53-55]. The listing of applications is given only to acquaint the reader to the vast number and types of determinations already carried out. For complete procedures and experimental details, the reader is referred to the original literature references to be found in the references used to compile Table IV.

P. Atomic Absorption Spectrometry with Nonflame Cells

In atomic absorption spectrometry, the sample must be atomized prior to the measurement. Flames have been almost exclusively used despite the numerous disadvantages mentioned previously. Nonflame cells of a variety of types have been used in atomic absorption studies. Since none of these nonflame cells have been used for routine studies, only a brief listing of some of the more useful and interesting nonflame cells will be given.

Gatehouse and Walsh[56-59] have used a hollow-cathode discharge tube as the sample cell. The sample was placed inside the hollow-cathode cylinder and after suitable degassing, an atomic vapor was produced using a dc discharge. Goleb and co-workers[58-60] have also used HCDT's as the absorption cell for trace element analysis and for isotopic analysis of uranium and lithium. L'vov[61] used a graphite furnace in which the sample was placed on a graphite rod and arced. The transient atomic vapor was measured using atomic absorption techniques. Nelson and Kuebler[61] thermally vaporized samples (heating fine filaments or thin strips of a solid in the optical path using an intense pulse of light from a capacitor discharge lamp). Kinetic absorption spectra were recorded using a second flame of light as the source.

Q. Atomic Fluorescence Flame Spectrometry

Atomic fluorescence flame spectrometry is a new method of chemical analysis. The method is based upon the absorption of radiation by an atomic vapor producing excited atoms and the measurement of the radiation emitted when a fraction of these atoms lose their energy by a radiational (called fluorescence) process. The method was first described by Winefordner and

TABLE IV

Some Applications in Which Atomic Absorption Flame Spectrometry is Used as the Measurement Method

Element	Samples
Agricultural applications	
Ca	Plants, soils
Cu	Fertilizers, plants, soils
Fe	Fertilizers, plants, soils
K	Plants, soils
Mg	Fertilizers, plants, soils
Mn	Fertilizers, plants, soils
Mo	Fertilizers, plants, soils
Na	Plants, soils
Rb	Plants
Sr	Plants, soils
Zn	Fertilizers, plants, soils
Biological applications	
Bi	Urine
Ca	Bone and teeth, cerebrospinal fluid, feces, saliva, serum, tissue, urine
Cr	Feces, urine
Cu	Serum, tissue, urine
Cd	Urine
Co	Urine
Hg	Urine
K	Bone and teeth, serum, tissue
Mg	Bone and teeth, cerebrospinal fluid, feces, saliva, serum, tissue, urine
Na	Bone and teeth, serum, tissue
Ni	Urine
Pb	Serum
Sr	Tissue, urine
Tl	Urine
Zn	Serum, tissue, urine
Food products applications	
Ca	Beer, diet foods
Cu	Beer, butter, milk, wine
Fe	Beer
K	Beer
Mg	Diet foods
Na	Beer
Pb	Wine
Sr	Fish and meat, vegetables, milk
Rb	Vegetables
Zn	Wine

Industrial applications

Al	Cement, glass
Ag	Lubricating oils
Ba	Lubricating oils, phosphorus
Ca	Cement, coal ash, phosphorus
Cr	Lubricating oils
Cu	Liquors, lubricating oils, plating solutions
Fe	Cement, lubricating oils, tungsten carbide
Hg	Isotopes
K	Cement, coal ash
Li	Isotopes
Mg	Cement, coal ash, fibers, glass, lubricating oils
Mn	Cement, fibers
Ma	Cement, coal ash, limestone, lubricating oils, phosphorus
Ni	Lubricating oils
Pb	Gasoline, lubricating oils, plating solutions
Sn	Lubricating oils
Sr	Coal ash
U	Isotopes
Zn	Liquors, plating solutions

Metallurgical applications

Ag	Aluminum alloys, gold, iron and steel, zinc alloys
Al	Copper alloys, iron and steel, zinc alloys
Ca	Magnesium alloys
Cd	Zinc alloys
Co	Iron and steel, nickel and nickel alloys
Cu	Aluminum alloys, copper alloys, gold, iron and steel, magnesium alloys, zirconium
Fe	Aluminum alloys, gold
Mg	Aluminum alloys, gold, iron and steel, nickel and nickel alloys, uranium
Mn	Copper alloys, iron and steel, magnesium alloys
P	Copper alloys
Pb	Copper alloys, gold, iron and steel
Pd	Gold
Se	Copper alloys
Si	Aluminum alloys, iron and steel
Zn	Aluminum alloys, copper alloys, gold, iron and steel, magnesium alloys, zirconium

Mining and geochemical applications

Ag	Ores
Au	Ores
Ga	Seawater, silicate rocks, water samples
Cd	Silicate rocks
Co	Silicate rocks
Cu	Brines, ores, petroleum crudes, seawater, silicate rocks, water samples
Fe	Petroleum crudes, seawater, silicate rocks, water samples
K	Seawater, silicate rocks, water samples

Mg	Seawater, silicate rocks, water samples
Mn	Seawater, silicate rocks
Na	Seawater, silicate rocks, water samples
Ni	Ores, seawater, silicate
Pb	Brines, ores
Rb	Seawater, silicate rocks
Sr	Seawater, silicate rocks
Zn	Ores, seawater, silicate rocks

co-workers[63-67]. The experimental system is similar to the one used in atomic absorption flame spectrometry and differs only in the entrance optics. The basic atomic absorption flame spectrometer shown in Fig. 1 can be converted to an atomic fluorescence flame spectrometer by placing the source flame cell perpendicular to the flame cell-monochromator system. A number of articles have been published recently concerning atomic fluorescence flame spectrometry, and the reader will be referred to them[63-72].

The major advantages of atomic fluorescence flame spectrometry appear to be its great sensitivity, e.g., elements like zinc and cadmium can be measured at concentrations below 1 ppb with ease, its great linearity, e.g., analytical curves are linear over a concentration range of 10^4 or more, and its possible general use, e.g., all elements having intense sources of excitation can be easily and simply measured. Since this chapter is concerned with atomic absorption flame spectrometry, no more discussion of atomic fluorescence flame spectrometry will be given. However, the reader is referred to the literature[63-72] for a more thorough discussion of the principles, instrumentation, results, and applications of this interesting, new analytical method.

REFERENCES

(1) J. Fraunhofer, Ann. Physik., 56, 264 (1817).

(2) G. Kirchhoff, Pogg. Ann., 109, 275 (1860).

(3) G. Kirchhoff and R. Bunsen, Pogg. Ann., 110, 161 (1860).

(4) G. Kirchhoff and R. Bunsen, Pogg. Ann., 113, 327 (1861).

(5) G. Kirchhoff and R. Bunsen, Phil. Mag., 22, 275 (1861).

(6) A. Walsh, Spectrochim. Acta, 7, 108 (1955).

(7) C. Th. J. Alkemade and J. M. W. Milatz, Appl. Sci. Res., 4B, 289 (1955).

(8) B. J. Russell, J. P. Shelton, and A. Walsh, Spectrochim. Acta, 8, 317 (1957).

(9) A. Unsöld, Physik der Sternatomospharen, Springer, Berlin, 1955.

(10) A. C. G. Mitchell and M. W. Zemansky, Resonance Radiation and Excited Atoms, University Press, Cambridge, 1934.

(11) J. D. Winefordner, W. W. McGee, M. L. Parsons, J. M. Mansfield, and K. E. Zacha, Anal. Chim. Acta, 36, 25 (1966).

(12) M. L. Parsons, W. J. McCarthy, and J. D. Winefordner, Appl. Spectr., 20, 223 (1966).

(13) S. S. Penner, Quantitative Molecular Spectroscopy and Molecular Emmissivities, Addison-Wesley, Reading, Mass., 1959.

(14) J. D. Winefordner and T. J. Vickers, Anal. Chem., 36, 1947 (1964).

(15) L. de Galan, W. W. McGee, and J. D. Winefordner, Anal. Chim. Acta, 37, 436 (1967).

(16) L. de Galan and J. D. Winefordner, Spectrochim. Acta, 23B, 277 (1968).

(17) G. L. Vidale, U.S. Dept. Comm., Office of Tech. Serv., P.B. Rept. 148, 206 (1960).

(18) J. D. Winefordner and T. J. Vickers, Anal. Chem., 36, 1939 (1964).

(19) J. D. Winefordner, C. T. Mansfield, and T. J. Vickers, Anal. Chem., 35, 1611 (1963).

(20) M. L. Parsons and J. D. Winefordner, Anal. Chem., 38, 1593 (1966).

(21) L. de Galan and J. D. Winefordner, J. Quant. Spectr. Radiat. Transfer, 7, 251 (1967).

(22) L. de Galan and J. D. Winefordner, Anal. Chem., 38, 1412 (1966).

(23) V. A. Fassel and D. W. Golightly, Anal. Chem., 39, 466 (1967).

(24) J. H. Gibson, W. E. L. Grossman, and V. D. Cooke, Analytical Chemistry (Proc. Feigl. Anniv. Symp.), Elsevier, 1962.

(25) V. A. Fassel, V. G. Mossotti, W. E. L. Grossman, and R. D. Kniseley, Spectrochim. Acta, 22, 347 (1966).

(26) W. B. Jones and A. Walsh, Spectrochim. Acta, 16, 249 (1960).

(27) W. Slavin and D. C. Manning, Appl. Spectr., 19, 65 (1965).

(28) J. V. Sullivan and A. Walsh, Spectrochim. Acta, 21, 721 (1965).

(29) R. K. Skogerboe and R. A. Woodriff, Anal. Chem., 35, 1977 (1963).

(30) B. M. Gatehouse and A. Walsh, Spectrochim. Acta, 16, 602 (1960).

(31) A. C. Menzies, Anal. Chem., 32, 898 (1960).

(32) W. Slavin, Atomic Absorption Newsletter, No. 10, 1963.

(33) W. Slavin and D. C. Manning, Anal. Chem., 35, 253 (1963).

(34) R. W. Tabeling and J. J. Devaney, Talk at "Symposium on Spectroscopy," Chicago, 1961.

(35) K. Fuwa and B. L. Vallee, Anal. Chem., 35, 942 (1963).

(36) S. A. Koirtyohann and C. Feldman, in Developments in Applied Spectroscopy, Vol. 3, Plenum Press, New York, 1964.

(37) W. W. McGee and J. D. Winefordner, Anal. Chim. Acta, 37, 429 (1967).

(38) D. J. David, Analyst, 85, 779 (1960).

(39) J. W. Robinson, Anal. Chim. Acta, 24, 451 (1961).

(40) J. W. Robinson, Anal. Chem., 33, 1067 (1961).

(41) R. N. Kniseley, A. P. D'Silva, and V. A. Fassel, Anal. Chem., 35, 910 (1963).

(42) J. D. Winefordner and C. Veillon, Anal. Chem., 36, 943 (1964).

(43) M. D. Amos and J. B. Willis, Spectrochim. Acta, 22, 1325 (1966).

(44) G. F. Box and A. Walsh, Spectrochim. Acta, 16, 255 (1960).

(45) J. P. Shelton and A. Walsh, Proceedings of 15th International Congress on Pure and Applied Chemistry, Vol. II, p. 403-409, Lisbon, Portugal, 1956.

(46) M. L. Parsons, W. J. McCarthy, and J. D. Winefordner, J. Chem. Ed., 49, 214 (1967).

(47) J. D. Winefordner, W. J. McCarthy, and P. A. St. John, J. Chem. Ed., 44, 80 (1967).

(48) J. D. Winefordner and C. Veillon, Anal. Chem., 37, 416 (1965).

(49) R. Herrmann and C. Th. J. Alkemade, Flammenphotometrie, 2nd ed., Springer, Berlin, 1960. Chemical Analysis by Flame Photometry (P.T. Gilbert, Jr. transl.), Wiley, New York, 1960.

(50) C. Th. J. Alkemade, Anal. Chem., 38, 1252 (1966).

(51) J. W. Robinson, Atomic Absorption Spectroscopy, Dekker, New York, 1966.

(52) T. H. Gibson, W. E. L. Grossman, and W. D. Cooke, Anal. Chem., 35, 266 (1963).

(53) Atomic Absorption Newsletter, No. 1, 1962 Perkin-Elmer Corp., Norwalk, Conn.

(54) Atomic Absorption Newsletter, No. 16, 1963, Perkin-Elmer Corp., Norwalk, Conn.

(55) W. Slavin, Atomic Absorption Newsletter, 4, 186 (1965).

(56) B. M. Gatehouse and A. Walsh, Spectrochim. Acta, 16, 602 (1960).

(57) B. M. Gatehouse and A. Walsh, Nature, 192, 929 (1961).

(58) J. A. Goleb and J. K. Brody, Anal. Chim. Acta, 28, 457 (1963).

(59) J. A. Goleb, Anal. Chem., 35, 1978 (1963).

(60) J. A. Goleb and Y. Yokoyama, Anal. Chem., 30, 213 (1964).

(61) B. V. L'vov, Spectrochim. Acta, 17, 761 (1961).

(62) L. S. Nelson and N. A. Kuebler, Spectrochim. Acta, 19, 781 (1963).

(63) J. D. Winefordner and T. J. Vickers, Anal. Chem., 36, 161 (1964).

(64) J. D. Winefordner and R. A. Staab, Anal. Chem., 36, 165 (1964).

(65) J. D. Winefordner and R. A. Staab, Anal. Chem., 36, 1367 (1964).

(66) J. M. Mansfield, J. D. Winefordner, and C. Veillon, Anal. Chem., 37, 1049 (1965).

(67) C. Veillon, J. M. Mansfield, M. L. Parsons, and J. D. Winefordner, Anal. Chem., 38, 204 (1966).

(68) R. M. Dagnall, T. S. West, and P. Young, Talanta, 13, 803 (1966).

(69) R. M. Dagnall, K. C. Thompson, and T. S. West, Anal. Chim. Acta, 36, 269 (1966).

(70) D. N. Armentrout, Anal. Chem., 38, 1235 (1966).

(71) D. W. Ellis and D. R. Demers, Anal. Chem., 38, 1943 (1966).

(72) J. D. Winefordner, M. L. Parsons, J. M. Mansfield, and W. J. McCarthy, Spectrochim. Acta, 23B, 37 (1967).

PART 2

Phase State and Structure

Chapter 5

X-RAY DIFFRACTION

R. W. Gould

Department of Metallurgical
and Materials Engineering
University of Florida
Gainesville, Florida

I. INTRODUCTION

X-ray diffraction has played a vital role in materials research over the past 30 years, and it would be impossible to summarize in one chapter the effect that x-ray diffraction has had upon the science of ceramic materials in this time period. Therefore, a definite subject area limitation must be placed when writing a single chapter designed to cover the entire field of x-ray diffraction.

In this chapter, we will assume that the reader is familiar with the simpler aspects of crystallography, has some knowledge of vector analysis, and has a basic knowledge of x-ray generation, detection, and diffraction theory. The approach taken in this chapter will be based upon a thorough use of the reciprocal lattice and its application to diffraction problems. Using this method, it will be possible to show qualitatively and quantitatively how various x-ray techniques can reveal information concerning the structure and properties of crystalline and noncrystalline materials. Several highly specialized techniques will be developed more fully in subsequent chapters.

In a recently published text Buerger(1) discusses the history of the reciprocal lattice and points out that it was developed, at least insofar as its geometrical features are concerned, as early as 1850 by a French mining engineer named N. A. Bravais. In 1913, P. P. Ewald showed that x-ray diffraction could be explained in terms of a reciprocal lattice whose translations were reciprocals of the translations of the crystal lattice and a sphere whose radius was equal to the wave number of the radiation, i. e., $1/\lambda$. In 1926, J. D. Bernal, pointed out that Ewald's reciprocal lattice and his sphere of reflection provided a very simple means of indexing photographs made by rotating crystal methods. The use of the reciprocal lattice, however, has not been popular because it is only a conceptual lattice not existing in real space. It was recognized early by crystallographers that the reciprocal lattice could aid immeasurably in visualizing diffraction effects from single crystals. In the last fifteen years the reciprocal lattice has been used to illustrate diffraction effects from imperfect crystals and it is a major purpose of this chapter to show how the reciprocal lattice, once understood, can be used to clearly visualize both the effects caused by various crystalline imperfections and the relationship of the x-ray diffraction technique to the type of information that it can provide. Before discussing the geometry of reciprocal space, it is necessary to look at the simple equations which relate the scattering intensity of an x-ray beam to the substance which is causing the scattering.

II. X-RAY DIFFRACTION

A. Scattering of X Rays

It is instructive to know in advance what direction you are taking when a subject is being presented. In this section on scattering we will discuss the intensity of radiation scattered by various assemblages of scattering centers, starting with a single electron and proceeding in steps to a "perfect" single crystal. It would require many pages to fully develop the theory, thus the reader will be directed to various texts for details. In general, we are approaching the subject from the direction of disorder toward order rather than the reverse.

B. Scattering by a Single Free Electron

The scattering of radiation by a free electric charge was first treated by J. J. Thompson who considered the interaction of x rays with the electron in a classic (nonquantum) sense. If we allow the primary x-ray beam whose intensity (Fig. 1) is I_o to be unpolarized (which in general it is), then the intensity (I_e) scattered by a single free electron per unit solid angle per second is

$$I_e = 7.90 \times 10^{-26} \left(\frac{1 + \cos^2 2\theta}{2} \right) I_o \tag{1}$$

Equation (1) shows that I_e is proportional to the scattering angle θ. This is due to the fact that when x rays are scattered they become polarized. Equation (1) is fundamental to all absolute intensity measurements with x rays

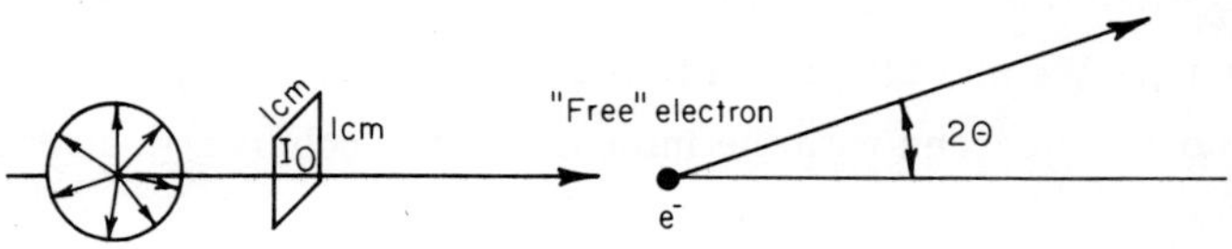

Fig. 1. Scattering of x rays by a single free electron.

because, as we will see, all subsequent intensity measurements will be expressed as ratios involving the scattering of a single free electron. Quantum mechanical calculations show that the free electron scattering occurs, but it always results in a wavelength shift. All of the scattered radiation will have its λ increased (energy lost) due to the Compton effect, the magnitude of which is θ dependent, as give by Eq. (2):

$$\Delta\lambda = 0.024\ (1 - \cos 2\theta)\ \text{Å} \tag{2}$$

Nevertheless, for x-ray wavelength normally used, the Thompson equation is valid within a few percent. Thus, we say that all of the scattering by a single free electron is incoherent (λ changed), and since the incoming beam and the scattered beam do not have the same wavelength there is no definite phase relationship between them and they will not interfere with each other, i. e., their intensities simply add.

C. Atomic Electron (Bound Electron)

The next logical extension is to consider the electron that is not a free space charge. Consider the hydrogen atom. Here, the single free electron is bound to the positive nucleus by a coulomb force. No longer can we consider electrons to occupy a fixed position in space but now we know from quantum mechanics that the electron occupies a volume surrounding the nucleus. This electron "cloud" is shown in Fig. 2. We now ask what happens when a monochromatic x-ray wave strikes the cloud? In order to evaluate this interaction, let us introduce some nomenclature that is used extensively throughout the x-ray literature. Let us designate the incoming monochromatic x-ray beam by a vector $\vec{S}_0$, whose direction gives the direction of the incoming beam. In the text we will consider $|\vec{S}_0|$ to be unity.*
The scattered beam is designated by $\vec{S}$ and the absolute value of $\vec{S}$ equals unity.

Quantum mechanical calculations show that we can expect both incoherent and coherent radiation (the electron is bound and will not always recoil with an energy loss). We are interested in both intensities, but as we shall see, we will be more concerned with the intensity of the coherently scattered

*The x-ray diffraction literature is not uniform in its choice of incident beam vectors. In some cases it is $\vec{S}_0$, sometimes $\vec{s}_0$ also $\vec{K}_0$.

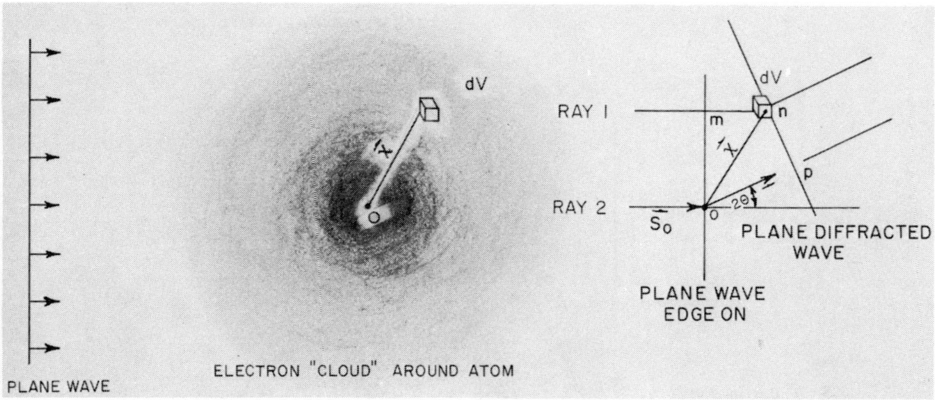

Fig. 2. Scattering of x rays by an electron cloud.

radiation. How do we calculate it? Consider the electron cloud of Fig. 2 to consist of many small volume elements dv. The charge density ρ (x) [charge per centimeter cubed] is a function of the vector $\vec{x}$ from the center of the atom. As the absolute value of $\vec{x}$ gets large the charge $\rho\ \overrightarrow{(x)}$ dv decreases. We can mentally assume that a single free electron is present at the origin and ask how each charge element ρdV scatters relative to the scattering ability of this single free electron. Let the scattering ability of the single free electron be given by unity. Figure 3 shows how we can "add up" the scattering from all such volume elements and allow for the phase relationship between the scattered waves. First, consider the path difference traveled by a wave scattered by the hypothetical free electron at the origin and the volume element dv. This path difference δ is conventionally given in vector notation by

$$\begin{aligned} \delta &= \overrightarrow{OP} - \overrightarrow{MN} \\ \delta &= \vec{x} \cdot (\vec{S} - \vec{S}_0) \end{aligned} \qquad (3)$$

The phase difference ϕ between waves scattered by these two scattering centers is given by the familiar relationship

$$\phi = \left(\frac{2\pi i}{\lambda}\right)\delta \qquad (4)$$

In order to sum up a group of waves, one must consider the wave amplitudes and the phase relationships between the waves. This is evident when adding two sine waves which are 90° out of phase. The same procedure is used

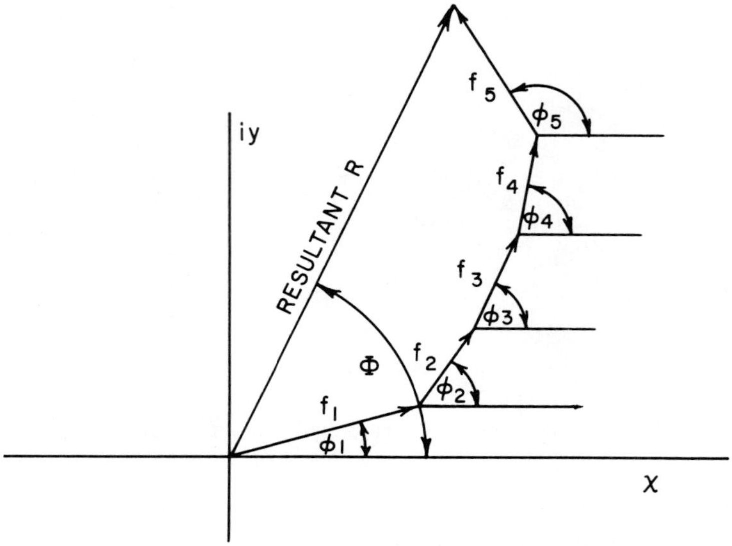

Fig. 3. Summing waves using vector $\vec{s}$.

here, only now there are many waves (a continuum). The method for summing waves of this type has been shown clearly by Buerger[2]. The resultant amplitude $\vec{R}$ from N scattering centers is given by

$$R = A \sum_{n=o}^{N-1} \exp\ (in\phi) \tag{4a}$$

Here each wave (exp inϕ) is represented in complex form, a feature which simplifies subsequent calculations. In Eq. (4a), A represents the scattering amplitudes of each of the scattering centers. For the electron cloud, A is given by ρ (x)dv/1 where the 1 in the denominator is the scattering ability of the single free electron which has been set equal to unity.

For a continuous distribution of scattering centers, the phase difference is given by Eq. (4b):

$$\phi = -2\pi i\ \vec{x} \cdot \left(\frac{\vec{S}-\vec{S}_o}{\lambda}\right) = -2\pi i\ \vec{x} \cdot \vec{s} \tag{4b}$$

where we let the vector $\left(\frac{\vec{S} - \vec{S}_o}{\lambda}\right)$ to be given by small $\vec{s}$. It will be shown that $\vec{s}$ bisects the angle between $\vec{S}$ and $\vec{S}_o$ and has a magnitude of 2 sinθ/λ. The total scattering by the electron cloud is then the sum of scattering of all the volume elements dv. Call this total scattering f:

$$f = \int_{vol} \rho(\vec{x}) \exp(-2 \ \vec{ix} \cdot \vec{s}) dv \tag{5}$$

At this point one generally makes an assumption that is more or less valid, namely, that the electron cloud is spherically symmetric, so that $\rho(\vec{x})$ can be replaced by $\rho(x)$ where x equals the absolute value of $\vec{x}$. This is valid since all radial directions away from the atom center are now equivalent. This greatly simplifies the mathematics. Thus, f for the hydrogen atom is given by

$$f = \int_{o}^{\infty} 4\pi x^2 \rho(x) \frac{\sin 2\pi sx}{2\pi sx} dx \tag{6}$$

The steps between Eqs. (5) and (6) are shown clearly by Azaroff[3]. Since f equals the amplitude scattered by a hydrogen atom divided by the amplitude scattered by a single electron, and the intensity (I) is given by amplitude times the complex conjugate of the amplitude, we have

$$f = \frac{\text{Hydrogen atom amplitude}}{\text{Single electron amplitude}}, \quad I = A \quad A^* \tag{7}$$

$$f^2 = \frac{I\ (\text{Hydrogen atom coherent})}{I_e}, \quad I^{\text{hydrogen}}_{\text{coherent}} = f^2 I_e$$

As stated earlier, quantum theory predicts that the total intensity I_T for the one-electron hydrogen atom is given by the classic Thompson formula (I_e) and is the sum of I_{coh} and $I_{Compton}$, so that

$$I^{\text{hydrogen}}_{\text{total}} = I^{\text{Thompson}}_{e} = I_{\text{coherent}} + I_{\text{Compton modified}} \tag{8}$$

$$I^{\text{hydrogen}}_{\text{comp}} = I_e - I_{\text{coherent}} = I_e(1-f^2_{\text{Hydrogen}}) \tag{9}$$

Before proceeding to a multielectron atom, one should examine the θ dependence of the I_e in Eq. (8). Note that it depends on the value of $\vec{s}$, which is a function of θ, the scattering angle. Figure 4 shows this angular dependence. Consider a hypothetical point atom, i. e., an atom where all of the charge is

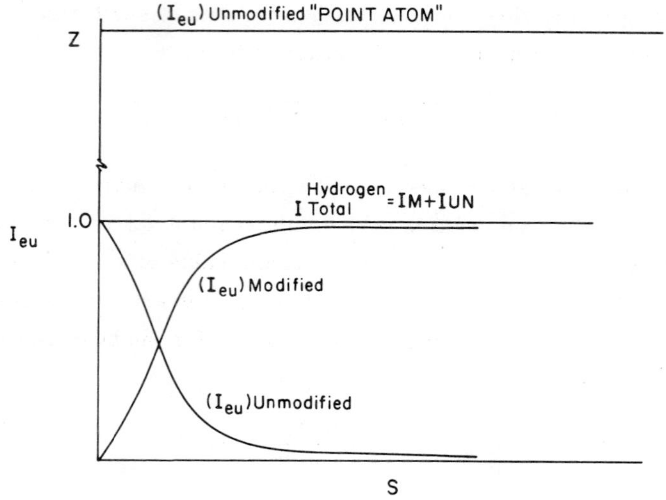

Fig. 4. Scattering of the hydrogen atom as a function of S.

located at a mathematical point so small that there can be no path differences for scattering between various volume elements of this point. Thus, since $\delta = 0$ in Eq. (3) $\phi = 0$ and Eq. (6) reduces to

$$f = \int_0^\infty 4\pi x^2 \rho(x)dx = Z \text{ (atomic number)} \tag{10}$$

which contains no angular dependent terms. Thus, I_e (unmodified point atom) would equal Z for all values of s. It can be shown that relationships similar to (7), (8), and (9) can be derived from multielectron atoms, so that in general

$$I_{coherent} = \left(\sum_{j=1}^{Z} f_j\right)^2 I_e \tag{11}$$

$$I_{Compton} = \sum_{j=1}^{Z} (1 - f_j^2)\, I_e \tag{12}$$

where the sum is over the Z electrons of the element of atomic number Z. Equations (11) and (12) are good approximations, and the reader is referred to James[4] for further information.

It is not difficult to extend the previous arguments to the case of a monatomic ideal gas. Here we will see that we have scattering by the electron cloud of each atom in the gas and interference between the scattered waves. The amplitude of this scattering from each atom (intraatomic scatter) has been shown above to be given by f_{atom}. In addition, we will also have interatomic scattering, i. e., interference between waves scattered by the various atoms. If the atoms in the assembly are very regularly positioned (as in a crystal), very interesting things occur. But first consider the present case of an ideal gas where the atoms are all of one type and are well separated. The gas atoms are in constant motion, and we can only visualize their positions at some short time instant t. Figure 5 shows such a hypothetical instant photograph of some gas atoms. Note that after some finite time increment, T (T>>t), which is approximately equal to the time necessary for a typical x-ray scattering experiment, every volume element available in Fig. 5 will have been occupied at one time by one of the atoms. Thus, we can treat the interatomic scattering in much the same way that we looked at the scattering from the "continuous" electron cloud of the single atom. Thus, the instantaneous amplitude A_t scattered by such an array as shown in Fig. 5 will be given by a scattering wave summation, where path difference and therefore phase difference is taken into consideration. This was done

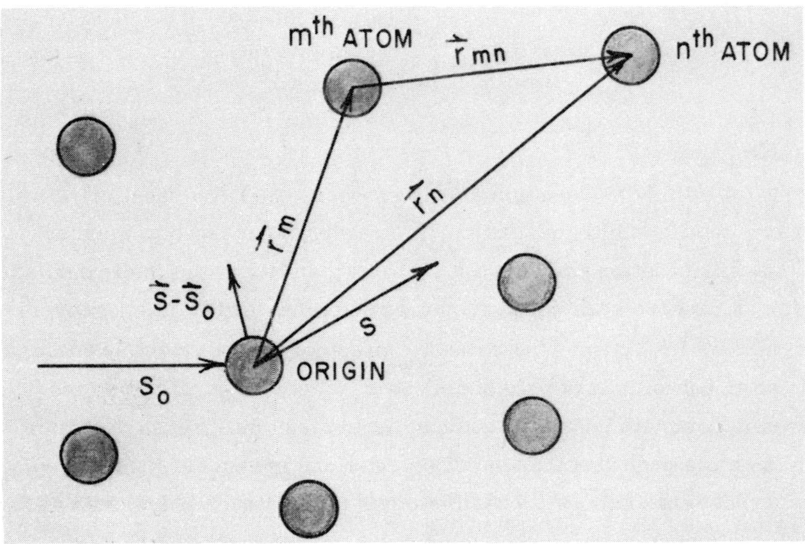

Fig. 5. Instantaneous location of some gas atoms.

for the electron cloud, and in a similar manner we can show that

$$A_t = \sum_m f_m \exp (2\pi i/\lambda) \ \overrightarrow{(S - S_o)} \cdot \overrightarrow{r}_m \tag{13}$$

Now the intensity is related to the amplitudes through

$$I_t = A_t A_t^* \tag{14}$$

where A* is the complex conjugate of A. Thus, the intensity scattered by the array shown in Fig. 5 at the time instant t will be

$$I_t = \sum_m f_m \exp (2\pi i/\lambda) \ \overrightarrow{(S\text{-}S_o)} \cdot \overrightarrow{r}_m \sum_n f_n \exp \left(-\frac{2\pi i}{\lambda}\right) \overrightarrow{(S\text{-}S_o)} \cdot \overrightarrow{r}_n$$

When these summations are combined into a double sum, realizing that $\overrightarrow{r}_m - \overrightarrow{r}_n = \overrightarrow{r}_{mn}$ (see Fig. 5), we have

$$I_t = \sum_m \sum_n f_m f_n \exp (2\pi i/\lambda) \ \overrightarrow{(S\text{-}S_o)} \cdot \overrightarrow{r}_{mn} \tag{15}$$

When we allow the gas atoms to take on all possible configurations and ask again what is the scattered intensity actually measured in some real experimental time T, we arrive at Eq. (16):

$$I_e = \sum_m \sum_n f_m f_n \frac{\sin k\overrightarrow{r}_{mn}}{k\overrightarrow{r}_{mn}} \tag{16}$$

where $k = 4\pi \sin\theta/\lambda$. Equation (16) is known as the Debye scattering equation. For more details on this derivation, the reader is referred to Warren[(5)]. Note that this equation contains the scattering ability of the inidividual atoms f (which as we have seen depends on the scattering angle θ). Moreover, and of much more importance, note that I_{eu} does not depend on the position of a single atom but on interatomic distances $\overrightarrow{r}_{mn}$ between a pair of atoms. This equation is completely general and can be applied to all forms of matter in which the atom positions take on all orientations in space. Figure 6 compares the coherent scattering of a monatomic gas (neon) with a polyatomic gas (carbon tetrachloride) in which each molecule has well-defined structure. Note that I_{eu} for carbon tetrachloride shows well-defined peaks. These peaks correspond to distances $|\overrightarrow{r}_{mn}|$ that occur with high probability in the carbon

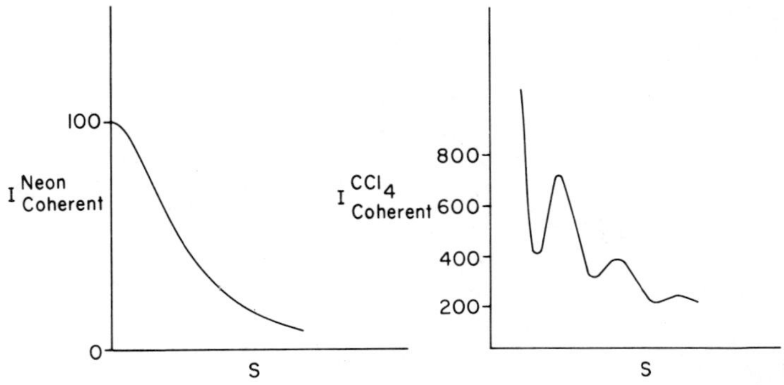

Fig. 6. Coherent scattering of neon and carbon tetrachoride (gas phase).

tetrachloride structure. Figure 6 (b) illustrates the concept that preferred interatomic distances, i.e., those that have a high probability of occurrence, will give rise to peaks in the x-ray scattering curve. If monochromatic x rays are scattered from most liquids (4) we find peaks similar to those shown in Figure 6 (b). The same is true of vitreous solids. These amorphous materials give rise to a number of broad peaks which indicate the presence of preferred $\vec{r}_{mn}$ distances. The structure of vitreous solids is receiving much attention in the literature owing to several recent improvements in experimental techniques and theory[6-9]. The reader who is interested in pursuing the subject should consult Warren's new text[5].

D. Crystalline Solids

In Eq. (15) the vector $\vec{r}_{mn}$ can have any magnitude greater than the diameter of the scattering atoms involved. In the case of a crystalline solid, the atoms are fixed in a precise periodic array, and there will be a fixed set of vectors $\vec{r}_{mn}$. In Fig. 7 a vector $\vec{R}_{m_1m_2m_3}$ is defined which locates the corner of any unit cell m_1a, m_2b, $m_3\vec{c}$ where m_1, m_2, and m_3 are integers and $\vec{a}$, $\vec{b}$, $\vec{c}$ are the unit-cell edges. The unit cell contains n atoms, and the vector ($\vec{r}_n$) in this unit cell locates any of these atoms. Thus, by choosing $\vec{R}_m$ and $\vec{r}_n$ any atom in the crystal can be located. Let

$$\vec{R}^{n}_{m_1m_2m_3} = m_1\vec{a}+m_2\vec{b}+m_3\vec{c}+\vec{r}_n \tag{17}$$

Thus,

$$\sum_{m_1}\sum_{m_2}\sum_{m_3}\sum_{n} R^{n}_{m_1m_2m_3}$$

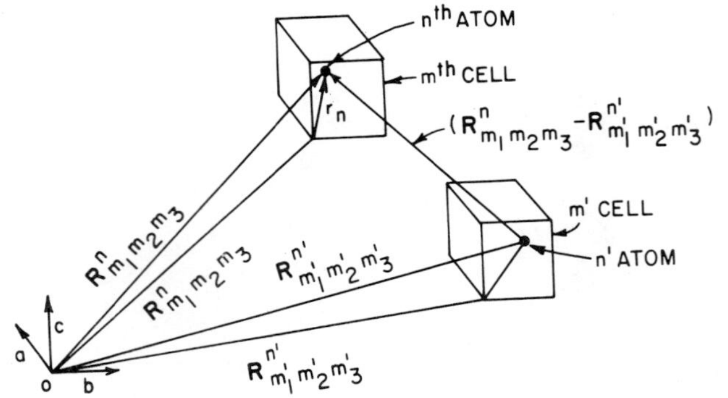

Fig. 7. Defining the unit cell locating vector $\vec{R}_{m_1 m_2 m_3}$.

will include all possible atoms in the structure. In Eq. (15), it was shown that the intensity expression contains a vector of the type $\vec{r}_{mn}$; that is, one which expresses interatomic separations rather than distance from the origin. Thus, for a crystal

$$I_{eu} = \left(\sum_{m_1} \sum_{m_2} \sum_{m_3} \sum_{n} f_n \exp\left(\frac{2\pi i}{\lambda}\right) (\vec{S}-\vec{S}_o) \cdot \vec{R}^n \right) \left(\sum_{m_1'} \sum_{m_2'} \sum_{m_3'} \sum_{n'} f_{n'} \exp\left(\frac{-2\pi i}{\lambda}\right) (S-S_o) \cdot \vec{R}^{n'} \right) \tag{18}$$

When Eq. (18) is multiplied out, a difference vector is obtained:

$$\vec{R}^n_{m_1 m_2 m_3} - \vec{R}^n_{m_1' m_2' m_3'} \qquad (19$$

$$I_{eu} = F^2 \sum_{m_1} \sum_{m_2} \sum_{m_3} \sum_{m_1'} \sum_{m_2'} \sum_{m_3'} \left[f_n f_{n'} \exp(2\pi i) \left(\frac{\vec{S}-\vec{S}_o}{\lambda} \right) \left(\vec{R}^n_{m_1 m_2 m_3} - \vec{R}^{n'}_{m_1' m_2' m_3'} \right) \right] \tag{20}$$

Equation (20), which at first appears complicated, simply represents a multiple sum over all atoms in all unit cells for which a distance $\vec{R}^n_{m_1 m_2 m_3} - R^{n'}_{m_1' m_2' m_3'}$ exists. If one substitutes Eq. (17) into Eq. (20), it is possible to show (see Ref. (5), Chapter 10; (3), Chapter 9) that

$$I_p = I_e F^2 \left\{ \frac{\sin^2\left[\frac{\pi}{\lambda}(\vec{S}-\vec{S_o})\cdot N_1\vec{a_1}\right]}{\sin^2\left[\frac{\pi}{\lambda}(\vec{S}-\vec{S_o})\cdot \vec{a_1}\right]} \frac{\sin^2\left[\frac{\pi}{\lambda}(\vec{S}-\vec{S_o})\cdot N_2\vec{a_2}\right]}{\sin^2\left[\frac{\pi}{\lambda}(\vec{S}-\vec{S_o})\cdot \vec{a_2}\right]} \times \frac{\sin^2\left[\frac{\pi}{\lambda}\,\vec{S}-\vec{S_o}\,N_3\cdot\vec{a_3}\right]}{\sin^2\left[\frac{\pi}{\lambda}(\vec{S}-\vec{S_o})\cdot \vec{a_3}\right]} \right\} \tag{21}$$

where I_p refers to the scattered intensity at some point P at a distance from the crystal. F is called the structure factor or structure amplitude, that is, the scattering ability of the unit cell relative to a single free Thompson electron and given by

$$F = \sum_n f_n \exp\left(\frac{2\pi i}{\lambda}\right)(\vec{S}-\vec{S_o})\cdot r_n \tag{22}$$

Note that Eq. (21) will predict zero diffracted intensity if F equals zero of if any of the three quotient terms is zero. These three quotient terms are of great interest in diffraction theory since it can be shown(2, 3, 5) that they lead directly to the three Laue equations which must all be satisfied simultaneously in order for diffraction to occur:

$$\frac{\vec{S}-\vec{S_o}}{\lambda}\cdot\vec{a} = h, \quad \frac{\vec{S}-\vec{S_o}}{\lambda}\cdot\vec{b} - k, \quad \frac{\vec{S}-\vec{S_o}}{\lambda}\cdot\vec{c} = 1 \tag{23}$$

Here, h, k, 1 are integers and are generally referred to as the Miller indices of the diffraction plane.

III. INTRODUCTION TO RECIPROCAL SPACE

Ewald in 1913 combined the reciprocal lattice with a sphere having a radius equal to the reciprocal x-ray wavelength, and found that this combination could be used to explain most diffraction effects. Since that time many clear explanations of difficult diffraction effects have been made utilizing the concepts of reciprocal space. When introducing reciprocal space concepts it is tempting to state without proof the geometric concepts involved and allow the student to accept these gradually before proceeding to the more mathematical aspects. It is then easy to show that a mathematical manipulation exists which will transform operations occurring in real space (the space in which the real crystal exists) into a fictitious or reciprocal space,

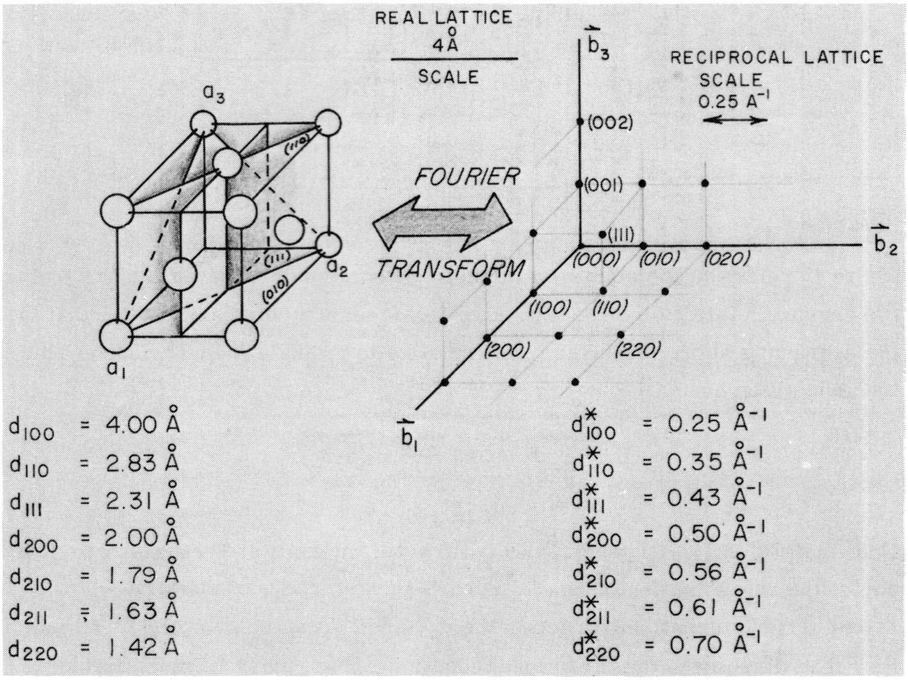

Fig. 8. Unit cell of fcc lattice and a portion of its reciprocal lattice.

a space which does not exist in the sense that one can feel it and touch it. This mathematical manipulation is called the Fourier transformation. Let us say then that the Fourier transformation is the key that allows us to open the door between real space and reciprocal space, and, once this key is obtained, the student can pass easily from one space to the other. The question now naturally arises, why should one want to go to reciprocal space from real space? It will be shown, as was shown over 50 years ago by Ewald, that diffraction space is reciprocal space.

A. Geometry of Reciprocal Space

Because of its familiarity let us consider a real, three-dimensional lattice and construct its reciprocal lattice. In Fig. 8, we see a unit cell of the fcc lattice with several planes shaded. The axial length $a_1 = a_2 = a_3$ is chosen for simplicity to be 4 Å. Note that each shaded plane really represents a complete set of parallel planes. The interplanar spacing for the planes shown can be calculated easily from the expression d_{hkl} =

$a/(h^2 + k^2 + \ell^2)^{1/2}$, and these values are shown in the figure. Let us move to the right-hand side of the figure and construct another lattice. Choose an origin call it (000). Now consider the "set of planes" (010). Draw a perpendicular to the real lattice planes (010) through the origin (000) of the reciprocal lattice. Now choose a scale for the reciprocal lattice as shown in the right-hand side of the figure. The scale is in units of reciprocal angstroms (Å^{-1}). Let $d^*_{hkl} = 1/d_{hkl}$ and calculate the various d_{hkl} values as shown on the figure. Note than $d^*_{100} = 0.25\ \text{Å}^{-1}$. If we lay off this distance ($0.25\ \text{Å}^{-1}$) along the normal to the (010) planes starting at the origin of reciprocal space (000), we will define the reciprocal lattice point (010). Now consider the set of planes (020). These are parallel to the (010) set, but $d^*_{020} = 0.50\ \text{Å}^{-1}$ since $d_{020} = 1/2\ d_{010}$. We can thus define the reciprocal lattice point (020). In a similar manner, we can define the reciprocal lattice point (100), (200), and (001), and (002). Now consider the set of planes (110) and see that the same construction holds. That is, the (110) reciprocal lattice point lies on a line (through the 000 point) perpendicular to the (110) face of the real lattice. The spacing of the (110) planes in a real cell is 2.83 Å. Thus, the distance from the (000) to the (110) point in the reciprocal lattice is $1/2.83\ \text{Å}^{-1}$ or $0.35\ \text{Å}^{-1}$. Note that the reciprocal lattice point (111) is on a line in reciprocal space passing through the (000) that is perpendicular to the (111) plane in the real cell. Now it is an easy matter to label all of the reciprocal lattice points in the reciprocal space. The interested reader can verify this by labeling the (101) and the (011) and the (210) reciprocal lattice points himself. These reciprocal lattice points will hereafter be referred to as reciprocal lattice nodes, or simply nodes. What does a node represent? Apparently, it represents a set of planes in real space, and from the position of the node with respect to the chosen origin (000) of the reciprocal lattice one can know the orientation and interplanar spacing of the set of real lattice planes. We have tacitly assumed that each node is a mathematical point but, as we shall see, the shape and size of the reciprocal lattice nodes are directly related to the physical state of the specific set of planes that the reciprocal node represents. Let us now look briefly at a two-dimensional real lattice and its geometrical reciprocal lattice. We will also obtain the Fourier transform of this two-dimensional lattice and show that the Fourier transform of the real lattice is identical with the geometrically constructed reciprocal lattice. We do this for two dimensions because the Fourier transform is easier

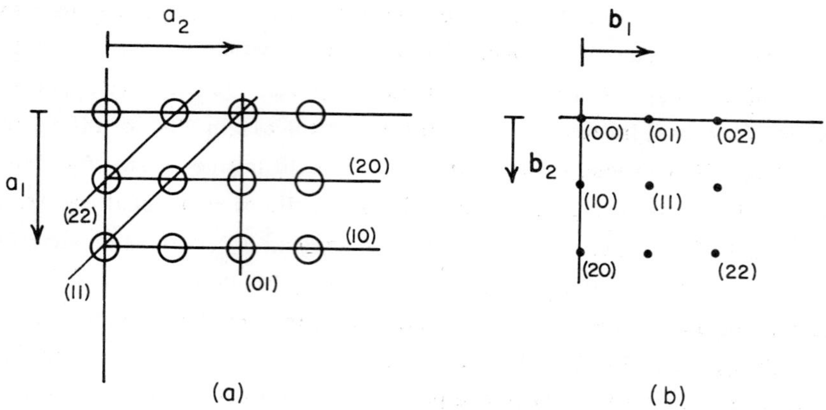

Fig. 9. Two-dimensional real plane lattice and its reciprocal lattice.

to visualize for this case. A real plane lattice (two dimensional) is shown in Fig. 9. We must speak of lattice rows, not lattice planes. Each row in the lattice can be indexed using two symbols such as h and k. In addition, the real lattice points can be defined to be at the end of a vector R directed from the origin of the real lattice where $\vec{R} = M\vec{a}_1 + N\vec{a}_2$ where m and n are the set of integers m, n = 0, 1, 2, 3 . . . The two-dimensional reciprocal lattice has been constructed on the right-hand side of the figure using the geometric methods discussed in the previous section. The reciprocal lattice points can be defined by a vector $\vec{H}_{hk} = h\vec{b}_1 + k\vec{b}_2$ where h, k are integers h, k = 0, 1, 2, 3 . . . and $\vec{b}_1$ and $\vec{b}_2$ are the vectors shown in Fig. 9. Thus, the vector $\vec{H}$ must always terminate on a reciprocal lattice point. The vector $\vec{s}$ is a general reciprocal lattice vector but at the nodes $\vec{s}=\vec{H}$. Let us now show how the Fourier transform method can transform the real lattice into the reciprocal lattice shown in Fig. 9 (b).

The Fourier transform is an integral transform. That is, to go from one side of the equation to the other, one has to integrate something. We will here state without proof the form of the Fourier transform and direct the interested reader to the literature[10–12]. For the particular case at hand, each point in the real lattice is considered to be identical. If $\vec{r}$ is any vector in real space (not necessarily a vector that terminates on a real lattice point) and $\vec{s}$ is a similar general vector in reciprocal space, we have the following relationship for the scattering amplitude[10]:

$$G(\vec{s}) = \int \rho(\vec{r}) \exp(2\pi i \vec{R} \cdot \vec{S})\, dv \tag{24}$$

where $\rho(\vec{r})$ is the scattering ability (relative to a single free electron) of the volume element dv at the end of the vector $\vec{r}$. Thus, $\rho(\vec{r})$ represents the spatial distribution of electron density in the crystal. In our case $\rho(\vec{r})$ is unity, that is, each lattice point in the real lattice is identical. Furthermore, $\rho(\vec{r})$ is not a continuous function in our case, but is zero except when $\vec{r} = \vec{R}$. We can, therefore, replace the integral with a summation (a double summation over m and n). Letting $\rho(\vec{r}) = 1$ and $\vec{r} = m\vec{a_1} + m\vec{a_2}$ we have

$$G(s) = \sum_{m=-\infty}^{\infty} \sum_{n=-\infty}^{\infty} (1) \exp (2\pi i) (\overrightarrow{ma_1 + na_2}) \cdot \vec{s} \tag{25}$$

which can be written as

$$G(\vec{s}) = \sum_{m=-\infty}^{\infty} \exp (2\pi i \; m\vec{a_1} \cdot \vec{s}) \sum_{n=-\infty} \exp (2\pi i \; n\vec{a} \cdot \vec{s}) \tag{26}$$

It is easy to show[2, 3] that each of the summations in Eq. (25) can be presented by a quotient of sine terms. Thus, we can obtain the amplitude G(s):

$$G(s) = \left[\frac{\sin 2\pi \; (2m+1) \; \vec{a_1} \cdot \vec{s}}{\sin \pi \; \vec{a_1} \cdot \vec{s}}\right]\left[\frac{\sin 2\pi \; (2n+1) \; \vec{a_2} \cdot \vec{s}}{\sin \pi \; \vec{a_2} \cdot \vec{s}}\right]$$

The first term in Eq. (26) is zero except when $\vec{a_1} \cdot \vec{s}$ equals an integer; that is, where $\vec{a_1} \cdot \vec{s} = h$, h equaling 0, 1, 2, 3.... The vector equation $\vec{a_1} \cdot \vec{s} = h$ mathematically represents a set of parallel equidistant planes perpendicular to the direction $\vec{a_1}$, having an interplanar spacing $1/|\vec{a_1}|$ [10, 12]. In a like manner $\vec{a_2} \cdot \vec{s} = k$ represents a set of parallel equidistant planes of spacing $1/|a_2|$ perpendicular to the vector $\vec{a_2}$. $G(\vec{s})$ is nonzero only when both equations are satisfied, i. e., where these two sets of planes intersect. The intersection of these planes is a set of infinitely long parallel lines perpendicular to the real lattice (see Fig. 10). If we section this set of parallel lines with a plane perpendicular to them, we obtain a two-dimensional reciprocal lattice projection consisting of points where $G(s) \neq 0$. Further, $G(s) \neq 0$ only when $G(s) = h(1/\vec{a_1}) + k(1/\vec{a_2}) = h\vec{b_1} + k\vec{b_2}$. It is easy to extend this argument[10] to three dimensions and see that the reciprocal lattice corresponding to the real lattice shown in Fig. 8 can be described by a vector:

$$\vec{H}_{hkl} = h(1/\vec{a_1}) + k(1/\vec{a_2}) + l(1/\vec{a_3}) = h\vec{b_1} + k\vec{b_2} + l\vec{b_3}. \tag{28}$$

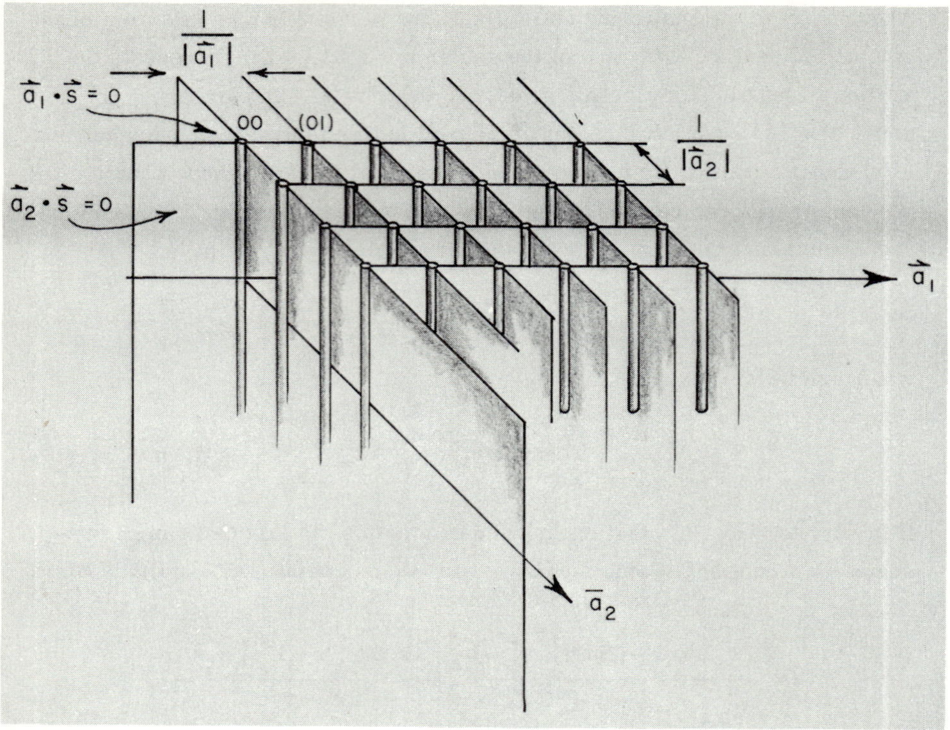

Fig. 10. Geometric interpretation of Fourier transform of two-dimensional lattice.

B. Vector Relationships between Real and Reciprocal Space

The real lattice vectors $\vec{a}_1$, $\vec{a}_2$, and $\vec{a}_3$ can easily be related to the reciprocal vectors $\vec{b}_1$, $\vec{b}_2$, and $\vec{b}_3$ in a general manner even when the real cell is nonorthogonal[13]. These relationships are given below

$$\vec{b}_1 = \frac{\vec{a}_2 \times \vec{a}_3}{\vec{a}_1 \cdot \vec{a}_2 \times \vec{a}_3} \qquad \vec{b}_2 = \frac{\vec{a}_3 \times \vec{a}_1}{\vec{a}_1 \cdot \vec{a}_2 \times \vec{a}_3} \qquad \vec{b}_3 = \frac{\vec{a}_1 \times \vec{a}_2}{\vec{a}_1 \cdot \vec{a}_2 \times \vec{a}_3} \tag{29}$$

Recall that the vector cross product $\vec{a}_2 \times \vec{a}_3$ produces a vector which is perpendicular to the plane of $\vec{a}_2$ and $\vec{a}_3$ and is directed away from that plane according to the familiar right-hand rule. In a similar manner, we can define $\vec{a}_1$, $\vec{a}_2$, and $\vec{a}_3$ in terms of $\vec{b}_1$, $\vec{b}_2$, and $\vec{b}_3$. Figure 11 illustrates the familiar case of Bragg diffraction by a single crystal having the planes (hkl) (whose interplanar spacing is d_{hkl}) parallel to the crystal surface. The

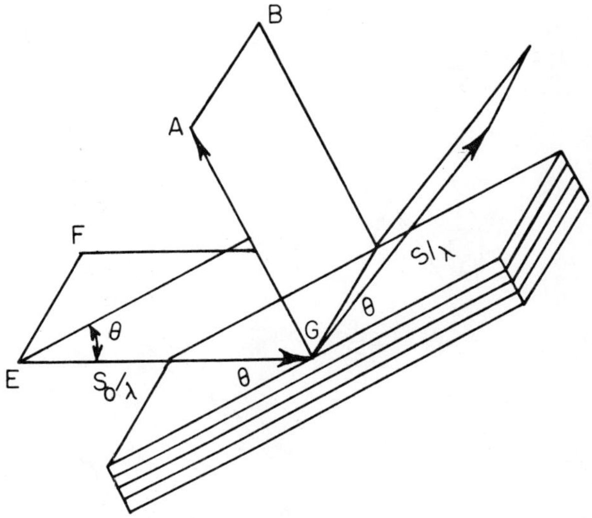

Fig. 11. Bragg diffraction from (hkl) planes of single crystals.

incident beam in this idealized case is considered to be monochromatic and line shaped. In addition, it is assumed that the incident beam is perfectly collimated, i. e., that it lies entirely in the plane as shown. The direction of the incident beam is given by the vector $\vec{S}_0/\lambda$, where S_0 is a unit vector ($|\vec{S}_0| = 1$) and λ is the wavelength of the incident radiation. The crystal is assumed to be ideally perfect so that the diffracted beam is confined to lie in a perfectly flat plane as shown. The direction of the diffracted beam is given by $\vec{S}/\lambda$ where $|\vec{S}/\lambda| = 1$. Bragg's law states that

$$\frac{2\sin\theta}{\lambda} = \frac{1}{d_{hkl}} \tag{30}$$

and by applying simple trigonometry to Fig. 25, it can be shown that

$$(\vec{S} - \vec{S}_0)/\lambda = 2 \sin\theta/\lambda \tag{31}$$

since $\vec{S}$ and $\vec{S}_0$ are unit vectors. The vector $(\vec{S} - \vec{S}_0)/\lambda$ is constructed in Fig. 11, and it is clearly perpendicular to the set of planes (hkl) from which diffraction is said to occur. By comparing Eq. (29) and (30), we see that the absolute value of $(\vec{S} - \vec{S}_0)/\lambda = 1/d_{hkl}$. Therefore, when diffraction occurs the vector $(\vec{S} - \vec{S}_0)/\lambda$ fulfills all of the requirements necessary for it to be equivalent to the reciprocal lattice vector $\vec{H}_{hkl}$, i. e., it is perpendicular to the

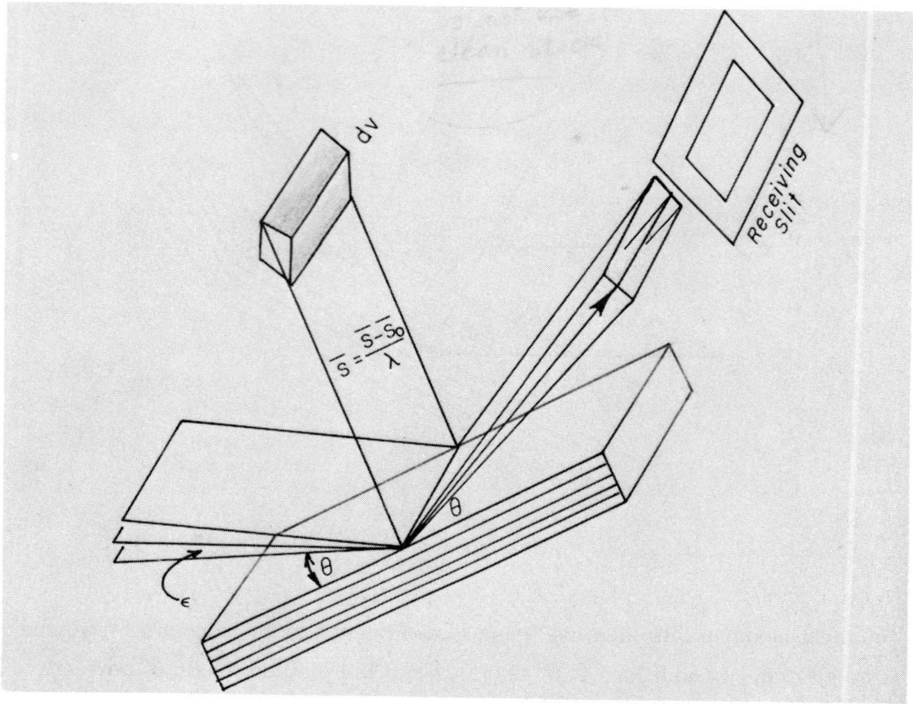

Fig. 12. General case of Bragg diffractions from (hkl) planes when incident beam is not perfectly collimated.

plane (hkl) and has a magnitude equal to $1/d_{hkl}$ (recall geometric construction shown in Fig. 8). We designate this reciprocal lattice vector by $\vec{s}$

$$\vec{s} = \frac{\vec{S} - \vec{S}_o}{\lambda} \tag{32}$$

and define $k = 2\pi s$. Note that $\vec{s}$ is equal to $\vec{H}_{hkl}$ only when diffraction occurs! At other points is reciprocal space $\vec{s}$ can have any value and is only limited by the wavelength of the x rays. That is, $\vec{s}_{max}$ equals 2.8 for MoK α radiation ($\lambda = 0.709$ Å), while $\vec{s}_{max}$ equals 1.30 for CuK_α radiation ($\lambda =$ 1.54 Å). As we shall see, all of reciprocal space is not available for diffraction investigation because $\vec{s}$ can't reach the most outlying nodes due to the above limitation of its maximum value.

Figure 12 illustrates a more general Bragg diffraction case than is shown in Fig. 15. Here we see that the incident beam is not perfectly collimated, i.e., there is some divergence ($\pm\epsilon$). In addition, the crystal is not perfect so that diffraction can take place over some finite angular region $\theta \pm \Delta\theta$ instead of precisely at θ, and the receiving slit can "see" more than just a narrow line. Thus, the diffraction vector $\vec{s}$ will not terminate on a line $\overline{ab}$, such as shown in Fig. 11, but will occupy a volume element dv as is shown in Fig. 12. So long as $\vec{s}$ lies in the volume (dv) the Bragg diffraction conditions are satisfied and diffraction will occur from the plane (hkl). We will see the significance of this reciprocal space volume element in the discussion that follows.

C. Diffraction and Reciprocal Space

Consider Fig. 8 and assume that we allow a monochromatic x-ray beam of wavelength 1.54 Å to be perpendicularly incident on the (010) face of a crystal. In reciprocal space this would be equivalent to a vector directed along the $\vec{b}_2$ reciprocal axis of length 1/1.54 Å -1. In Fig. 13 we show this vector with its head touching the (000) node in reciprocal space. Let us now draw the vector $\vec{S}/\lambda$ which defines the direction of the diffracted beam if a diffracted beam if a diffracted beam should occur*, $\vec{S}/\lambda$ has the same magnitude as $\vec{S}_0/\lambda$. The direction of $\vec{S}_0/\lambda$ and $\vec{S}/\lambda$ defines the direction of small $\vec{s} = (\vec{S} - \vec{S}_0)/\lambda$, as we have seen. Thus, let us mentally allow $\vec{s}$ to wander around reciprocal space and ask the question, when will a diffraction event take place? As we saw earlier when $(\vec{S} - \vec{S}_0)/\lambda = 1/d_{hkl} = \vec{H}_{hkl}$ Braggs' law is satisfied and diffraction is said to occur. Since $\vec{H}_{hkl}$ is a vector joining the origin of reciprocal space with any node (hkl), we can say that diffraction will occur if the vector $(\vec{S} - \vec{S}_0)/\lambda$ ends on a node since if it does $\vec{S} - \vec{S}_0/\lambda = H_{hkl}$. Ewald showed that a very simple geometric construction could be used to find all such nodes that meet the condition $\vec{s} = \vec{H}_{hkl}$ stated above. A good question to ask yourself at this point is the following. If you mentally set up the experiment we are describing, i.e., a stationary single crystal of aluminum with a monochromatic copper x-ray beam perpendicular to the cube face, would diffraction occur from many sets of planes? The answer will appear in the text. In Fig. 14, we have translated $\vec{S}/\lambda$ so that its end coincides with $\vec{S}_0/\lambda$ at point p. The vector $\vec{s}$ then joins the heads of these two vectors. If a sphere is drawn with the point p as its center having

---

*A node touching the Ewald spere is a necessary condition for diffraction. If, however, the structure factor for the set of planes represented by that node is zero then the diffracted intensity will be zero, e.g., the (100 node in a fcc structure such as aluminum.

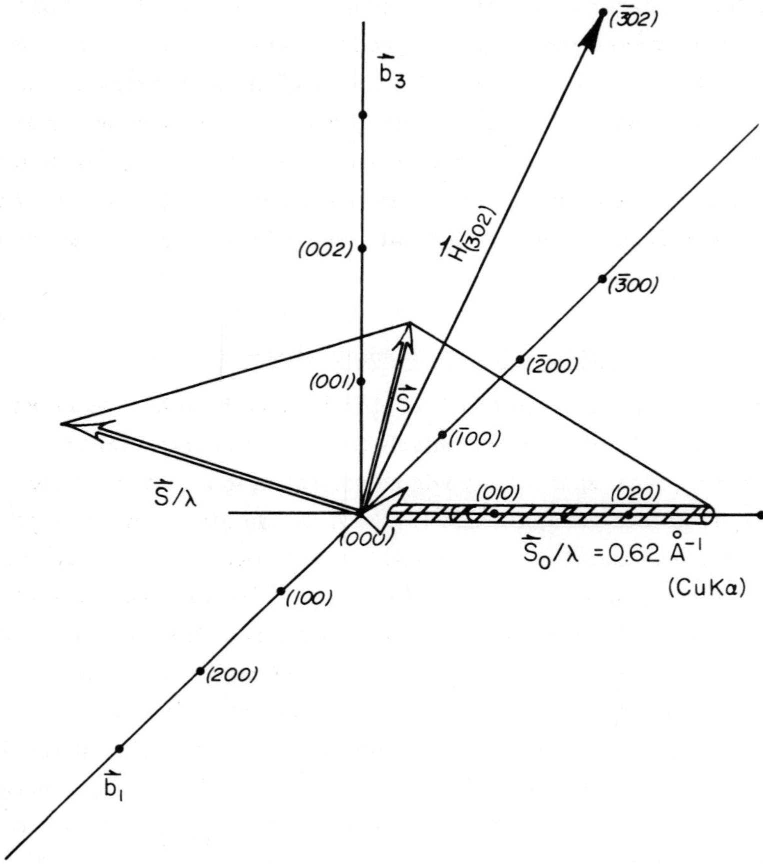

Fig. 13. Diffraction and the reciprocal space vector S and S₀.

a radius of $1/\lambda$ Ewald showed that $\vec{s}$ equals $(\vec{S} - \vec{S}_o)/\lambda = \vec{H}_{hkl}$ whenever the surface of the sphere touches a node. It is difficult to draw this in three dimensions, but if we only consider the plane of $\vec{b}_1$ and $\vec{b}_2$, then the trace of the Ewald sphere becomes a circle as shown in Fig. 15. Since $\overrightarrow{S_o/\lambda}$ and $\overrightarrow{S/\lambda}$ have the same magnitude and their ends originate at p, their heads must always lie on the circumference of the circle. Thus, whenever the circle touches a node, say at ($\bar{2}$10) or at (210), $\vec{s}$ equals $\vec{H}_{hkl}$ and diffraction will occur. In Fig. 15 the drawing was made in such a manner so as to force some of the nodes to touch this circle in order to illustrate a point. Normally, under the conditions given it is unlikely that many, if any, nodes would touch the surface of the Ewald sphere. It is a good idea to stop at this point and examine the construction of Fig. 14 and 15 in detail. Note that our

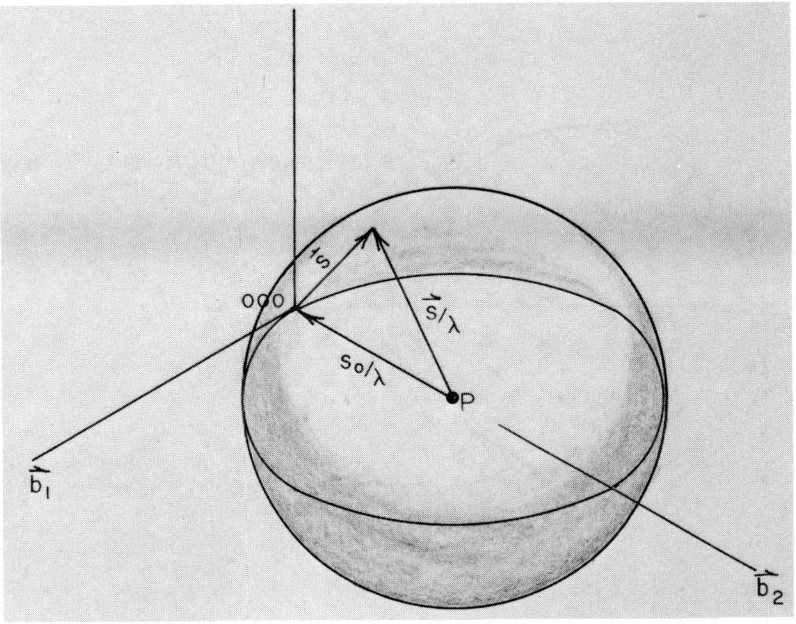

Fig. 14. Defining the vector $\vec{s} = (\vec{S} - \vec{S_0})/\lambda$.

Ewald sphere has a paper-thin skin, i.e., we assumed that $\vec{s}$ ends on a mathematically small point. Recall what we showed in Fig. 12. That is, that our experimental setup, i.e., collimation, etc., produces a volume element at the end of the vector $\vec{s}$. If we now think of this volume element existing at the end of vector $\vec{s}$, we can approximate this situation by allowing the Ewald sphere to have a "thick skin." The probability (Fig. 15) that a node such as (150) will touch a thick-skinned Ewald sphere is greater than the probability that this node will touch a thin-skinned Ewald sphere. What about the nodes, are they really small points? In Figs. 14 and 15, we thought of these nodes as being mathematical points, i.e., infinitesimally small points which might exist for a perfect crystal of large size containing no defects. In general, however, the nodes are not infinitesimal spherical points but may in the case of small crystals (that is, less than 1000 Å in diameter) become large balls, having a diameter up to one quarter of the distance between adjacent nodes. Thus, the larger the nodes, the higher the probability that a node-Ewald sphere intersection will occur. In the section that follows, we will examine how the probability of node-sphere intersections, i.e., a diffraction event, can be increased and also look at the shape of the node-sphere intersection to see what information it can provide

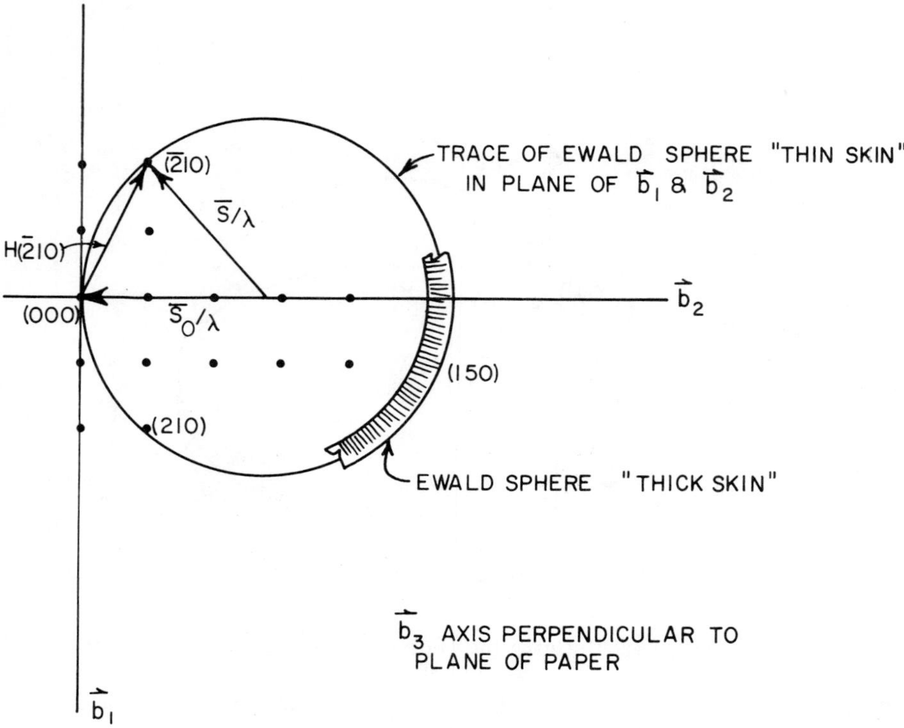

Fig. 15. The intersection of the Ewald sphere with the plane $\vec{b}_1$, $\vec{b}_2$.

with regard to the nature of the diffracting crystal. In truth, the nodes may be football-shaped, disc-shaped, etc., depending upon the shape of the small crystallites and/or the strain associated with each of the coherently diffracting domains.

D. Increasing the Probability of Diffraction

As we have seen, a stationary, perfect crystal has little likelihood of giving rise to diffraction when struck by a monochromatic x-ray beam. If we insist that the radiation must remain monochromatic, what are the conditions that can be relaxed to increase the probability that diffraction will occur? Perhaps the simplest way to accomplish this is to rotate the crystal. When the crystal rotates with some angular velocity ω (degrees per second) around a crystallographic axis then the reciprocal lattice also rotates with an angular velocity ω around a parallel axis. Let us examine this further. Consider Fig. 12; if we rotate the real lattice on the left counterclockwise about

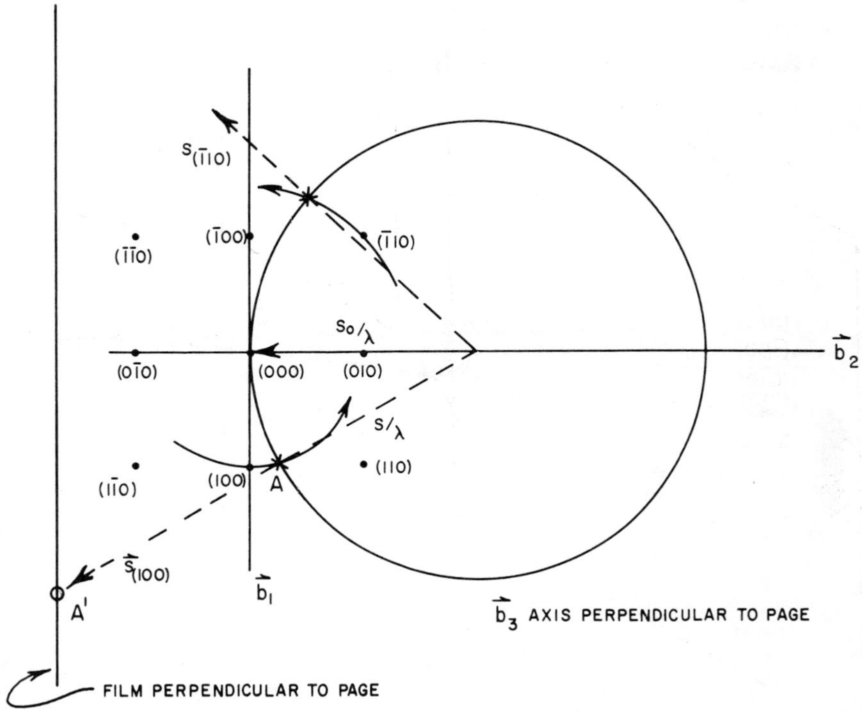

Fig. 16. Rotation of nodes through Ewald sphere.

the $\vec{a}_3$ axis, then it is easy to see that the reciprocal lattice would be rotated about the $\vec{b}_3$ axis in the same direction and with the same velocity. Now consider Fig. 16. Nodes such as (110) which would not touch the Ewald sphere when the crystal is stationary can be made to do so by this rotation. Figure 16 illustrates this in the $b_1 b_2$ plane of the reciprocal lattice. Assume that the real crystal is being rotated as described above. Note that, after a few degrees of rotation, the (100) node pierces the sphere. This gives rise to a diffracted beam $\vec{S}_{100}$. Further rotation would bring $(\bar{1}10)$ node through the sphere, giving rise to the $\vec{S}_{\bar{1}10}$. In a similar manner, all of the nodes shown in Fig. 16 will penetrate the sphere. Note that a complete 360° rotation will cause the (100), (010), and $(\bar{1}00)$, and the $(0\bar{1}0)$ all to pass through the same point A on the Ewald sphere. Thus, the diffraction spot a A' on the film contains superimposed information about the (100), (010), $(\bar{1}00)$, and $(0\bar{1}0)$ crystallographic planes. The method we have described is called the rotating crystal method. How can we reduce the number of planes in the real crystal that give rise to diffraction effects at A'? One method would be simply

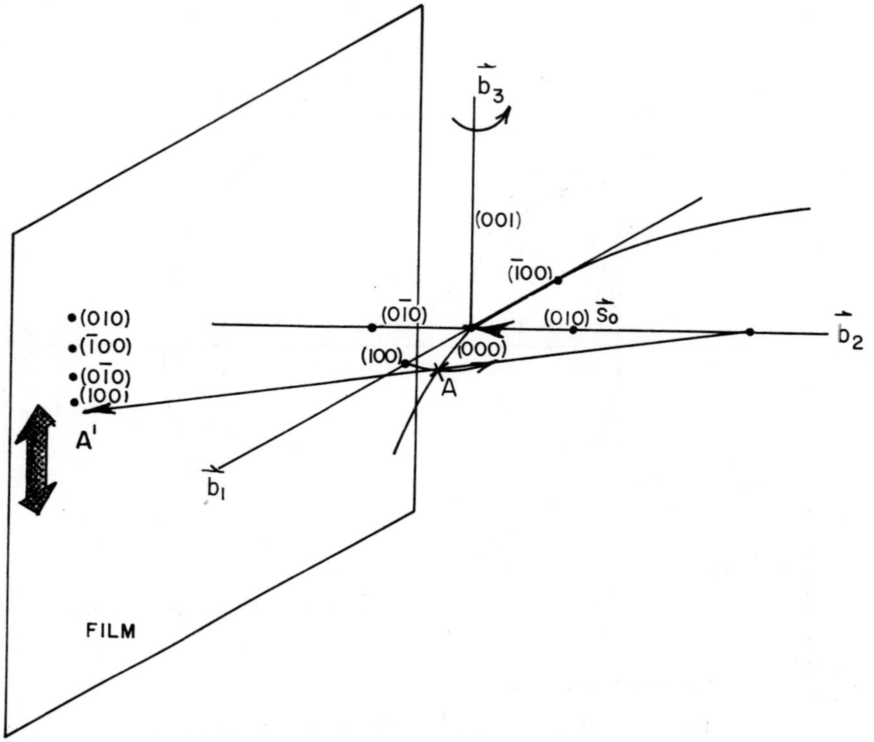

Fig. 17. Moving film technique, a schematic illustration.

to oscillate the real crystal back and forth about the b_3 through some small angle β large enough to allow the (100) node to pierce the Ewald sphere but not so large that the (010) can pierce. There is yet another method to accomplish the same effect. Suppose, as is shown in Fig. 17, we move the film (in the direction shown by the large arrow) while we rotate the crystal 360° about the $\vec{a}_3$ axis. In this way we would spread out the superimposed diffraction effects occuring at A'. If the film motion is correctly correlated with the rotational motion of the crystal (Weissenberg camera) we can cause the diffraction events from each of the four planes (100), ($\bar{1}$00), (010), and (0$\bar{1}$0) to reoccur at the same place on the film each time.

At this point, one can easily notice the effect that changing the wavelength will produce. In Figs. 14 and 17 the radius of the Ewald sphere is $\vec{S}_0/\lambda$, where the absolute value of $\vec{S}_0$ is 1. Thus, as λ becomes shorter the radius of the Ewald sphere increases. In Fig. 18, we can see the curvature

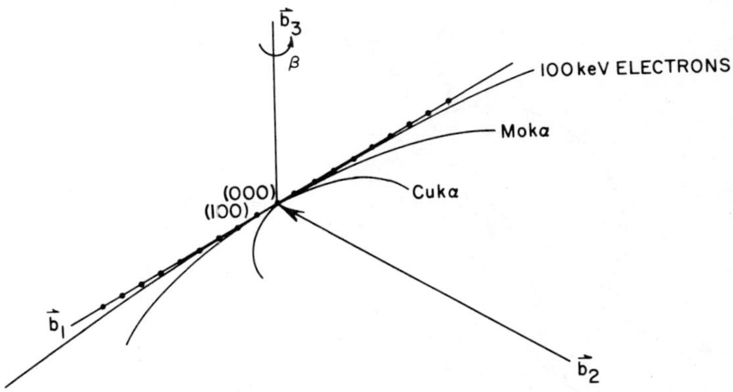

Fig. 18. Curvature of Ewald sphere for Cu and Mo x-radiation and 100 keV electrons.

of the Ewald sphere in the b_1b_2 plane when the incident x-ray beam is CuK_α and MoK_α, respectively. Thus a node, such as (100), will hit the MoK_α Ewald sphere with a smaller rotation angle β than for the copper Ewald sphere. For the sake of completeness, consider the wavelength of the electron beam in the electron microscope operated 100 kV (approximately 0.02 Å). Thus, in the case of electron diffraction the Ewald sphere is nearly planar and no rotation would be necessary for the (100) node to penetrate the sphere*.

E. Powder (Polycrystalline) Method and Reciprocal Space

We have up until now considered only single crystals. What happens when we have our sample in the form of a one-component powder (a large assembly of nearly identical particles)? Naturally, we will no longer have a picture such as shown in Fig. 8 because we no longer have one stationary crystal. What we have is a very large number of small "single crystals," each having its own reciprocal lattice. Even though the reciprocal lattice of each single-crystal particle is identical with all others, they are not oriented in the same way because the particles themselves do not have the same orientation in the powder. Consider for a moment the reciprocal lattice vector $\vec{H}_{(100)}$ for each

*In transmission electron diffraction the specimen is in the form of a very thin foil. Thus due to the small crystal effect the reciprocal lattice nodes are elongated in a direction perpendicular to the foil surface. These nodes become rods and have been called rel rods in the older literature.

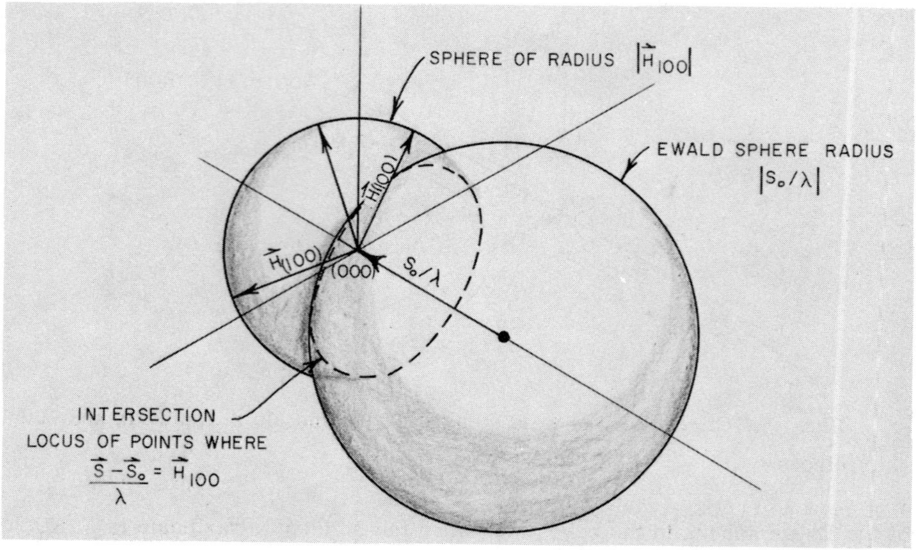

Fig. 19. The powder method and the reciprocal lattice.

particle. This vector has the same length in all of the various reciprocal lattices but, because these particles are oriented at random, the vectors will also be oriented at random. If we put the ends of all of the $\vec{H}_{(100)}$ vectors at (000), as shown in Fig. 19, we see that these vectors will have their heads lying on a sphere whose radius is $|\vec{H}_{(100)}|$. The condition for powder diffraction can easily be constructed from this information. If we now superimpose the Ewald sphere onto Fig. 24, we see that $\vec{H}_{100} = (\vec{S} - \vec{S})/\lambda$ only at the intersection of the small sphere of radius $\vec{H}_{100}$ with the Ewald sphere. The intersection of two interpenetrating thin-skinned spheres is a circle. The diffracted beam then takes the form of a circle as shown in Fig. 20.

F. Crystalline Defects and the Reciprocal Lattice Nodes

In a previous section, it was stated that the size of the nodes of the reciprocal lattice is inversely related to the size of the real crystals (if no strain is present). That is to say, very small crystals, i.e., less than 500 Å in diameter, will give rise to nodes which are considerably larger than those associated with 10,000 Å crystals. Equation (26) can give a clue as to why this should be so. For a three-dimensional real crystal the amplitude G (s) is given by Eq. (33):

$$G(s) = \left[\frac{\sin 2\pi N_i \overrightarrow{a_1} \cdot \overrightarrow{s}}{\sin \pi \overrightarrow{a_1} \cdot \overrightarrow{s}}\right]\left[\frac{\sin 2\pi N_2 \overrightarrow{a_2} \cdot \overrightarrow{s}}{\sin \pi \overrightarrow{a_2} \cdot \overrightarrow{s}}\right]\left[\frac{\sin 2\pi N_3 \overrightarrow{a_3} \cdot \overrightarrow{s}}{\sin \pi \overrightarrow{a_3} \cdot \overrightarrow{s}}\right] \tag{33}$$

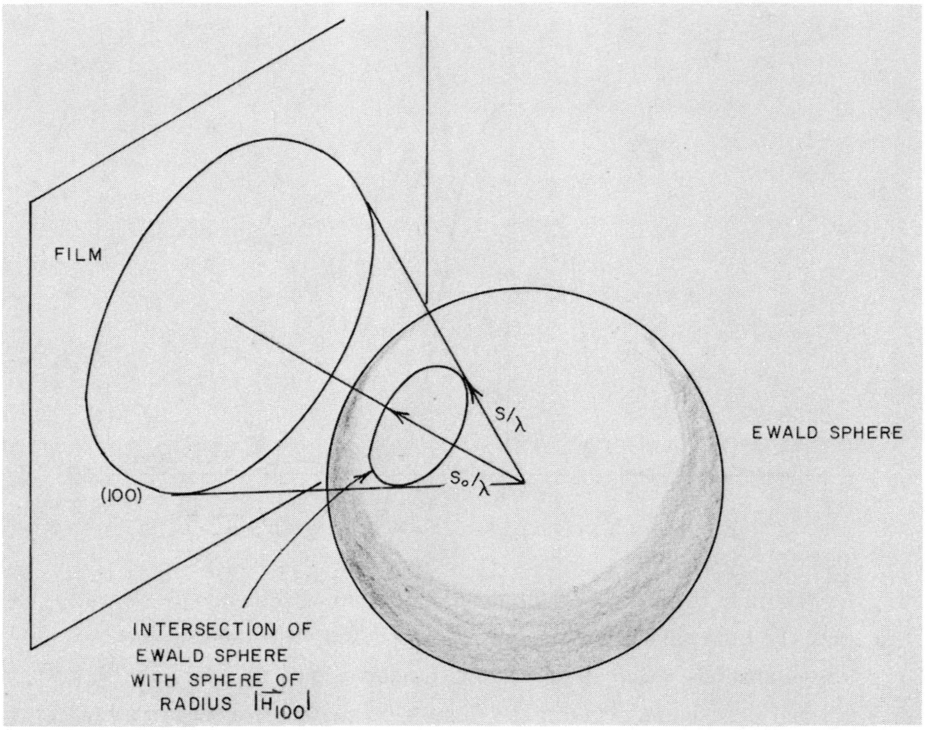

Fig. 20. The diffracted beam shape in the powder method.

The corresponding intensity $I = G(s) \cdot G^*(s)$ is

$$I = \frac{\sin^2 2\pi N_1 a_1 \cdot s}{\sin^2 \pi \vec{a}_1 \cdot \vec{s}} \; \frac{\sin^2 2\pi N_2 \vec{a}_2 \cdot \vec{s}}{\sin^2 \pi \vec{a}_2 \cdot s} \; \frac{\sin^2 2\pi N_3 \vec{a}_3 \cdot \vec{s}}{\sin^2 \pi \vec{a}_3 \cdot \vec{s}} \tag{34}$$

Each of the sine quotients if of the form $\sin^2 2\pi N_1 \vec{a}_1 \cdot \vec{s}$ divided by $\sin \pi \vec{a}_1 \cdot \vec{s}$, where N_1 refers to the number of planes in a stack of planes all of which are perpendicular to the a_1 axis. Thus, N_1 is a crystal size parameter. If N_1 is small in the a_1 direction then the crystal is correspondingly small in this direction; a graph of $\sin^2 2\pi N_1 \vec{a}_1 \cdot \vec{s}$ divided by $\sin^2 \pi \vec{a}_1 \cdot \vec{s}$ as shown in Fig. 21. The little maxima at the foot of the large peak can be neglected since the intensity of the largest of these peaks is only of the order of 1% of the main peak. It is not difficult to show that the width of the main peak is inversely proportional to $N_1 |a_1|$. Thus, as $N_1 a_1$ increases the peak becomes narrower. This peak represents one "diameter" of a reciprocal lattice node. Consider the small single crystal shown in Fig. 22. If we assume that it has a cubic crystal structure, then $\vec{a}_1 = \vec{a}_2 = \vec{a}_3 = \vec{a}$. If during the growth of this crystal

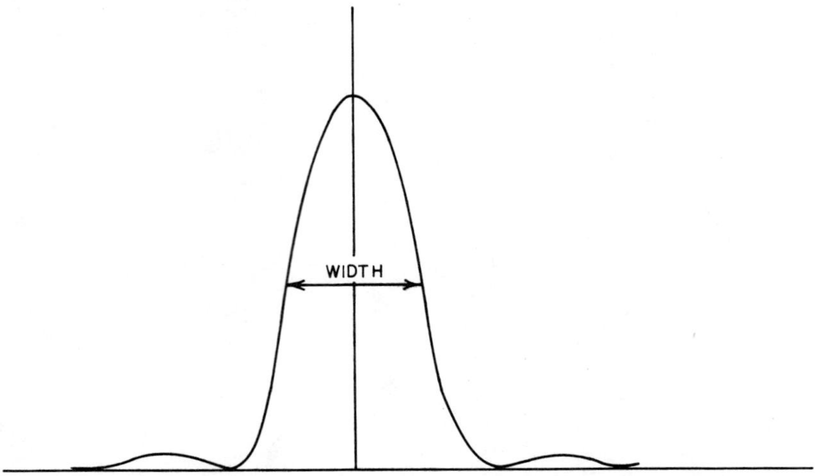

Fig. 21. A graph of $\sin^2 2\pi N_1 \vec{a}_1 \cdot \vec{s} / \sin^2 \pi \vec{a}_1 \cdot \vec{s}$.

it grew faster in the $\vec{a}_3$ direction, it might end up with the shape shown in Fig. 22. For the sake of discussion, let $N_1 = 100$, $N_2 = 200$, and $N_3 = 500$ and let the absolute value of $|\vec{a}| = 4$ Å. Since we know N_1, N_2, and N_3 and $\vec{a}_1$, $\vec{a}_2$, and a_3, we can evaluate the three $\sin^2$ quotient terms in Eq. (33). Then the diameter of any reciprocal lattice node will be very small (1/2000 Å)

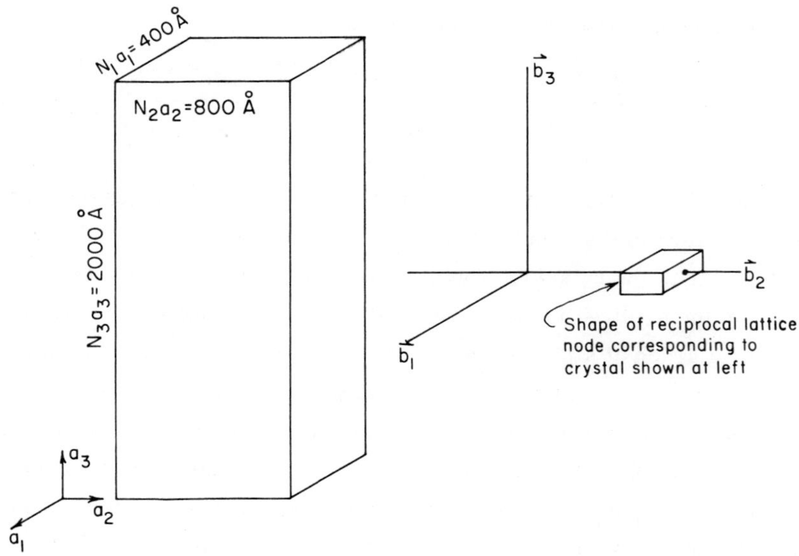

Fig. 22. A small single crystal having a parallelopipidon shape.

in the $\vec{b}_3$ direction, moderately large in the $\vec{b}_1$ direction (1/400 Å) and proportional to 1/800 Å in the $\vec{b}_2$ direction. Thus the reciprocal lattice node for the small crystal shown in Fig. 22 is not a point but rather a flat disc.

We can see from the above argument that the size and shape of the reciprocal lattice nodes are strongly related to the size and shape of the real crystal. It can be shown that all reciprocal lattice nodes are in general identical in shape and size, see Guinier [13]. If the crystal gets too large, that is greater than ~ 2000 Å in average diameter, we can no longer measure a corresponding change in reciprocal lattice node size due to the fact that instrumental broadening occurs. It should also be pointed out that strains, stacking faults, and other lattice imperfections can give rise to asymmetrical and symmetrical broadening of reciprocal lattice nodes.

From the preceding discussion, we can see why a small crystal gives rise to x-ray diffraction line broadening. If the small crystal is rotated about the a_3 axis, the intersection of the flat disc mentioned above with the Ewald sphere will not be a point as in the case of a large perfect crystal but rather a large area. Thus, as is seen in Fig. 27 (zero layer only) the conditions for diffraction are satisfied for many vectors $\vec{s}$ giving rise to a large diffraction spot. The reader should realize that a polycrystalline array of randomly oriented small crystals such as is shown in Fig. 22 will necessarily give rise to reciprocal lattice nodes that have nearly spherical symmetry (Fig. 23). Measurement of the diffraction spot size (or diffraction profile) width can thus be used to obtain a measure of an average particle size (see Refs. 13, 14, and 5).

IV. APPLICATION OF X-RAY TECHNIQUES TO MATERIALS RESEARCH

A. Amorphous Solids

A wide variety of amorphous materials are in common use today. Even though these materials lack the perfect periodicity of crystals, x-ray diffraction can reveal details about their structural makeup. It is beyond the scope of this chapter to describe the x-ray radial distribution analysis technique [5, 4] that is used to interpret diffraction effects from amorphous solids and liquids. Suffice to say, it is possible to obtain information regarding the number of neighbors and the radii of the first three or four coordination shells as well as information regarding bond angles. This information can provide valuable

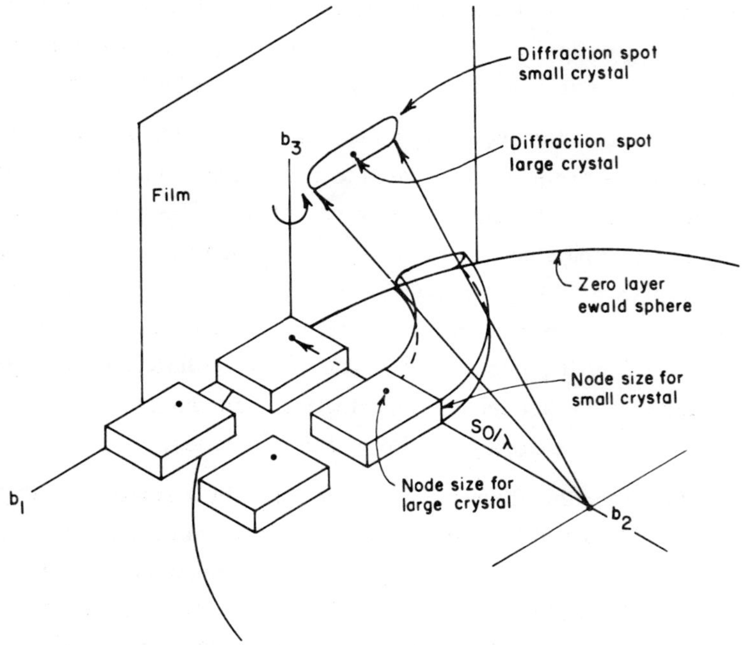

Fig. 23. Reciprocal lattice (zero layer) for small crystal shown in Fig. 22.

insight into the physical properties of these materials. Figure 24 illustrates the appearance of an x-ray diffraction pattern for several amorphous-type materials. Notice the absence of sharp diffraction maxima. The broad peak in the pattern on the right is predicted by Eq. (16) (Debye equation) on the basis of a few preferred interatomic distances r_{mn}. For a more complex amorphous solid, such as vitreous silica, the x-ray diffraction pattern (Curve A, Fig. 25) shows several peaks. The first strong peak can be shown to correspond to silicon-oxygen bonds having a length of 1.62 Å where four oxygen atoms surround each silicon atom in a tetrahedral array. Such structural information as is described by the radial distribution method can be used to study the evolution of the structure in amorphous solids during crystallization. The reader is directed to Warren's text [5] for futher details.

B. Particle Size and Shape

As was seen in the previous section, x-ray diffraction techniques are capable of revealing size and shape information. Size is related to the size

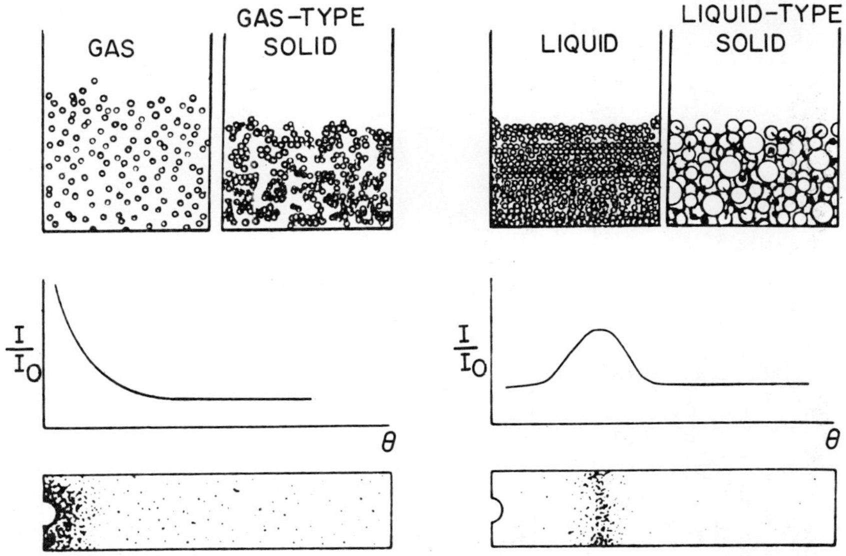

Fig. 24. X-ray diffraction pattern from several amorphous-like materials: (a) Scattering at small angles, small, loosely bound molecular aggregates with high percentage of free space, differences in electron density, bituminous coal, silica gel, activated charcoal. (b) Diffuse ring, dense packing anthracite, glass, vitreous selenium, explosive antimony, chemically deposited nickel, unstretched rubber, vacuum-evaporated arsenic.[20]

of the diffraction spots through the size of the reciprocal lattice nodes. Shape is also directly related to the shape of the reciprocal lattice nodes. It is easy to see (Fig. 23) how the particular x-ray technique might strongly affect the appearance of the diffraction spots, i.e., the shape of the node may be distorted by the geometry of the x-ray diffraction technique used. When the powder method is used much shape information is averaged out and the resulting x-ray diffraction ring (Figs. 19 and 20) contains average shape and size information which is still very useful. Figure 26 (a) illustrates the appearance of a powder x-ray diffraction pattern of small particles (approximately 1000 Å in diameter). When the same particles are well separated [Fig. 26 (b)] we will see small angle x-ray scattering in addition to the broadened diffraction line. When the particle size is too large, i.e., greater than 30 - 40μ incomplete diffraction rings will result (see Fig. 27). Recall that complete diffraction ring comes as a result of having a random array of particles so that the ends

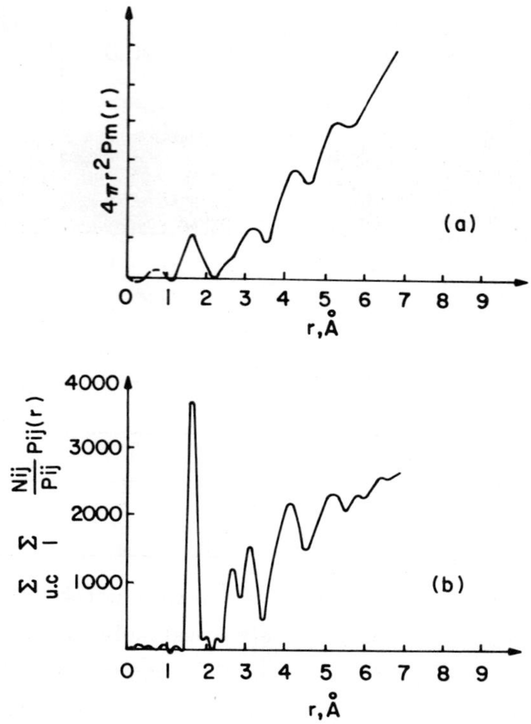

Fig. 25. The structure of vitreous silica (SiO_2) using the older x-ray radial distribution method (a) and with the newer method of Warren[25] (b).

of the vector $\overrightarrow{H}_{hkl}$ completely cover the surface of a sphere. If the number of particles is limited, i.e., by being too large, the surface of the sphere whose radius is $\overrightarrow{H}_{hkl}$ has "holes" similar to cheesecloth. When such a sphere (containing holes) contacts the Ewald sphere, an incomplete or spotty contact results, and the diffraction ring will be discontinuous.

Many naturally occurring mineral species have highly anisotropic shapes: asbestos fiber, platy clay minerals, etc. The same is true for many synthetic compounds. X-ray diffraction can reveal this shape anisotropy. Figure 28 shows schematically how the shape information is registered on a powder diffraction film. When crystallites have a random layer structure, for example a layered structure in which the stacking sequence is random rather than ordered, the very thin layers behave like a two-dimensional crystal; i.e., certain reciprocal lattice nodes are severely elongated perpendicularly to the

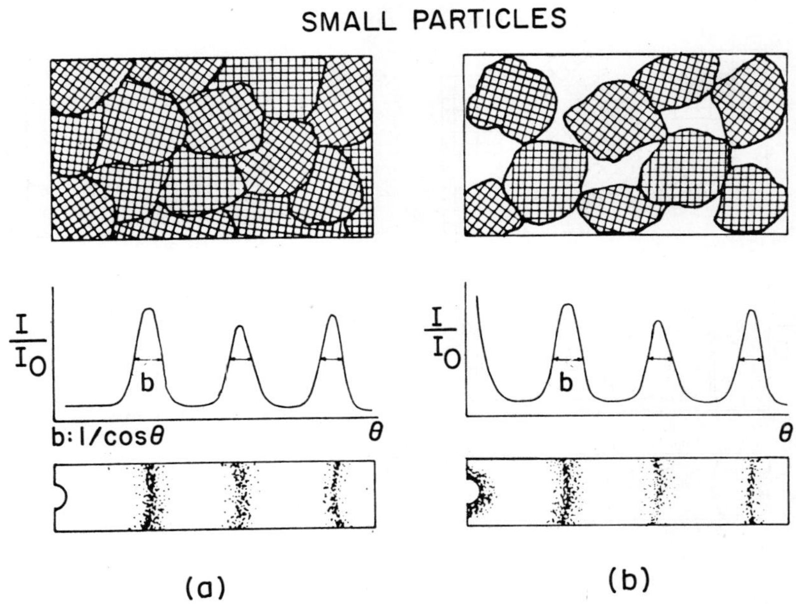

Fig. 26. Powder pattern from a sample containing small particles: (a) Line broadening; large, secondary particles; size of primary particles 20 - 1500 Å; products of incomplete topochemical reactions; surface films caused by reaction with atmosphere. And (b) Line broadening and small angle scattering; primary particles separated by free spaces; thermal decomposition of solids containing volatile materials; vapor-deposited substances.[20]

thin dimension of the stacking layers (Fig. 29) while other nodes may be small and approximately spherical. The powder diffraction pattern thus consists of sharp lines as well as broad asymmetrical peaks having their tails extending toward large angles. The reader should try to visualize and construct the intersection of the diffraction sphere with the Ewald sphere for the powder case. The results are very interesting.

In most fabricated metal parts the individual grains or particles are not oriented at random. Their relative orientations are the result of "preferred" crystallographic slip directions as a result of mechanical deformation and/or recrystallization. This "preferred" grain orientation is also found in many ceramic materials and may come about as a result of preferred particle packing during a pressing operation or, again, as a recrystallization process.

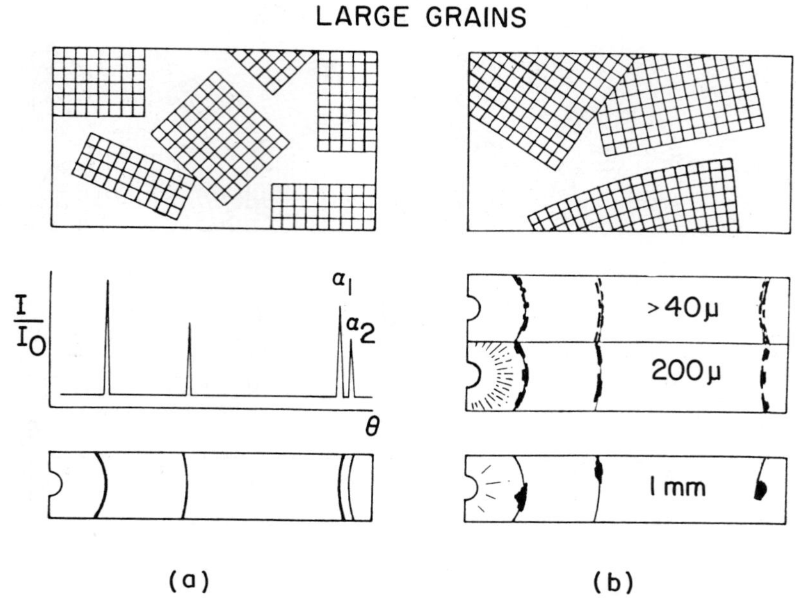

Fig. 27. Powder pattern from a sample containing large particles D 30, -40: (a) smooth lines grain 0.2-20μ, inorganic pigments, ground substances; and (b) spotty lines, asterism grain size $\geq$ 40μ, quartz sand, metals after recrystallization. [20]

Preferred orientation gives rise to x-ray diffraction results that are very unique. Figure 30 shows schematically how this preferred orientation is manifested on the x-ray diffraction patterns. It is left to the reader to visualize the reciprocal lattice construction for the preferred orientation case.

C. X-Ray Diffraction Investigations of Atomic Substructure

This field has received considerable attention in recent years[15-17]. In this section, we will briefly discuss only one technique, i.e., the Lang method, which is used to observe lattice imperfections in nearly perfect crystals. This method has found wide application in the electronics industry for characterizing highly perfect silicon and germanium single crystals which are used extensively in the manufacture of semiconductor devices and integrated circuits. The presence of lattice imperfections, such as dislocations, precipitates, etc., in the wrong places can be detrimental to the proper operation of such electronic devices and can limit their effective lifetimes. Figure 31 is a schematic

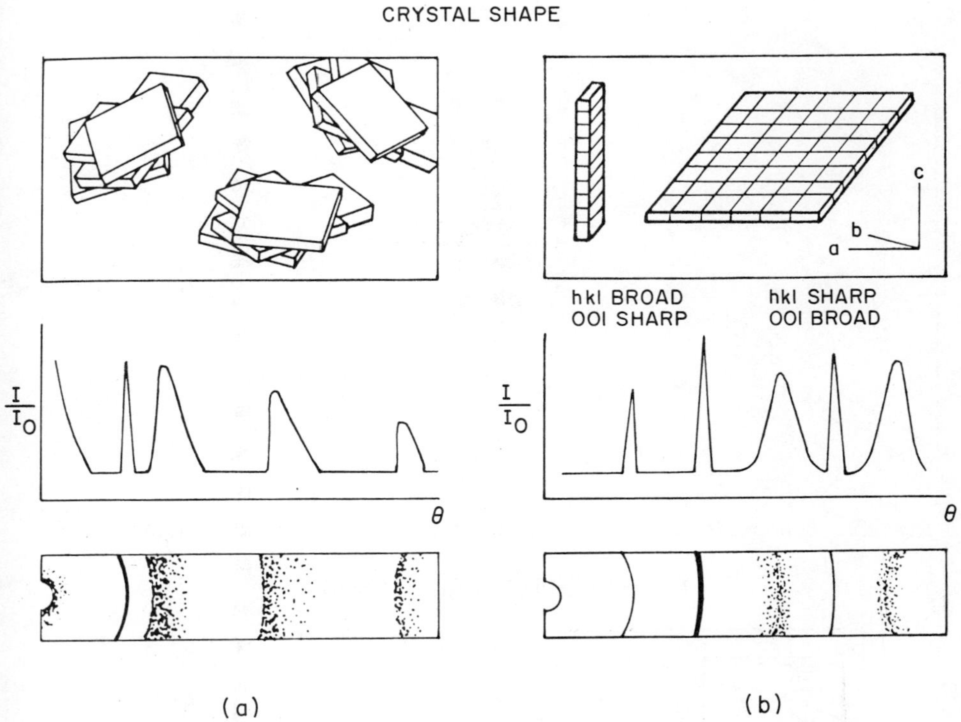

Fig. 28. Schematic illustration of particle shape and the effect on the diffraction pattern pattern: (a) saw-toothed line shape, two-dimensional crystals, random layer lattices, regular interposition of amorphous material, displacement of whole lattice layers, clay, carbon, precipitated magnesium silicate (talc); and (b) selective line broadening, needlelike or platelike crystals having large dimensions in one direction and small in the other, nickel hydroxide, precipitated wollastonite (calcium silicate); stacking disorder.[20]

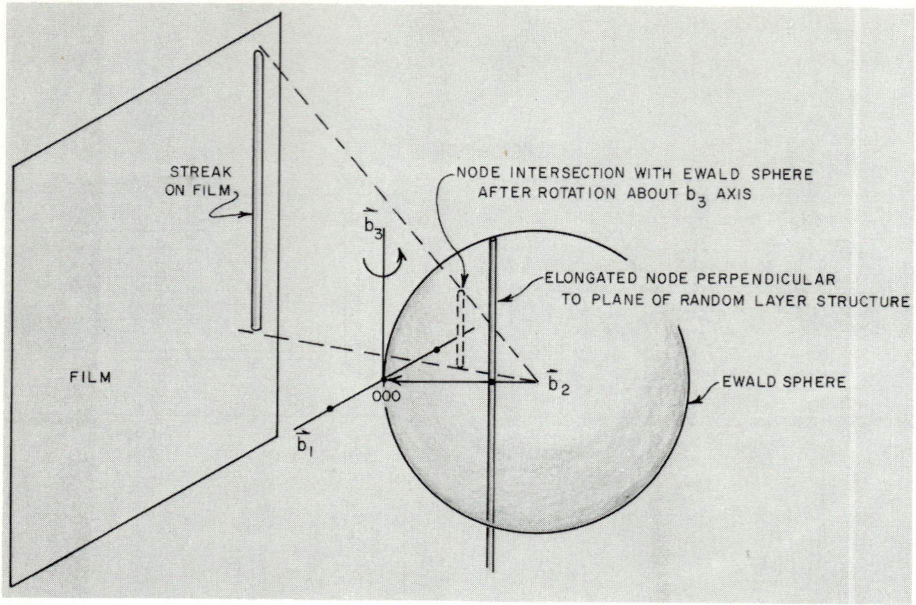

Fig. 29. Schematic diffraction results from a random layer structure.[20]

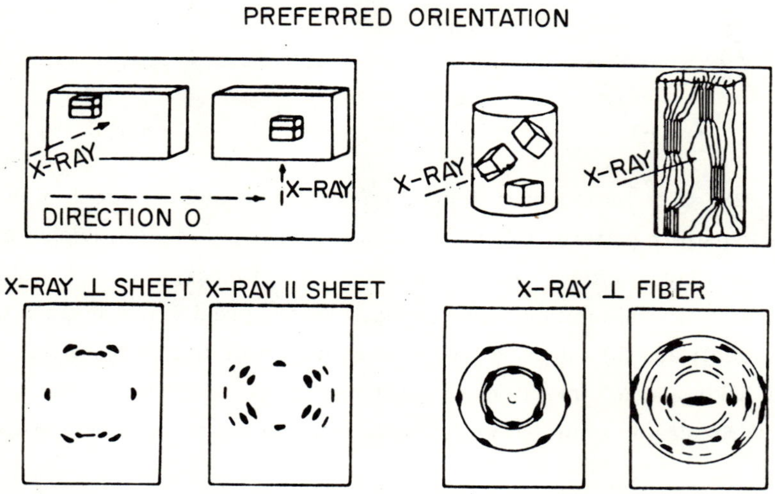

Fig. 30. The effect of preferred orientation or the diffraction pattern: (a) orientation in two directions higher orientation in rolled sheets, stretched rubber; (b) orientation in one direction, aluminum wire x-ray⊥ fiber axis; and (c) Ramie fiber, orientation fiber axis, high polymers small angle scattering shows particles elongated ॥ fiber axis.[20]

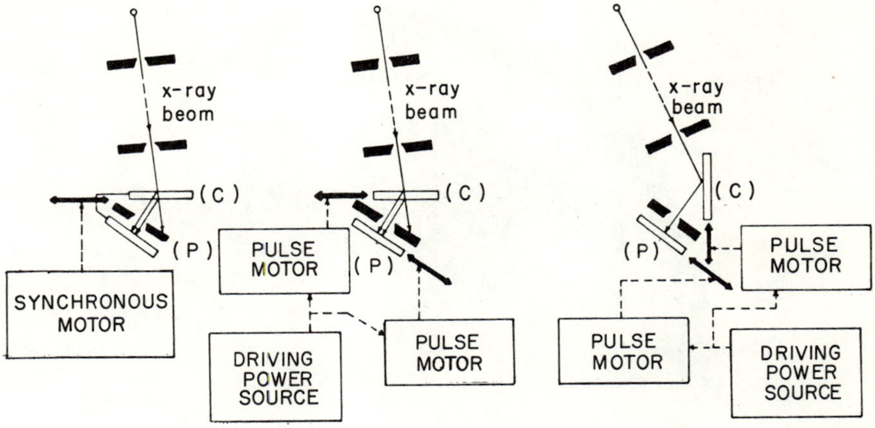

Fig. 31. Lang x-ray topographic camera (schematic drawing).

drawing of a Lang x-ray topographic camera. A narrow line-shaped polychromatic beam of x rays is incident on a single-crystal sample. A transmitted diffracted beam of characteristic $K_{\alpha 1}$ and $K_{\alpha 2}$ radiation passes through the slit p and strikes a very high resolution photographic plate. The intensity of the diffracted beam can be shown to be a direct function of the positions of various crystalline imperfections in the incident beam x-ray path. The sample and films can be moved slowly in unison, thus obtaining diffraction data from the entire sample. Imperfections such as dislocations give rise to contrast effects on the photographic image. Figure 32 shows a boron diffused silicon wafer [19]. The boron was diffused through the triangular windows which are approximately 5 mm on a side. Figure 33 shows the detailed dislocation substructure near a corner of one of the triangles [19]. The round black spots are precipitates and the dark curved and straight lines represent dislocations. Note that this method looks at the details in a single diffraction spot. Thus the node in reciprocal space for such a nearly perfect crystal is not a uniformly intense spherical ball! A normal diffraction camera would not be able to reveal this fact owing to the lack of resolution. It is only when we use a very high resolution camera such as the Lang device that we can see details in the reciprocal lattice nodes which arise from atomic substructural details.

V. SUMMARY

In this chapter we have introduced the reciprocal lattice concept and showed how it may be readily used to explain nearly all x-ray diffraction effects

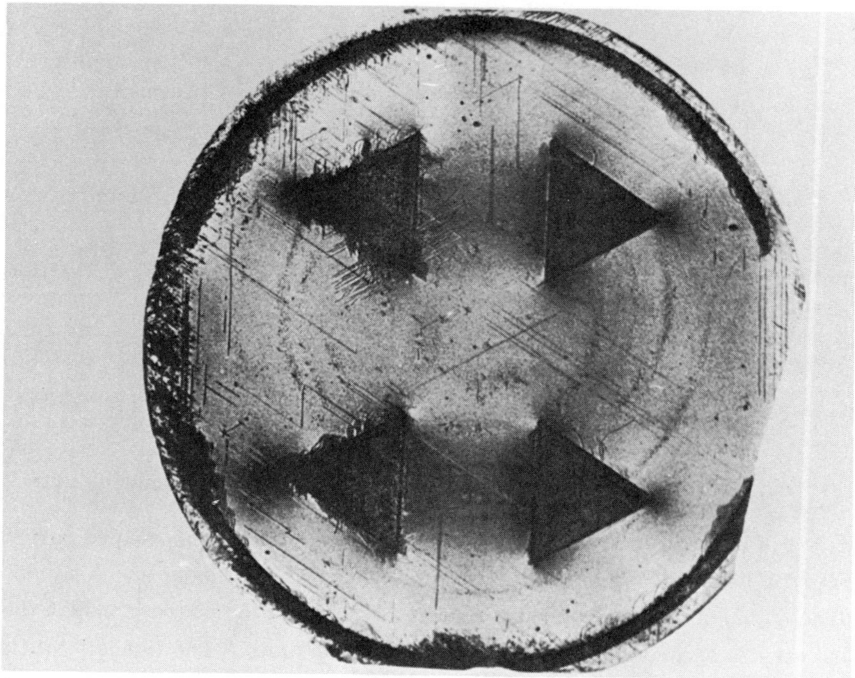

Fig. 32. Lang topograph of a boron diffused silicon wafer.

from the familiar Debye-Scherrer method to the more complicated methods such as Lang topography.* The approach has been geometric with the hope that the student will achieve a facility for thinking of x-ray diffraction in terms of reciprocal space. In a subsequent chapter, we will discuss in greater detail two specialized x-ray diffraction methods which are finding much application in current ceramic science research. It is hoped that this chapter will serve as an appropriate introduction to this later material.

*Dynamical diffraction effects are a noted exception.

Fig. 33. Detail dislocation substruction of boron diffused silicon wafer.

REFERENCES

(1) M. J. Buerger, Introduction to Crystal Geometry, 1970, McGraw Hill, New York.

(2) M. J. Buerger, Crystal Structure Analysis, Wiley, New York 1960.

(3) L. V. Azaroff, Elements of X-ray Crystallography, McGraw-Hill, New York, 1968, p. 155-157.

(4) R. W. James, The Optical Principles of the Diffraction of X-rays, Cornell Univ., 1965.

(5) B. E. Warren, X-Ray Diffraction, Addison Wesley, Reading, Mass., 1969, p. 116.

(6) B. E. Warren and G. Mavel, "Elimination of the Compton Component in Amorphous Scattering," Rev. Scient. Ins., [36], 2, 196-197, 1965.

(7) B. E. Warren and R. L. Mozzi, "Multiple Scattering of X-rays by Amorphous Samples," Acta Crystallographica 21, 459 (1966).

(8) R. L. Mozzie and B. E. Warren, "The Structure of Viteous Silica," to appear in J. App. Crystallography, 1970.

(9) D. B. Dove, "Structure of Amorphous Films Using Direct Recording Electron Diffraction" Proc. EMSA 26th Annual Meeting, p. 396, (1968).

(10) H. Lipson and C. A. Taylor, Fourier Transforms and X-Ray Diffraction, London G. Bells, 1958.

(11) I. N. Sneddon, Fourier Transform, McGraw-Hill, New York (1951), p. 23-27.

(12) Mary L. Boas, Mathematical Methods in the Physical Sciences, Wiley, New York (1966), p. 601.

(13) A. Guinier, X-Ray Diffraction, W. H. Freeman (1963), p. 88.

(14) B. D. Cullity, Elements of X-Ray Diffraction, Addison Wesley (1956), p. 261.

(15) V. K. Bonse, M. Hart, and J. B. Newkirk, "X-Ray Diffraction Topography," Adv. X-Ray Analysis, 10, 1.

(16) A. R. Lang, "Some Recent Applications of X-Ray Topography," Adv. X-Ray Analysis, 10, 91.

(17) J. Auleytner, X-Ray Methods in the Study of Defects in Single Crystals, Pergamon (1967).

(18) A. R. Lang, J. Applied Physics, 30, 7748 (1959).

(19) G. H. Schwuttke, "Semi-conductor Junction Properties as Influenced by Crystallographic Imperfections," AFCRL-66-192, March, 1966.

(20) F. Schossberger, Proceedings 6th Annual Denver X-Ray Conference, August 1957.

Chapter 6

TRANSMISSION ELECTRON MICROSCOPY AND ELECTRON DIFFRACTION

Charles F. Tufts

The Bayside Research Center of General Telephone &
Electronics Laboratories Inc.
Bayside, New York

I. INTRODUCTION

The methodologies of high-energy, transmission electron microscopy and electron diffraction that are used to characterize materials have expanded significantly during the last ten years. From the beginning of general practice about 1935 until about 1960, replication of surface geometry on thick specimens was the principal method of electron microscopy. The surfaces examined could be "as fractured"; "as formed" by chemical and physical processes; or "as prepared" by critical techniques of cutting, polishing, and etching to reveal structural character within the specimen bulk. During this period, electron diffraction was used in the incident mode to study surface layers "in situ" on thick specimens; the transmission mode was used to study surface layers isolated from thick specimens by chemical or electrolytic treatment; and the transmission mode was employed also to analyze thin deposits on glasslike substrates that were relatively transparent to high-energy electrons. Some correlation was made between surface geometry, as resolved by the replica, and crystal structure, as identified from the incident electron diffraction pattern. A number of analyses were reported that correlated transmission electron diffraction results with those from transmission electron microscopy when the specimens were thin deposits on glasslike substrates.

During the last ten years the significant advance in methodology has been based on the ability to prepare thin crystal sections (in the order of 1000 Å* thick) and to use the electron microscopical transmission mode with electrons of about 100-keV energy in order to form microscopical images that can be interpreted on the basis of electron diffraction theory. This method is called transmission electron diffraction contrast microscopy (TEDCM) and the image is formed by electron diffraction-contrast effects that result from the interaction of incident, monoenergetic electrons with the atomic architecture of the thin section. Electron diffraction methods that result in diffraction pattern formation have been integrated with techniques of TEDCM. Replication electron microscopy (REM) remains as a fundamental part of current methodology because it is the only technique by which the three-dimensional topography of a surface plane can be resolved to better than 100 Å.

This expansion in methodology demands the application of considerably more basic theory about image formation and diffraction effects than was required for the interpretation of REM. The purpose of this chapter is to discuss some nonrigorous concepts and describe some methods that are

---

*An angstrom unit (symbolized by Å) is equivalent to 10^{-10} m. A micrometer (symbolized by μ m) is the preferred notation for 10^{-6} m; another notation for this length is the micron (symbolized by μ).

required for routine practice of high-energy electron microscopy and electron diffraction when the task is analysis for materials science.

II. ELECTRON BEAM

Electron optical methods that can be used to characterize materials are summarized in Table I. All of these methods employ a beam of high-energy electrons (de Broglie waves) contained in a vacuum better than 10^{-4} Torr. The electrons are classified as a high-energy type because the energy falls in the range of 20,000 electron volts (20 keV) and 100,000 eV (100 keV). In the former case, a potential of 20,000 V (20 kV) is used to accelerate the electrons and in the latter, a potential of 100,000 V (100 kV). In the incident beam it is assumed that the electrons are monoenergetic (hence monochromatic) and travel at uniform velocity and with constant wavelength. The electron energy range for REM is usually between 20 and 100 keV; 100 keV is used for TEDCM and the diffraction methods. The electron beam is characterized by a dual wave-particle behavior in which electrons (carriers of negative charge) can exhibit wave properties. Extending this concept, it can be said that all matter is composed of particles which may be charged or uncharged, and these particles behave in some events as particles and in other events as waves. Particulate concepts are as old as the corpuscular theories of Rene Descartes (1596-1650) and Sir Isaac Newton (1642-1727); the wave description originates in the theory of Christian Huygens for light (1629-1695).

Wave character for nonrelativistic cases* can be described by classical theory using parameters of frequency, velocity, wavelength, amplitude, and phase relationship. The wavelength is related to wave velocity and frequency in classical theory by:

$$\lambda = \frac{v}{f} \tag{1}$$

where λ is wavelength (in meters); v the wave velocity (in meters/second) and f, wave frequency (in cycles/second). Wave frequency is determined by the frequency of oscillations in the source. Wave velocity is determined by

---

* Cases where wave velocity is considerably less than the speed of light (3.00 x 10^8 m/sec).

TABLE I

Methods of Analysis

Method	Acronym	Mode of examination and basis for image formation	Type of data obtained	Location sampled by method
Replication electron microscopy (bright field)	REM	Transmission through a replica of the specimen surface; replica < 200 Å thick. Microscopical image contrast formed by differences in the electron scattering cross section[a] produced by variation in mass thickness within the replica.	Surface topography of specimen is resolved to 20 Å for routine work using two-stage or one-stage carbon replicas shadowed with Pt metal. Resolutions to < 10 Å obtained for special cases using one-stage unshadowed carbon replicas.	Surface of thick specimen
Incident electron diffraction[b]	IED	Incident diffraction from a surface layer < 200 Å thick. Image is diffraction pattern formed primarily by elastic scattering of electrons.	Specific crystal system and space lattice identified when crystal size range is 1000 Å to 10 Å; crystallographic orientation determined for crystals in this size range and for large single crystals.	Surface layer of thick specimen
Transmission electron diffraction	TED	Transmission through a thin section of specimen or through individual particles in a dispersion; specimen usually < 1500 Å thick. Image is diffraction pattern formed by effects similar to those of IED.	It is the same as IED except precision of data is higher.	A zone in the specimen that is always larger than the incident beam diameter; zone geometry is a function of electron scattering cross section[a]

Transmission electron diffraction contrast microscopy (bright field and dark field)	TEDCM	Transmission through a thin section of specimen or through individual particles in a dispersion; specimen usually < 1500 Å thick. Microscopical image formed by electron diffraction-contrast effects. If diffracted beam is removed from final image by an aperture placed in back focal plane of objective lens, bright field image is formed; if diffracted beam is imaged and direct beam (undiffracted beam) is removed by objective aperture, dark field image is formed.	Gives microscopical images of crystal defects such as dislocations, stacking faults, and boundaries; resolution can be better than 400 Å; this is the only method for obtaining such data. Images crystal defects in individual particles in a dispersion. Dark field method can differentiate glasslike particles from crystalline particles in a dispersion. Dark field method can determine regions of a crystal that contribute to diffraction pattern.	Same as TED
Selected area electron diffraction and selected area electron microscopy	SAED and SAEM	Transmission through a thin section of specimen or through individual particles in a dispersion; specimen thickness is usually < 1500 Å. A variable aperture is placed in image plane of first projector lens; aperture selects part of diffraction image which is always formed at back focal plane of objective lens; by reducing strength of first projector lens the selected part of the diffraction image is projected on final viewing screen as a diffraction pattern; if strength of first projector lens is increased, a microscopical image is obtained showing zone of specimen that produced diffraction pattern.	Used extensively with TEDCM to relate diffraction pattern information to the microscopical image. Zone sampled can be as small as 0.2 x 0.2 x 0.1 μm. It determines crystallographic orientation for thin sections and crystal defects, identifies crystal structure, and determines crystallographic orientation for crystals in a dispersion where size ranges from 1000 Å to 10 Å.	Same as TED

TABLE I (Continued)

Method	Acronym	Mode of examination and basis for image formation	Type of data obtained	Location sampled by method
Extraction replication electron microscopy	EREM	Transmission through individual particles attached to unshadowed carbon replica. Particles are extracted from thick specimens or thin films by electrolytic, chemical, or mechanical stripping techniques. Replica is analyzed by TED, TEDCM, SAED, and SAEM.	Permits analysis of inclusions without interference effects from specimen matrix. This is the most sensitive electron optical method for analysis of inclusions in the size range of 1000 Å to 10 Å.	Same as TED

[a] The electron scattering cross section is that area projected on the atomic nucleus within which an electron collision will result in a scattering angle of θ or larger. In electron scattering, this cross section increases with atomic number (Z) and decreases with the potential used to accelerate the electron.

[b] Also called reflection electron diffraction.

the properties of the medium through which the wave travels and varies inversely as the refractive index of the medium. When frequency and velocity are established, wavelength is fixed.

Although the behavior of large particles traveling at relatively low velocities can be described by classical mechanics, the rigorous description for the behavior of particles in the order of atomic size traveling at velocities approaching the speed of light requires the assistance of quantum, relativistic mechanics. Limitations of classical principles are removed in part by applying relativistic correction. In quantum theory, energy is composed of elemental units called quanta which can be visualized as "granules," "bundles," or "corpuscles". Emission of radiation takes place discontinuously in "jumps" or quanta according to rigorous quantum rules. These quanta of energy can be emitted only by a quantum transition which occurs when the source changes from one of its quantized energy states to another such energy state. As long as the source remains in one of these quantized states (stationary state) it neither emits nor absorbs energy. In this system a source cannot emit just any energy but only quantized energy that conforms to quantum rules according to the general relation:

$$E = hf \tag{2}$$

where E is a general term for energy; h Planck's constant; and f the frequency of the source or oscillator. If the energy quantity (E) is replaced by the energy of a single light quantum (photon), Eq. (2) becomes the relationship that Professor Albert Einstein (1879-1955) developed to explain the photoelectric effect. Electron energy (eE) can be described by:

$$eE = 1/2\ mv^2 \tag{3}$$

where v is the electron velocity that determines the value of electron mass m. The quantity ($1/2\ mv^2$) is a general term for kinetic energy (K). This being so (eE) is kinetic energy that is specific for the electron. Now (K) and (eE) can be described also in the following way:

$$K = eE = mc^2 - m_oc^2 = (m - m_o)c^2 = \Delta mc^2 \tag{4}$$

where m is the relativistic mass which is a function of velocity, c the speed of light, m_o the mass of an electron at rest (rest mass), Δm the increase in mass resulting from motion, m_oc^2 the rest energy, and mc^2 the energy at very high velocity approaching the speed of light. The relativistic mass (m) requires additional explanation which has been stated by Einstein in his redefinition of classical mass:

$$m = \frac{m_o}{\left[1 - v^2/c^2\right]^{1/2}} \tag{5}$$

where m_o is rest mass; v the velocity, and c the speed of light. To summarize the information in Eqs. (3), (4), and (5) it can be said that the kinetic energy (eE) of an electron in the incident beam is a product of the acceleration potential (E) and the electron charge (e). The kinetic energy is equal to the product of the speed of light squared (c^2) and the increase in mass (Δm) that results when velocity approaches the speed of light. This is equivalent to the celebrated equation of Einstein:

$$E = \Delta mc^2, \tag{6}$$

by which he predicted that the mass of a particle is just one form of energy.

In 1924 Louis-Victor de Broglie (1892-) postulated that the electron wavelength (λ) could be described by:

$$\lambda = h/mv \tag{7}$$

where h is Planck's constant; m the electron mass that is a function of electron velocity v, and mv which is electron momentum. This relationship is nonrelativistic. It can be made relativistic by using the relationship, $eE = mc^2 - m_oc^2$, from Eq. (4) in which the m is relativistic mass as defined in Eq. (5). With these parameters it can be shown that $(m + m_o)\,eE = m^2v^2$. If substitution for mv is made in Eq. (7), it can be shown also that the relativistic electron wavelength (λ) is expressed by:

$$\lambda = \frac{h}{\left[2m_oeE\left(1 + eE/2m_oc^2\right)\right]^{1/2}} \tag{8}$$

where h is Planck's constant, m_o electron rest mass, E the acceleration potential for the electron, e the electron charge, eE the electron kinetic energy, c the speed of light. The quantity $\left[1 + (eE/2m_oc^2)\right]$ in Eq. (8) is the relativistic correction factor which is very significant in general practice. If the correction factor is not used, the error in wavelength amounts to 2% for 50-keV electrons ($\lambda = 0.0536$ Å) and 5% for 100-keV electrons

($\lambda = 0.0370$ Å). If numerical values are substituted for constants in Eq. (8), a relation between electron wavelength (λ) and acceleration potential (V expressed in volts) can be written as:

$$\lambda = \frac{12.26}{V^{1/2}\,(1 + 0.9788 \times 10^{-6} V)^{1/2}}\ \text{Å} \tag{9}$$

This completes a description of the high-energy incident electron beam. Equations (7) and (8) illustrate clearly the dual wave-particle nature of the de Broglie waves which are characterized by a wave parameter (wavelength) and particulate parameters (energy and momentum).

III. INTERACTION OF THE INCIDENT ELECTRON BEAM WITH MATTER IN THE SPECIMEN

When the high-energy incident electrons interact with matter in a specimen, there is produced a complex, three-dimensional pattern of events containing signals that supply information about the chemical and structural character of the specimen. Some effects and signals resulting from this interaction are summarized in Table II. The primary result of interaction is that incident electrons, which are characterized by high kinetic energy, lose some of this energy to the specimen as a result of encounters with atoms of specimen material. If an incident electron loses all of its kinetic energy in one event, the frequency of the quantum produced is given by the relation shown in Eq. (2). Usually the total energy of the incident electron is not converted to a single quantum in one event but is dissipated during a series of encounters that produce heat within the specimen and quanta of lower frequency than that obtained by a single-event conversion. About 2% of the incident electron energy is converted into x-ray quanta, and most of this appears as a continuous x-ray spectrum or continuum (white radiation). During these events additional spectra are excited that range from the x-ray region into the infrared region. Some of the balance of x radiation appears as a high-intensity line spectrum that is superimposed on the continuum.

It should be evident from Table II that the signals used for transmission electron microscopy and electron diffraction are only part of the complicated pattern of events resulting from electron beam interaction with the specimen. In these methods the signals that reach the recorder (usually a photographic plate or film) are contained in an incident beam that has been modified by the

TABLE II

Interaction of High-Energy Electrons with Matter in a Specimen

Effects	Signals	Applicability to transmission electron microscopy and electron diffraction
Elastic scattering of electrons	Electron diffraction maxima; diffraction patterns; microscopical images for special cases	If scattering is produced by a crystal, a qualitative description of the diffraction is given by Bragg's law and the Laue equations. Signals are the basis of image formation in TEDCM and diffraction pattern formation in IED, TED, SAED.
Inelastic scattering of electrons	Diffuse background; electron diffraction maxima; diffraction patterns; microscopical images for special cases	Contributes to image contrast in TEDCM. Contributes also to Kikuchi line and band formation and to Laue spot formation in IED, TED, SAED when scattering is within a crystal.
Excitation of x radiation by electrons	Continuous x-ray spectra	Not applicable
	Characteristic x-ray spectra	Not applicable
Excitation of x radiation by x-ray quanta	Characteristic x-ray spectra	Not applicable
Coherent scattering of x rays	X-ray diffraction maxima; diffraction patterns; microscopical images for special cases	Not applicable
Incoherent scattering of x rays (modified x radiation described by the Compton effect)	Diffuse background; x-ray diffraction maxima; diffraction patterns	Not applicable
Excitation of Auger electrons by x-ray quanta or electrons	Characteristic Auger spectra	Not applicable

Excitation of electroluminescence[a] by electrons	Emission of ultraviolet, light, and infrared quanta; microscopical and macroscopical images	Not applicable
Generation of electrical effects	Variation in current induced between different zones of the specimen	Not applicable
	Currents induced between specimen and ground	Not applicable
	Variation in potential at specimen surface	Not applicable
Heating within specimen cross section		Serious effects (including phase transformation and melting) are possible unless specimen is cooled during examination. Specimen stages that contain cooling facilities are used routinely in TEDCM and SAED.
Radiation damage		Damage from 100-keV incident electrons should not be significant for inorganic crystals but effects from ionic contamination of high energy traveling in the incident beam can cause detectable effects. Electrons and ions of 100-keV energy can produce damage in organic specimens.

[a] Also called cathodoluminescence which is a partial misnomer; "electron beam" is the preferred notation for a wave of particles carrying negative charges rather than the notation, "cathode rays."

interaction. It is an important point that these signals are basically different from signals excited in the specimen such as characteristic x-ray spectra or Auger spectra which reach the recorder as new events. Electron scattering is the primary effect that produces image formation in transmission electron microscopy and pattern formation in electron diffraction. During interaction an incident electron is scattered by the potential field that surrounds an atomic nucleus in the specimen. This field is a resultant of potential contributed by the positively charged nucleus as modified by screening effects from the potential distribution of negatively charged electrons which form the clouds that surround the nucleus. The scattering is described as elastic if the path direction of the incident electron changes but the velocity of the electron remains constant. If both the path direction and the velocity change, the scattering is described as inelastic. Elastically scattered electrons are said to be coherent and inelastically scattered electrons, incoherent with respect to the incident electrons. Elastic scattering of electrons is called diffraction. Heating effects in the specimen and radiation damage to the specimen can be significant, as indicated in Table II. Methods of analysis must include specimen cooling, sufficiently clean and sufficiently high vacuum conditions, and also the use of appropriate electron energies in order to minimize radiation damage.

IV. METHODS OF ANALYSIS

Instrumental resolution (in the ideal or Gaussian sense) for electron and light microscopy can be described by a mathematical relationship developed from the Rayleigh criterion which says that if a point on the object is imaged by a lens, the point will appear as a diffraction pattern. SAED and SAEM make good use of this effect. The mathematical relationship that defines limit of resolution is:

$$d = \frac{0.61\,\lambda}{n\,\sin\alpha} \tag{10}$$

where d is the limit of resolution or the minimal distance between two points on the object that appear separated in the image, λ is wavelength of the illuminating radiation, n the index of refraction for the medium in which observation is made, and α is one-half the angular aperture or the angle between the undiffracted wave (direct beam) and the diffracted wave from a periodic structure. The quantity $n \sin \alpha$ is called numerical aperture (NA). If oil immersion methods are used with light microscopy, approximate values for λ, n, α, and NA are 5000 Å, 1.5, 65°, and 1.4, respectively. This makes d approximately 2200 Å, which is a realistic value for general practice. In electron optics the effect of spherical aberration must be reduced by making the aperture of the objective lens small to the order of 10^{-2} rad (0.573°). If 50-keV electrons are used, approximate values for λ, n, α, and NA are 0.054 Å, 1.0, 34 min, and 0.01, respectively. This makes d about 3.3 Å which can be obtained experimentally for special cases. It is primarily the significant reduction in wavelength that makes practical instrumental resolution in electron microscopy about 500 times better than that obtained by conventional light optics using a wavelength in the order of 5000 Å.

Figure 1 shows the basic replication techniques used for REM. There is an extensive literature that reports the applications of various replica types. Carbon, silica, and aluminum oxide have been used most frequently for the single-stage method. In recent years the carbon replica method, which was introduced by D. E. Bradley, has been improved and applied to many problems by numerous workers. For the two-stage method many plastics have been used to make the negative impression including Parlodion and Formvar. Carbon has been the most popular material used for the evaporated film of the two-stage method. Resolution limit in general practice is about 20 Å for single-stage or two-stage carbon replicas shadowed with platinum metal.

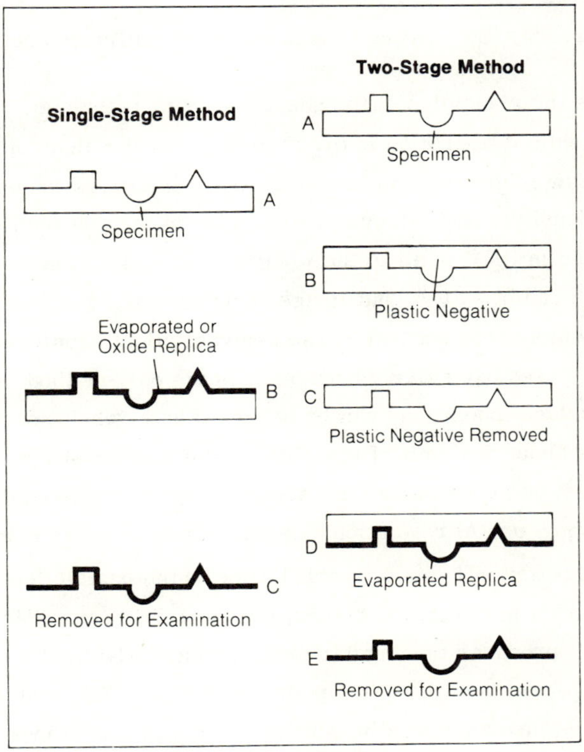

Fig. 1. Basic replication methods.

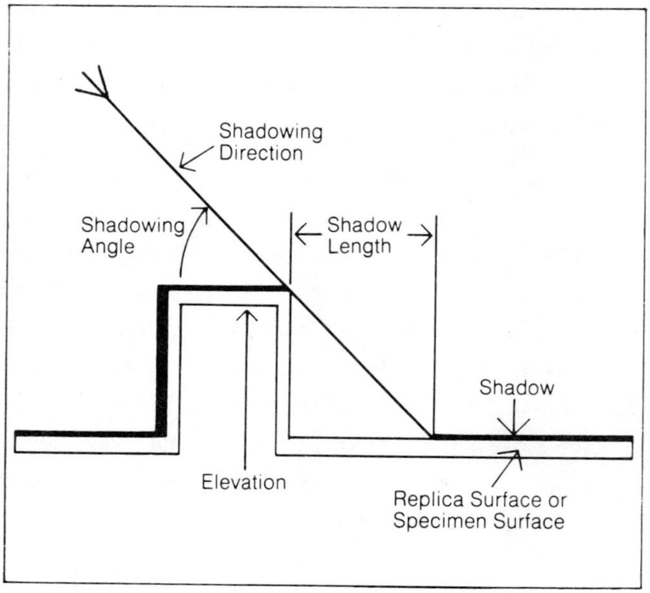

Fig. 2. Shadow-casting to enhance electron image contrast.

Figure 2 shows the principles of the shadow-casting technique. The shadow is added by vacuum evaporation in order to improve the electron image contrast by increasing the variation in replica mass thickness (physical thickness multiplied by density). It is the particulate size in the platinum shadow that limits resolution in REM to about 20 Å. Practical resolution for unshadowed carbon replicas can be <10 Å, but image contrast usually is low and interpretation can be complicated because phase contrast effects contribute to the image. REM is used extensively because it is the only method by which three-dimensional surface topography can be observed with resolution better than 100 Å. This is about the limit of practical resolution for scanning electron microscopy when using secondary electrons and when image contrast is adequate. If it is not, useful resolution can deteriorate to more than 1000 Å.

Electron diffraction from a crystal lattice is the basic effect that produces diffraction contrast in the microscopical image of TEDCM. It is also the basic effect that produces the angular distribution of elastically scattered electrons which form the diffraction patterns of IED, TED, and SAED. A crystal can be defined as a solid in which an array of atoms in space forms a periodic pattern in three dimensions. Any atomic array found in any crystal can be related arbitrarily to a three-dimensional net of straight lines called a space-lattice in which the atoms occupy the points of intersection formed by the lines of the net. These points of intersection are called lattice points. Every atom at a lattice point is surrounded by an identical configuration of other atoms in the space lattice. Only 14 ways have been found to arrange lattice points so that each atom has identical surroundings. These are the 14 space lattices of crystallography and are used in association with the seven systems of axes that constitute the basic geometries for the seven crystal systems (triclinic, monoclinic, orthorhombic, tetragonal, cubic, hexagonal, and rhombohedral or trigonal). The lines through the lattice points divide the space lattice into parallelepipeds called unit cells of the crystal structure. Each unit cell is equivalent to all other unit cells, and the crystal structure is built by repeating this fundamental unit. Grain size of a single crystal can extend from centimeters (in the case of semiconductors such as silicon) to about 10 Å which is the order of size for two adjacent unit cells in numerous crystal structures. This combination of two unit cells is the smallest grain size in which crystallinity can exist. With only one unit cell, the repetition or periodicity of the fundamental unit disappears and the crystallinity is zero.

A glass can be defined as a noncrystalline solid. In some parts of the literature it is called amorphous and in other parts quasi-amorphous structure because it lacks the three-dimensional periodicity of a crystal. The term amorphous should not be used to describe glass or liquids because it connotes too strongly a state that contains no order, form, or periodicity. Such a state does not exist in glass or liquids. These can be classified according to some degree of order based on the periodicity of their atomic array. Although this ordering can be of very low degree and approach the amorphous state as a limit, some detectable periodicity will always be present. This argument is substantiated by experimental work in electron diffraction using high energy electrons. The transmission patterns formed by glass structures contain from one to several diffuse, concentric rings that are the images of diffuse diffraction maxima produced by elastic scattering from an atomic array in which the periodicity along one or all dimensions is relaxed but not absent. If glass were amorphous, no diffraction maxima would appear; the pattern would contain only diffuse background. Taking these facts into account, a glass should be defined as a structure in which relatively long-range periodicity can be present in two, one, or none of the dimensions; it never assumes the geometry of a crystal because it always lacks periodicity in the third dimension; the range of periodicity in the other two dimensions can vary from the relatively long to the very short which approaches the amorphous state. When glass devitrifies or changes to the crystalline state, periodicity in all dimensions is made long when atoms rearrange during annealing below the melting point. Even liquids can contain some order. Liquids have less periodicity than glasses because all dimensions experience relaxation in the regularity of the atomic array. It is interesting that detectable periodicity, as evidenced by the formation of diffuse diffraction maxima, has been observed in oils by elastic scattering of high energy electrons.

Adequate models for the routine practice of electron diffraction can be based on Bragg's law and the Laue equations. Both assume, according to kinematical theory, that the space lattice is static. The Bragg diffraction equation, originally developed for the x-ray case, considers the crystal to be a diffraction grating composed of a family of planes that consist of a systematic array of atoms. Each atom acts as a scattering center. The simple, elegant relation is:

$$n\lambda = 2\, d_{(hkl)} \sin\theta_{(hkl)} \tag{11}$$

where n is always an integer called the order of diffraction, λ the wavelength of incident and diffracted electrons, $d_{(hkl)}$ the distance or interplanar spacing between consecutive, parallel, diffracting planes whose Miller indices are (hkl), and $\theta_{(khl)}$ the angle between the direction of the incident electron beam and the diffracting planes whose Miller indices are (hkl). $\theta_{(hkl)}$ is also the angle between these planes and the diffracted beam. The diffraction angle (the angle between the diffracted beam and the undiffracted beam) is $2\theta_{(hkl)}$. The Bragg relation says that diffraction maxima will form only when the path difference ($2d_{(hkl)}\sin\theta_{(hkl)}$ between two diffracted waves equals an integral number of electron wavelengths (λ). The pattern that records these effects is composed of individual diffraction maxima called Laue spots. When relative intensity factors are associated with a system of $d_{(hkl)}$ values, the crystal structure that produced the system can be identified. Equation (11) for Bragg diffraction does not require a space lattice for the proof. The Laue equations, however, describe diffraction from a space lattice composed of atoms at the lattice points:

$$\begin{aligned} &\text{Row A: } a\,(\cos\alpha - \cos\alpha_o) = h\lambda \\ &\text{Row B: } b\,(\cos\beta - \cos\beta_o) = k\lambda \\ &\text{Row C: } c\,(\cos\gamma - \cos\gamma_o) = l\lambda \end{aligned} \tag{12}$$

A set of coordinate axes is chosen and the origin placed at one of the lattice points. The set is one of the seven systems of axes for the seven crystal systems. When incident electrons interact with a row of atoms having equal spacing, the electrons are diffracted from each atom. This incident wave makes angles α_o, β_o, and γ_o with the atoms in row A (the a axis), B (the b axis), and C (the c axis), respectively; the diffracted waves make angles α, β, and γ with A, B, and C, respectively. The distance between atoms in row A is a; in row B, b; and c in row C. The system of equations (12) gives the path difference traversed by diffracted electrons in each of the three rows. In the case of row A, for example, path difference is $[a(\cos\alpha - \cos\alpha_o)]$. Now according to Bragg's argument, the path difference must be an integral number of wavelengths if the elastically scattered electrons are to be in phase and produce diffraction maxima. That is, $h\lambda$, $k\lambda$, and $l\lambda$ must be integral. The λ designates electron wavelength; h, k, and l are integers and designate the order of diffraction. In addition, they are the Miller indices for the diffracted beam. When diffraction maxima appear, all of the requirements as stated by the three Laue equations must be satisfied simultaneously. When conditions are satisfied, the diffracted waves form in space a set of concentric cones around each atom row; the row acts as axis for its set; since angles α, β, and γ are the angles between the diffracted beams and the atoms in row A, B, and C, respectively, the diffraction angles (angle between the diffracted beam and the undiffracted beam) are 2α, 2β, and 2γ for row A, B, and C, respectively; each cone in a set represents one order of diffraction. If a photographic plate is placed normal to the direction of the incident beam; and if the crystal causing the diffraction is a simple cubic structure with atoms at cube corners only; and if the incident beam is parallel to a cube edge (a axis) — then atom row B (b axis) normal to the direct beam and parallel to the plate will produce on the plate one family of hyperbolae; atom row C (c axis) which is normal to the beam, parallel to the plate and normal to atom row B will produce on the plate a second family of hyperbolae with axis normal to that of the first family; atom row A (a axis) which is parallel to the incident beam will produce on the plate a family of concentric circles. Since electron wavelength is very short, the hyperbolae appear on the plate as approximately straight lines. The spot on the plate made by the undiffracted beam (direct beam) is the geometric center for this symmetrical array. Figure 3 shows such an array diagrammatically. The complex requirements stated by the Laue equations limit significantly the number of diffraction maxima that can

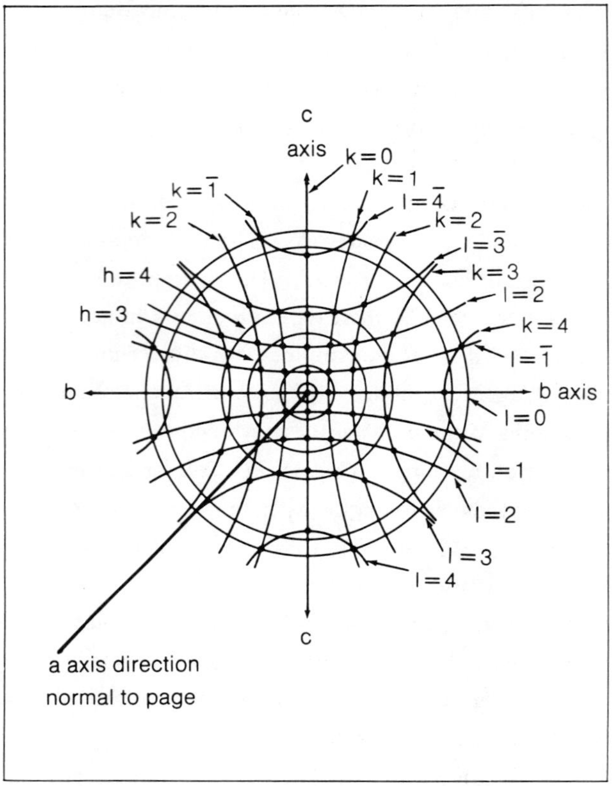

Fig. 3. Geometry of diffraction effects.

appear. For a dispersion of randomly oriented crystals with size in the order of 200 Å, all three Laue conditions are satisfied, and the diffraction pattern obtained by TED and SAED consists of a family of concentric rings that vary in intensity. If the specimen is a section that is thin in the dimension parallel to the direction of the direct beam, then Laue conditions are relaxed in that direction and no maxima appear. The pattern will consist of two families of hyperbolae as shown in Fig. 3; the pattern on the plate will be a cross-grating of Laue spots. This is the pattern obtained from a thin section of single crystal.

To calculate the $d_{(hkl)}$ values for general practice, a relationship can be derived from the Bragg equation (11). The new relationship is:

$$d_{(hkl)} = L\lambda / r \tag{13}$$

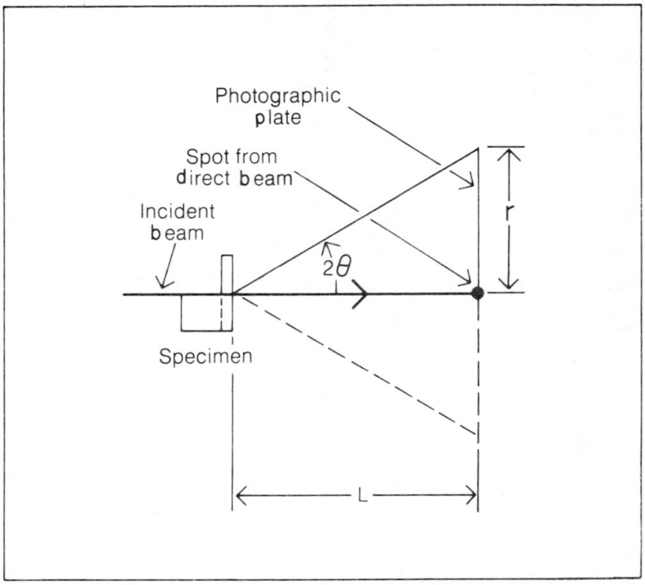

Fig. 4. Electron diffraction.

where L is the camera length or the physical distance between the specimen and the photographic plate that records the pattern; r is the radius of a diffraction ring on the pattern. This relaxed relationship can be justified because the short electron wavelength (λ) makes $\theta_{(hkl)}$ a small value. $\theta_{(hkl)}$ would be equal to 1.1° using 100-keV electrons when $d_{(hkl)} = 1.00$ Å; n = 1 (first order of diffraction); and $\lambda = 0.0370$ Å. In the diffraction camera geometry, as expressed in Eq. (13), the tan $2\theta_{(hkl)} = r/L$. For small angles in the order of 1.0°, 2 sin $\theta_{(hkl)}$ (from Bragg's equation) is a good approximation for the tan $2\theta_{(hkl)}$. The schematic diagram of Fig. 4 shows diffraction camera geometry for IED and TED. Relation (13) can be employed unless very high precision is required. If such is the case, precision of one part in 10^4 can be obtained by using:

$$rd_{(hkl)} = L\lambda\left(1 + \frac{3r^2}{8L^2}\right) \tag{14}$$

in which the terms have the same meanings as used for Eq. (13).

When specimen thickness for TEDCM is about 1000 Å, multiple scattering events enhance incoherent scattering; chromatic aberration increases and resolution in the image deteriorates. These cases require dynamical theory of electron scattering to explain the mechanisms. The incident electrons experience not only complex interactions with potential fields about atomic nuclei in the space lattice of the specimen, but also interactions with diffracted beams and these diffracted beams interact with each other. In brief, dynamical theory attempts to explain diffraction intensity produced by a dynamic space lattice that contains potential distribution not only within the lattice but also at the crystal surface. The theory attempts to account for the experimental observation in which diffracted beams have the same intensity as the direct beam. Diffraction patterns containing Kikuchi lines also represent a dynamical case although S. Kikuchi used a geometrical argument to explain the effects. In kinematical theory, diffraction takes place in a static space lattice; intensity of the diffracted beam is assumed to be negligible with respect to direct beam intensity. For rigorous interpretation, dynamical theory should be used for most cases where specimen thickness is greater than several hundred angstroms. Since the dynamical theory is still in development it is fortunate that kinematical theory can interpret many of the image contrast effects observed in TEDCM and the intensity of diffraction maxima observed in electron diffraction patterns. Further discussion of dynamical theory is beyond the scope of this chapter.

In practice, the surface topography of individual particles in a powder specimen can influence significantly the behavior of the powder. An additional critical parameter for such materials is the architecture of aggregates. This cannot be determined when aggregates are imaged as silhouettes on the electron micrograph — a condition that occurs when particles with a thickness greater than about 2000 Å are used as objects on the specimen grid of the microscope. Figures 5 through 7 illustrate the application of REM to the study of particulate geometry. The general method of specimen preparation for microscopy includes dispersing the powder on a microscope slide using minimal viscous shear and a liquid in which the powder is insoluble; adding a platinum metal shadow by vacuum evaporation; covering the shadow with a carbon film using vacuum deposition; covering the carbon with a backing film of plastic to increase replica strength during handling and subsequent treatment; scoring the replica to form squares while still on the microscope slide; transferring the squares to the surface of a liquid solvent for the specimen material; removing the material from the replica by slow chemical attack; mounting a square of replica on a specimen grid and removing the plastic

Fig. 5

backing with solvents. This leaves a one-stage carbon replica preshadowed with platinum for examination in the electron microscope. In this case, elevations on the specimen cast shadows. In the three micrographs used for illustration the shadow details are white.

Figure 5 shows the surface of a phosphor of the ZnS-type whose emission behavior is affected by surface properties that include topography. The powder specimen was dispersed on a microscope slide using amyl acetate; the angle for platinum shadowing was $\tan^{-1} 1/2$; the shadow thickness was 10 Å; the carbon thickness was 200 Å and a 10% solution of Formvar in ethylene dichloride was used to make the backing. The replica was scored to form a pattern of squares having sides of 1/8 in; the squares were released from the slide by brief treatment in a 10% solution of HCl in distilled water containing

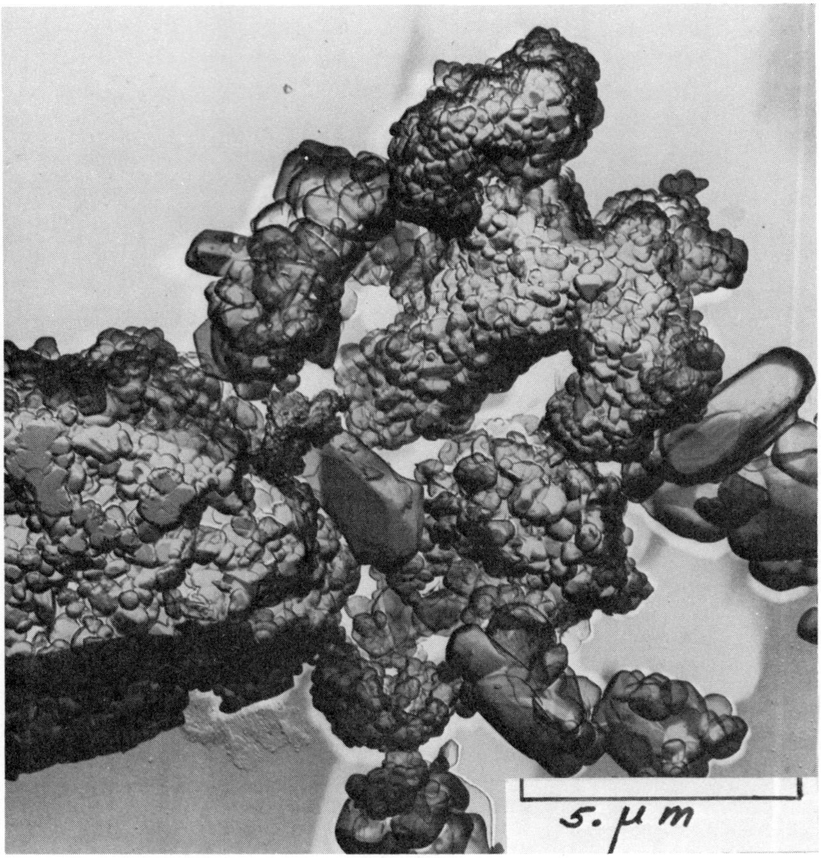

Fig. 6

several drops of hydrofluoric acid (48%). Specimen material was removed by chemical treatment for 15 hr in a 10% solution of HCl in distilled water; the plastic backing was removed by dipping in ethylene dichloride.

Figure 6 shows an aggregate of ZnS crystals that resulted from a thermal threatment at 1000°C for 2 min. Aggregate geometry is very complex and cannot be interpreted on the basis of silhouette images. For replication, amyl acetate was used to disperse the powder on a microscope slide; platinum shadow thickness was 10 Å and shadow angle was $\tan^{-1} 1/2$; carbon was 200 Å thick and was deposited while the specimen was rotating; the backing was made from a solution of 0.25% Parlodion in equal volumes of amyl acetate and ethyl acetate. After scoring, squares of replica were released onto the surface of distilled water; specimen material was removed from the replica by chemical attack for 15 hr in 50% HCl in distilled water and the Parlodion backing was removed by dipping in a bath containing equal volumes of amyl acetate and ethyl acetate.

Figure 7 shows crystallographic features on the surface of a calcium fluoride phosphate crystal used in fluorescent lamp technology. The crystal is known also as fluorapatite [$Ca_5F(PO_4)_3$] and has a hexagonal crystal structure. The crystal had a tabletlike habit; the length of the longest dimension was in the order of 10 μm. For replica preparation the powder was dispersed as in the previous case and the platinum shadowing and carbon deposition were equivalent; the backing film was made from a solution of 0.5% Parlodion in equal volumes of amyl acetate and ethyl acetate; squares of replica were floated onto the surface of distilled water. Specimen material was dissolved by chemical treatment for 2 hr in a 10% solution of HCl in distilled water. The Parlodion backing film was removed by flow-washing with a mixture of equal volumes of amyl acetate and ethyl acetate supplied from a burette.

It is well known that the electrical behavior of a semiconductor crystal such as germanium, silicon or gallium arsenide is influenced strongly by the population of crystallographic and impurity defects that are present in the crystal. Crystallographic defects include dislocations, stacking faults, twins, and strains induced during growth; impurity defects include precipitates, inclusions, and zones of segregation. In many cases, precipitates and inclusions decorate crystallographic defects. It is axiomatic that these defects affect performance of a device fabricated from the crystal. During fabrication additional defects can be generated at crystal surfaces and propagate into crystal bulk as a result of mechanical cutting, lapping, and polishing procedures. In order to understand and control device performance it is necessary that defects be described as quantitatively as possible and that these descriptions be correlated with electrical behavior. The methods of IED, REM, TEDCM, SAEM, and SAED can be coordinated into a scheme of analysis capable of supplying significant information about defects.

Figures 8 and 9 show, respectively, an electron diffraction pattern obtained by IED and an electron micrograph obtained by REM. The specimen

Fig. 7

was undoped, single-crystal silicon in the form of a wafer; the crystal was grown by Czochralski methods; the average intrinsic resistivity was 20 ohm-cm and the surface of the wafer was a $\{111\}$ plane of the silicon crystal. The wafer was cut to a thickness of 3 mm by a diamond saw of the inside-diameter type. Both wafer surfaces were lapped mechanically using garnet (a silicate of iron and aluminum) in a vehicle of organic oil. A layer 50 μm thick was removed in order to produce the surface that gave the pattern shown in Fig. 8.

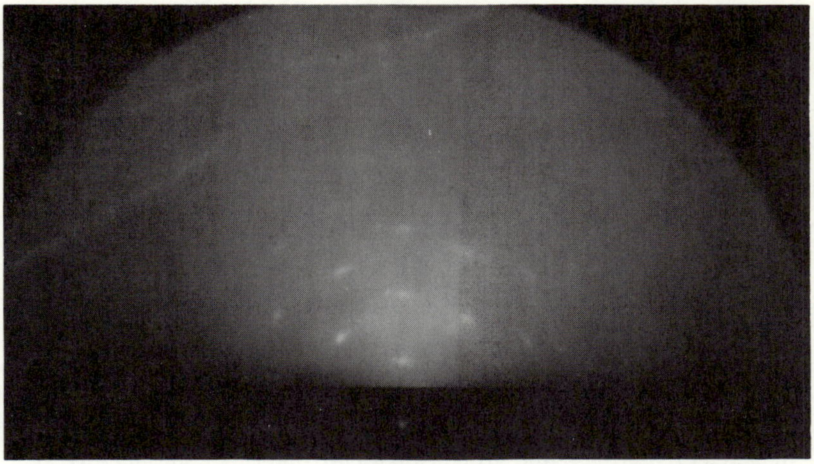

Fig. 8

Fig. 9

To obtain the micrograph in Fig. 9, this surface was replicated using the two-stage Parlodion-carbon method that included platinum shadowing at $\tan^{-1} 1/2$. The topography shown in Fig. 9 represents a flatness to the order of 3000 Å as determined by interference fringe measurement using an optical flat in sodium light.

The pattern in Fig. 8 is composed primarily of discrete Laue spots arranged in a geometry that approximates a cross-grating array. The pattern indicates that the surface has been modified by lapping and although it has retained the general character of a single crystal, it has lost the high degree of perfection that can be expected for a silicon surface relatively free of defects. Figure 10 shows a pattern obtained from a $\{111\}$ surface of another wafer prepared from the same silicon material. In this case lapping was followed by mechanical-chemical polishing. Population of defects is considerably lower on the polished surface as evidenced by the intensity and sharpness of Kikuchi lines which cannot be imaged unless the degree of crystal perfection with respect to crystallographic orientation is high. In addition to the cross-grating array of Laue spots in Fig. 8, there is a pattern of very weak, continuous rings some of which pass through Laue spots. These rings indicate the presence of silicon fragments produced by lapping. The fragments are oriented at random and represent particulate sizes less than about 500 Å. There are also very weak rings that do not pass through Laue spots. Although these reflections could not be identified it is probable that they were produced by small particles of abrasive and/or other contamination embedded in the surface during lapping.

Figure 9 shows fragments and small particles and in addition a high population of conchoidal markings that can be associated with fracture mechanisms. If such a surface is polished by mechanical-chemical procedures in which silicon is removed concurrently by mechanical cutting and by chemical attack, it is possible to obtain the relatively high degree of perfection indicated in Fig. 10. In this case the lap speed during polishing was critical and never exceeded 65 rpm. Even these slowly polished surfaces may contain particles of abrasive and other contamination introduced during the polishing. If contamination is crystalline, IED is a very sensitive method for establishing its presence. Such surfaces usually contain a low population of crystallographic defects that can be detected qualitatively by etch pit methods using Dash

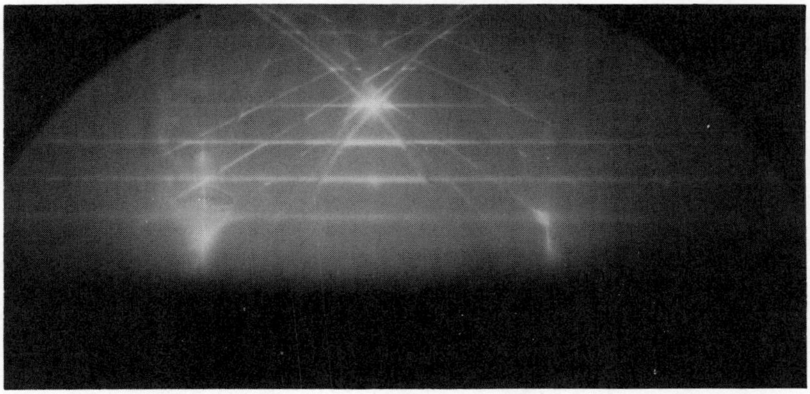

Fig. 10

reagent [1 part hydrofluoric acid (48%) — 3 parts nitric acid (concentrated) — 8 to 12 parts acetic acid (glacial)]. Treatment time in this reagent can be as long as 15 hr. REM and light microscopy are used to observe the etch pit geometry and population. Crystallographic defects and contamination introduced by surface preparation can be removed by chemical polishing in various concentrations of hydrofluoric acid in nitric acid. This treatment tends to reduce flatness of surface topography. Surface preparation can generate defects in thick wafers to a depth of 50 to 100μ m. When wafers are less than about 250μ m thick a network of defects can form throughout the entire cross section.

Figure 11 is a micrograph prepared from a two-stage, Parlodion-carbon replica shadowed with platinum at $\tan^{-1}$ 1/2. The field shows a $\{111\}$ surface of a silicon wafer composed of material that was similar to that used to illustrate the lapping procedure (Figs. 8 and 9). The wafer used for Fig. 11 was prepared by lapping followed by mechanical-chemical polishing; then the wafer was polished chemically using a solution containing 3% hydrofluoric acid (48%) in concentrated nitric acid. The chemical polishing and subsequent washing in distilled and deionized water followed by methyl alcohol were done in an atmosphere of chemically pure nitrogen gas. The thickness removed by the treatment was about 360μ m. According to analysis by IED this surface was free of defects introduced by preparation.

The SAED and IED methods are not always capable of detecting all defects in a thin section or on a specimen surface. Figure 12 is a micrograph made

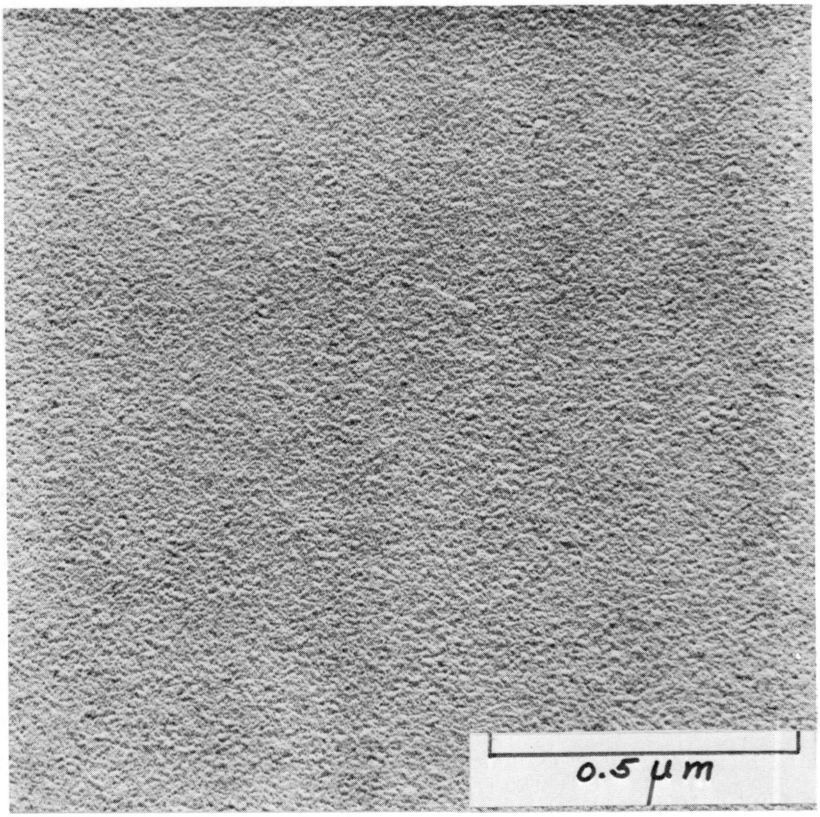

Fig. 11

by TEDCM; it shows a thin section (1500 Å thick) prepared from a wafer of the same material shown in Figs. 8 and 9. The wafer of Fig. 12 was lapped and then polished by mechanical-chemical methods. A thin section composed of crystal material that was near and below the polished surface was prepared by chemical polishing using the hydrofluoric acid-nitric acid solution. A disk of specimen (3 mm in diameter) was formed by using Teflon masks and chemical polishing in the acids. The thin section was developed at the center of this disk using the same acids and a different mask configuration. Figure 12 shows that a network of crystallographic defects was present in the crystal. When a SAED pattern of this zone was photographed only Laue spots and Kikuchi lines were observed as shown in Fig. 13. It is an important point that Kikuchi lines in the diffraction pattern do not always indicate that the thin section is free of defects. If a description of the three-dimensional

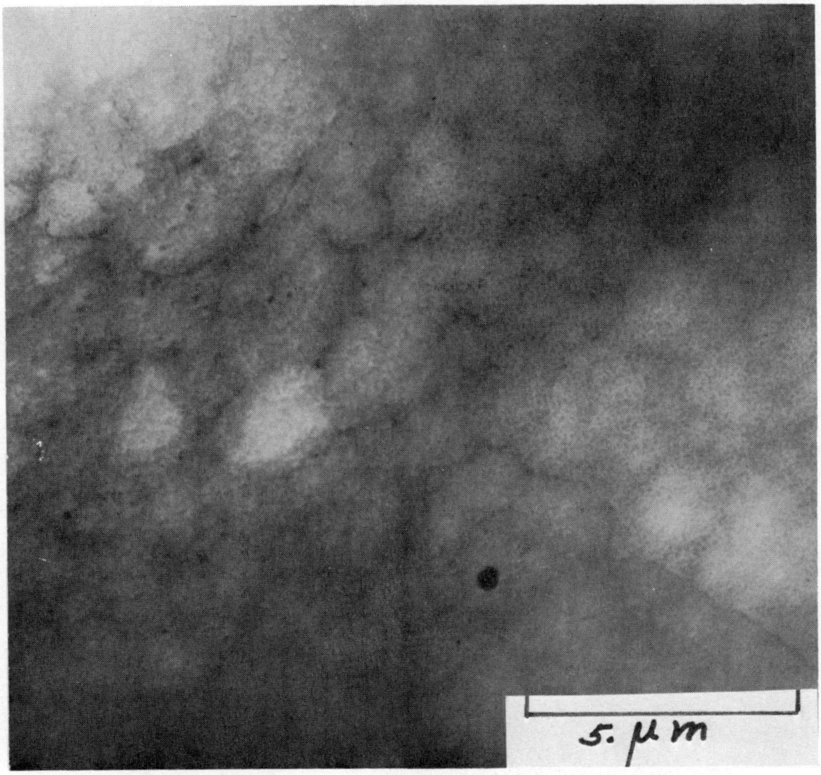

Fig. 12

geometry of crystallographic defects is required with resolution better than 2000 Å, TEDCM is the only method that is applicable.

Epitaxy is the preferred orientation of crystals in a deposit that has grown on a substrate. The texture or type of orientation is influenced strongly by the chemistry, topography, crystal character and temperature of the substrate, and also by the crystal character and chemistry of the deposit. In many cases the crystal size in a deposit is less than 1000 Å, making it necessary to use electron diffraction methods for the identification of crystal structure. Deposits can be monocrystalline, polycrystalline with crystals arranged in preferred orientation, or polycrystalline with random orientation.

Figures 14 through 18 illustrate the application of REM, IED, TEDCM, and SAED to the analysis of silicon carbide (SiC) deposits on single-crystal silicon. In all cases the substrates were $\{110\}$ surfaces of high-purity, n-type silicon grown by Czochralski methods; resistivity ranged from 120 to

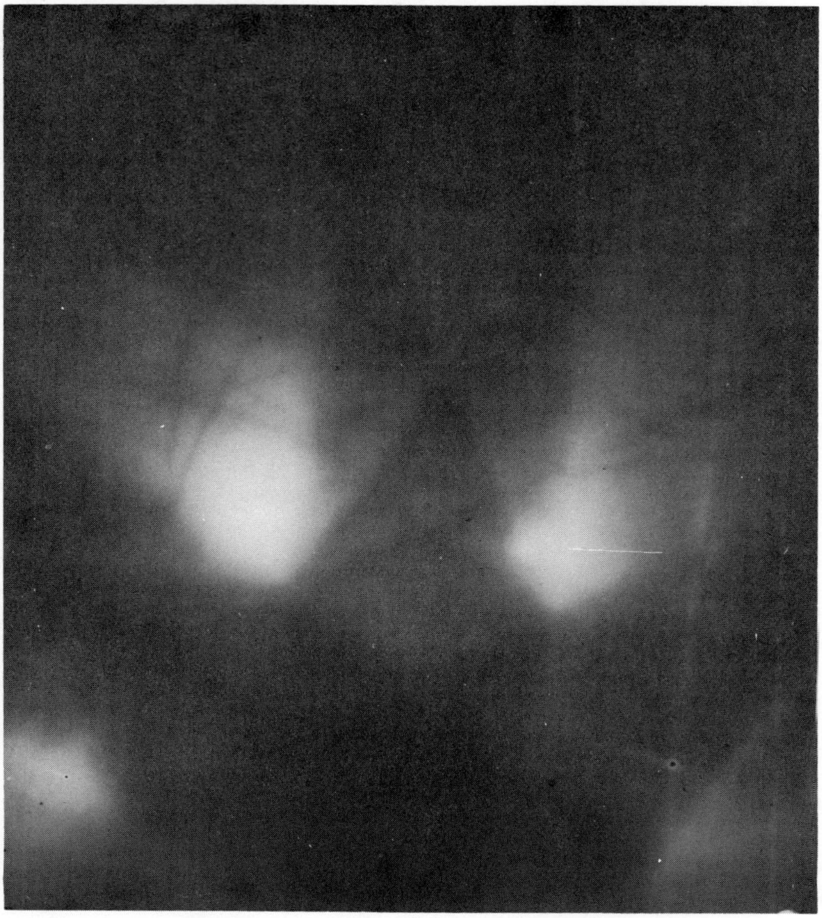

Fig. 13

250 ohm cm; average dislocation density was less than 10^4 cm^{-2}. The substrates were 10 mm x 3 mm x 0.75 mm thick and were prepared by the same methods of cutting, lapping, mechanical-chemical polishing and chemical polishing that were used for the surface shown in Fig. 11. The surface reaction that produced epitaxy lasted 2 hr at 900°C in acetylene (C_2H_2) with partial pressure of 5 x 10^{-7} Torr. Before the introduction of C_2H_2, substrates were annealed for 20 min at 900°C in vacuum of 1.6 x 10^{-9} Torr. Average "as calculated" thickness of deposits was about 1000 Å. The substrates were analyzed before deposition and the deposits were analyzed "in situ" by IED using a diffraction camera length of 50 cm and a double condenser lens system.

Figure 18 shows a typical pattern obtained from a substrate prior to deposition. As in the case of Fig. 10 the Kikuchi lines indicate a high

Fig. 14

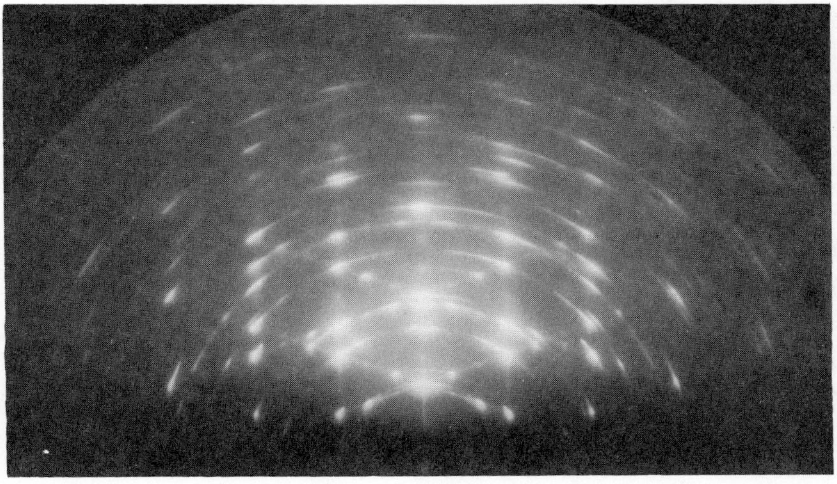

Fig. 15

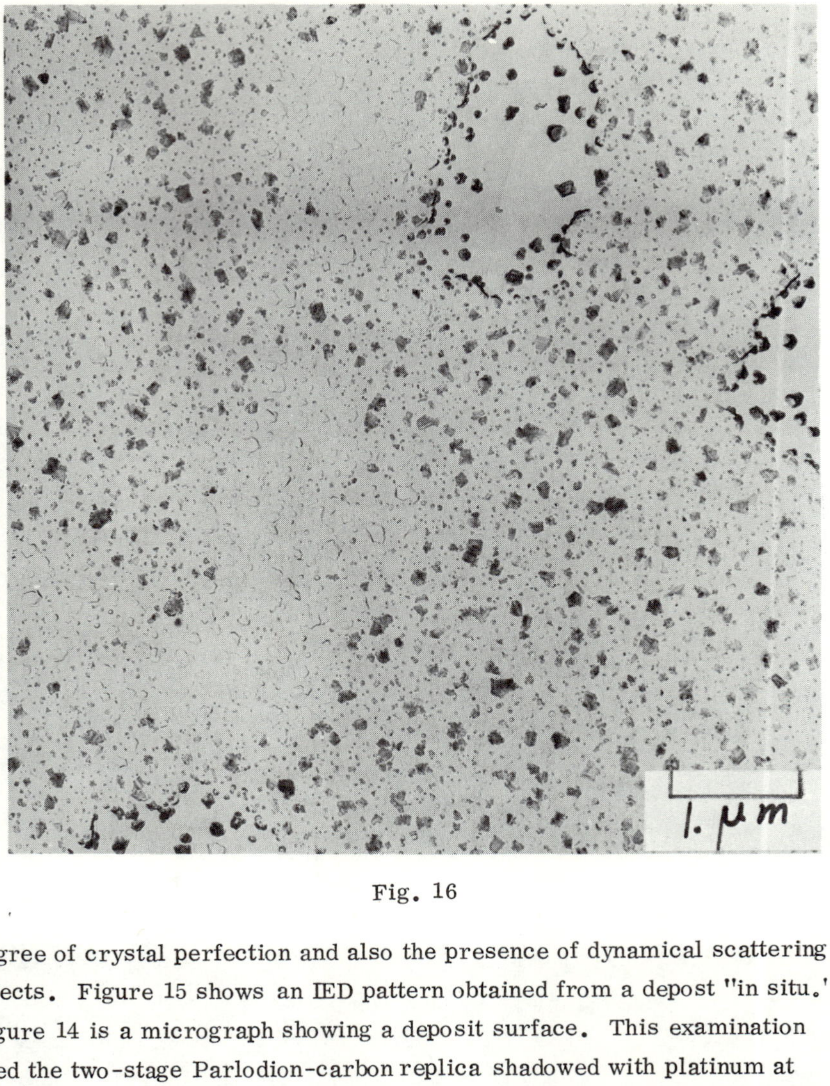

Fig. 16

degree of crystal perfection and also the presence of dynamical scattering effects. Figure 15 shows an IED pattern obtained from a depost "in situ." Figure 14 is a micrograph showing a deposit surface. This examination used the two-stage Parlodion-carbon replica shadowed with platinum at $\tan^{-1} 1/2$. The large crystallographic markings are etch pits in the substrate and were developed by heating during the deposition process. These pits appear as elevations on the micrograph because a two-stage replication process was employed. After REM the deposits were covered with a film of glasslike carbon (100 Å thick) using vacuum deposition. The carbon film reinforced the deposit while it was being removed from the substrate by chemical treatment in 4% hydrofluoric acid (48%)-nitric acid (concentrated). After removal, deposits were analyzed by TEDCM and SAED.

Fig. 17

Figure 16 is a micrograph obtained by TEDCM; a typical SAED pattern of an isolated deposit is shown in Fig. 17. IED and SAED identified the major phase in all deposits as β SiC (cubic; space group: $T_D^2 - F\bar{4}3M$; a_o= 4.358 Å). The IED established that the deposits were polycrystalline. Some crystals were oriented at random; many crystals formed preferred orientation textures with (111) and (110) planes parallel to the (110) plane of the substrate; a few crystals were oriented with (001) or (311) planes parallel to the substrate surface. Reflections from (220) and (440) planes of silicon were detected also indicating the possibility that small fractions of substrate had migrated to the deposit surface. Twinning about the [111] axis of β SiC was established. Some deposit did not remain attached to the carbon film when it was isolated, but remained on the substrate. Such cases appear in Fig. 16 as replicated shapes instead of the diffraction-contrast images produced by βSiC crystals. Twinning effects in these crystals are resolved in Fig. 16. Information obtained from SAED patterns such as Fig. 17 substantiated the preferred orientation textures and twinning effects established by IED.

Fig. 18

Another application of IED to the analysis of semiconductor material has to do with identification of βSiC precipitates formed in silicon single crystals as a result of annealing. The specimen was a p-type $\{111\}$ silicon wafer grown by Czochralski methods; resistivity ranged from 50 to 60 ohm cm. Etch pit analysis before annealing established that the wafer was composed of zones with dislocation density varying from zone to zone over the range of 0 to 10^5 cm^{-2}. The wafer was prepared by the same methods of cutting, lapping, mechanical-chemical polishing, and chemical polishing as used for the crystal of Fig. 11. A layer 100 - 300 μm thick was removed from the surface by the chemical polishing. IED established that, prior to annealing, the crystal perfection of the surface was high and that defects introduced by mechanical preparation were absent. β SiC was not detected by IED prior to annealing. The wafer was annealed for 16 hr at 1200°C in a quartz capsule evacuated to 10^{-6} Torr. The wafer was polished chemically in several stages to a depth of 11 μm using 3% hydrofluoric acid (48%)-nitric acid (concentrated). The IED analysis was made at each stage; the β SiC structure was detected at most stages.

Figure 19 shows an IED pattern for a case where 6.4 μm of surface layer had been removed. The ring pattern indicates reflections from βSiC with most crystals oriented at random; average crystal size was about 200 Å as determined from diffraction pattern geometry. Population of crystals larger than about 1000 Å was very low. The Kikuchi lines and Laue spots in Fig. 19 indicate reflections from the silicon single crystal.

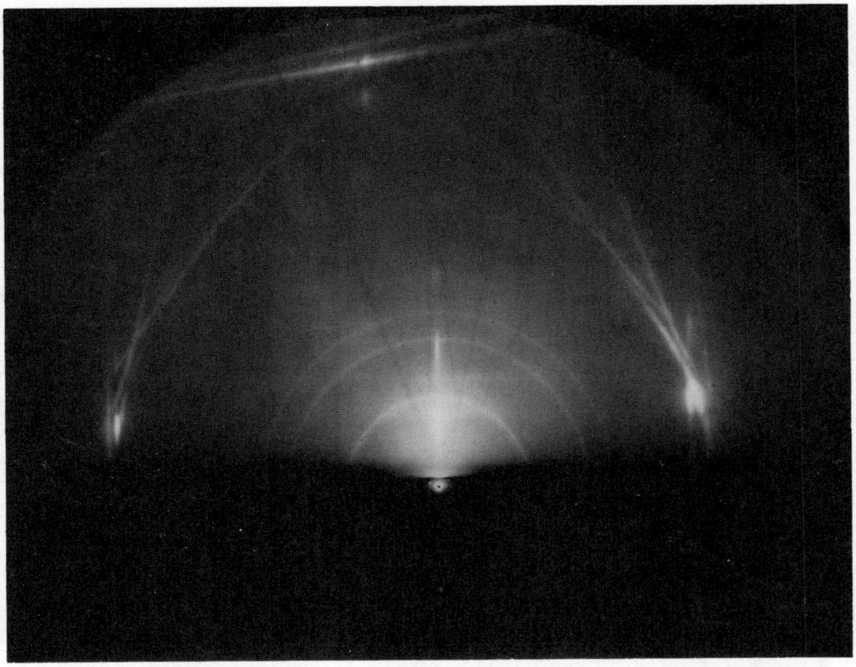

Fig. 19

Devitrifying solder glasses are used to form glass-to-glass seals in such applications as the encapsulation of electronic components or the sealing of faceplates to funnel sections in color television tubes. These solders make good seals in the mechanical sense; they can be fired at a relatively low temperature in the order of 425°C and they can be used in service at temperatures up to 450°C. The solder is applied to the sealing area as a suspension of powdered glass in a low viscosity vehicle such as nitrocellulose in amyl acetate. This "green seal" is baked thoroughly and then fired. During firing crystalline phases nucleate and grow in the solder to produce a devitrified glass seal. The strength and hardness of this seal are related to the population, distribution, and morphology of crystalline phases developed during firing. REM is a useful method for studying this mixture of glass and crystals in which morphological features, less than 500 Å in size, can be of significance.

Figure 20 is a micrograph prepared by REM; it shows a field from the cross section of a seal. The specimen was polished automatically using diamond (0.25 μm diameter) and then etched chemically in a solution of 5%

Fig. 20

nitric acid in water; the etching developed the crystal geometry illustrated in the micrograph. The replica was prepared by the two-stage Parlodion-carbon method with the negative impression being made from a solution of 6% Parlodion in equal volumes of amyl acetate and ethyl acetate; the negative was removed from the specimen by mechanical stripping and shadowed on the replica side with platinum at $\tan^{-1} 1/2$; the carbon film was vacuum-deposited on top of the shadow and was 200 Å thick. Deposition of the carbon was made while the replica surface was rotating under the carbon source which consisted of two spectrographic carbon rods (3 mm diameter); one rod was reduced at the tip to form a cylinder (one mm diameter and 6 mm long). This tip was the amount used for one deposition and two depositions were made to form the replica. Washing in equal volumes of amyl acetate and ethyl acetate removed the plastic negative from the carbon film containing the shadow. Figure 20 shows crystals that developed in the seal during processing. The smooth zones represent glasslike phases. Cracks in one of these phases appear at about 3:00 o'clock on the micrograph. The cracks terminate when they encounter the crystal habit shown at the center of the micrograph.

Adequate characterization of a glass system should include not only description of the glass phases that compose the matrix but also delineation of the crystallites that may be dispersed throughout the matrix as minor phases. The geometry, structure, perfection, population, and distribution of such crystallites can affect not only the mechanical properties of the glass but also its electrical, magnetic, and optical character as well. Such crystallites are basic in one of the mechanisms by which glass can devitrify or change to a stable, crystalline state. In this mechanism, crystallites that have specific composition and structure form as a dispersion throughout the glass matrix as a minor phase. During early devitrification history crystal growth progresses in a size range from angstroms to micrometers. Description of crystals in such systems should include sizes less than 1000 Å. This information can be obtained by analyzing thin sections (<2000 Å thick) using TEDCM and SAED. If surfaces can be prepared that represent bulk character, IED can be used as a supplementary tool. Preparation of thin glass sections and surfaces by chemical methods or by mechanical polishing usually introduces serious artifacts.

Figures 21 and 22 show micrographs prepared from thin sections that were made by fracturing alumino-borosilicate glasses containing iron oxide. Mole composition was 34% SiO_2, 25% B_2O_3, 8% Al_2O_3, 21% CaO, and 12% Fe_2O_3. To obtain thin sections, a specimen is fractured in a drybox containing an inert gas such as nitrogen or argon. Nitrogen was used for the work illustrated in Figs. 21 and 22; the dew point of this gas was lower than $-50^{\circ}C$ and the specimen was protected at all times by the gas when not under vacuum. After fracture the thin sections were collected by an unshadowed extraction replica technique that used Parlodion in equal volumes of amyl acetate and

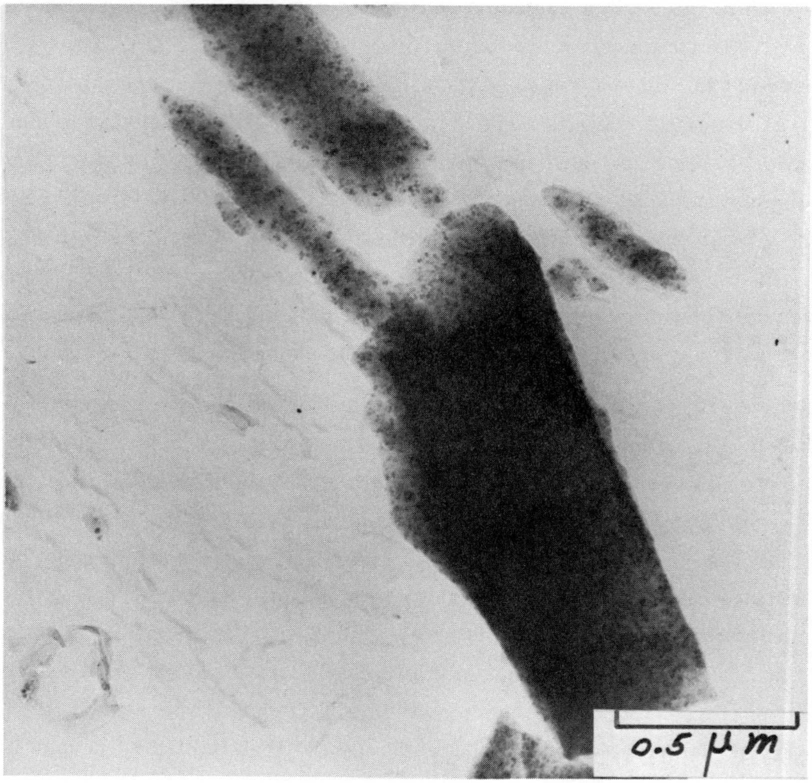

Fig. 21

ethyl acetate to make the negative plastic impression; this negative was removed from the fracture surface by physical stripping; roughness of the fracture surface dictated negative thickness. This replication process removed from the place of fracture thin glass sections varying in thickness from about 200 Å to more than 5000 Å. The extraction replica was coated (on the side that contained thin sections) with a carbon film made by vacuum deposition; thickness of carbon was 100 Å. The plastic negative was removed by washing in equal volumes of amyl acetate and ethyl acetate with washing being done in a drybox containing nitrogen. The sections were analyzed in an electron microscope whose specimen chamber was filled with nitrogen gas when not evacuated; during analysis the specimen was cooled to about 0^oC by a liquid nitrogen cooling device attached to the goniometer specimen stage of the microscope.

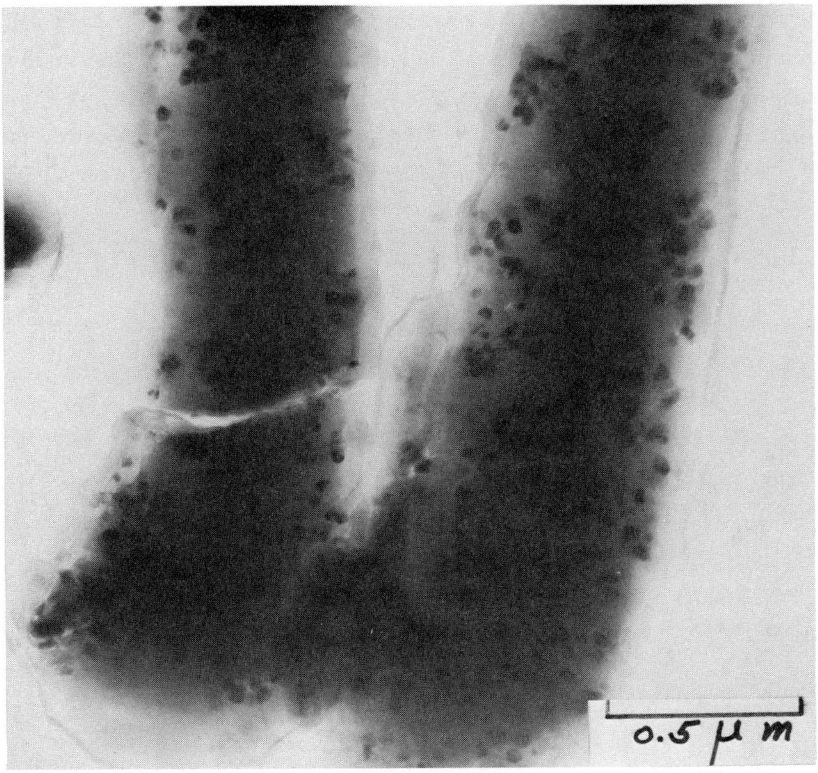

Fig. 22

The specimen used for Fig. 21 had been processed by annealing at 700°C for 20 min. and the specimen used for Fig. 22 by annealing at 750°C for 2 hr.

The SAED identified crystals of Fe_3O_4 and/or gamma Fe_2O_3 in both specimens. Because of similarity between the $d_{(hkl)}$ values and their intensities for these structures, discrimination cannot be made by electron diffraction methods. TEDCM permitted determination of size and shape for aggregates and individual crystals; crystal growth as a result of annealing was established; defects in crystals of the specimen that had been annealed at 750°C were resolved by TEDCM.

ACKNOWLEDGMENTS

The following persons have assisted substantially in the experimental work discussed in Section IV, this chapter: Thomas A. Emma, Ruth S.

Letendre, and Fred Grieco (photographic printing). The author thanks Dr. Rene Simon of this staff for his critical review of the manuscript. The author thanks also Mr. Richard N. Rose, Managing Editor, for granting permission to reprint the part of this chapter which appeared in the "Norelco Reporter" (Vol. 16, No. 1, 1969, published by Philips Electronic Instruments, Mount Vernon, New York).

BIBLIOGRAPHY

A. General Principles and Applications of Electron Optics, Electron Microscopy, and Electron Diffraction

R. Rudenberg, "The Early History of the Electron Microscope," J. Appl. Phys., 14, 434-436 (1943).

V. K. Zworykin, G. A. Morton, E. G. Ramberg, J. Hillier, and A. W. Vance, Electron Optics and the Electron Microscope, Wiley, New York, 1945.

R. W. G. Wyckoff, Electron Microscopy, Interscience, New York, 1949.

V. E. Cosslett, Electron Optics, Oxford, Clarendon, (Second Ed.), 1950.

V. E. Cosslett, Practical Electron Microscopy, Academic, New York, 1951.

R. B. Fischer, Applied Electron Microscopy, Indiana Univ. Press, Bloomington, Indiana, 1953.

C. E. Hall, Introduction to Electron Microscopy, McGraw-Hill, New York, 1953.

R. D. Heidenreich and C. J. Calbick, "Electron Microscopy," in Physical Methods in Chemical Analysis (W. G. Berl, ed.) Academic, New York, 1960, p. 547.

F. A. Hamm, "Electron Microscopy," in Physical Methods of Organic Chemistry (Third Ed.) (A. Weissberger, ed.) Vol. I, Part 2, Interscience, New York, 1960, p. 1561.

The Encyclopedia of Microscopy, (G. L. Clark, ed.) Reinhold, New York, 1961.

M. E. Haine and V. E. Cosslett, The Electron Microscope. The Present State of the Art, Interscience, New York, 1961.

S. Wischnitzer, Introduction to Electron Microscopy, Pergamon, New York, 1962.

Modern Developments in Electron Microscopy, (B. M. Siegel, ed.) Academic, New York, 1964.

G. G. Cocks, "Electron Microscopy," in Treatise on Analytical Chemistry, (I. M. Kolthoff and P. J. Elving, eds.) Part 1, Vol. 6, Chapt. 73, Interscience, New York, 1965, p. 4113.

J. A. Belk and A. L. Davies, Electron Microscopy and Microanalysis of Metals, Elsevier, New York, 1968.

B. Principles and Applications of Transmission Electron Diffraction Contrast Microscopy

G. Thomas, Transmission Electron Microscopy of Metals, Wiley, New York, 1962.

Electron Microscopy and Strength of Crystals, (G. Thomas and J. Washburn, eds.), Interscience, New York, 1963.

R. D. Heidenreich, Fundamentals of Transmission Electron Microscopy, Interscience, New York, 1964.

P. B. Hirsch, A. Howie, R. B. Nicholson, D. W. Pashley, and M. J. Whelan, Electron Microscopy of Thin Crystals, Butterworth, Washington, D.C., 1965.

C. Principles and Applications of Electron Diffraction

G. P. Thomson and W. Cochrane, Theory and Practice of Electron Diffraction, Macmillan, London, 1939.

Z. G. Pinsker, Electron Diffraction, Butterworths Scientific, London, 1953.

W. C. Bigelow, "Electron Diffraction," in Physical Methods in Chemical Analysis, (W. G. Berl, ed.) Academic, New York, 1960, p. 619.

L. O. Brockway, "Electron Diffraction by Gases," in Physical Methods of Organic Chemistry, (3rd Ed.) (A. Weissberger, ed.), Vol. I, Part 2, Interscience, New York, 1960, p. 1739.

J. H. Reisner, "Electron Diffraction and Scattering Techniques," in Treatise on Analytical Chemistry, (I. M. Kolthoff and P. J. Elving, eds.), Part 1, Vol. 6, Chapt. 68, Interscience, New York, 1965, p. 3765.

D. Techniques of Electron Microscopy and Electron Diffraction

"The Practice of Electron Microscopy", (D. G. Drummond, ed.) J. Roy. Microscopical Soc., London, Series III, Vol. LXX, Part 1, March, 1950.

W. J. McG. Tegart, The Electrolytic and Chemical Polishing of Metals in Research and Industry, Pergamon, London, 1956.

Techniques for Electron Microscopy, (D. H. Kay, ed.) (Second Ed.), Davis, Philadelphia, 1965.

I. S. Brammar and M. A. P. Dewey, Specimen Preparation for Electron Metallography, American Elsevier, New York, 1966.

K. W. Andrews, D. J. Dyson, and S. R. Keown, Interpretation of Electron Diffraction Patterns, Plenum, New York, 1967.

Inorganic Index to the Powder Diffraction File, 1969, (Compiled by the Joint Committee on Powder Diffraction Standards), Publication PDIS-19i, American Society for Testing and Materials, Philadelphia, Pennsylvania, 1969 (revised yearly).

Fink (Six-Entry) Inorganic Index to the Powder Diffraction File, 1969, (Compiled by the Joint Committee on Powder Diffraction Standards, Publication PDIS-19F, American Society for Testing and Materials, Philadelphia, Pennsylvania, 1969 (revised yearly).

E. Light Microscopy

C. P. Shillaber, Photomicrography, Wiley, New York, 1944.

E. M. Chamot and C. W. Mason, Handbook of Chemical Microscopy, Vol. I, (3rd Ed.), Wiley, New York, 1958.

E. E. Jelley, "Light Microscopy," in Physical Methods of Organic Chemistry, (3rd Ed.), (A. Weissberger, ed.) Vol. I, Part 2, Interscience, New York, 1960, p. 1347.

W. C. McCrone and L. B. McCrone, "Chemical Microscopy," in Treatise on Analytical Chemistry, (I. M. Kolthoff and P. J. Elving, eds.) Part I, Vol. 6, Chapt. 72, Interscience, New York, 1965, p. 4009.

F. Crystallography

M. J. Buerger, X-ray Crystallography, Wiley, New York, 1942.

C. W. Bunn, Chemical Crystallography, Oxford Univ. Press, London, 1945.

C. S. Barrett, Structure of Metals, McGraw-Hill, New York, 1952.

A. N. Winchell, Elements of Optical Mineralogy, Part I, (5th Ed.), 1958; Part II, (4th Ed.), 1959; Part III, (2nd Ed.), 1957; Wiley, New York.

B. D. Cullity, Elements of X-ray Diffraction, Addison-Wesley, Reading, Massachusetts, Second Printing, 1959.

N. H. Hartshorne and A. Stuart, Crystals and the Polarising Microscope, (3rd Ed.), Edward Arnold, London, 1960.

A. Taylor, X-ray Metallography, Wiley, New York, 1961.

E. E. Wahlstrom, Optical Crystallography, (4th Ed.), Wiley, New York, 1969.

Chapter 7

ANALYSIS OF MICROSTRUCTURAL DEFECTS

R. W. Newman

Florida Institute of Technology, Melbourne, Florida

I. INTRODUCTION

Field ion and electron microscopy are complementary tools used for the study of defects in crystalline solids. This chapter discusses the theory and application of these two techniques to the study of the various types of defects normally encountered in materials investigations. The basic experimental techniques associated with transmission electron microscopy are, however, covered in Chapter 6, and it is suggested that the reader familiarize himself with that chapter before reading the following material.

II. DEFECTS IN CRYSTALS

When we speak of a perfect crystal, we mean that the atoms occupy fixed sites in a regular array. Real crystals, however, are not perfect, but contain regions where the regular pattern breaks down and some atoms are not properly surrounded by their neighbors. Defects in crystals take on many forms but can be categorized as (A) point defects, (B) line defects, (C) planar defects, or (D) volume defects.

A. Point Defects

Point defects result when one, two, or perhaps, three atoms are "out of place." A foreign atom or an atom occupying a nonlattice site are considered as such, since the atoms in the immediate vicinity will be displaced from their proper position. An equilibrium number of these imperfections are present in all crystals; however, they are too small to be seen in the electron microscope and are visible only via field ion microscopy. This is a technique, to be described later, which permits one to "see" the positions of the atoms on the surface of a small hemispherical crystal. Figure 1 is a field ion micrograph of a platinum crystal containing vacant lattice sites.

B. Line Defects

Line defects (i.e., dislocations) are not equilibrium defects, but almost always occur in real crystals. They are extremely important in that plastic deformation of a crystalline body can be attributed to and described by the motion of these imperfections within the crystal. In the simple picture of slip, the crystal is envisioned as deforming by shearing the planes of atoms over one another much like a deck of cards. In real crystals slip does not occur simultaneously over the entire slip plane, but only over part of it (see Fig. 2). A dislocation separates the slipped from the unslipped region. The mechanism by which the motion of a single dislocation produces a displacement of the upper half of the crystal with respect to the lower half,

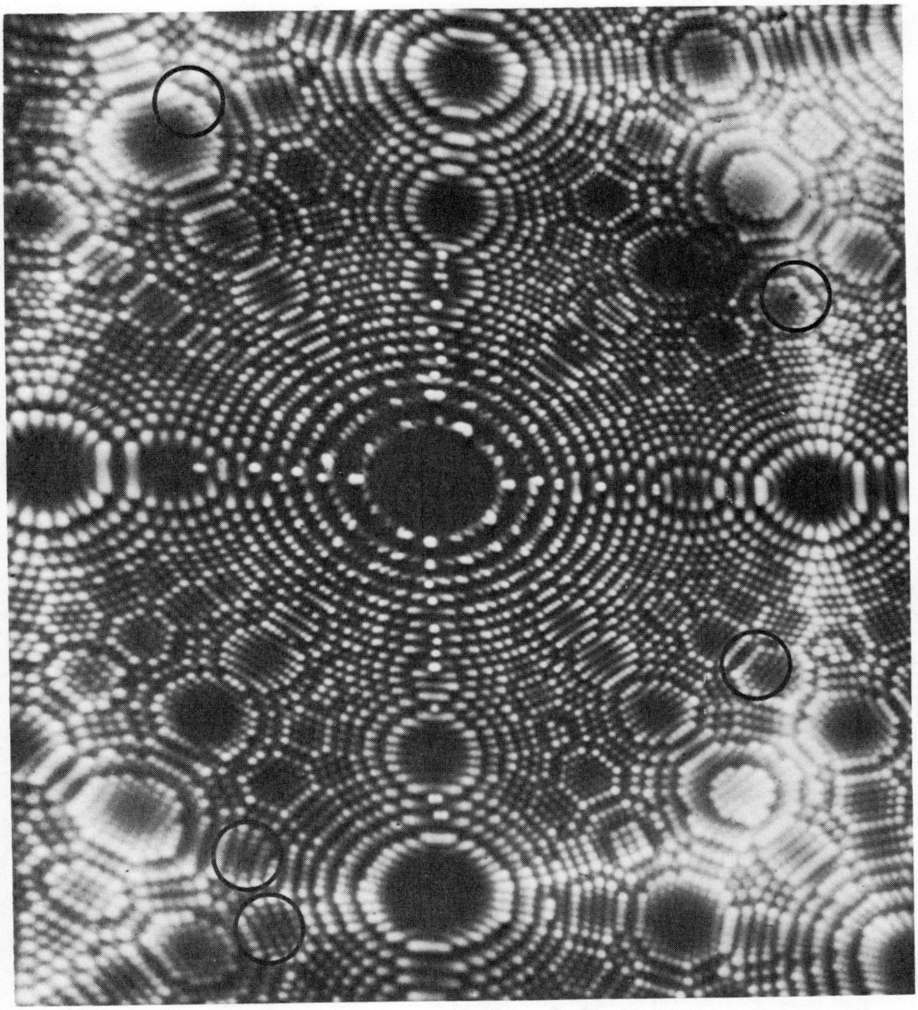

Fig. 1. Field ion micrograph of platinum showing point defects.

across the slip plane, is analogous to the way one usually relocates a large rug in a living room. It's simply not sensible to grab one end of the rug and drag it several inches over the floor. Instead, one introduces a hump (i. e., a dislocation) at one end and pushes the hump to the other end with relative ease, as shown in Fig. 3.

The general dislocation is quite difficult to describe geometrically, but may be more easily understood as a mixture of two limiting cases, the edge

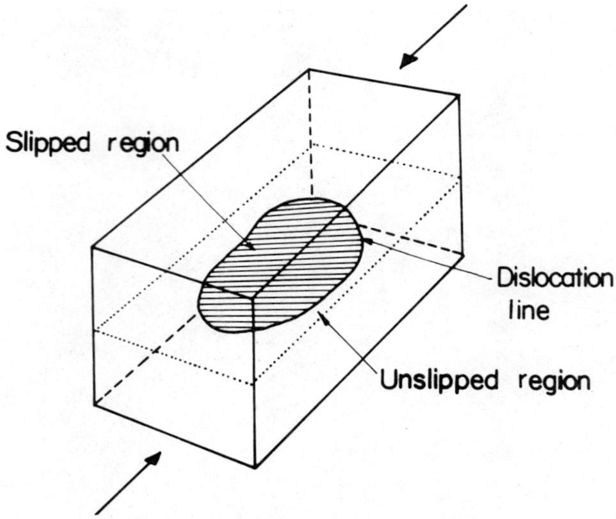

Fig. 2. The dislocation model of plastic deformation. When a shearing stress is applied to the crystal, slip begins in a limited region on the slip pane. The slipped region is bounded by a dislocation line.

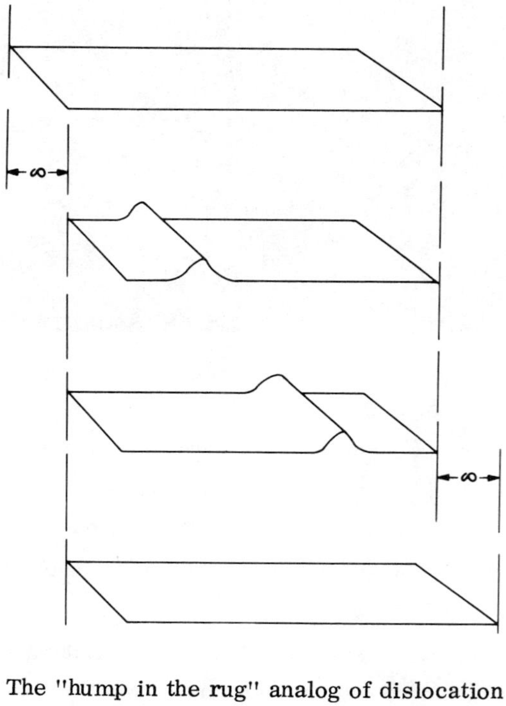

Fig. 3. The "hump in the rug" analog of dislocation motion in a solid.

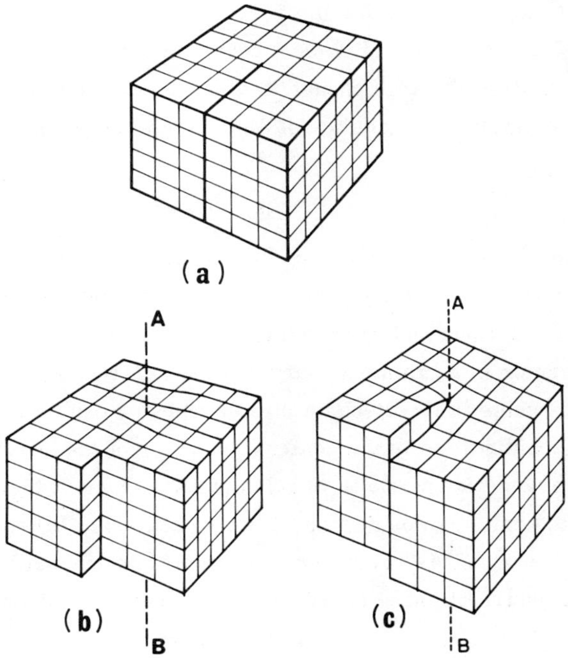

Fig. 4. (a) Schematic cubic crystal cut along the heavy line. (b) Sheared to produce an edge dislocation along A-B. (c) Sheared to produce a screw dislocation along A-B.

and the screw. The characteristics of these two can best be explained by artificially introducing them into a perfect body. Let a sharp cut be made part way into the schematic crystal as shown in Fig. 4a. Then let a force be applied such that one-half moves one atomic spacing relative to the other half, and let the two surfaces be rejoined (Fig. 4b). The result is an edge dislocation lying along the line A-B. A screw dislocation is produced by making the same cut on the perfect crystal (Fig. 4a), but applying a force such that the material on one side of the cut is displaced by one atomic spacing relative to the other as in Fig. 4c. When the atoms are again joined across the boundary, the crystal contains a screw dislocation along A-B.

The density of dislocations in a real crystal may vary between 10^4 and 10^{12} lines/cm^2, depending on its mechanical and thermal history. (Since a dislocation cannot terminate within a crystal, this density is a measure of the number of dislocations which intersect each square centimeter of surface.) This high population of defects coupled with the fact that the local atom displacements are large enough to alter the diffraction conditions, makes electron microscope studies of dislocation arrays quite feasible.

C. Planar Defects

Planar defects are characterized by a disruption in the order of atomic packing at a surface within the solid. The interface between two crystals of different orientations and/or crystal structures clearly falls in this category. The former is a grain boundary and the latter is a phase boundary. They are observable in the electron microscope and can be readily recognized because of discontinuities in the contrast behavior of the adjacent crystals. The relative orientation of the two grains and the plane of the boundary (on a macroscopic scale) can be deduced from an analysis of their respective diffraction patterns. It is sometimes possible to obtain additional information about the nature of phase boundaries if interfacial dislocations are present.

In many instances it is energetically favorable for a dislocation to split into two or more partial dislocations. The region between the partials is called a stacking fault and, if the separation is more than a few atom diameters, the fault can certainly be classified as a planar defect. Adjacent regions (domains) of ordered crystals which have nucleated on different sublattices interface at a domain boundary. Unlike the grain boundary, however, there is no atomic disorder at the boundary, but only an alteration in the directionality or type of the atomic bonds. Stacking faults and domain boundaries usually appear as alternating dark-light fringes in the electron microscope.

D. Volume Defects

Point defects are often found to be thermodynamically more stable in an agglomerated state. Vacancies and interstitials migrate to form so-called secondary defects, and impurity atoms may form clusters or precipitates. These volume defects are important from many standpoints, but of particular interest is their effect on the deformation behavior of solids. It is well known that the response of a material to an applied stress is governed by the motion of dislocations. Obstacles, such as volume defects, interfere with this motion and may either improve or adversely affect the mechanical properties of the material. A large number of theoretical and experimental investigations have been directed toward a better understanding of these interactions in materials, and electron microscopy is perhaps the foremost technique.

III. KINEMATICAL THEORY OF DIFFRACTION CONTRAST

A. Introductory Remarks

Diffraction of electrons by the crystal lattice is the principle contrast mechanism in thin foil transmission microscopy. In this section we will develop, from first principles, the equations necessary to explain the observed contrast behavior of perfect and imperfect crystals. (This treament is taken mainly from Refs. 9 and 14.)

De Broglie postulated that matter particles have an associated wavelength give by

$$\lambda = h/p = h/mu \tag{5}$$

The kinetic energy of an electron accelerated to a classical velocity u by an impressed potential E is

$$Ee = \frac{mu^2}{2} = \frac{h^2}{2m\lambda^2} \tag{6a}$$

or

$$\lambda = h/(2mEe)^{1/2} \tag{6b}$$

For 100 kV electrons

$$\lambda = \frac{150}{E(\text{volts})^{1/2}} = 0.04\ \text{Å} \tag{6c}$$

This is considerably shorter than the wavelengths normally employed in x-ray diffraction, and has some consequences that will become apparent later. Relativistic effects require only a 5% correction for 100-kV electrons and can be ignored for all but the most exacting microscopy.

The scattering of x rays by atoms arises from an interaction with the electron cloud, whereas electron scattering is primarily affected by the nucleus. The electron-electron interaction contributes mainly to the inelastic component. This is revealed in the expression for the atomic scattering amplitude

$$f(\theta) = \frac{m_o e^2}{2h^2} \left(\frac{\lambda}{\sin\theta}\right)^2 \left[Z - f_x\right] \tag{7}$$

where m_o = electron rest mass

e = electronic charge

h = Planck's constant

θ = scattering angle

Z = atomic number of the scatterer

f_x = x-ray scattering factor

The first bracketed term is due to Rutherford scattering from the nucleus and the second is due to scattering from the electron cloud. The equivalent expression for x-ray scattering is

$$f(\theta) = \frac{e^2}{m_o c^2} f_x \tag{8}$$

For a typical low-order reflection in the electron microscope, $\sin\theta/\lambda \sim 0.2\,\text{Å}^{-1}$, and the value of f_x is such that

$$\frac{f(\theta)}{(e^2/m_o c^2)\, f_x} \sim 10^4$$

Thus, the scattering of electrons at low angles is much greater, and also less sensitive to the atomic number of the scattering material (i.e., $f_x \rightarrow Z$ as $\sin\theta/\lambda \rightarrow 0$ for x rays).

This discussion to follow will be limited entirely to the kinematical theory of diffraction. It will be assumed that the scattering potential is small enough to allow us to ignore the change in the intensity of the incident beam, even though it is partially scattered by each plane of atoms traversed. Consequently, the intensity of the scattered wave is small, and it follows that the crystal must not be precisely oriented at a Bragg angle. It is further assumed that the scattered beam does not undergo a secondary scattering event.

B. Scattering from Atoms

Assuming the potential inside a metal foil to be constant, the Schrodinger equation has plane wave solutions of the form

$$\psi = A \exp (2\pi i\, x/\lambda) \tag{9}$$

for a beam traveling in the x direction. At all points on the wavefront the beam is in phase. As shown in Fig. 5, this can best be expressed in vector notation

$$\psi = A \exp (2\pi i\, \vec{k} \cdot \vec{r}) \tag{10}$$

where k is the incident wave vector of modulus $1/\lambda$ and $\vec{r}$ is a point on the wavefront.

We will use the Fresnel half-period zone construction to calculate the amplitude of the scattered wave at a point P due to the interaction of the incident wave $\vec{k}$ with a slice of material dx thick. The plane in Fig. 6a is divided into annular rings such that the perimeters are at distances x, $x + \lambda/2$, $x + \lambda$, $x + 3\lambda/2$, etc., from the point P. The contribution from each successive zone is one-half wavelength out of phase with the previous zone and is decreased in magnitude because the distance traveled is increased. Thus, the resultant amplitude-phase diagram takes the form of a spiral

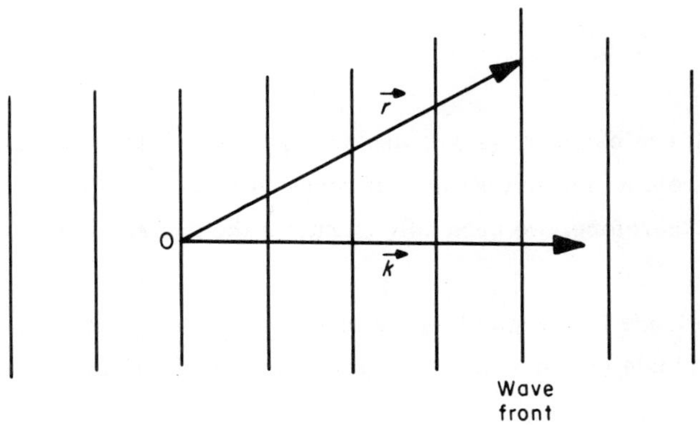

Fig. 5. Schematic illustration of a plane wave showing the wave vector $\vec{k}$ and a vector $\vec{r}$ to a point on the wave front.

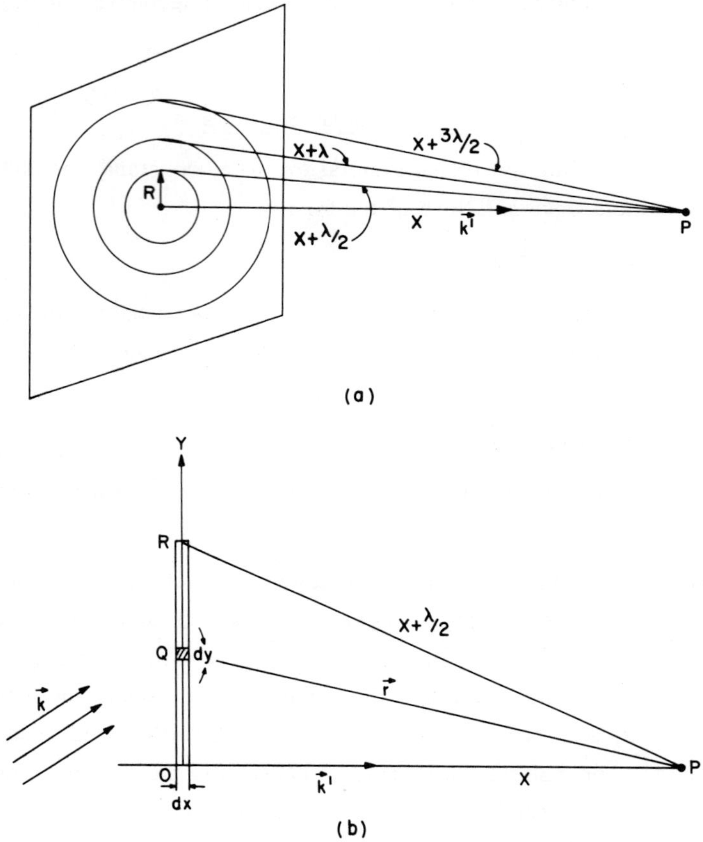

Fig. 6. (a) Fresnel zones in a plane; (b) cross section of first Fresnel zone.

(Fig. 7). It can be shown (e. g., Jenkins and White*) that the resultant amplitude from an infinite number of zones is just one-half that from the first zone, therefore, we need only calculate the contribution from the first zone.

The distance R in Fig. 6b is the radius of the first zone. The contribution of any single vector from the annular segment dy to the amplitude from

*F. A. Jenkins and A. E. White, Fundamentals of Optics, McGraw-Hill, New York, 1957.

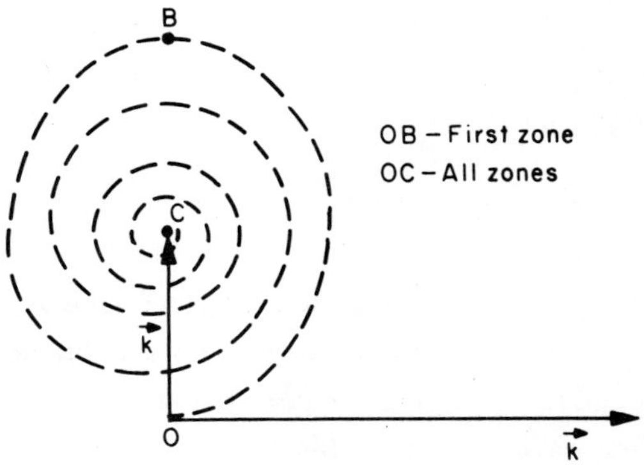

Fig. 7. Amplitude phase diagram for n Fresnel zones.

the first zone is equal to the projection of that vector onto OB, viz., its magnitude times the sine of the phase angle ϕ, where

$$\phi = \frac{2\pi\Delta}{\lambda} = \frac{2\pi(r-x)}{\lambda} \tag{11}$$

The length of each vector is proportional to the area of the respective annular segment and inversely proportional to the distance r (the intensity falls off as $1/r^2$, so the amplitude is proportional to $1/r$). Hence, for a plane wave ψ_o incident upon a volume of material dx thick and having a scattering power f_p per unit area,

$$dA = f_p \frac{2\pi y dy}{r} \sin \frac{2\pi(r-x)}{\lambda} dx\,\psi_o \tag{12}$$

For $y \ll x$, we assume $r = x$ in the first term, and in the sine term we take

$$r = (y^2 + x^2)^{1/2} = x(1 + y^2/x^2)^{1/2} \simeq x + y^2/2x$$

Thus,

$$A_{dx} = \int_{y=o}^{R} f_p \frac{2\pi y dy}{x} \sin \frac{\pi y^2}{\lambda x} dx\,\psi_o \tag{13}$$

With a change of variable

$$A_{dx} = f_p \lambda dx \psi_o \int_o^{(\lambda x)^{1/2}} \sin q\, dq = 2 f_p \lambda\, dx\, \psi_o \tag{14}$$

for the first zone. Writing this in wave form for the whole slice, including the 90° phase change due to scattering, the disturbance at P resulting from scattering from the plane is:

$$d\psi_s = i f_p \lambda\, dx\, \psi_o \exp(2\pi i \vec{k}' \cdot \vec{r}) \tag{15}$$

Each of the atoms in this volume element gives rise to a scattered wavelet. The amplitude of the scattered wave $d\psi_s$ is obtained by summing the amplitudes from each scattering center, but we must account for the phase of the incident wave ψ_o for angles $\theta \neq 0$ (i. e., $k' \neq k$). The amplitude of the scattered wave at a point P where $|r| >> |r_i|$ is (Fig. 8)

$$d\psi_s = i f_i(\theta) \lambda\, dx \exp(2\pi i \vec{k'} \cdot \vec{r}) \exp(-2\pi i \vec{K} \cdot \vec{r_i}) \tag{16}$$

where $\vec{K} = (\vec{k'} - \vec{k})$ and $f_i(\theta)$ is the scattering amplitude for the ith atom located at $\vec{r_i}$. Clearly, the phase factor ϕ is no more than a measure of the difference in path length for a wavelet scattered from the ith atom relative to one scattered from the origin. If we sum over the atoms in a unit cell

$$\psi_s = i \lambda \exp(2\pi i \vec{k'} \cdot \vec{r}) \sum_i f_i(\theta) \exp(-2\pi i \vec{K} \cdot \vec{r_i}) \tag{17}$$

where

$$F(\theta) = \sum_i f_i(\theta) \exp(-2\pi i \vec{K} \cdot \vec{r_i})$$

is the structure factor of the unit cell for scattering at an angle .

C. Extinction Distance

So far it has been assumed, as an inherent feature of the kinematical theory, that the amplitude of the incident wave is unaffected by the scattering process since ψ_s is small. It is instructive to calculate the crystal

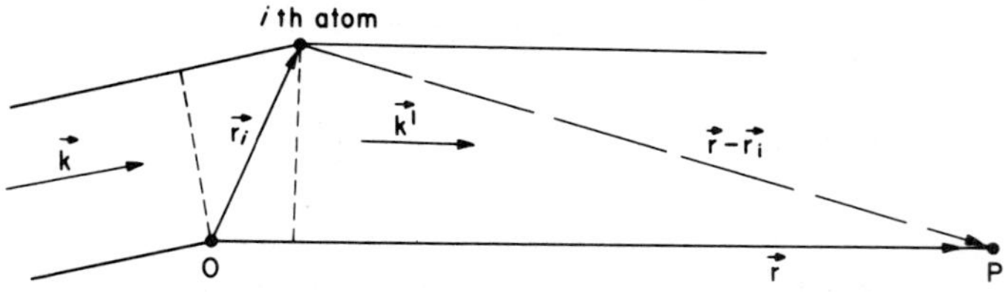

Fig. 8. The amplitude of the scattered wave at a point P.

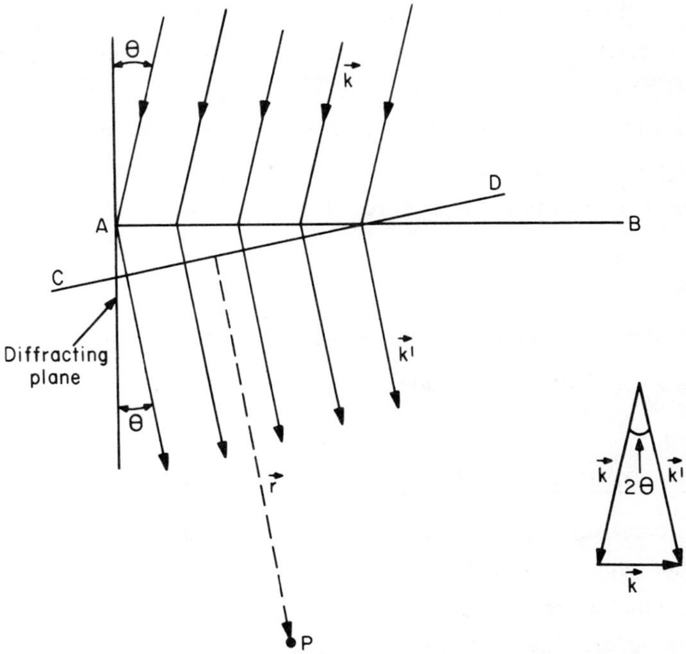

Fig. 9. A crystal oriented for Bragg reflection from a set of planes normal to AB.

thickness for which this assumption remains valid. Let the crystal in Fig. 9 be oriented for Bragg diffraction from the set of planes normal to AB. The scattered plane wave is in phase along CD. For unit amplitude incident at CD the scattered amplitude at P is, from Eq. (15):

$$d\psi_g = i\lambda((f_p dx) \exp(2\pi i \vec{k'} \cdot \vec{r}) \tag{18a}$$

where $(f_p dx)$ is the scattering amplitude per unit area of plane CD. However, the actual scatterers are the unit cells on AB and $f_p dx = nF_g/\cos\theta$; where F_g is the structure factor for the Bragg reflection $\vec{g}$, n is the number of unit cells per unit area, and θ is the scattering angle. Thus

$$d\psi_g = \frac{in\lambda F_g}{\cos\theta} \exp(2\pi i \vec{k'} \cdot \vec{r}) \tag{18b}$$

Substituting typical values in the pre-exponential factor one finds that for unit incident amplitude, the scattered amplitude approaches unity after the incident beam has traversed a mere 25 layers (contrasted with 10^4 layers for x rays). We must conclude that the kinematical theory of electron diffraction is strictly valid only for very thin crystals or when the orientation deviates significantly from the Bragg angle. At the Bragg orientation the amplitude scattered per layer is $n\lambda F_g / \cos\theta$. The extinction distance is defined as twice the distance occupied by the number of layers required to reduce the incident amplitude to zero. It follows that

$$\xi_g = \frac{\pi a \cos\theta}{n\lambda F_g} \tag{19}$$

where a is the interplanar spacing. Hence, Eq. (18b) becomes

$$d\psi_g = \frac{i\pi a}{\xi_g} \exp(2\pi i\, \vec{k'} \cdot \vec{r}) \tag{20}$$

D. Electron Diffraction Pattern and Reciprocal Space

It is clear from Eq. (17) that the scattered intensity is a maximum when $(\vec{K} \cdot \vec{r_i})$ is an integer. That is, constructive interference obtains when the path difference between scattering centers is an integral number of wavelengths. In order to get Bragg scattering from a row of atoms along the direction $\vec{r_1}$ (Fig. 10), we require that

$$OP - QR = a(\cos\alpha_2 - \cos\alpha_1) = h\lambda$$

or

$$(\vec{k'} - \vec{k}) \cdot \vec{r_1} = \vec{K} \cdot \vec{r_1} = h\lambda \tag{21a}$$

Similarly, for rows of atoms along $\vec{r_2}$ and $\vec{r_3}$

$$b(\cos\beta_2 - \cos\beta_1) = \vec{K} \cdot \vec{r_2} = k\lambda \tag{21b}$$

$$c(\cos\gamma_2 - \cos\gamma_1) = \vec{K} \cdot \vec{r_3} = 1\lambda \tag{21c}$$

These three relations are known as the Laue equations. The criterion for the formation of a diffraction spot from a three-dimensional array of atoms rows (i.e., a crystal) is that all three be satisfied simultaneously.

If one squares the equations in the trigonometric form and sums them, assuming the axes are orthonormal, one obtains

$$\frac{4a^2 \sin^2\theta}{h^2 + k^2 + 1^2} = \lambda^2 \tag{22}$$

where 2θ is the angle between the incident and diffracted beams and h, k, 1

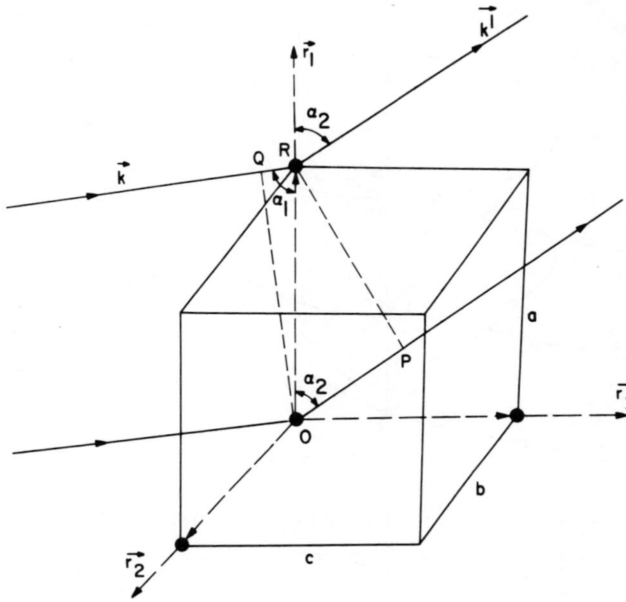

Fig. 10. Conditions for Bragg scattering from a row of atoms along direction r.

are integers which have no physical meaning as yet. For a cubic lattice the interplanar spacing is

$$d^2\,(hkl) = \frac{a_o^2}{h^2 + k^2 + l^2} \tag{23}$$

where a_o is the lattice parameter. Again h, k, l are integers, but they have meaning in terms of the Miller notation for crystal planes. Substituting into Eq. (22), we have

$$4d^2\,(hkl)\,\sin^2\theta = \lambda^2$$

or

$$2d\,\sin\,\theta = \lambda \tag{24}$$

which is the well-known Bragg law. What we are saying in a physical sense is that the integers in Eqs. (21) and (22) define a set of planes in real space, such that the Laue conditions are simultaneously met along the three orthogonal directions in the crystal.

How do we extend this to explain the diffraction pattern and the concept of the reciprocal lattice? If we rotate the crystal in Fig. 10 through all

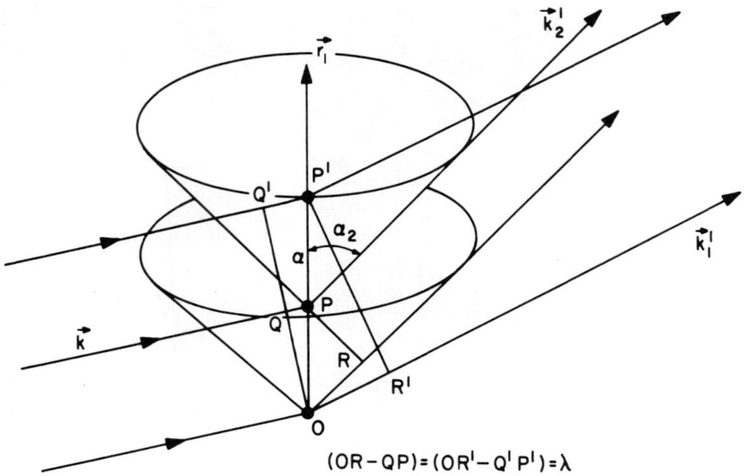

Fig. 11. A family of diffraction cones caused by Bragg scattering from a row of atoms having a separation distance a.

possible orientations in space, we would find Eq. (21a) satisfied only for those orientations where the scattered beams $\vec{k}'$ defined a cone with an apex angle $2\alpha_2$. For an infinite row of atoms along $\vec{r}_1$ we would obtain a family of cones, one for atoms a apart, one for atoms 2a apart, etc. Figure 11 is a schematic illustration of this point. Orthogonal rows of atoms comprising an infinite crystal would give rise to orthogonal families of cones. The intersection of any three cones defines a direction $\vec{k}'$ which satisfies the Laue equations (Fig. 12 illustrates this for two families of cones). Since $\vec{k}$ is fixed by the experimental conditions, $\vec{K} = \vec{k}' - \vec{k}$ is also determined. From Figs. 9 and 10 it is clear that $\vec{K}$ is normal to the set of diffracting planes (hkl). Further, $\vec{K}$ is equal to the reciprocal of the interplanar spacing, since

$$\vec{K} = \frac{2 \sin \theta}{\lambda} = \frac{1}{d\,(hkl)} \tag{25}$$

For Bragg scattering we define $\vec{K} = \vec{g}\,(hkl)$, a vector in reciprocal space. The so-called reciprocal lattice is the point lattice defined by the termini of all possible vectors $\vec{g}\,(hkl)$. The direction of the vector is normal to the (hkl) planes in real space and the length is equal to the reciprocal of the interplanar spacing. Figure 13 demonstrates this construction for several planes in a simple cubic cell.

The Ewald sphere construction for determining which beams will be visible in any one experiment is discussed at length elsewhere in this book

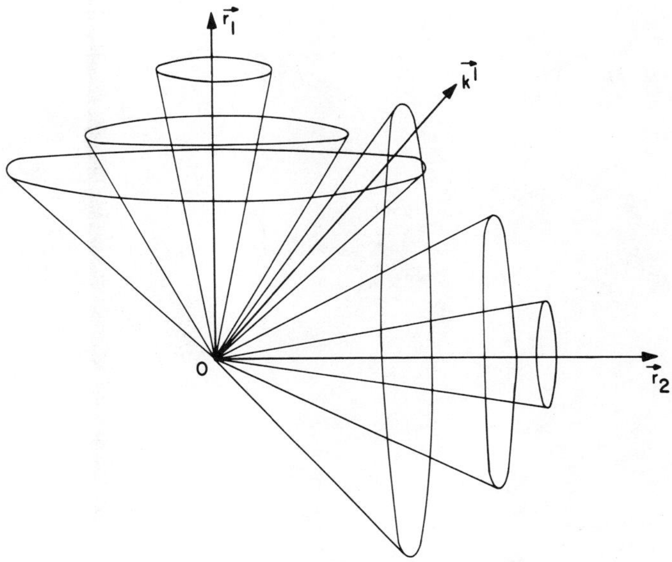

Fig. 12. Intersection of two families of cones.

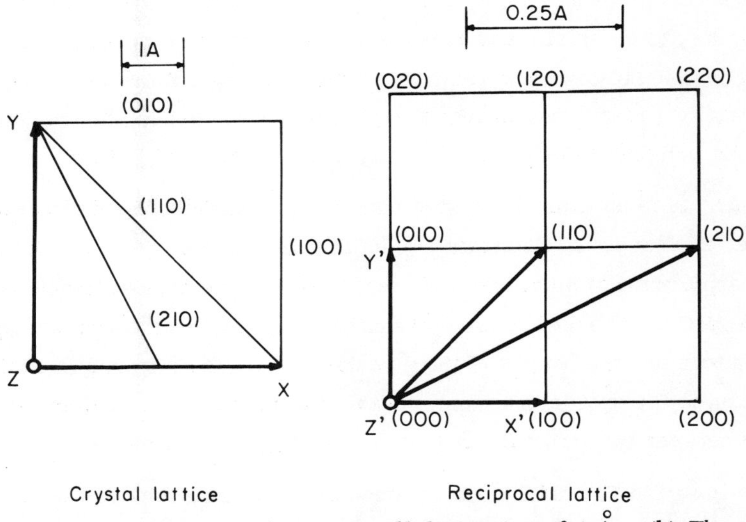

Fig. 13. (a) A crystal having a cell dimension of 4 Å. (b) The reciprocal lattice of the crystal in (a). To those familiar with the Miller system of indexing, the use of parentheses to label the reciprocal lattice points may seem strange, particularly since these points are at the terminus of vectors. However, parenthesis signify planes regardless of whether they are planes as in real space, or points (i. e., vectors) in reciprocal space.

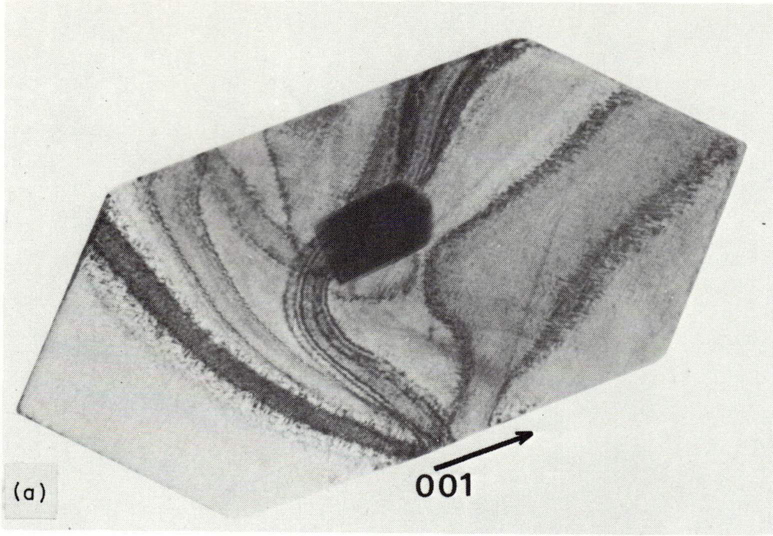

Fig. 14. (a) Transmission micrograph of a MoO_3 crystal (20 KX).

and will not be presented here. Suffice it to say that because the electron wavelengths are so small, and because the Laue condition is relaxed in the thin dimension (analogous to particle size broadening), the Ewald sphere is essentially a planar cut through reciprocal space. Hence, the resulting diffraction pattern exhibits the symmetry of the reciprocal lattice.

Figure 14 is an example of what can be done with electron diffraction patterns, and serves to illustrate a point that is very important to the successful application of diffraction contrast theory. The crystal is orthorhombic MoO_3. The diffraction spots can be indexed from the symmetry of the pattern and the lengths of the $\vec{g}$ vectors (i.e., the distance from the main beam to the spot in question). In this instance there is a clear relationship between the structure and the morphology of the crystal.

The image in Fig. 14 (c) was obtained by increasing the diffraction lens current above that required for the diffraction pattern. In this defocused image we can see regions of the crystal where various sets of planes are at

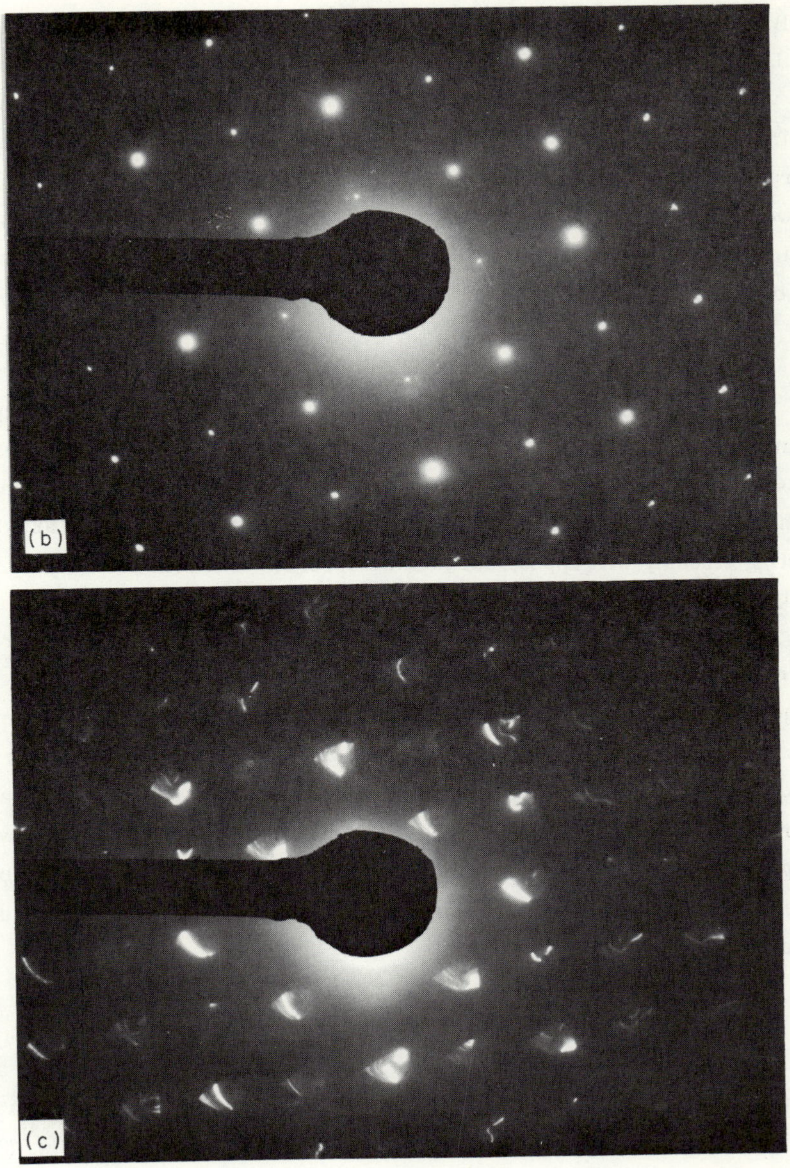

Fig. 14. (b) Electron diffraction pattern of the crystal oriented to correspond to the orientation of the image. (c) Defocused diffraction pattern showing which regions of the crystal contribute to each of the diffracted beams.

the Bragg orientation and are, therefore, scattering strongly. In other words, the diffraction vector $\vec{K}$ in that region is equal to the reciprocal lattice vector $\vec{g}$ for the (hkl) plane defined by that diffraction spot. Knowing the indices of the diffraction spots we can deduce the set of planes that are diffracting strongly in that region of the crystal. Regions exhibiting contrast in two of the enlarged diffraction spots do so because both sets of planes are sufficiently close to the Bragg orientation. These multiple dark-field images can be extremely useful in defect analyses.

E. Column Approximation

The normal procedure for viewing a thin crystal foil (~1000 A thick) is to intercept all diffracted beams with an aperture, allowing only the transmitted beam $\vec{k}$ to pass through the lens system to be magnified and projected on the fluorescent screen. Thus, any regions of the specimen that are diffracting strongly will appear dark on the screen. (The foil is rarely perfectly flat, so local undulations may cause one region to be rotated into a Bragg orientation while adjacent regions are not. The resulting dark areas are called bend contours.) Defects or precipitates causing a displacement of the lattice planes will affect the diffraction conditions locally, and thereby give rise to contrast in the image. It is also possible to view the foil in the dark-field mode, where all but a single diffracted beam $\vec{k'} = \vec{k} + \vec{g}$(hkl) are intercepted by the aperture. The kinematical theory predicts that the two images should be complementary. Indeed, all of our calculations have been in terms of the scattered amplitude; the transmitted amplitude ψ is assumed to be given by

$$\psi = 1 - \psi_g$$

To calculate the amplitude ψ_g, we assume that the wave emitted at any point on the bottom surface of the crystal is due only to diffraction from within a narrow column of material above the point. According to Eq. (20), the disturbance at P due to scattering at A in Fig. 15 is

$$d\psi_g = \frac{i\pi a}{\xi_g} \exp\left[2\pi i \vec{k'} \cdot (\vec{r} - \vec{r}_n)\right] \exp(2\pi i \vec{k} \cdot \vec{r}_n) \tag{26}$$

where the first exponential is the phase of the scattered wave at P and the second is the phase of the incident wave at A. It follows that

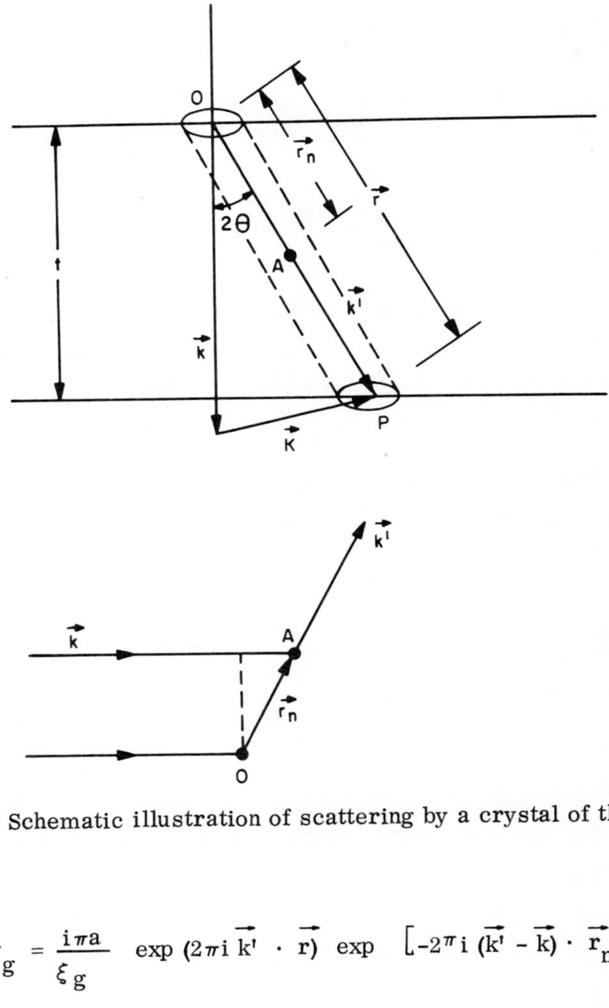

Fig. 15. Schematic illustration of scattering by a crystal of thickness t.

$$d\psi_g = \frac{i\pi a}{\xi_g} \exp(2\pi i\, \vec{k}' \cdot \vec{r}) \exp\left[-2\pi i\, (\vec{k}' - \vec{k}) \cdot \vec{r}_n\right]$$

$$= \frac{i\pi a}{\xi_g} \exp(2\pi i \vec{k}' \cdot \vec{r}) \exp(-2\pi i \vec{K} \cdot \vec{r}_n) \tag{27}$$

The phase change in the first exponential is constant for all columns and will be neglected in this treatment.

For a perfect crystal oriented for Bragg diffraction $\vec{K} = \vec{g}$, a reciprocal lattice vector, and we have

$$d\psi_g = \frac{i\pi a}{\xi_g} \exp(-2\pi i\, \vec{g} \cdot \vec{r}_n) \tag{28}$$

where $\vec{r}_n$ is a real lattice vector defining the position of the nth unit cell. Now:

$$\vec{g} \cdot \vec{r}_n = (ha^* + kb^* + lc^*) \cdot (u\hat{a} + v\hat{b} + w\hat{c}) = n \qquad n = 0, 1, \ldots$$

Therefore,

$$\exp(-2\pi i\, \vec{g} \cdot \vec{r}_n) = 1$$

and the amplitude of the scattered wave has its maximum value. This is equivalent to the statement that the Ewald sphere intersects a reciprocal lattice node when the diffraction vector $\vec{K}$ is a reciprocal lattice vector.

We next inquire as to the amplitude of a wave scattered in directions deviating slightly from a Bragg direction. This is equivalent to stating that the crystal is rotated so that the Ewald sphere passes near, but not through, the reciprocal lattice point. Hence, $\vec{K} = \vec{g} + \vec{s}$ and

$$d\psi_g = \frac{i\pi a}{\xi_g} \exp\left[-2\pi i(\vec{g} + \vec{s}) \cdot \vec{r}_n\right]$$

$$= \frac{i\pi a}{\xi_g} \exp(-2\pi i\, \vec{g} \cdot \vec{r}_n) \exp(-2\pi i\, \vec{s} \cdot \vec{r}_n) \qquad (29)$$

For small scattering angles $\vec{r}_n$ is nearly parallel to z, the foil normal, and $\vec{s}$ is small and parallel to z. The first exponential is again unity and, therefore the scattering per plane for planes spaced a apart is

$$d\psi_g = \frac{i\pi a}{\xi_g} \exp(-2\pi i\, sz) \qquad (30)$$

A thickness dz contains dz/a planes and scatters and amplitude

$$d\psi_g = \frac{i\pi}{\xi_g} \exp(-2\pi i\, sz)\, dz$$

The total amplitude scattered at P from a thickness t is

$$\psi_g = \frac{i\pi}{\xi_g} \int_0^t \exp(-2\pi i\, sz)\, dz$$

$$= \frac{i\pi}{\xi_g} \frac{\sin \pi ts}{\pi s} \exp(-\pi i\, st) \qquad (31)$$

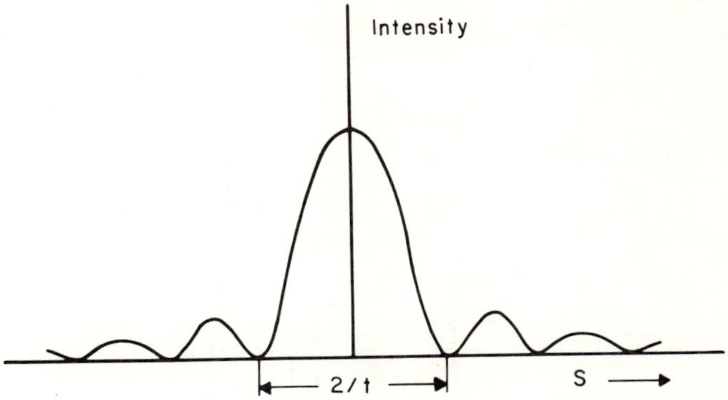

Fig. 16. A qualitative plot of diffracted intensity versus distance from the reciprocal lattice point. The lattice point is located at $\vec{s} = 0$; t is the thickness of the crystal. Note that the extent of the streaking varies inversely as the foil thickness.

Since we measure and see intensity rather than amplitude, we have

$$I = \psi_g^* \psi_g = \frac{\sin^2 \pi ts}{(\xi_g s)^2} \tag{32}$$

This function is plotted in Fig. 16.

Several pieces of information can be extracted from this equation. First, the intensity of the scattered radiation varies sinisoidally with foil thickness. Thus, a wedge-shaped foil will exhibit alternating dark-light fringes (Fig. 17). Second, the depth periodicity of these of these oscillations is 1/s and approaches a limiting value at a Bragg orientation, i. e.,

$$\lim_{s \to o} \frac{1}{s} = \xi_g$$

Third, Fig. 16 represents the intensity distribution in the thickness direction of the reciprocal lattice nodes.

Our ultimate aim is to derive an analytical expression for the scattered amplitude from a crystal containing defects. Let $\vec{K} = \vec{g} + \vec{s}$ and $\vec{r}'_n = \vec{r}_n + \vec{R}$, where $\vec{r}_n$ is the position of the nth unit cell in the perfect crystal and $\vec{R}$

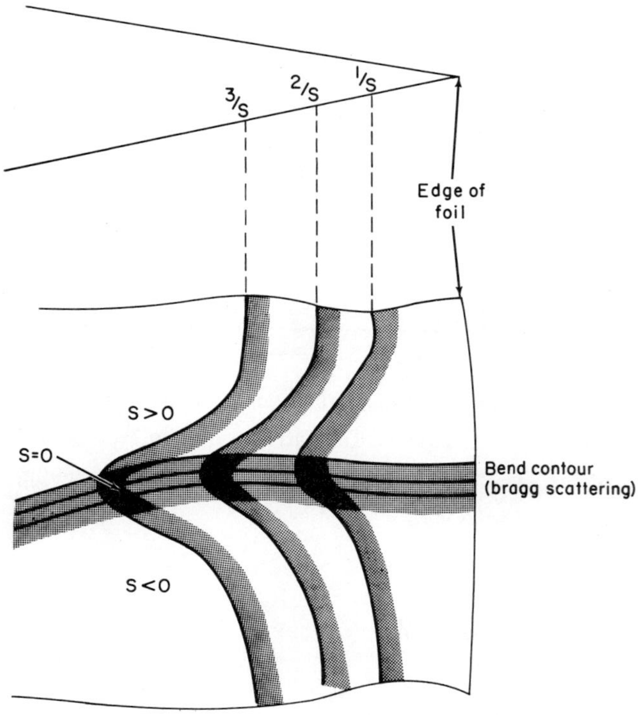

Fig. 17. Thickness fringes from a wedge-shaped foil.

describes the displacement of the cell due to the presence of the defect. Then, from Eq. (28),

$$\begin{aligned} d\Psi_g &= \frac{i\pi a}{\xi_g} \exp(-2\pi i\, \vec{K} \cdot \vec{r}_n') \\ &= \frac{i\pi a}{\xi_g} \exp\left[-2\pi i\, (\vec{g} + \vec{s}) \cdot (\vec{r}_n + \vec{R})\right] \end{aligned} \tag{33}$$

Because $\vec{g} \cdot \vec{r}_n$ is an integer and $\vec{s} \cdot \vec{R}$ is negligible,

$$d\Psi_g = \frac{i\pi a}{\xi_g} \exp(-2\pi i\, \vec{g} \cdot \vec{R}) \exp(-2\pi i\, \vec{s} \cdot \vec{r}_n)$$

and

$$\Psi_g = \frac{i\pi}{\xi_g} \int_0^t \exp(-2\pi i\, \vec{g} \cdot \vec{R}) \exp(-2\pi i\, sz)\, dz \tag{34}$$

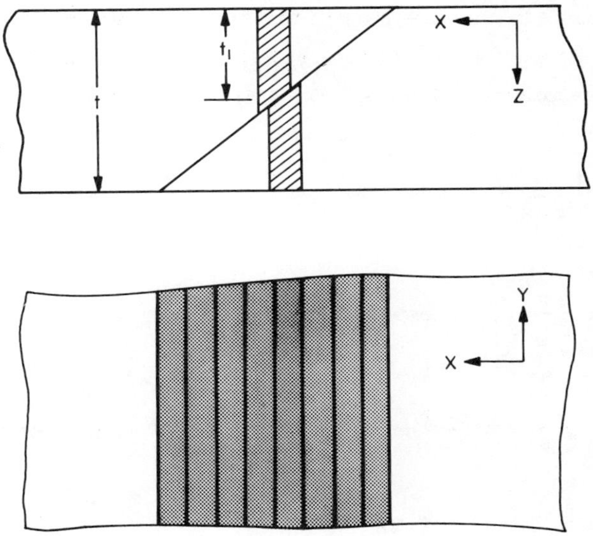

Fig. 18. Schematic illustration of stacking fault fringes.

The first exponential contains an additional phase change $\alpha = 2\pi \vec{g} \cdot \vec{R}$ introduced by the defect. This term simply modulates the scattering amplitude from the perfect crystal as given by the second exponential. This equation is the basis for the calculation of scattering in the presence of defects of any kind.

F. Defect Analysis

Let us now apply Eq. (34) to ascertain the contrast at some simple defects. Consider a stacking fault inclined to the surface and extending through the foil, as illustrated in Fig. 18. We would naturally expect some kind of contrast, since the defect displaces potential diffracting planes, relative to one another, above and below the fault plane. For an fcc material one can conceive of two means by which this fault could be generated: collapse of a vacancy cluster to form a Frank loop (R = 1/3 111), and splitting of a dislocation into partials ($\vec{R}$ = 1/6 [112]). We will treat the former. The phase factor is

$$\alpha = 2\pi \vec{g} \cdot \vec{R} = 2\pi (ha^* + kb^* + 1c^*) \cdot \frac{1}{3} (\hat{a} + \hat{b} + \hat{c})$$

$$\alpha = \frac{2\pi}{3} (h+k+1) = \pm \frac{2\pi n}{3}, \quad n = 0, 1, 2, \ldots$$

(The reader ought to convince himself that the phase factor is identical for the fault between partial dislocations.) Then, since the fault lies at a depth t_1 for the column of interest, Eq. (34) can be written as the sum of two integrals:

$$\psi_g = \frac{i\pi}{\xi g}\int_o^{t_1} \exp(-2\pi i\ sz)\,dz$$

$$+\frac{i\pi}{\xi_g}\int_{t_1}^{t} \exp(-2\pi i\ sz)\exp(-i\alpha)\,dz$$

$$\psi_g = \frac{i}{\xi_g s}\exp\left(\frac{-i\alpha}{2}\right)\exp(-\pi ist)\left\{\sin\left(\pi ts+\frac{\alpha}{2}\right) -\sin\frac{\alpha}{2}\exp\left[2\pi is\left(\frac{t}{2}-t_1\right)\right]\right.$$

and the intensity $I = \psi_g^* \psi_g$ is

$$I = \frac{1}{(\xi_g s)^2}\left[\sin^2\left(\pi ts+\frac{\alpha}{2}\right)+\sin^2\frac{\alpha}{2} -2\sin\frac{\alpha}{2}\sin\left(\pi ts+\frac{\alpha}{2}\right)\cos\pi sz\right] \tag{35}$$

where $z = t/2-t_1$ is the distance from the center of the foil to the fault plane. The contrast is in the form of fringes parallel to the line of intersection with the surface and is symmetrical with the center of the foil (from the cos $2\pi sz$ term). It also follows that the depth periodicity for a constant t is 1/s, reaching a maximum spacing equal to the extinction distance ξ_g in a region oriented for Bragg diffraction. The fault between partials would be identical in appearance, but the contrast at the dislocations surrounding the faults would necessarily be different.

Dislocation contrast is computed in the same manner, although the atomic displacements are more complex. We will treat the simpler screw dislocation for illustrative purposes. The displacement function is

$$\vec{R} = \vec{b}\frac{\phi}{2\pi} = \frac{b}{2\pi}\tan^{-1}\left(\frac{z-y}{x}\right)$$

where b is the Burgers vector; the other parameters are as illustrated in Fig. 19. The phase factor is

$$\alpha = 2\pi\vec{g}\cdot\vec{R} = \vec{g}\cdot\vec{b}\tan^{-1}\left(\frac{z-y}{x}\right)$$

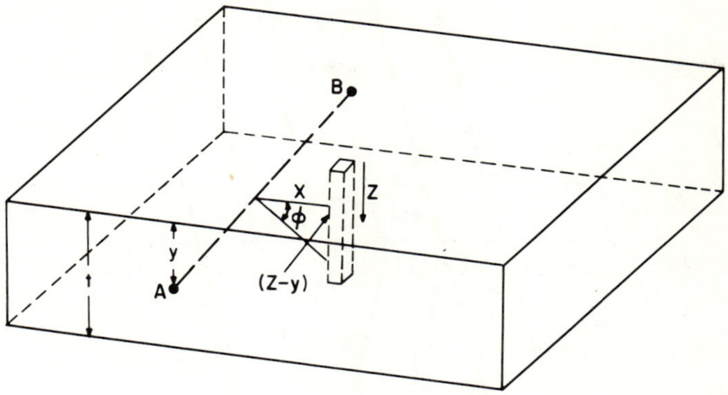

Fig. 19. Calculation of dislocation contrast by a screw dislocation.

Since $\vec{b}$ is a real lattice vector for total dislocations, $\vec{g} \cdot \vec{b}$ can be integral or zero. In general,

$$\psi_g = \frac{i\pi}{\xi_g} \int_0^t \exp\left[-in \tan^{-1}\left(\frac{z-y}{x}\right)\right] \exp(-2\pi i\, sz)\, dz \tag{36}$$

Now, since $\vec{R}$ is a continuous function of position, the additional phase angle smoothly subtracts from $2\pi sz$ for columns on one side of the dislocation and smoothly adds to it on the other (i. e., as the parameter x changes sign). Consequently, we observe strong contrast on one side of the dislocation and essentially perfect crystal contrast on the other. This behavior can be understood on an intuitive basis using the pure edge dislocation pictured in Fig. 20. Let us assume that the planes outside the region $E_1 E_2$ would satisfy the Bragg condition for diffraction if the foil were rotated slightly counterclockwise about the normal to the paper. The presence of the dislocation at E effects this rotation for the region E_1, but the reverse holds for the region E_2. Hence, the image of the defect appears to the side of the true position of the dislocation line. Note, however, that if the dislocation crosses a bend contour, $\vec{s}$ changes sign and the image will flip to the opposite side of the dislocation line.

A special case obtains where $\vec{g} \cdot \vec{b} = 0$. Analytically, $\alpha = 0$ and $\exp(-i\alpha) = 1$. Thus

$$\psi_g = \frac{i\pi}{\xi_g} \int_0^t \exp(-2\pi i\, sz)\, dz \tag{37}$$

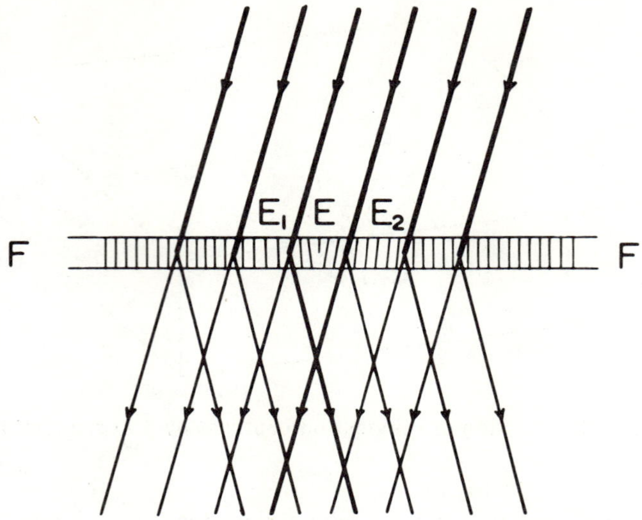

Fig. 20. Diffraction of electrons in a foil containing an edge dislocation at E. In E_2, less intensity is diffracted away than in the perfect part while more is diffracted away in E_1. (After Amelinckx, 1964.)

which is identical to Eq. (32), the amplitude scattered by a perfect crystal. We conclude that if the displacement vector lies wholly within the diffracting planes, it cannot induce a distortion normal to the planes, and the defect must be invisible. This is, in fact, a powerful tool for determining the Burgers vector of dislocations and the displacement function of defects in general. By finding two sets of diffracting planes for which there is no contrast (i. e., $\vec{g} \cdot \vec{b} = 0$), we know that the Burgers vector must lie along the line of intersection of these two planes.

It is worth mentioning that it is quite possible for the contrast in a bright field image to be due to scattering from more than one set of planes. One must take care to tilt the foil so that one $\vec{g}$ vector operates much more strongly than the others, in order to make meaningful $\vec{g} \cdot \vec{b}$ analyses. An alternative is to employ dark-field techniques where one can be certain that the contrast is arising from the diffracted beam alone.

Provided one knows the displacement function of a defect it is possible, in principle, to determine the contrast behavior expected. However, the kinematical theory can only qualitatively account for some of the contrast effects; to obtain truly quantitative information, one must resort to dynamical theory. The latter allows for the fact that ψ_g may not be negligible compared to ψ_o, and that the diffracted beam may undergo secondary scattering events. For most purposes, though, the kinematical theory is found to be quite satisfactory, provided one is aware of the shortcomings.

IV. FIELD-ION MICROSCOPY

Field-ion microscopy has yet to enjoy the popularity of other metallographic tools, primarily because of present limitations on the types of materials that can be studied. However, continuing advances in experimental technique have considerably extended the range of materials suitable for this kind of analysis, and it is probable that ceramic materials will be routinely imaged in the not too distant future (see Fig. 21). It is for this

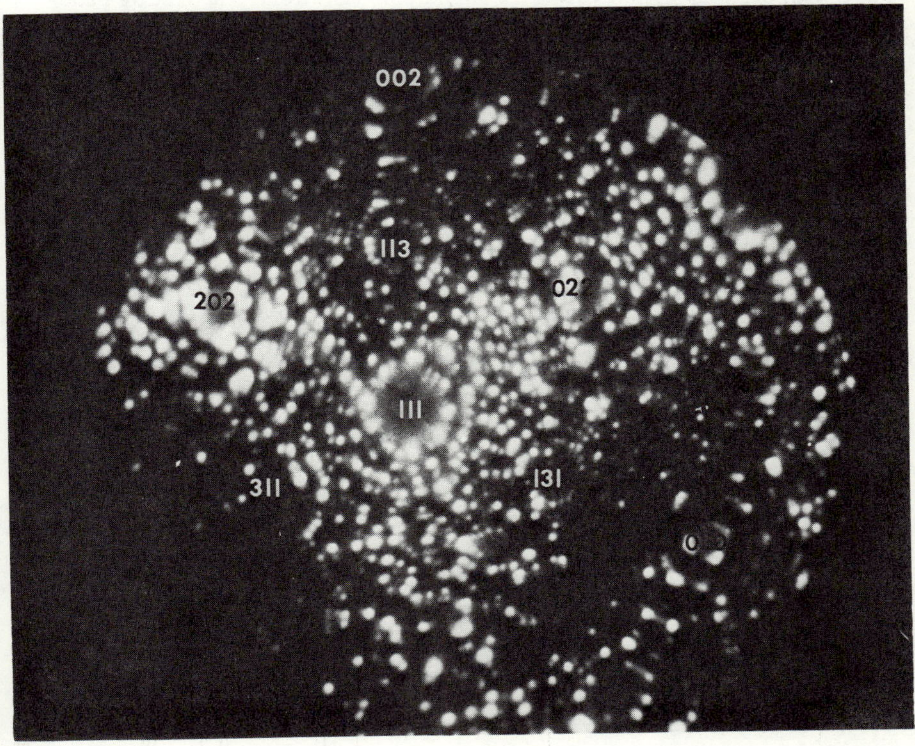

Fig. 21. Helium field-ion image of tantalum carbide ($TaC_{0.73}$) taken at 78° K. The crystallographic planes are appropriately indexed. (Courtesy N. V. Durai Raghavan.)

reason that the author felt it worthwhile to include a discussion of the fundamentals of the technique and of applications to structure analysis.

A. Basic Principles

The field-ion microscope has resolving power and magnification capabilities enabling the direct observation of the atomic lattice. In contrast to electron microscopy and x-ray diffraction, FIM does not depend upon an interference phenomenon to form an "image" of the internal structure of a crystal; instead, it utilizes an inert gas as a vehicle to transmit information about the specimen surface to a phosphor screen. However, only 5 to 10% of the atoms are visible at any one time. To a first approximation, those atoms having fewer than half their bulk number of nearest neighbors can be considered as imaging surface atoms. These surface atoms can be removed in a controlled fashion to reveal a new set of surface atoms by simply raising the applied voltage. Employing this field evaporation process, the specimen can be microdissected, such that the structure of the bulk can be deduced in atomic detail.

The current generation of field ion microscopes is quite complex in design, but all have the same basic components. The specimen is a short fine wire polished to produce a sharp, nearly hemispherical tip several hundreds of angstroms in radius. The specimen is thermally shorted to a cryogenic fluid and connected to a stable, dc power supply. Facing the tip a few centimeters away is a flat phosphor screen, which is generally an integral part of the vacuum chamber enclosing the specimen as illustrated in Fig. 22. The system is evacuated to at least 10^{-7} Torr, backfilled with high purity helium gas to a pressure of about 1μ, and a positive potential of up to, say, 30 kV is impressed upon the tip relative to the screen. The necessary voltage is primarily a function of the tip radius. The strength of the resulting field at the specimen surface is on the order of 5 V/Å, which is sufficient to cause field ionization of the imaging gas.

A neutral atom of the helium imaging gas experiencing the field around the tip becomes an oriented dipole and accelerates toward the specimen. When the polarized gas atom is in the vicinity of the tip, the potential well is severely distorted by the tremendous electric field at the surface (Fig. 23). Thus, the atomic electron no longer sees the symmetric potential

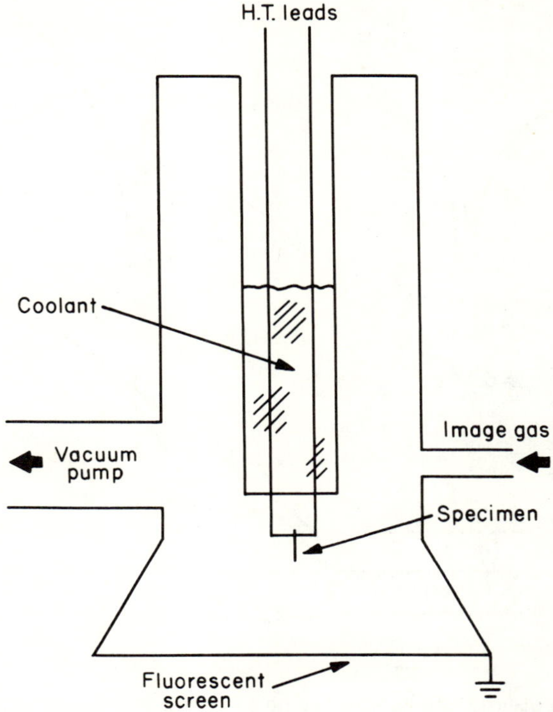

Fig. 22. Schematic drawing of a field ion microscope.

well of the helium nucleus as in free space but, rather, an asymmetric barrier that reaches a maximum and has a finite width on the side nearest the specimen. According to wave mechanics, ionization is possible by tunneling of the atomic electron through the barrier into the metal tip, provided there is an available energy level. At distances greater than some critical distance x_c, the ionization probability falls off rapidly because of the increasing width of the barrier; at lesser distances the ionization probability is negligible due to the absence of available energy levels. The ionization zone, then, can be envisioned as a thin disc (on the order of 0.1 Å thick) positioned approximately 3-4 Å above each surface atom. Since the transition rate is dependent upon the time the atom spends in the zone, the helium atom is thought to hop over the surface, losing energy with each impact. As the velocity is reduced the atom spends increasing lengths of time in the ionization zone. Calculations indicate that the probability of an ionization event is maximized after approximately one hundred

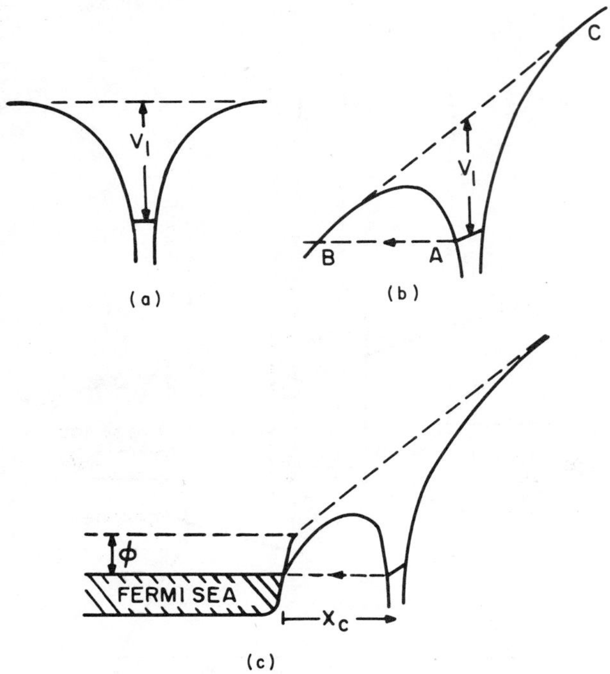

Fig. 23. Potential diagrams for an electron in field ionization. (a) the potential due to the parent ion in zero field; (b) in an applied field; (c) in an applied field near the surface of a metal of work function ϕ .

such hops. (This process is schematically illustrated in Fig. 24. The resulting helium ions are then repelled by the positive field at the tip and travel to the screen along a trajectory defined by the electrostatic lines of force. Under imaging conditions, the flux of the beam from a single surface atom is about 10^5 ions/sec. The field ion image is, therefore, a sensitive contour plot of the surface field distribution. Bright spots on the screen, corresponding to regions of high local field fluctuations (i. e., surface atom positions), are arranged in a very regular pattern which can be interpreted directly in terms of the atomic lattice.

This then is a simplified explanation of how the field ion microscope is capable of producing a radial projection image of a hemispherical surface with a magnification of some one million diameters. For further discussion of the theory of image formation, the reader is referred to the suggested references listed at the end of the chapter.

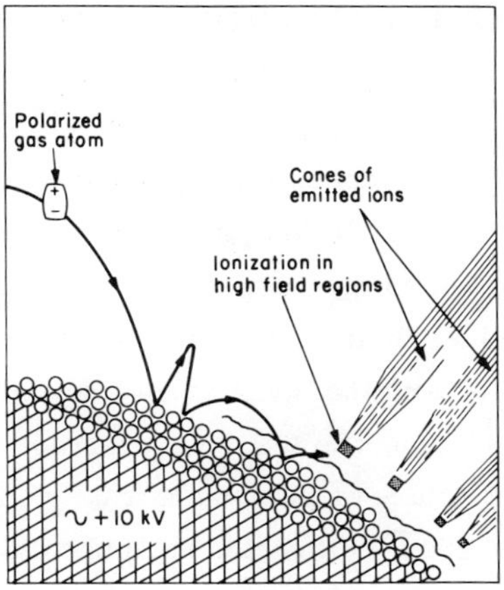

Fig. 24. Detailed schematic of the ionization process illustrating the relationship between the atom positions and the cones of emitted ions.

B. Defect Analysis

It seems appropriate to briefly discuss the methods of defect analysis with this micrscope. It would seem initially that point defect analysis would be reasonably straightforward using FIM. After all, the microscope affords atomic resolution, so one ought to be able to conclude that a missing image point infers a missing atom. Unfortunately, this is not necessarily the case. It is rare that two atom species in a single specimen will image with the same intensity; therefore, that which appears to be a vacancy may simply be an impurity atom. Without question, there will be instances when a missing image point is due to a vacant lattice site, but even then one cannot be sure that it is an equilibrium vacancy since it is well known that artifact vacancies can be created on the surface as a result of the imaging conditions (i. e., field stress, gas impact, etc.). * The difficulties are similar for the interpretation of interstitials regardless of whether they are self- or impurity interstitials. Consequently, it is desirable to limit oneself to the study of point defect clusters, as the data are statistically more reliable.

Observation and interpretation of images of dislocated crystals has been very successful. Dislocations were recognized as such during the early

*See the papers by D. G. Brandon and E. W. Mueller in Ref. 17 .

years of FIM, but quantitative interpretations were not possible unitl 1966, when Ranganathan developed a geometrical analysis which was subsequently verified with computer simulation (Sanwald et al.)** Simply stated, the theory predicts that the emergence of a dislocation will displace the normally concentric rings of a particular pole on a field ion image to that of a spiral configuration, provided the Burgers vector of the dislocation is nonparallel to the set of planes defining the pole. The complexity of the image distortion is a function of both the magnitude of the Burgers vector and the point of emergence with respect to the crystallography of the specimen. Furthermore, the exact nature of this configuration will be given by the scalar product $\vec{g} \cdot \vec{b}$, where $\vec{g}$ is a vector normal to the set of planes disturbed by the presence of the dislocation and $\vec{b}$ is the Burgers vector of that dislocation. Since $\vec{g}$ and $\vec{b}$ are geometrical in nature, the scalar product is simply a measure of the displacement component in the direction normal to the set of planes denoted by $\vec{g}$. For a total dislocation the two parameters are lattice vectors, the dot product of which is always an integer that determines the number of leaves in the spiral array. Although difficult to picture in the mind's eye, this holds true regardless of the character of the dislocation (i.e., edge, screw, or mixed). The sense of the Burgers vector determines the direction of rotation of the spiral for a given pole. Figure 25 clearly illustrates this behavior.

Having this information at our disposal, we are ready to attempt an analysis of defect structures comprised of one or more dislocations. A dislocation loop is defined by a single Burgers vector (it is implicit in this definition that the reference coordinate system move around the perimeter of the closed loop). If the loop should be intersected by a free surface it would be difficult to distinguish it from a dislocation dipole (i.e., two individual dislocations having opposite Burgers vectors). To interpret such a defect in terms of the $\vec{g} \cdot \vec{b}$ criterion, it need only be assumed that the loop is large enough to render as negligible the strain field of the loop on the two points of emergence of the defect. We may then treat the loop as two separate dislocations for which the $\vec{g} \cdot \vec{b}$ criterion predicts spirals of the opposite sense.

**See the paper by J. J. Hren in Ref. 17 .

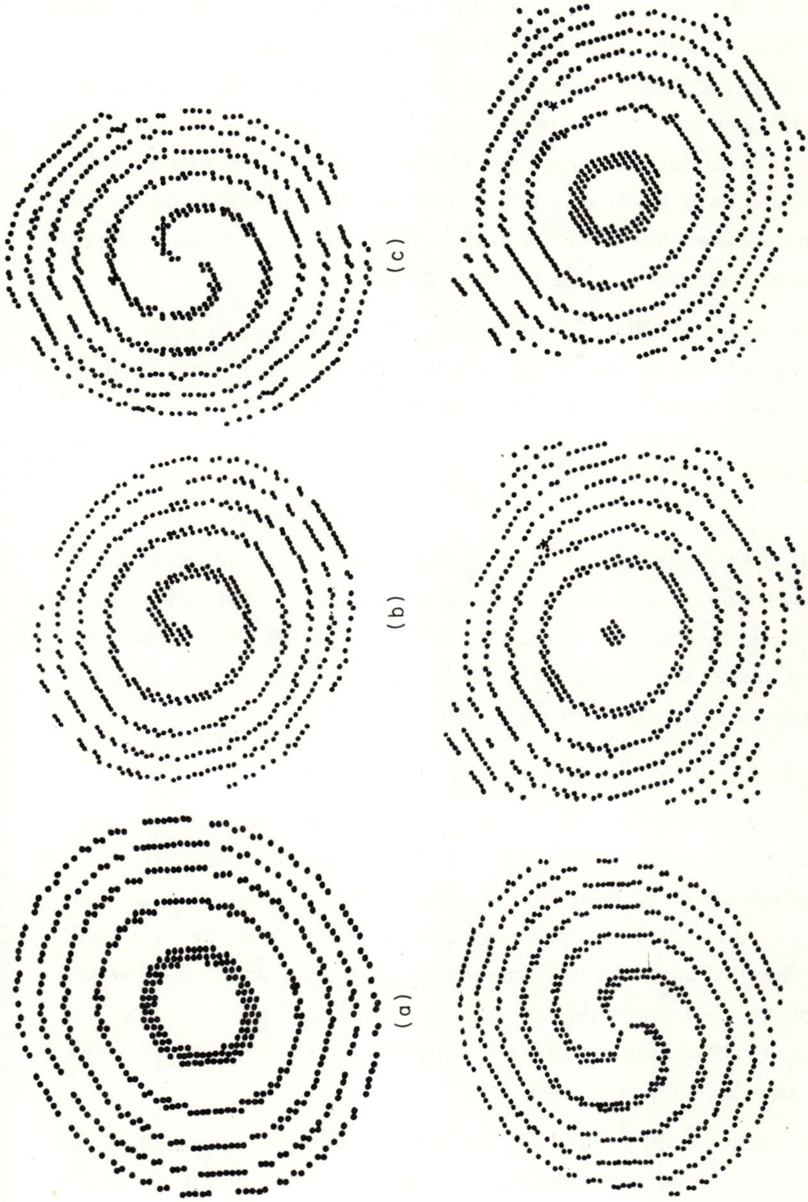

Fig. 25. Computer simulation of the effects of emerging dislocations on field-ion micrographs.

Faulted loops present a problem since the line bounding the fault is a partial dislocation by definition. This in turn means that $\vec{g} \cdot \vec{b}$ is not necessarily an integer, and a fraction of a spiral is both difficult to detect experimentally and to understand geometrically. Partial dislocations have been found by computer simulation to cause various types of disturbances depending upon the location of the defect. For instance, net plane facets appear to be split or sheared, while a pinching-type distortion occurs in the regions between the poles. However, the $\vec{g} \cdot \vec{b}$ criterion does not seem to yield as much information as would be desired.

An experimental difficulty which will undoubtedly have an effect on defect interpretation is the stress generated by the applied electric field. Mueller pointed out as early as 1958 that the normal stress is given by:

$$\sigma_n = F^2/8\pi = 44.3F^2$$

where F is the field in volts/Å and σ_N is in kg/mm^2. A spherical specimen of an isotropic material will experience a uniform dilation and the shear stresses would therefore be zero. For a hemispherical cap attached to a cylindrical shank, the tensile stress in the shank will be approximately σ_N, so that the maximum shear stress τ will be:

$$\tau = \sigma_n/2$$

These shear stresses are large enough to cause plastic deformation in the bulk material and might have an effect on glissile dislocation configurations in the FIM specimen. Add to this the fact that the tip is small enough so that dislocation image forces are appreciable, the tip is never truly hemispherical, and that no material is completely isotropic, and it is clear that the sum of these effects could have a complex bearing on the final defect distribution. These factors must be kept in mind when analyzing the defect structure of a field-ion specimen.

In the realm of planar defects, one might reasonably include the interface between a second phase particle and the matrix, but relatively little has been accomplished in field-ion studies of this type because of inherent experimental difficulties. On the other hand, FIM has been extremely useful in characterizing the grain boundary. The dislocation structure of the low-angle boundary has been further substantiated, and many investigators have shown that the width of a high-angle boundary is, indeed, less than 10 Å, although the exact nature of the interface itself is not presently known.

This discussion is by no means complete; it is only intended to illustrate that FIM, properly employed, can be useful for the characterization of microstructures. Improvements in experimental technique and instrumentation, and advances in image interpretation will surely allow application to some of the microstructural analyses confronting ceramists.

V. SUMMARY

What we have tried to accomplish in this chapter is to introduce the reader to electron and field-ion microscopical techniques for the characterization of microstructures. To this end we have derived the basic equations of the kinematical theory of electron scattering theory; demonstrated by example how the theory may be applied to defect analysis; and illustrated the use of electron diffraction patterns for crystallographic analysis.

The purpose of the discussion of field-ion microscopy is twofold. FIM is a viable tool, and those concerned with microscopy should be aware of its potential, although present use is fairly limited to metallic systems. Extensive application to ceramics will develop only when those involved in the area require the capability, and subsequently develop the necessary experimental techniques.

REFERENCES

(1) C. E. Hall, Introduction to Electron Microscopy, McGraw-Hill, New York, 1953.

(2) G. Pinsker, Electron Diffraction, Butterworths, London, 1953.

(3) Gareth Thomas, Transmission Electron Microscopy of Metals, Wiley, New York, 1962.

(4) B. K. Vainshtein, Structure Analysis by Electron Diffraction, Pergamon, New York, 1964.

(5) R. D. Heidenreich, Fundamentals of Electron Microscopy, Wiley, New York, 1964.

(6) S. Amelinckx, The Direct Observation of Dislocations, Academic, New York, 1964; (Solid State Physics, Suppl. 6).

(7) Modern Developments in Electron Microscopy (Benjamin M. Siegal, ed.), Academic, New York, 1964.

(8) Techniques for Electron Microscopy (Desmond H. Kay, ed.), Davis, Philadelphia, 2nd Ed., 1965.

(9) P. B. Hirsch, A. Howie, R. B. Nicholson, D. W. Pashley, M. J. Whelan, Electron Microscopy of Thin Crystals, Butterworths, London, 1965.

(10) R. E. Smallman and K. H. G. Ashbee, Modern Metallography, Pergamon, New York, 1966.

(11) I. S. Brammer and M. A. P. Dewey, Specimen Preparation for Electron Metallography, American Elsevier, New York, 1966.

(12) J. M. Cowley, Crystal Structure Determination by Electron Diffraction, Pergamon, New York, 1967; (Prog. in Mat'ls. Sci., Vol. 13, No. 6).

(13) K. W. Andrews, D. J. Dyson, and S. R. Keown, Interpretation of Electron Diffraction Patterns, Adam Hilger, London, 2nd Ed., 1968.

(14) M. J. Makin, "The Theory of Image Contrast in Electron Microscopes," in Metallography, Vol. 1, 1968, p. 109.

(15) L. E. Murr, Electron Optical Applications in Materials Science, McGraw-Hill, New York, 1970.

(16) R. Gomer, Field Emission & Field Ionization, Harvard Univ. Cambridge, Mass., 1961.

(17) Field-Ion Microscopy (J. J. Hren and S. Ranganathan, eds., Plenum, New York, 1968.

(18) E. W. Mueller and T. T. Tsong, Field-Ion Microscopy Principles & Application, American Elsevier, New York, 1969.

(19) Applications of Field-Ion Microscopy in Physical Metallurgy & Corrosion (R. F. Hochman, E. W. Mueller, and B. Ralph, eds., Georgia Tech., Atlanta, Georgia, 1969.

(20) K. M. Bowkett and D. A. Smith, Field-Ion Microscopy, North-Holland, Amsterdam, 1970; (Defects in Crystalline Solids, Vol. 2, S. Amelinckx, R. Gevers, and J. Nihoul, eds.).

Chapter 8

PETROGRAPHIC ANALYSIS

V. D. Fréchette

State University of New York
College of Ceramics at
Alfred University
Alfred, New York

I. INTRODUCTION

Microscopy, of all analytical techniques the most broadly useful in the ceramic industry, is not easy to discuss in terms that are readily understandable. The problem lies in the fact that microscopic methods do not rest on any one principle. They capitalize on a great number of optical, mechanical, and chemical phenomena by means of which minerals can be (1) identified and recognized in mixtures, (2) their chemical compositions inferred, and (3) their grains characterized with respect to size, shape, preferred orientation in aggregates, and abundance in mixtures. The techniques by which such principles are reduced to practical procedures are as various as might well be imagined from reflection on the considerable span of years over which they have been developed.

The present discussion will begin, therefore, with brief descriptions of the most relevant instruments and the simplest possible statement of their operation and application.

The microscope is usually thought of, first of all, as a device for seeing objects too small to be studied with the naked eye. The value of the microscope in this respect is surely enormous. In some cases it permits identification of tiny mineral grains merely by reason of the resemblance of their magnified images to the appearance of hand specimens. It permits convenient measurement of grain sizes and characterization of grain shapes. It enables distinction between free-milling ores and those from which no amount of grinding can free the valuable constituent from the gangue. But magnification (or better to say resolution and magnification of fine details) is by no means the only function of the modern microscope and indeed much of the success of modern microscopy rests on other, additional grounds. Such quantities as refractive index, birefringence, dispersion, optic character and sign, pleochroism, optic angle, optic orientation, angle of extinction, and sign of elongation characterize a mineral so that it may be identified from published description. Measurements of one or more of these may also be used as a basis for inferring chemical composition and such textural characteristics as preferred orientation of the crystalline components of an agglomerate. The following sections will outline the nature of such measurements and will suggest some typical applications and some procedures of particular use in the characterization of ceramics.

II. INSTRUMENTS

A. Stereoscopic Microscope

Using twin optical systems, one for each eye, the stereoscopic microscope gives three-dimensional vision at magnifications up to about 150X. With it a mineral fragment appears exactly like one having a diameter up to 150 times as great, viewed with the same type of illumination, and it can be examined for size, shape, surface contamination (adhering grains and stains), and included minerals.

Illumination is extremely important. If the grains are to resemble a hand specimen in appearance, they should be lighted from above and the aperture of illumination should be very large. For most purposes a spot flood lamp of the type commonly used in outdoor illumination is excellent, and its heat may be absorbed by interposing a Florence flask filled with a dilute solution of copper sulfate, acidified, for permanence, with a little sulfuric acid.

Transmitted light may also be useful, especially in conjunction with polarizing accessories. A sheet of Polaroid is placed beneath the specimen and a second sheet, with polarization direction 90° from the first, is supported above it. In the absence of a specimen no light reaches the eye. For such examination the specimens should be immersed in liquid in a shallow dish. Water is adequate for minerals of low refractive index but liquids of higher refractive index than water (mineral oil, monochloronaphthalene, or diiodomethane in extreme cases) should be used for the high-index minerals.

The "working distance" separating the object from the objective lenses of the stereoscopic microscope is ample to permit manipulation of the specimen. Many tests may be performed to explore properties which may be of practical interest or aid in recognition. Response to probings with a sharp needle and a bar magnet, behavior in contact with chemical reagents, tendency to sink or float in calibrated heavy liquids, and possibly wetting behavior in selected liquids may be observed directly on the mineral constituents of a mixture.

Grains of special interest may be selected from a granular mixture in air by simply touching them with an oily needle (they may be released by dipping the loaded needle tip in a drop of oil) so that they may be studied separately under the petrographic microscope, by x-ray diffraction or fluoroscopy, by spectroscopy or by other means.

To facilitate particle size measurement, the stereoscopic microscope can be fitted with an eyepiece micrometer whose scale divisions can be calibrated for each of the objectives by comparison with a standard scale placed for the purpose on the object stage. A mineral grain diameter can thus be converted from eyepiece scale divisions to absolute units by applying the calibration factor. An even more useful arrangement is a scale drawn by the microscopist to suit his needs. A blackened photographic plate illuminated from below is placed in a horizontal position beside the microscope under a camera lucida (Abbe drawing camera). Lines scratched through the blackened emulsion are visible as bright lines superimposed

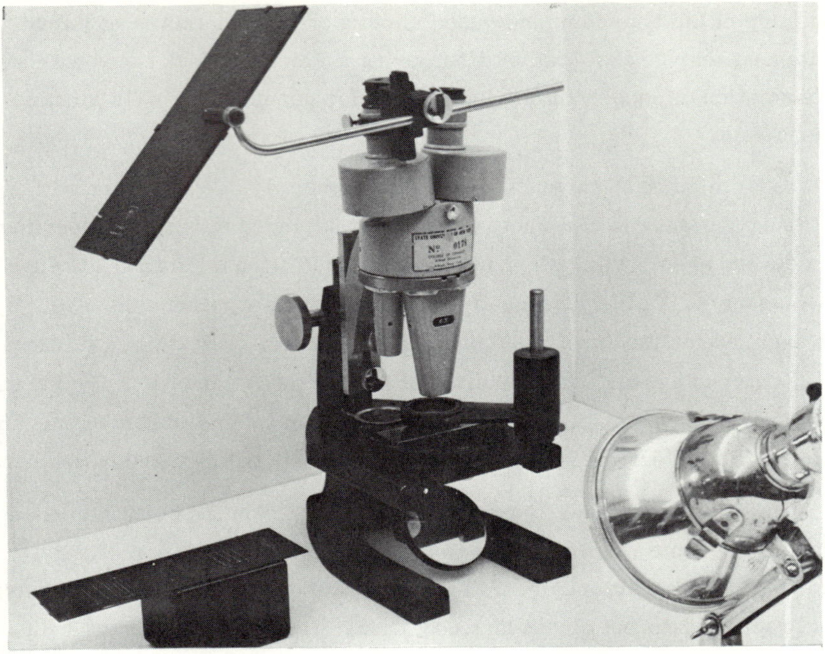

Fig. 1. Stereoscopic microscope equipped for particle size measurement with camera lucida which superimposes a lighted scale (lower left) on the field of the microscope.

on the field of view of the microscope. They can be ruled in the form of a linear scale reading directly in English or metric units, a grid, or any convenient reference scheme (Fig. 1).

Comparison of a specimen powder with reference photomicrographs constitutes the simplest and fastest of all particle-size checks and can save time and labor in control of milling operations, for example. A series of photomicrographs covering the range within the control limits is prepared and printed as transparencies. Half of the microscope field is than blacked out by placing a black sheet of paper over the object stage above the specimen powder. In this blacked-out portion of the field of view the image of the reference transparency, resting on its illuminated screen, is brought into view by the camera lucida. Such external reference scales and reference photomicrographs can be useful with all types of microscopes and at all magnifications.

B. Polarizing Microscope

The polarizing microscope developed by F. E. Wright especially for the study of minerals by transmitted light is capable of magnifications as high as 1500X. Special features include polarizing elements in the condenser and above the objective lens, rotating stage and a swing-out condenser element to provide convergent (conoscopic) illumination. A retractable auxiliary lens is provided immediately below the eyepiece to focus on the back lens of the objective where, under conoscopic illumination, an interference figures is formed. Only a few of the observations possible with this extraordinarily versatile instrument will be indicated here. Details are available in standard texts.

1. Preparation of Specimens

When viewed in air, the light scattered from grain surfaces is apt to obscure internal details and make measurement of optical properties intrinsic to the mineral impossible. Liquid immersion gives the possibility of comparing the refractive index of the mineral with that of the immersion liquid to provide a means for measuring mineral refractive indices; to this end the mineral is mounted successively in a series of liquids of known refractive indices until its own index is bracketed within desired limits.

In cases where it is desirable to preserve for study the textural details of a ware, a thin section may be cut with a diamond saw, ground smooth on one side, cemented on a glass slide smooth side down, thinned by grinding to an appropriate thickness ($30\,\mu$ is standard but not invariable) and capped with a cover slip.

Figures 2 and 3 show sections of Bavarian porcelain prepared by splintering. The technique involves pinching a specimen of the material between the jaws of a hydraulic press and selecting from the resulting splinters those which are suitable by examination under the stereoscopic microscope. In Fig. 2 a concentration of randomly oriented needlelike mullite crystals can be seen on the site of a feldspar grain. The surrounding gray grains are quartz particles rounded by reaction with the clay and with alkali which has diffused out of the feldspar. The clay appears as unresolved masses in which, however, a directional characteristic has resulted from shear during the forming process. Figure 3 shows the glaze of the same porcelain, containing quartz grains rounded by reaction and rimmed with cristobalite formed by inversion of the quartz.

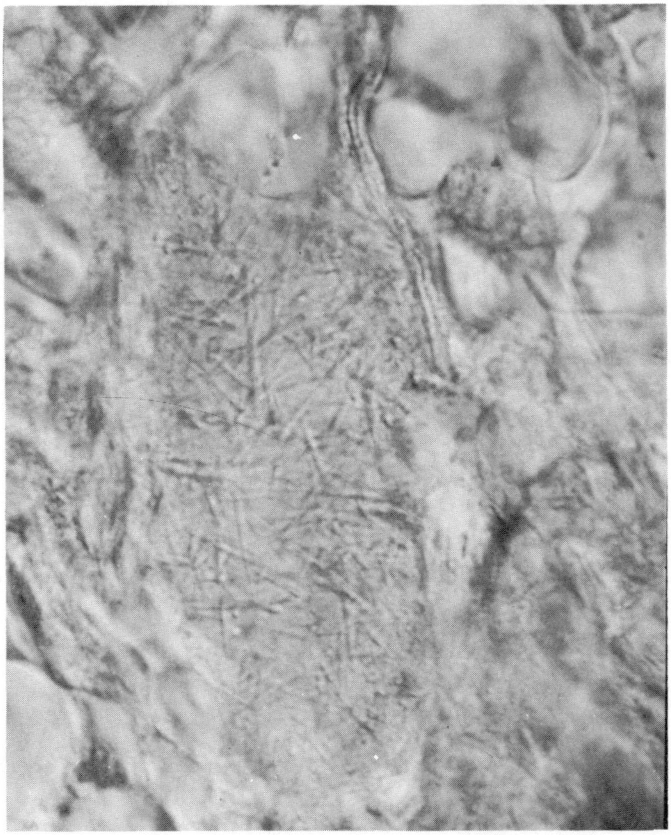

Fig. 2. Bavarian porcelain body showing a nest of randomly oriented mullite needles in glass on the site of a feldspar grain (center), surrounded by rounded quartz grains and (unresolved) clay masses oriented in a flow structure. Transmitted light, inclined; 1000X.

2. Optical Measurements

Most useful of all optical measurements is refractive index, n, according to which minerals are listed in determinative tables and by means of which the members of a mixture may be easily distinguished from one another in textural studies and quantitative analysis. Several phenomena indicate the relationship between the refractive index of a mineral and that of the liquid in which it is immersed (Table I). Figure 4 illustrates the method of apparent relief. Quartz grains stand out in strong positive relief, i.e., they appear as hills, while albite shows very low negative relief and

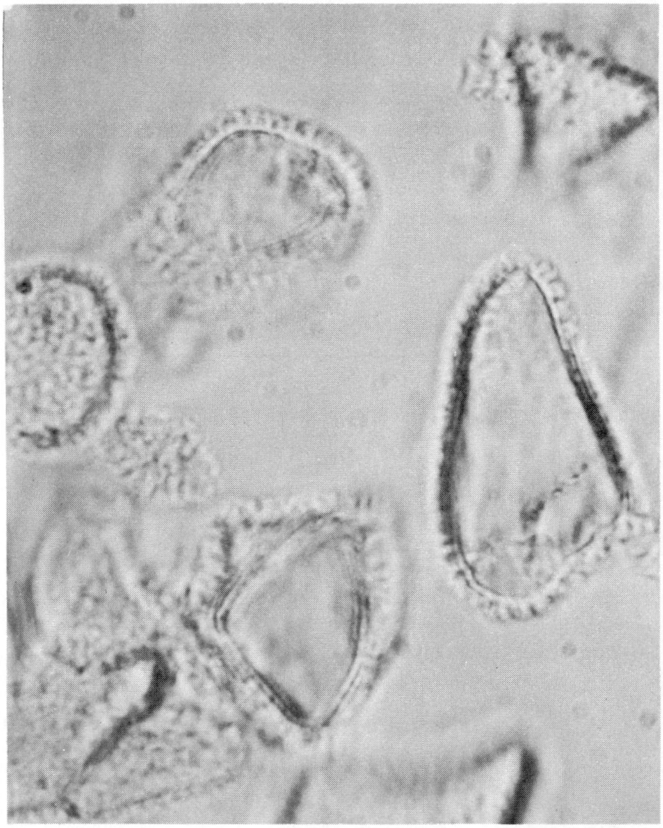

Fig. 3. Bavarian porcelain glaze showing rounded quartz grains rimmed with cristobalite in glass matrix. Transmitted light; 1000X.

microcline moderate negative relief. Because of the relatively steep dependence of refractive index of liquids on the wavelength of light, a mineral may appear higher in index than the liquid for red light and lower for blue light. In such a case the grains appear red on one side and blue on the other. The phenomenon is useful in "tagging" a particular mineral in a mixture, choosing for the purpose an immersion liquid of high dispersion. Blends between di-n-butyl succinate (n_D = 1.43) and transcinnamaldehyde (n_D = 1.62) are convenient. Especially sensitive discrimination among a group of minerals by this method is to be obtained by use of a Heine condenser set at the boundary between light and dark field conditions. Figure 5 shows the effect in a mixture of apatite and quartz immersed in an oil of

TABLE I

Comparison of Refractive Indices

Manipulation	Observation
Mineral grains in sharp focus, illumination oblique (decentered)	Grains appearing in positive relief are higher in refractive indices than the liquid in which they are immersed.
Mineral grains are lowered slightly from sharp focus	Grains which brighten are higher, those which darken are lower in index than surrounding liquid.

Fig. 4. Quartz grains (high positive relief), microcline (moderate negative relief), and albite (very low negative relief) in oil having refractive index 1.526. Oblique transmitted light; 150X.

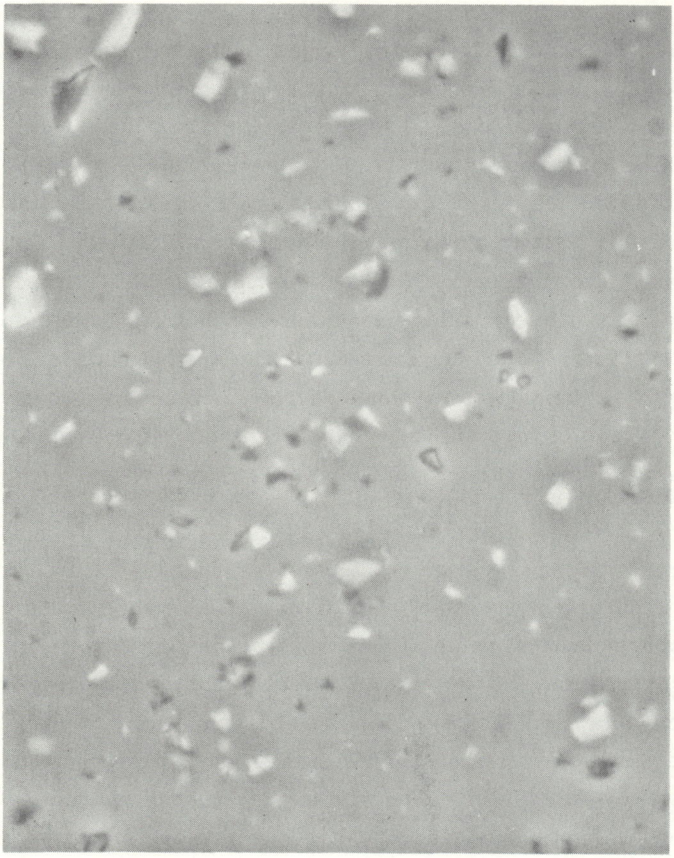

Fig. 5. Dispersion "tagging" of minerals. Apatite (dark) and quartz (bright) in a high-dispersion liquid having refractive index 1.590. Boundary phase contrast; 800X. Halos around the grains and poorly resolved boundaries are typical.

refractive index 1.590. Under these conditions the apatite grains appear dark while the quartz grains appear bright. A drawback of the method is the poor resolution at grain boundaries and the halos which surround the grains.

Minerals may be classified into two main groups, the optically isotropic and anisotropic. Isotropic minerals, which are those of the cubic system, carry light at the same velocity irrespective of direction and so show no variation of refractive index with orientation. Light is split by an anisotropic mineral into two rays, in general, and these have different refractive indices depending on the direction in which they are traveling. Those of the tetragonal and hexagonal systems are termed uniaxial because there exists in them a single direction along which only one refractive index is observed. (This direction, the optic axis, coincides with the crystallographic c axis in tetragonal and hexagonal minerals.) At a small angle to the optic axis only a small difference in refractive indices is observed between the two rays and this increases to a maximum for light traveling perpendicular to the optic axis. The maximum difference is termed the birefringence of the mineral and it constitutes another parameter by which minerals can be characterized. Under the polarizing microscope the birefringence may be measured by reason of the retardation which develops between the two rays, traveling at velocities inversely proportional to their refractive indices, and increasing with the thickness of the grain. Thus

$$\text{Retardation} = \text{thickness} \times \text{birefringence}$$

The effect is observed between crossed Nicols, i.e., between polarizing elements whose planes of polarization are perpendicular to one another. Under these conditions the two rays are brought into interference, forming interference colors which vary with the retardation exactly as soap-bubble colors, the result of interference between the first- and second-surface reflections of the bubble wall, vary with the thickness of the film. The Michel-Levy scale of birefringence, reproduced in several texts, relates the colors to retardation and provides graphical solutions to the birefringence equation.

Upon rotation of the stage, the mineral grain becomes dark every 90°. Observation of the grain at such an extinction position characterizes it further. If its cleavage or crystal face directions are seen to be parallel to the vibration directions of one of the polarizing elements (North-South and East-West for the lower and upper elements, respectively) it is said to have parallel extinction. If two cleavages or faces are arranged symmetrically with respect to the vibration directions it has symmetrical extinction. If neither applies but the face or cleavage is inclined to the vibration direction it has inclined extinction and the angle of inclination is called the angle of extinction. Tetragonal and hexagonal minerals never show inclined extinction, triclinic minerals show only inclined extinction. Any of the three extinctions may occur in monoclinic minerals.

Elongated minerals are easily characterized with respect to sign of elongation depending on whether the slow or the fast ray is more nearly paralel to the elongation direction. The length-slow case is defined as positive elongation. Determination of the sign of elongation is made by aligning the direction of elongation 45° clockwise from the North-South and inserting an appropriate accessory plate with Nicols crossed. The gypsum plate (first-order red plate) is most often used. If the first-order gray portion of the mineral grain (this is almost always present at a thin edge) is seen to rise to blue upon insertion of the gypsum plate, the slow ray of the grain is concluded to be parallel, or more nearly parallel, to the slow ray of the plate. If the first-order gray falls to yellow, the fast ray of the grain is parallel to the slow ray of the plate. The elongation sign can be inferred correspondingly.

By means of interference figures, uniaxial minerals are characterized with respect to sign depending upon whether the faster or the slower of the rays is observed when light travels along the optic axis.

Biaxial minerals, i.e., those of the orthorhombic, monoclinic, and triclinic systems, have two optic axes and their sign depends on whether the slowest or the fastest of all rays can be observed when light travels along the direction bisecting the acute angle between the optic axes. The angle between them is measured as the optic angle. Biaxial minerals vary also in the relationship between these optic directions and their crystallographic axes. The determinations involving interference figures require lengthy description which can be found elsewhere. Powerful as they are, especially for mineral identification, much useful work can be done in ceramic laboratories without their use.

C. Metallographic Microscope

Instead of illuminating the specimen by light transmitted through it along the microscope axis, as in the case of the polarizing microscope, or by light incident on one side, as in the case of the usual stereoscopic microscope, the metallographic instrument brings light to bear perpendicular to the polished specimen surface. If a mirror were placed on the object stage the field of view (through the eyepiece) would be bright. Mineral surfaces under these conditions exhibit a brightness proportional to their reflecting power which in turn is related to their refractive index n by the equation:

$$\text{Fraction of light reflected} = \left(\frac{n - 1}{n + 1}\right)^2$$

Zirconia grains in a matrix of aluminum oxide, therefore, appear as bright grains in a gray field. In a few laboratories photometric attachments are used with the microscope to determine the reflecting power of an unknown mineral for purposes of identification, but the equipment is expensive and the procedure is exacting. It is usually far more practical to supplement qualitative estimation of brightness with observations on the polishing hardness and the effects of chemical etches and stains.

Methods of specimen preparation vary widely in their details. Typically the specimen is sliced with a diamond saw and if porous is impregnated with an epoxy resin and cured. The surface is flattened and smoothed and its edges beveled by grinding on a series of two or three abrasive papers of successively finer grits. Polishing is accomplished with diamond pastes on photographic paper until no scratches show at the magnifications to be used in examination. The choice of polishing abrasive (diamond, Al_2O_3, Cr_2O_3, CeO_2, SnO_2, Fe_2O_3, MgO, etc.), of liquid (water, alcohols, oils), of surface (hard or soft fabric, bronze, paraffin, lead), of holding method (hand, inertial, planetary), and of movement (hand, vibratory, rotary), depends on the nature of the specimen, on the volume of work, and on the personal preferences of the microscopist.*

Etching and staining procedures are highly specific to the particular application. In principle an etchant is chosen for its specific action on a particular mineral: (1) simple solution, resulting in a depression in the level of the mineral surface relative to the surrounding minerals, (2) pitting of the mineral surface, (3) formation, by reaction or deposition, of a surface film providing subdued, enhanced, or colored reflection, (4) modification of the surface to receive subsequent staining, or (5) no action whatever, although adventitious minerals are changed. Etchants may be pure liquids, solutions,

*Helpful recommendations on polishing practice are given in The Metal Analyst: A Review of Selected Apparatus and Equipment for Metallurgical and Mineralogical Testing, Buehler, Evanston, Illinois, 1962.

or gases. They may be applied at temperatures from zero to several hundred degrees and for times from seconds to hours. No formal guide to the selection of etchants is possible at present except in the light of trials. An open-minded approach to etching procedure is recommended by the success of such unconventional approaches as etch-polishing (after preparation of a relief-free surface polish, the surface is repolished using a softer medium which leaves the harder minerals standing in positive relief above their softer neighbors), thermal etching (simple heating of the specimen to accentuate grain boundaries or to degrade certain minerals by oxidation or decomposition) and ion bombardment.

Figure 6 illustrates the ease with which the constituents of portland cement clinker can be recognized by polishing and etching with HF vapor. The euhedral grains are triacalcium silicate and the rounded, pock-marked grains are dicalcium silicate (These are the desirable constituents of portland cement. Tricalcium silicate releases more heat on setting but it hardens more rapidly than the dicalcium silicate.) The light colored background includes tricalcium aluminate and the somewhat brighter tetracalcium aluminoferrite.

Staining formulations are also based largely on the basis of trials. Excellent procedures are available for the rhombohedral carbonates, for the feldspars, and for the clay minerals. With their aid the several components of a rock may be recognized at a glance and their relative abundance, size, shape, and orientation defined.

III. SEPARATIONS

It must be emphasized that microscopy is always most powerful when used in conjunction with other analytical methods and separation techniques. The results of chemical analysis, if available, are always helpful in mineral identifications. Separation of mineral constituents of a mixture by means of heavy media, magnetic separators, electrostatic separators, froth flotation, particle-size classification by sieving, sedimentation or elutriation, or chemical separation are all useful in conjunction with the microscope in characterizing mineral mixtures and in inspecting for possible contamination.

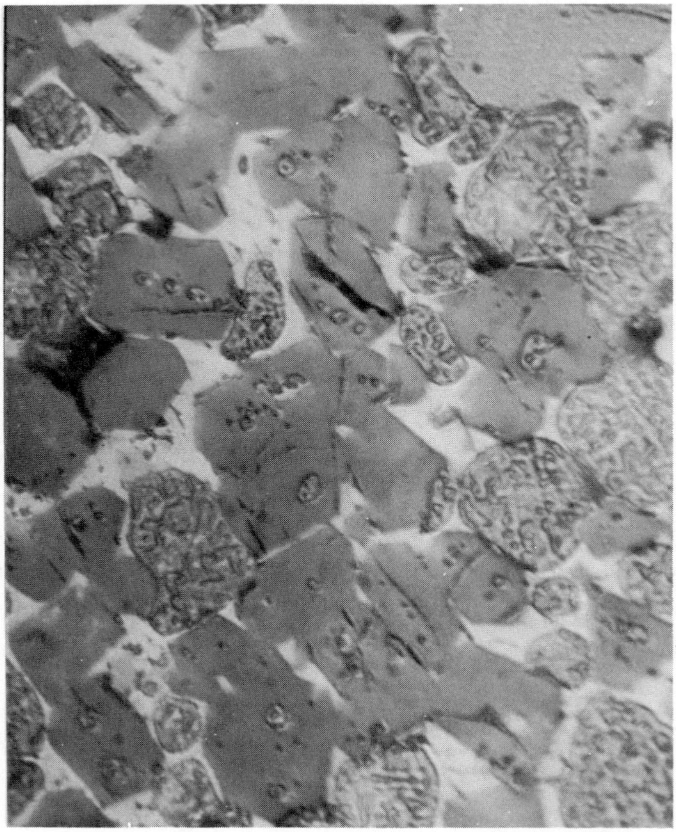

Fig. 6. Portland cement clinker, polished and etched with HF vapor. Large smooth euhedral crystals are $3CaO \cdot SiO_2$; rounded, pock-marked grains are $2CaO \cdot SiO_2$. The interstices are filled with bright $3CaO \cdot Al_2O_3$ and brighter $4CaO \cdot Al_2O_3 \cdot Fe_2O_3$. Reflected light; 600X.

Among analytical techniques, microscopy is peculiarly a catch-as-catch-can game. Establishment of routine procedures must always take account of the special needs of a particular operation lest time and money be wasted in acquiring trivial data on the one hand or lest important points be missed on the other. By careful attention to the precise requirements together with imaginative exploitation of the optical peculiarities of the combination of minerals of interest, great gains in speed and economy can be achieved over a general approach. The design of appropriate procedures requires expert attention, but such procedures when carefully specified may be followed by technicians with minimal training. Recommended procedures for a number of determinations often required are available in the Standards of the American Society for Testing and Materials. These include:

Waterformed deposits, Method for Examination by Chemical Microscopy D1245.

Practice for Air-Void Content, Specific Surface and Spacing Factor of Air-Void System in Hardened Concrete D457.

Definition of Terms E175.

Qualitative Examination of Mineral Filler and Mineral Coating of Paper D686

Particle Size Distribution of Subsieve Size Particulate Substances E20.

Petrographic Examination of Aggregates for Concrete C295.

Symposium on Light Microscopy STP No. 143.

These make good starting points for a new microscopy laboratory, but because they are designed to fit general needs they can be made shorter, faster, and more productive, in most cases, by modifications to fit the specific applications of the inidividual laboratory.

REFERENCES

(1) F. Donald Bloss, An Introduction to the Methods of Optical Crystallography, Holt, Rinehart and Winston, New York, 1961.

(2) Buehler Ltd., The Metal Analyst, A. I. Buehler, Inc, Evanston, Illinois, 1962.

(3) E. M. Chamot and C. W. Mason, Handbook of Chemical Microscopy, Vol. 1, Wiley, New York, 3rd Ed., 1958.

(4) N. H. Hartshorne and A. Stuart, Practical Optical Crystallography, American Elsevier, New York, 1964.

(5) H. Insley and V. D. Fréchette, Microscopy of Ceramics and Cements, Academic, New York, 1955.

(6) B. M. Siegel, Modern Developments in Electron Microscopy, Academic, New York, 1964.

(7) A. N. Winchell and H. Winchell, Elements of Optical Mineralogy Part II Descriptions of Minerals, Wiley, New York, 4th Ed., 1951.

(8) A. N. Winchell and H. Winchell, The Microscopic Characters of Artificial Minerals, Academic, New York, 1964.

Chapter 9

THERMAL ANALYSIS

Robert K. Ware

Owens-Illinois
Toledo, Ohio

I. INTRODUCTION

Since about 1960 various analytical procedures have converged into what is now a distinct field of study known as thermal analysis. Until recently, even those who utilized some thermal method usually used only one of these methods, such as differential thermal analysis or thermogravimetry. Now, however, as more is made known about these various techniques, and as a wider range of complementary methods have become available, the idea of using several complementary methods of analysis has spread. Obviously, more intelligent use can be made of these methods if each method itself and its advantages, disadvantages, and possible uses are fully understood. Without this understanding, it may well be that we end up treating an ailment with aspirin where penicillin is indicated. Indeed, we cannot know whether we need aspirin, penicillin, or both until we know the function of both drugs and the disease that must be treated. Each thermal method has an analytical function which it seves within the framework of the whole field of thermal analysis.

What is "thermal analysis?" If we measure the change in any property as a material is being heated or cooled (usually at a constant rate), we are using a "thermal analysis" method. Weight (thermogravimetry) and heat content (differential thermal analysis) are examples of such properties which have been frequently measured in the past. While size (dilatometry)[7] has also been frequently measured, more often than not, the measurement was an end in itself rather than a research tool for analyzing a material. Usually, it is not considered a thermal analysis method when the measurement is an end in itself. Thermal analysis methods involve a wide range of such properties. The following list gives additional examples of techniques included in the field of thermal analysis:

1. Gas evolution or absorption (evolved gas analysis)
2. Electrical resistance[13]
3. X-ray diffraction[6, 28]
4. Color[30]
5. Luminescence[1]
6. Radioactive emission[4]
7. Sound[25]
8. Light microscopy[29]
9. Pressure[3]
10. Internal friction[10]

In principle, the measurement of any property under dynamic temperature conditions could constitute a thermal analysis method. Some properties are easier to measure and some give more useful information than others. Only a few property measurements have gained widespread use for thermal analysis.

Why measure a property under dynamic temperature conditions? There are two principal reasons in addition to the measurement for its own sake: (1) to study the reactions which occur on heating, and (2) to characterize or identify a material of interest.

Knowledge of the reactions which occur on heating is frequently useful for understanding and improving a high-temperature manufacturing process. For example, Wilburn et al.[32] used differential thermal analysis and high temperature microscopy to study reactions between raw materials used in manufacturing glass. An improved understanding of the high-temperature reactions, which occur during the glass melting process, resulted from this study. Although the application of thermal analysis to high-temperature manufacturing processes constitutes an important field, this chapter will be concerned primarily with the characterization and identification of materials. The basic instrumental principles and theory are similar for all types of applications, however.

As a material is heated, its properties change in various ways; the size of most materials increases continuously due to thermal expansion. However, if melting occurs, an abrupt change in size usually occurs as shown in Fig. 1. This abrupt change can be used to study the melting reaction. However, reactions, other than melting, can cause abrupt changes in size. A more useful thermal analysis method might be high-temperature microscopy, where liquid formation and flow can be observed visually. Knowledge of the changes in properties of a material during heating can be used to identify the reactions and the conditions under which these reactions occur. It is somewhat like a murder mystery. The problem is to solve the mystery by determining what reaction would cause the change in properties which are observed. To arrive at a solution of the mystery, one must utilize clues from every source: information gleaned, quite often, from more than one method of analysis. After the particular reaction is identified, it may then prove necessary, or at least valuable, to study the influence of such variables as pressure, heating rate,

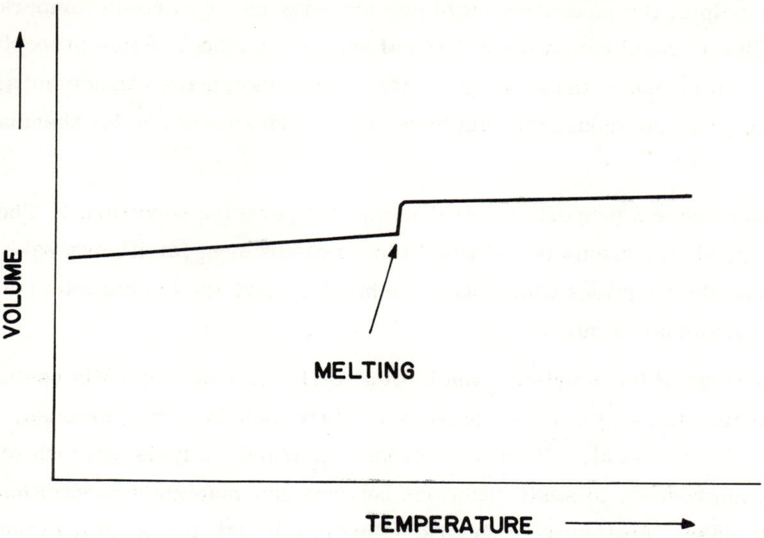

Fig. 1. Change in size during heating of a sample which melts.

or chemical additives; or the influence of these variables may have to be studied in order that the initial identification can be made. The mystery may be solved by different approaches, but each additional clue helps toward the final solution.

As discussed previously, abrupt changes in the properties of a material are caused by various reactions which occur during heating. For a constant set of sample preparation and heating conditions, a given material will always exhibit the same reactions and, hence, the same property changes. Consequently, these property changes may be used as a "fingerprint" to identify the material. However, the measurement of only one property usually cannot be relied upon for positive identification of a material, just as the color of a person's hair is not sufficient information for identification of a given individual. Several complementary thermal analysis methods must often be used to completely characterize a material. Indeed, other methods, such as chemical analysis, may be needed also. Thermal analysis can frequently improve the characterization of a material, even though chemical, x-ray diffraction, and microscopy methods have already been used.

Basically, there are three requirements for any thermal analysis method: (1) some particular property must be measured, (2) the material must be heated or cooled at controlled rates, and (3) the conditions which may influence the reactions, such as pressure and sample preparation, must be controlled. Unfortunately, neglect of the last requirement by some workers has produced undesirable results in the past. More will be said about these variables in the section on Reaction Rates. Basically, both the second and third requirement are similar for all thermal analysis methods; although special problems often arise because of the need to measure a particular property.

As indicated above, there are several thermal analysis methods which have been used in the past. Some of these have enjoyed only very limited use to date. Rather than trying to cover all of them, only the three most widely used methods will be described. These are: (1) thermogravimetry (T. G.), (2) evolved gas analysis (E. G. A.), and (3) differential thermal analysis (D. T. A.). The variables involved in their use will be discussed and their applications illustrated. Basic principles rather than specific instrument designs will be described.

II. THERMOGRAVIMETRY

Thermogravimetry (T. G.) is simply a method of measuring the weight of a material as it is heated or cooled. A large number of studies using "thermobalances" have been done at constant temperatures rather than under dynamic temperature conditions. Frequently this is the case when precise reaction rate studies are attempted. This chapter will be concerned with dynamic temperature conditions.

Basic methods of measuring weight are well-known and need not be discussed here; however, some requirements of the measuring equipment which are peculiar to thermogravimetry will be discussed. Most weight measurements used in this application involve the detection of the mechanical force needed to support a sample. Thus, all forces on the sample with components in a vertical direction should be taken into account. Any such forces which are neither controlled nor measured will tend to interfere with the weight measurement.

What does this mean in practical terms? It means that as far as possible the sample should be mechanically isolated from everything except the measuring device itself. This means that great care should be exercised in making thermocouple or other connections to the sample holder itself. Any such connections must be designed to exert as little force as possible on the sample holder. If a null-type balance is used so that the position of the sample is maintained constant, the force exerted by such a connection will normally be constant. Thus, the force will not interfere with measuring weight changes, except that it may decrease the sensitivity which the balance would have otherwise.

A major source of unwanted forces on the sample holder is the gas surrounding the sample. When runs are made in vacuum, most of these forces become negligible. Each type of force exerted by the surrounding gas will be described separately below.

A. Buoyancy

An upward buoyant force is exerted on an object suspended in a gas, just as when an object floats on water[21]. The upward force is equal to the weight of displaced gas. What is of primary interest is the change in buoyant force during a run, rather than the total buoyant force. The change in the buoyant force is due, primarily, to the change in gas density during heating. This causes an apparent increase in sample weight, since the gas density decreases and, hence, the upward buoyant force decreases. Although this effect is small, it is nevertheless significant and an appropriate correction must be made when attempting quantitative measurements. The common procedure is to simply subtract a "blank run" made on an inert sample having a volume similar to that of the sample under investigation. This procedure also tends to correct for other sources of error, some of which are described below.

B. Convection

When a gas is heated by a vertical surface, convection currents are produced[21, 5, 16]. This convection usually causes gas flow down across the sample as shown in Fig. 2. The gas flow shown exerts a downward force on the sample holder and tends to cancel the upward buoyant force. The error due to convection is strongly dependent on the equipment design. The diameter of the sample holder, the inside diameter of the furnace, and the extent to which the top of the furnace is closed, all have large effects on the convection. If a forced gas flow is used in order to control the atmosphere, this also will change the gas flow past the sample holder and must be considered

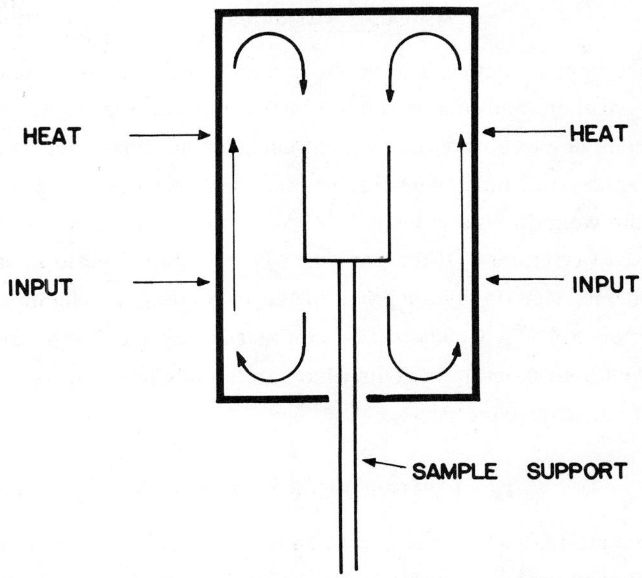

Fig. 2. Convection currents producing downward force on sample holder.

where quantitative results are desired. Proper design of the apparatus and procedure, plus the use of a blank run as mentioned above all minimize errors due to these forces.

C. Momentum

A small but sometimes significant force is exerted on the sample due to the evolution of gas from the sample[8]. This force results from the conservation of momentum law and is analogous to the recoil force experienced when a projectile is fired from a gun. This is usually a downward force or an apparent increase in weight since the gases are normally emitted from the top of the sample. The magnitude of this effect depends on the velocity of the emitted gas and the rate of weight loss according to the following relationship:

$$F = av\,dw/dt$$

where F = apparent force on sample

a = constant

v = velocity of emitted gas

dw/dt = rate of loss in weight

D. Other Variables

There are various sources of error in addition to the ones mentioned above. Most of them are minimized with good apparatus design. Effects of magnetic forces due to electrical current used for heating must be avoided. A problem can arise if heat from the furnace is allowed to increase the temperature of the weighing instrument[21]. Many instruments are susceptible to temperature variations. If the gas evolved from the sample is easily condensible, condensation on colder parts of the measuring mechanism or support rod may occur[26]. Condensation can usually be avoided by flushing the colder parts with an inert gas. This also has the advantage of protecting the mechanism from corrosive gases.

III. EVOLVED GAS ANALYSIS

This thermal analysis method is at an earlier stage of development than either thermogravimetry or differential thermal analysis. Various names, such as gas evolution analysis (G. E. A.)[31], effluence analysis[9], effluent gas analysis (E. G. A.)[2], and mass spectrometric thermal analysis (M. T. A.)[15] have all been used by various people to describe this method. Some of the early excursions into this area involved the periodic sampling of the gas evolved during differential thermal analysis runs[20]. The gas samples were then analyzed individually by such techniques as gas chromatography and mass spectrometry. While this approach has the advantage of a more complete analysis, it has the disadvantage that samples are obtained only at certain temperatures. Reactions which occur during a short temperature interval can be missed altogether.

Most investigators now use continuous sampling methods. If the runs are made at atmospheric pressure, a carrier gas is continuously passed over the sample and conveyed to a detector. The carrier gas is frequently, but not always, an inert gas such as nitrogen or helium. Gases evolved from the sample are "picked up" by the carrier gas and conveyed to the detector for analysis.

A. Gas Detectors

The detector can be of several types. Frequently, detectors originally developed for gas chromatography, such as the differential density detector, are used. This detector measures changes in density of the carrier gas.

When the sample evolves gas, the density of the carrier gas changes and a deflection is produced in the recorded curve. For example, if a nitrogen carrier gas "picks up" carbon dioxide from the sample, the density increases, since the molecular weight is greater for carbon dioxide (44 g-moles) than for nitrogen (28 g-moles). Thus, a peak in the curve occurs corresponding to the carbon dioxide evolution. If water is evolved, the density is decreased and a peak in the opposite direction occurs. However, if carbon monoxide is evolved, no peak occurs since the molecular weight (and density) is the same as nitrogen. If runs on a sample are made in different carrier gases, the molecular weight of the evolved gas can sometimes be bracketed. Often this is considerable help in identifying the evolved gas. Another identification technique is to put materials which will selectively absorb certain gases in the gas stream ahead of the detector.

A more specific analysis can be made using a mass spectrometer. For runs at 1 atm pressure and above, a portion of the carrier gas is continuously leaked into a mass spectrometer[31]. Usually, the mass spectrometer is tuned to a peak corresponding to a specific gas and the output signal recorded while the sample is heated. Thus, the evolution of a specific gas is detected during a run. A separate run is made for each gas which might be evolved. However, it is not always possible to choose a mass peak corresponding to only one gas. In this case, it is sometimes possible to selectively absorb the interfering gas from the carrier gas so that only the gas of interest is detected. In this manner, a complete analysis of the evolved gases is obtained.

When vacuum runs are desired, it is possible to actually heat the sample inside the mass spectrometer itself[15]. In this way, all of the evolved gas is analyzed rather than only a portion of it. This method is particularly useful when the evolved gas is very reactive or easily condensible. For example, the vaporization of metallic sodium could be detected with this method, whereas with the other method, the sodium would react with the walls of the apparatus before the carrier gas could transport the sodium to the detector. This method has two disadvantages, however: (1) the sample is heated in a vacuum (usually below 10^{-5} Torr) and, hence, the results will not correlate with 1 atm pressure data, and (2) the pressure in the mass spectrometer will be unstable due to the gas evolved from the sample[11].

B. Condensation

There are several problems which can arise during evolved gas analysis. Condensation of the evolved gas on parts of the apparatus which are at lower temperatures has already been mentioned. This problem is more serious in evolved gas analysis than in thermogravimetry. Any condensation or reaction of the evolved gas prior to its reaching the detector will introduce an error. In fact, the evolution of some gases may not be detected at all. Usually, however, in thermogravimetry only the gas condensation on the sample support rod or weighing mechanism is serious. This problem can be particularly serious where oxygen evolution occurs at high temperature. For example, the initial sample holder used in the author's laboratory was made of Inconel. Oxygen reaction with the sample holder prevented detection of oxygen evolution above 900°C. The use of high purity aluminum oxide permits oxygen detection at higher temperatures. A related problem which sometimes occurs is the outgassing of the sample holder assembly during heating. Fortunately, such outgassing usually decreases after a few runs have been made.

C. Flow Rate

The flow rate of the carrier gas should be controlled accurately if quantitative results are desired. Any variation in flow rate will affect the apparent sensitivity of virtually all of the evolved gas analysis methods. Another factor which is peculiar to those methods which use a carrier gas is the time lag which occurs between gas evolution and detection. In order to keep this lag time to a minimum, the volume of carrier gas between the sample and the detector should be as small as possible. A wide variety of sample holder designs are possible which minimize the above problems. Figure 3 shows a schematic diagram of the apparatus used in the author's laboratory.

Of the two methods used to study reactions involving a gas, thermogravimetry and evolved gas analysis, each exhibits its own advantages and disadvantages. The choice between these two methods will depend upon the study at hand. Perhaps the most important difference is that thermogravimetry measures the total weight change, whereas evolved gas analysis is more specific. Consequently, where more than one gas is involved, the use of evolved gas analysis may be indicated. Thermogravimetry, on the other

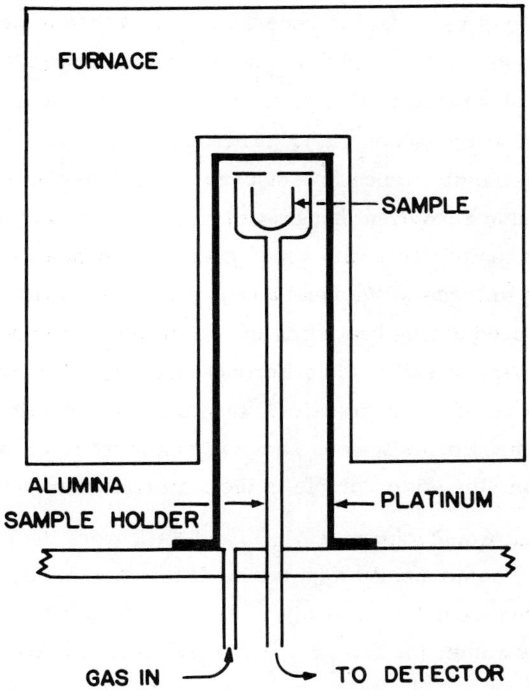

Fig. 3. Sample holder arrangement for evolved gas analysis.

hand, is dictated where the controlled atmosphere and the gas involved in the reaction are identical.

Thermogravimetry has the advantage that it is usually a cumulative measurement rather than a rate measurement. Therefore, reactions which last for hours rather than minutes can best be studied with this method.

IV. DIFFERENTIAL THERMAL ANALYSIS

Differential thermal analysis was one of the first thermal analysis methods to achieve widespread use, because of the early development of thermocouples and sensitive galvanometers for measuring very small temperature differences. The principle of operation for differential thermal analysis is simple. The interpretation of the results is complex. Compared to differential thermal analysis, the interpretation of the results obtained with evolved gas analysis and thermogravimetry is simple, as shall become apparent shortly.

A certain rate of heat input is necessary to increase the temperature of a sample at some given rate. This is due to the heat capacity of the sample. As long as the heat input and the heat capacity do not change, the temperature will increase at a constant rate. If heat is put into two identical samples at the same rate, the difference in temperature between the centers of the samples will remain zero even though both samples are increasing in temperature. If this temperature difference (zero in this case) is recorded during heating, a differential thermal analysis has been made. Thus, the "property" measured during heating is the difference in temperature between two samples; more correctly, the difference between two thermocouple junctions. If the samples are identical, then the curve obtained gives no useful information. Normally, one sample is an inert reference material, such as Al_2O_3, and the other sample is the material under investigation.

Consider what would happen if Al_2O_3 and NaCl were the two samples. As heating began, some small temperature difference would be established because of differences in heat capacity. As long as the relative heat capacities remained the same, the ΔT (temperature difference) would remain the same. On reaching 800°C, however, the NaCl melts. When this occurs, additional heat is required over and above that due to the heat capacity in order to raise the temperature further. The additional heat corresponds to the heat of melting. Because of the additional heat required, the temperature of the NaCl will lag further behind the Al_2O_3. This lag is only temporary, however. After melting is completed, the initial ΔT would be reestablished because of heat transfer between the two materials. If the heat capacity is different after melting, a slightly different ΔT will be reestablished. The above explanation gives the basic concept of differential thermal analysis, although it is somewhat oversimplified. An actual curve for Al_2O_3 versus NaCl is shown in Fig. 4.

The variables which are important in differential thermal analysis but not in other thermal methods of analysis are discussed below.

A. Sample-Holder Design

Differential thermal analysis is a comparative method. Consequently, the heating of the two samples must be symmetrical. The slightest asymmetry in the fabrication of the sample holder and furnace can produce a temperature difference even though identical samples were placed in both sides.

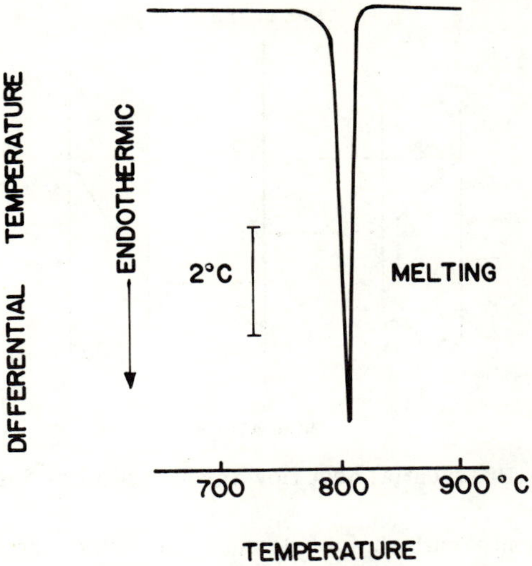

Fig. 4. Differential thermal analysis of NaCl.

Although this is a fairly obvious requirement, it is almost impossible to achieve completely. In practice, an adjustment in thermocouple position or some other variable is usually made to cancel the effect of asymmetries. This is particularly important when high sensitivity and high heating rate runs are being made.

A wide variety of results can be obtained depending on the sample holder-furnace assembly design. In order to understand this, let us consider in more detail the cause of the differential temperature during a reaction.

In order for heat to flow through a material, one side must be at a higher temperature than the other. The rate of heat flow is proportional to the temperature difference. Consider the two similar materials illustrated in Fig. 5. If these materials are heated uniformly, one side of each material will be at temperature T_o. The heat flow Q_1 through material 1 will be proportional to $T_o - T_1$. Similarily, the heat flow Q_2 through material 2 will be proportional to $T_o - T_2$. The difference in heat flow through the two materials $(Q_1 - Q_2)$ would be proportional to $(T_o - T_1) - (T_o - T_2) = T_2 - T_1$. Thus, the temperature difference at the colder sides provides a measure of the difference in heat flow. If samples which absorb heat are placed against the cold sides of the "heat flow meters," the differential temperature will provide a measure of the difference in heat flow into the samples. Thus, a differential thermal analysis provides a record of the relative heat flow into the samples.

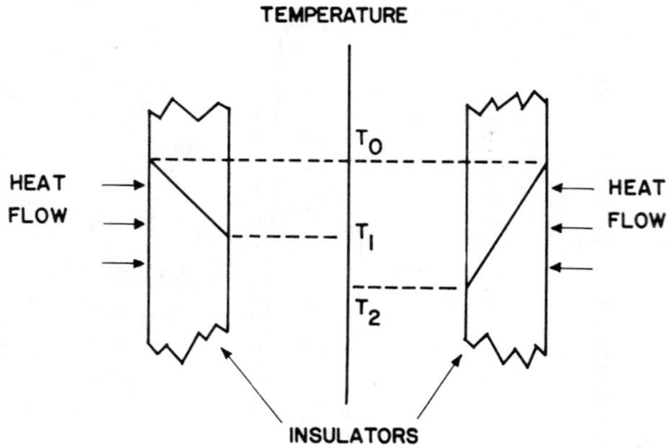

Fig. 5. Differential heat flow through similar insulators.

In many common sample-holder designs, however, the differential temperature is measured at the centers of the samples. Thus, the relative heat flow is measured by differences in temperature within the samples themselves. Under these conditions, the samples serve two purposes:

1. To absorb heat
2. To act as heat flow meters

In the above illustration, we said that the heat flow is proportional to the temperature drop across the material. This is true as long as the apparent thermal conductivity of the material is constant. However, any change in conductivity will also change the temperature drop even though the heat flow remains constant. Therefore, the differential temperature changes when either heat flow or conductivity changes.

It is sometimes desirable to detect changes in sample conductivity. For example, the "softening point" of glass can be detected from changes in apparent thermal conductivity as glass particles begin to fuse together[33]. When this is desired, the differential temperatures should be measured at the center of the samples.

More commonly, the actual heat absorption (or evolution) is desired. In this case it is necessary to minimize the influence of changes in conductivity. This is particularly true when quantitative determinations of heats of reaction or quantities of each component are to be determined. This can be done by separating the heat flow detection from the sample as shown in Fig. 6[22]. With this arrangement the metal block distributes the heat evenly

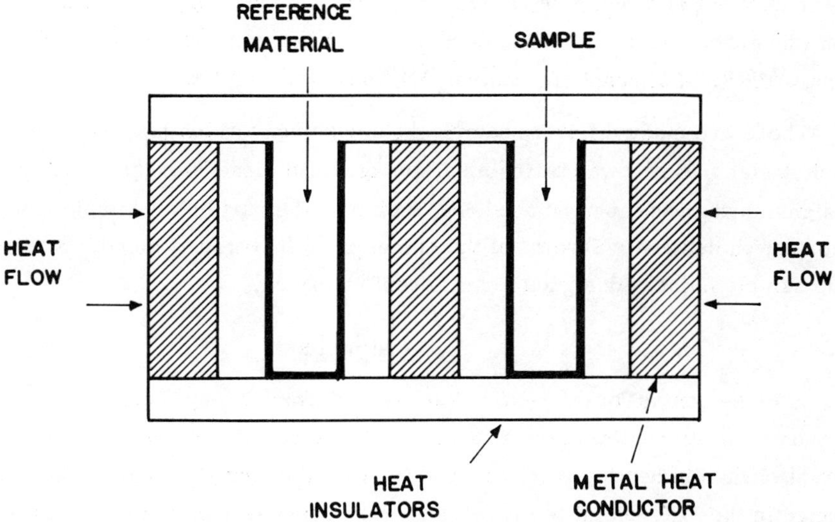

Fig. 6. Sample holder for quantitative differential thermal analysis. After Ozawa[22].

so that the temperature at the outside of both insulators is maintained the same. The differential temperature is measured at the inside of the insulators and, thus, provides a measure of the relative heat flow into the samples. The samples themselves are usually encased in a metal container, which distributes the heat to the sample uniformly.

An alternative method of minimizing the effects of thermal conductivity differences consists of diluting the unknown sample with the reference sample material[24]. When about 25% of the unknown material is mixed with 75% of of the reference material (usually Al_2O_3), the overall apparent thermal conductivity is essentially that of the reference material. Therefore, differences in the apparent thermal conductivity of the unknown sample material have little effect on the results. Incidentally, the presence of gases evolved from the unknown material can significantly change the overall conductivity, even when the sample is diluted. This is particularly true when helium is used as the control atmosphere. This method has two principal disadvantages: (1) a smaller quantity of unknown material is used and, hence, the size of the resulting peak is decreased, and (2) a spurious reaction may occur between the unknown material and the reference material used for dilution.

The effect of changes in sample conductivity is an important consideration when choosing a sample-holder design. Other considerations, such as sample size and atmosphere control, will be discussed later.

There are many different sample-holder designs which have been used. Each design has its own particular advantages and disadvantages. Different designs must sometimes be used, depending on the particular sample and information desired. Several of the commercial instrument suppliers provide sample holders designed for different purposes.

B. Heating Rate

Precise control of the heating rate is important in any thermal analysis method. In differential thermal analysis it is extremely important. An abrupt change in heating rate will tend to upset the heat flow and produce a change in the differential temperature. Such changes are easily mistaken for sample reactions. What is most important is the instantaneous heating rate rather than the average heating rate over a 10- or 20-min interval. A slight interruption of the furnace power can produce an effect on the differential temperature curve, even though only a fraction-of-a-degree change occurs in the furnace temperature. The magnitude of this effect will depend on the sensitivity of the instrument, sample-holder design, and other experimental conditions. Thus, "on-off" temperature controllers are not usually suitable for differential thermal analysis.

C. Electrical Interference

Most instruments use thermocouples for the differential temperature measurement. The magnitude of differential temperature detected is sometimes less than $0.01^\circ C$. For a Pt, Pt-10% Rh thermocouple, this corresponds to an electrical signal of about 0.0001 mV. Since the signal of interest is so small, spurious electrical signals from other sources can cause considerable "noise" in the resultant curve. Induced or leaked voltage from the furnace power circuit is a frequent source of trouble, particularly at high temperatures where the insulating materials may become semiconducting. In addition, spurious thermoelectric effects involving the insulating refractories can be generated which produce "no load" potentials as high as 1 V. Often these sources of electrical interference are aggravated by contamination of the insulating materials by vapors emitted by the sample under investigation.

V. SIMULTANEOUS MEASUREMENTS

Recently a trend toward making more than one type of measurement on a sample during a single experiment has developed[18,23]. For example, both differential thermal analysis and thermogravimetry curves are sometimes obtained on a single sample during a single run. Such combinations have several possible advantages: (1) more information is obtained when only enough sample for one run is available, (2) more efficient use is made of equipment and manpower, and (3) the results may be compared more easily than if they were obtained on different samples, especially when the conditions are not well-controlled or the sample is not uniform.

While much can be said for simultaneous measurements, these advantages are not obtained automatically just by making simultaneous measurements. Usually compromises in apparatus design and technique are necessary in order to make simultaneous measurements. Indeed, results may not correlate perfectly even when obtained on the same sample at the same time[23].

VI. REACTION RATES

Basically, an investigation using thermal analysis is a study of reaction rates, although the immediate purpose may be to determine the mineral components present in a sample. The changes recorded during the heating of a sample are directly related to the reaction rates. Therefore, a basic understanding of reaction rates is necessary in order to interpret thermal analysis data.

A. Homogeneous Reactions

Homogeneous reactions, i.e., reactions occurring with no phase change, can be described by an "order of reaction," usually denoted by n. Such reactions usually obey the equation given below when held at constant conditions such as temperature, pressure, atmosphere, etc.:

$$dc/dt = -kc^n$$

where c = concentration of the reacting substance
n = reaction order number
t = time
k = rate constant

In other words, the rate of reaction is an exponential function of the concentration of remaining reactant.

However, in thermal analysis, the temperature is not constant. Usually the value of k in the above equation varies with temperature according to the well-known Arrhenius equation as follows:

$$\log k = -E/RT + \text{constant}$$

where
E = activation energy
R = ideal gas constant
T = absolute temperature

Thus, where both time and temperature vary, these two equations can be combined to give a theoretical description of the reaction rate curve obtained from thermal analysis:

$$dc/dt = Ae^{-E/RT}c^n$$

where A = constant

If a linear heating curve is used, the temperature is a simple function of time:

$$T = T_o + rt$$

where
T_o = temperature at time zero
r = heating rate
t = time

Thus, the shape of the curve for a homogeneous reaction is described by

$$dc/dt = Be^{-D(T_o + rt)}c^n$$

where B and D are constants

As can be seen, the rate of reaction (dc/dt) is influenced by the heating rate r. The other constants are characteristic of the sample under study. Generally this is true of thermal analysis. The heating rate influences both the shape and temperature of most reaction peaks. This applies equally to heterogeneous and homogeneous reactions. Figure 7 shows differential thermal analysis curves of $CaCO_3$ for two different heating rates.

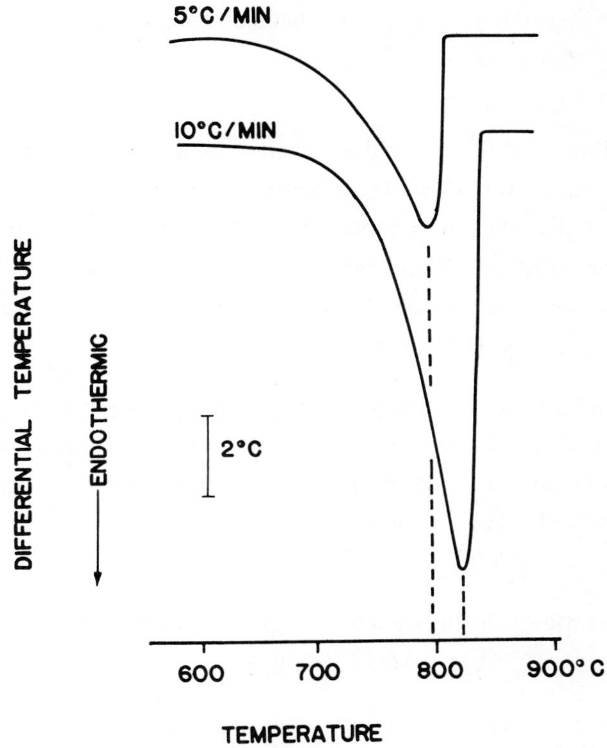

Fig. 7. Differential thermal analysis showing effect of heating rate on calcite decomposition in flowing N_2 atmosphere.

B. Heterogeneous Reactions

Normally, heterogeneous reactions (reactions involving more than one phase) cannot be described in terms of the concentration of the reacting substance. These reactions may be described more adequately by such terms as: (1) the area of the interfacial surface between the starting material and the reaction product, and (2) the thickness of the layer of the reaction product.

For example, the dissociation of calcite ($CaCO_3 \rightarrow CaO + CO_2$) has been shown to be proportional to the interfacial surface area for a given temperature and atmosphere[14]:

$$dv/dt = ka$$

where v = quantity of unreacted substance remaining

k = rate constant

a = interfacial area

The controlling mechanism for this reaction is attributed to the diffusion of CO_2 through an activated CaO layer having a constant thickness. As in homogeneous reactions, the rate constant varies with temperature according to the Arrhenius equation. Thus, for thermal analysis (where temperature is not constant), the rate constant is replaced by an exponential term as before:

$$dv/dt = Be^{-D(T_o + rt)}a$$

The interfacial area (a) is a function of the geometry of the sample. If the geometry is simple, the area can be related to the volume (v) and the equation can be solved. For example, the area of a sphere is related to its volume by the following equation:

$$a = 4.836v^{2/3}$$

Thus, the reaction rate equation for a sphere of $CaCO_3$ is

$$dv/dt = 4.836Be^{-D(T_o + rt)}v^{2/3}$$

The equation for a cube would be

$$dv/dt = 6Be^{-D(T_o + rt)}v^{2/3}$$

It is assumed that the shape of the interfacial surface will remain a cube throughout the reaction. This is true only to a first approximation. The surface area of a cube is not too different from that of a sphere having the same volume; consequently, the reaction rate equations are not too different.

A sample in the shape of a thin plate provides an example quite different from that of a sphere. Cleavage samples having this shape could be obtained easily from such materials as mica. For such a shape, the reaction surface area would remain essentially constant during the reaction.

The equations for a sphere, a cube, and a plate were solved with an electronic computer using the constants for $CaCO_3$ dissociation[14]. The resulting curves are shown in Fig. 8. The shapes of the curves do not change, however. This is true also for different heating rates when the curve is plotted versus time rather than temperature. When we are only interested in the area under the curve, which is a measure of the quantity of sample present, the sample shape is unimportant. Only where the shape of the curve

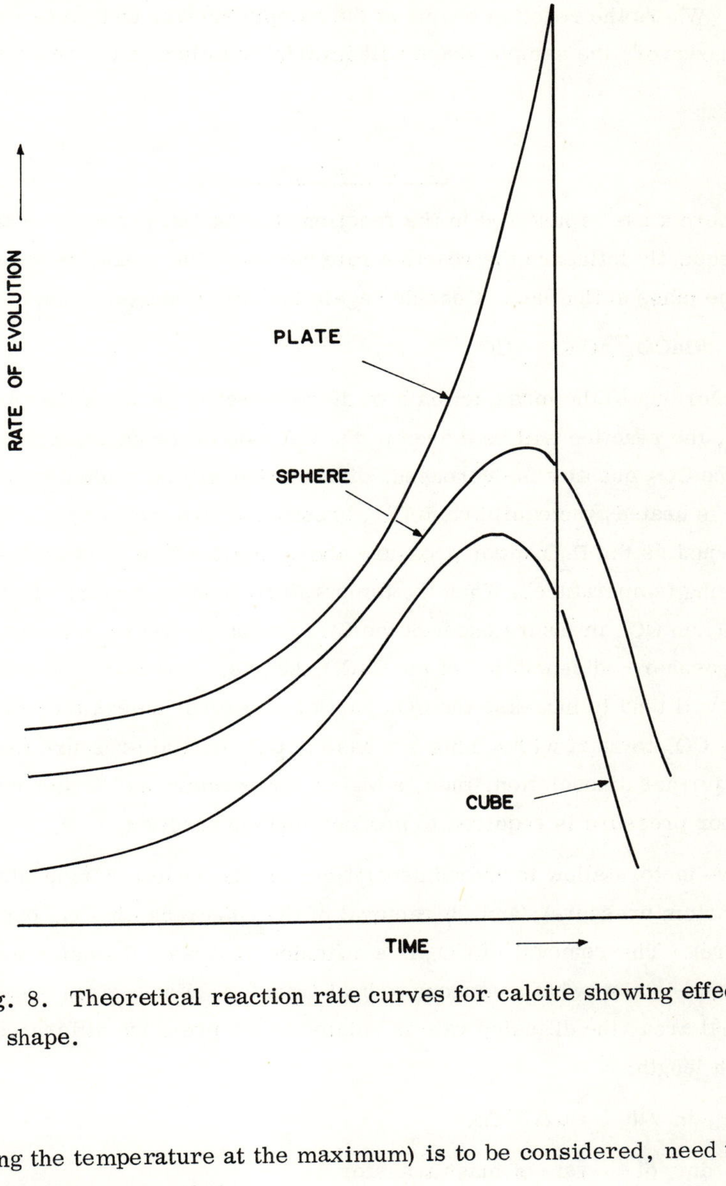

Fig. 8. Theoretical reaction rate curves for calcite showing effect of sample shape.

(including the temperature at the maximum) is to be considered, need the sample shape be controlled.

Many heterogeneous reactions are controlled by diffusion of some component to or away from the reaction surface in a more complex manner than for the $CaCO_3$ reaction discussed above. The reaction rate equations are frequently more complex than the ones given above and are frequently not known. Where the reaction begins at the sample surface and proceeds inward, however, the sample shape will invariably influence the reaction rate curve.

C. Atmosphere

Where a gas is involved in the reaction, the partial pressure of that gas will frequently influence the reaction rate curve. This is understandable from the mass action law. Consider again the calcite dissociation:

$$CaCO_3 \rightleftarrows CaO + CO_2$$

According to the mass action law, if we increase the quantity (pressure) of CO_2, the reaction will be driven to the left side of the equation. Conversely, if the CO_2 quantity is decreased, dissociation will be promoted. As $CaCO_3$ is heated, the equilibrium CO_2 pressure above the sample will increase just as the H_2O vapor pressure above liquid water increases with increasing temperature. When the temperature reaches a point where the equilibrium CO_2 pressure exceeds the CO_2 partial pressure in the surrounding atmosphere, dissociation of the $CaCO_3$ begins. However, the evolution of CO_2 will tend to increase the CO_2 partial pressure, unless the atmosphere is 100% CO_2 to start with. This increase in CO_2 partial pressure tends to retard further dissociation; thus, a higher temperature and higher equilibrium vapor pressure is required to produce further reaction.

Two factors allow further dissociation: (1) increase in temperature (equilibrium pressure), and (2) removal of CO_2 (decrease in CO_2 partial pressure). The removal of CO_2 is a diffusion process. Diffusion through a material is somewhat analogous to heat transfer. For a given cross-sectional area, the diffusion rate is related to the pressure difference and the path length:

$$dm/dt = \alpha \Delta P/\Delta x$$

where dm/dt = rate of mass transfer

ΔP = CO_2 pressure difference

Δx = material thickness (length of diffusion path)

For some reactions, the path length controls the diffusion rate which, in turn, controls the reaction rate. The controlling path length for diffusion is the thickness of the reaction product layer for many reactions, although this appears not to be the case for calcite dissociation; thus, the equation which describes the reaction rate might be:

$$dv/dt = \alpha A \Delta P / \Delta x$$

where α = diffusion constant

A = interfacial area

Δx = controlling path length

ΔP = equilibrium vapor pressure minus partial pressure in surrounding atmosphere

In this case, both the equilibrium vapor pressure and the diffusion constant would usually increase with increasing temperature. In the case of $CaCO_3$, the CO_2 pressure, which surrounds the particles, exerts a large influence on the reaction. This is illustrated in Fig. 9 which shows differential thermal analysis curves of $CaCO_3$ under various gas pressures. The path length and interfacial area are influenced by sample geometry. The same factors influence a reaction involving the absorption of a gas, such as the reverse of $CaCO_3$ dissociation:

$$CaO + CO_2 \rightarrow CaCO_3$$

Here we are only considering the net reaction and are ignoring the fact that both reactions occur simultaneously in a reversible reaction, such as this one.

D. Heat Transfer

The heat of reaction of the calcite dissociation is quite significant (40,000 cal/mole). At higher temperatures, the rate of heat transfer into the sample can become more important than the rate of diffusion of CO_2 out of the sample. Under these conditions, the reaction rate is controlled by heat transfer rather than mass transfer. As pointed out earlier, heat and mass transfer are analogous and follow similar physical laws; consequently, sample geometry should influence the reaction rate in either case.

An important limitation of most thermal analysis methods is that a temperature gradient exists through the sample. When the sample is heated by conventional means, the outside is hotter than the middle. If this were not so, heat would not flow into the sample and the average sample temperature would not be increasing. Consequently, it is impossible to ascribe a single

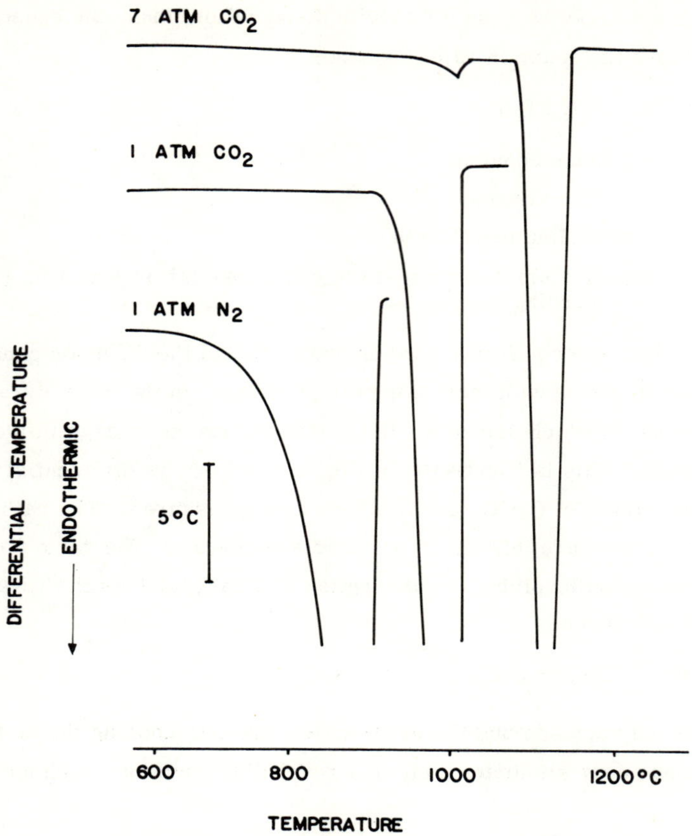

Fig. 9. Differential thermal analysis curves of calcite showing effect of CO_2 pressure.

temperature to the sample at any given time during thermal analysis. The measured property is frequently plotted vs. temperature rather than time. What temperature is to be used? Two alternatives are usually considered. The temperature at (1) the center of the sample, or (2) some point outside the sample may be used. Frequently, the temperature measured by a thermocouple placed near the sample is used both for controlling the furnace power and indicating the "temperature of the sample." The temperature difference depends on four factors: (1) heat transfer coefficient, (2) heat capacity, (3) sample thickness, and (4) heating rate. The temperature difference decreases as: (1) the heat transfer coefficient increases, and (2) the heat capacity, thickness, and rate decrease. This temperature gradient has the effect of spreading a peak out and, indeed, adjacent peaks may begin to overlap. This effect may be kept quite small by using a very small sample (a few milligrams) and/or using low heating rates. However, the sensitivity of the instrument must be high for this to work.

The variation in sample temperature due to the heat of reaction is even more important than the temperature gradient discussed above. Due to the heat of reaction, the temperature of the sample will lag behind or go ahead of the "furnace temperature." This, of course, is the property of interest in differential thermal analysis. When temperature is measured in the center of the sample, the temperature vs. time curve will differ from a curve where temperature is measured elsewhere. The shape of the reaction rate curve depends on the thermocouple location if it is plotted vs. temperature rather than time. Figure 10 shows differential thermal analysis curves of the $CaCO_3$ dissociation where the temperature was measured (1) at the outside surface of the reference material, and (2) at the center of the $CaCO_3$. The magnitude of this effect is influenced by the sample size and heating rate.

E. Reaction Rate versus Measured Property

The particular property measured during thermal analysis is not always a simple function of the reaction rate curve discussed above. Of the commonly used methods, thermogravimetry and evolved gas analysis come closer to providing a direct measure of the reaction rate; consequently, these methods frequently are used for quantitative studies. Obviously these methods are limited to reactions involving a gas.

With a proper sample-holder design and calibration, differential thermal analysis provides a measure of the rate of heat absorption or evolution. This is not related as directly to the reaction rate as is gas evolution or absorption, although an understanding of reaction rates is, nevertheless, the basis for interpreting differential thermal analysis results. Remember that the heat of reaction may change with temperature. Furthermore, the heat capacity of the products is frequently different from that of the reactants. Both of these factors will change the shape of the differential thermal analysis curve as compared to the reaction rate curve. If the sample itself provides the temperature drop used to indicate heat flow, changes in the curve due to changes in apparent thermal conductivity of the sample must also be considered.

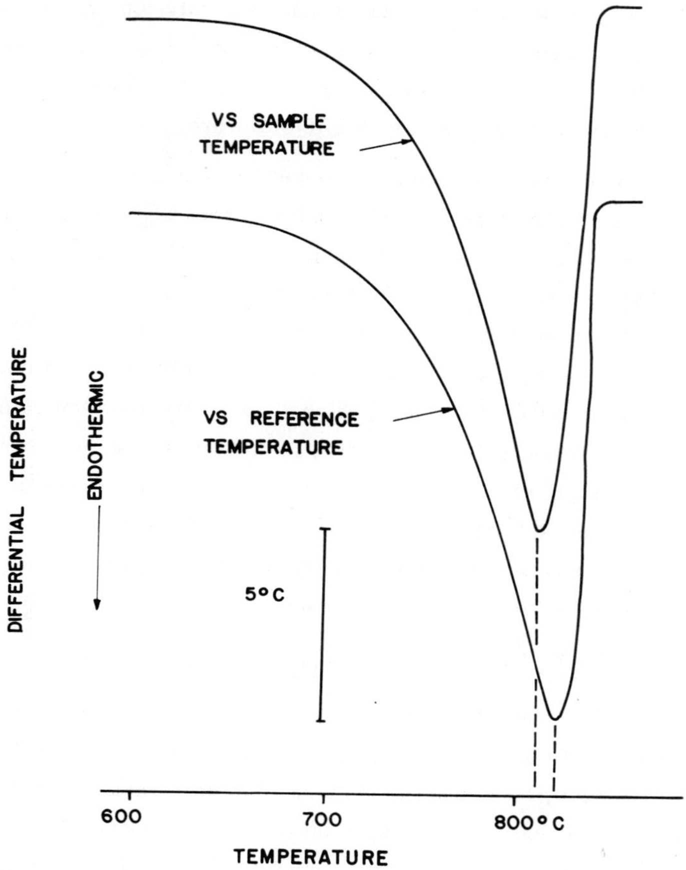

Fig. 10. Differential thermal analysis curves of calcite showing effect of thermocouple location.

These variables must be considered in addition to reaction rate curves if valid interpretations of differential thermal analysis data are to be made. The presence of such variables is not necessarily a disadvantage, however. For example, the variation in the curve due to changes in apparent thermal conductivity can provide useful information.

VII. APPLICATIONS

Now that the fundamentals of thermal analysis have been considered, a few applications of these methods shall be discussed.

One of the early examples of the use of differential thermal analysis for obtaining mineralogical analyses was in the area of clay minerals[17]. Because of the extremely small particle sizes of these materials (0.001 to 1μ)[12] and extensive solid solution, light microscopy and room temperature x-ray diffraction were not very useful analytical methods. Thermal analysis has been used widely for analyses of clay minerals. Although differential thermal analysis has been the method most widely used in this area, evolved gas analysis and thermogravimetry are useful also for studying clay minerals, since all of these minerals evolve water on heating.

Certain cations in the clay mineral, montmorillonite, are present in positions where they exert a strong influence on the water of hydration present between structural sheets. Cations producing stronger bonds tend to make the water of hydration more stable in the structure. Thus, the dehydration provides an indication of the cations present in the structure[19].

Figure 11 shows differential thermal analysis curves of montmorillonite containing Li, Na, or K. The K and Na samples exhibited a single endothermic peak in the 100-200°C range corresponding to dehydration; but, the Li sample exhibited two peaks in this range, indicating the greater water content. As is frequently the case, sample preparation was important in this

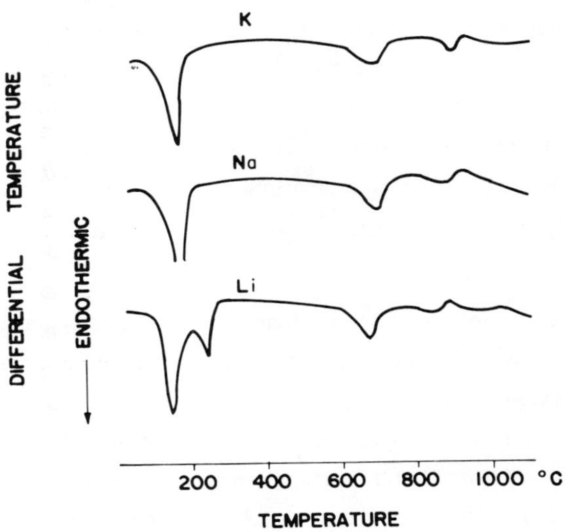

Fig. 11. Differential thermal analysis curves of montmorillonite samples having different base exchange cations. After Mielenz et al.[19].

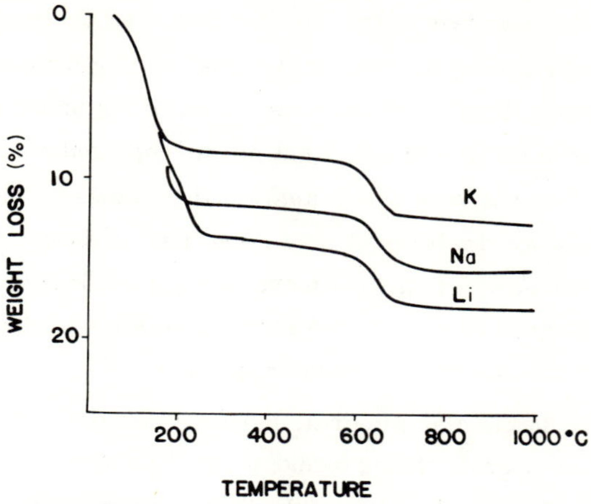

Fig. 12. Thermogravimetric curves of montmorillonite samples having different base exchange cations. After Mielenz et al. [19].

study. The samples were stabilized at 52% relative humidity prior to the runs. Meaningless results would be obtained in this case if the sample preparation were not controlled.

Since a single thermal analysis method is usually insufficient to give a complete picture, thermogravimetry also were obtained on these samples (Fig. 12). These results showed an excellent correlation between quantity of weight loss in the 100 - 300°C range and the cation size. Furthermore, the thermogravimetric curves indicate a weight loss corresponding to the 600-700°C endotherm, due to evolution of water present as hydroxyl groups in the structure. It should be noted that the heat effects at about 900°C do not involve a weight loss. High-temperature x-ray diffraction could be used to show that the samples recrystallize into more stable structures at this temperature.

In addition to information on the cations present and the water evolution, the entire curves serve to "fingerprint" the mineral montmorillonite and, thus, indicate its presence in a sample. Other clay minerals give different curves. The clay minerals usually occur in combination with other minerals, such as quartz and calcite. Quartz undergoes a polymorphic transition at about 573°C (Fig. 13). The temperature and characteristic shape of the endothermic peak caused by this reaction is characteristic of quartz. Where the influence of sample thermal conductivity is eliminated, the magnitude of

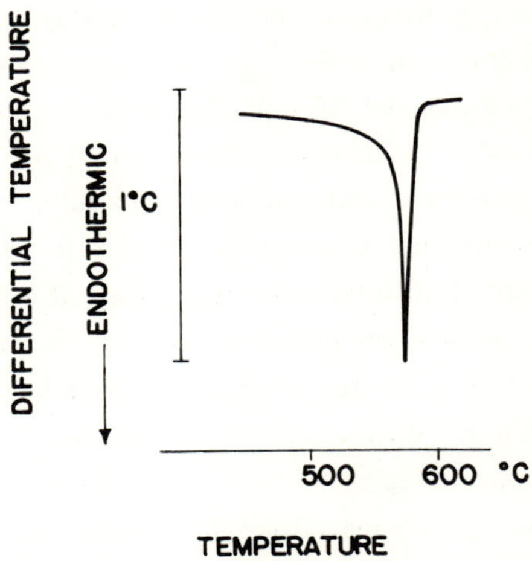

Fig. 13. Differential thermal analysis curve of quartz.

this peak can provide a measure of the quantity of quartz present. Unfortunately, the 600°C peak of montmorillonite will frequently mask the much smaller quartz peak, a common occurrence where a clay mineral is present. This is a case where our knowledge of reaction rates is particularly valuable. Recalling that changing the pressure will sometimes significantly change the temperature of a reaction when the reaction involves a gas, the quartz peak is exposed if the clay sample is heated in elevated water pressures, thus shifting the dehydroxylation peak to higher temperatures[27].

The presence of calcite ($CaCO_3$) in a clay sample could also present difficulties caused by overlapping peaks. Under some conditions the CO_2 evolution from calcite would overlap the 600°C water evolution from the clay mineral. Here again, the peaks might be separated by controlling the atmosphere. Heating in 1 atm pressure of CO_2 would shift the calcite dissociation to about 900°C, providing the water vapor was not allowed to displace the CO_2 surrounding the calcite particles. This would serve to identify the calcite peak as a reaction involving CO_2 as well as exposing the clay dehydroxylation peak. A more straightforward method would be to make evolved gas analysis runs using a mass spectrometer tuned to H_2O for one run and CO_2 for the other.

The above provides an example of the information that can be obtained on clay samples using thermal analysis. Not only may the various minerals which are present be identified, but the particular clay mineral may be identified and information concerning its composition obtained. Such information may, in some cases, be difficult or even impossible to obtain by other analytical methods. Chemical analyses usually are valuable but do not distinguish between mineral combinations having similar chemical compositions. While x-ray diffraction is usually valuable for mineral analyses, the patterns may be difficult to interpret when several minerals are present which produce overlapping x-ray peaks. Furthermore, evolved gas analysis can detect much smaller quantities of minerals, such as calcite and pyrite, than is possible with x-ray diffraction. For example, the author has detected as little as 0.02 ppm pyrite using evolved gas analysis. Microscopy is a powerful tool in mineral analyses; although for materials having very small particle sizes, such as clay minerals, the light microscope is useless. Furthermore, quantitative analyses are usually rather tedious. Thus, there are applications where thermal analysis is uniquely useful. The list is growing as instrumentation is improved and the techniques become more sophisticated. Today, the field of thermal analysis is advancing rapidly.

A word of warning: "the coin has two sides." For many purposes, thermal analysis is completely inapplicable. Some minerals exhibit no distinct changes in properties when heated to easily obtainable temperatures (e.g., corundum). Others may exhibit reactions which are not unique enough to serve as a "fingerprint." In the above discussion, the tacit assumption was made that the reactions exhibited by a mineral in a sample are always the same as when the mineral is heated alone. This is not always true, particularly when one or more component undergoes melting. In such cases, reactions at higher temperatures frequently are changed completely.

In some cases, interaction between component minerals also can occur as solid state reactions. This possibility should never be overlooked. Also, the condition of the sample can be most important. As discussed earlier, particle size and shape can influence the reaction rates and, hence, the curves obtained.

These words of warning are not intended to discourage the use of thermal analysis, for these are powerful tools and are becoming more so each day. As in other techniques, however, the results are no better than the person interpreting the data.

ACKNOWLEDGMENT

The author wishes to thank Warren H. Turner and Virginia L. Ware for their review of the manuscript and Owens-Illinois for permission to publish this work.

REFERENCES

(1) G. E. Ashby and R. C. Kellagher, "An Apparatus for the Study of Thermoluminescence from Minerals, "Am. Min., 43, 695-706 (1958).

(2) W. R. Bandi, et al., "Differential Thermal Analysis — Effluent Gas Analysis Method for Identification and Determination of Nitride in Steel," Anal. Chem., 38 [10] 1336-41 (1966).

(3) S. Bog and T. Rosenquist, "A High-Temperature Manometer and the Decomposition Pressure of Pyrite," Trans. Faraday Soc., 55, 1565-1569 (1959).

(4) P. Busiere, B. Claudel, J. P. Renouf, Y. Trambouze, and M. Prettre, "Study of Solid Phases By the Simultaneous Application of Thermogravimetry, Differential Thermal Analysis and Radio-active Emission," J. Chem. Phys., 58, 668-674 (1961).

(5) L. Cahn and H. Schultz, "Aerodynamic Forces in Thermogravimetry," Anal. Chem., 35 [11] 1729-1731 (1963).

(6) W. J. Campbell, S. Stecura, and C. Grain, "High Temperature Furnaces for X-ray Diffractometers," Bureau of Mines Rept. of Investigations 5738 (1961).

(7) D. W. A. Forbes, "Dilatometric Analysis as a Control Tool," Ceramic Age, 81 [9] 38-41 (1965).

(8) H. L. Friedman, "Error in Vacuum Thermogravimetry," Anal. Chem., 37 [6] 768-9 (1965).

(9) P. D. Garn and J. E. Kessler, "Effluence Analysis As An Aid to Thermal Analysis," Anal. Chem., 33, 952 (1961).

(10) J. K. Gillham, "Thermomechanical and Thermal Analysis of Fiber-Forming Polymers," Appl. Poly. Sym., 2, 59-75 (1966).

(11) R. S. Gohlke and H. G. Langer, "Thermal Analysis by Mass Spectrometry," Anal. Chem., 37 [10] 25A-29A (1965).

(12) R. E. Grim, Clay Mineralogy, p. 106-125, McGraw-Hill, New York (1953).

(13) H. Hoerster and E. Kauer, "Investigation of the Kinetics of Gas-Solid Reactions By Means of Electrical Measurements," Z-Elektrochem, 66 [8/9] 667-671 (1962).

(14) T. R. Ingraham and P. Marier, "Kinetic Studies on the Thermal Decomposition of Calcium Carbonate," Canadian J. Chem. Eng., 41, 170-173 (1963).

(15) H. G. Langer and R. S. Gohlke, "Mass Spectrometric Thermal Analysis (MTA)," Anal. Chem., 35 [9] 1301-2 (1963).

(16) G. M. Lukaszewski, "Accuracy in Thermogravimetric Analysis," Nature, 194 [4832] 959-961 (1962).

(17) R. C. Mackenzie, ed., The Differential Thermal Investigation of Clays, Mineralogical Society, London (1957).

(18) H. G. McAdie, "Simultaneous Thermal Analysis Using the Open-Pan Type of Sample Holder," Anal. Chem., 35, 1840-1844 (1963).

(19) R. C. Mielenz, N. C. Schieltz, and M. E. King, "Effect of Exchangeable Cation on X-ray Diffraction Patterns and Thermal Behavior of a Montmorillonite Clay," Proc. Third National Conf. on Clays and Clay Minerals, pp. 146-173, Publication 395, Nat. Acad. Sci. – Nat. Rech. Council, Washington, D. C. (1955).

(20) C. B. Murphy, et al., "Differential Thermal Analysis and Simultaneous Gas Analysis," Anal. Chem., 32, 1374-1375 (1960).

(21) A. E. Newkirk, "Thermogravimetric Measurements," Anal. Chem., 32 [12] 1558-1563 (1960).

(22) T. Ozawa, "A New Method of Quantitative Differential Thermal Analysis," B. Chem. Soc. Japan, 39 [10] 2071-2085 (1966).

(23) A. Reisman, "Isobaric Dissociation Studies of Alkali-Metal Carbonate Hydrate Using Simultaneous Differential Thermal Analysis-Thermogravimetric Analysis," Anal. Chem., 32, 1566-1574 (1960).

(24) G. Sabatier, "Measure of the Heats of Transformation by Differential Thermal Analysis," Bull. Soc. Franc. Mineral., 77, 953-968 (1954).

(25) H. S. Scott, "The Decrepitation Method Applied to Minerals with Fluid Inclusions," Econ. Geol., 43, 637-654 (1948).

(26) J. R. Soulen and I. Mockrin, "Improved Methods of Using the Thermobalance to Determine Thermal Stabilities," Anal. Chem., 33, 1909 (1961).

(27) R. L. Stone, "Differential Thermal Analysis of Kaolin Group Minerals Under Controlled Partial Pressures of H_2O," J. Am. Ceram. Soc., 35 [4] 90-99 (1952).

(28) F. M. Wahl, R. E. Grim and R. B. Graf, "Phase Transformations in Silica As Examined By Continuous X-ray Diffraction," Am. Min., 46 [1 and 2] p. 196-208 (1961).

(29) J. H. Welch, "A Simple Microscope Attachment for Observing High-Temperature Phenomena," J. Sci. Inst., 31, 458-462 (1954).

(30) W. W. Wendlandt, P. H. Franke, Jr., and J. P. Smith, "High Temperature Diffuse Reflectance Spectroscopy," Anal. Chem., 35 [1] 105-7 (1963).

(31) W. W. Wendlandt and T. M. Southern, "An Apparatus for Simultaneous Gas Evolution Analysis and Mass Spectrometric Analysis," Anal. Chim. Acta, 32, 405-10 (1965).

(32) F. W. Wilburn, S. A. Metcalfe, and R. S. Warburton, "Differential Thermal Analysis, Differential Thermogravimetric Analysis and High Temperature Microscopy of Reactions Between the Major Components of A Sheet Glass Batch," Glass Tech., 6 [4] 107-114 (1965).

(33) Akira Yamamoto, "Determination of Characteristic Points on Commercial Glasses by Differential Thermal Analysis," Bunseki Kagaku, 14 [8] 692-696 (1965).

SELECTED BIBLIOGRAPHY

C. Duval, Inorganic Thermogravimetric Analysis, 531 pp., Elsevier, New York (1953). A pioneer study of the weight loss during heating of precipitates proposed for inorganic gravimetric analysis. Several hundred curves are reproduced.

R. C. Mackenzie, The Differential Thermal Investigation of Clays, 456 pp., Mineralogical Society, London (1957). Although directed toward clays, a considerable number of curves on accessory minerals, such as carbonates and silica are included. A well-written and valuable book, although the sections on apparatus and theory are now becoming outdated.

W. W. Wendlandt, Thermal Methods of Analysis, 424 pp., Interscience , New York (1964). Diagrams and photographs of most commercially available apparatus are reproduced. The principles of operation and variables of a wide range of thermal methods of analysis are described. Applications of the several techniques are illustrated. Although the book is directed primarily to the analytical chemist, it is useful to others also.

P. D. Garn, Thermoanalytical Methods of Investigation, 606 pp., Academic, New York (1965). Primarily, this book is written for the researcher who is beginning or planning to use thermal analysis. Differential thermal analysis and thermogravimetry are emphasized, with brief coverage of other methods. The variables involved in the design and use of thermal analysis methods and in interpretation of results are discussed in detail.

H. G. McAdie, "Recommendations for Reporting Thermal Analysis Data," Anal. Chem., 39, 543 (1967). A discussion of the minimum information which should be included when publishing thermal analysis data.

Saul Gordon, "Thermoanalysis," Lab. Man., 3, 23 (1965); W. W. Wendlandt, "Thermoanalysis," Lab. Man., 3, 26 (1965); and Jen Chiu, "Thermoanalysis," Lab. Man., 3, 29 (1965). A series of articles directed to research managers who are considering the addition of thermal analysis to their laboratory's capabilities.

Chapter 10

POINT DEFECT ANALYSIS

William J. James and Gordon Lewis

Departments of Chemistry and Ceramic Engineering
and
Graduate Center for Materials Research
University of Missouri-Rolla
Rolla, Missouri

I. INTRODUCTION

It is commonly accepted practice to use the integral (stoichiometric) formula to represent compounds. This makes it convenient numerically to formulate chemical reactions employing the concept of oxidation number and the assignment of an atom to each possible position in a chemical structure. The problem, however, is that compounds seldom behave as simply as described by the integral formula. This is indeed fortunate because television sets, transistor radios, and tape recorders, to name a few, would not be current market items otherwise. These items use devices that incorporate compounds containing imperfections, that is, deviations exist in the ratio of the elements in the compound from the integer values, or small amounts of an impurity are present which may lead to the same thing.

To limit the scope of this chapter, the defect types considered will be point defects; impurities, unoccupied structural sites (vacancies), and atoms or ions at special sites not normally occupied in the structure (interstitial sites). The term impurity is extended beyond its normal meaning to include any atom occupying the structural site of another atom. On this basis, a cation (of the compound) occupying an anion site constitutes an impurity.

By associating each site in a space lattice with a particular chemical species, the ideas of structure and compound can be tied together in the sense that the ratio of sites or of atoms can each be expressed by integers.

Accordingly, an ideal compound is defined as one with a one-to-one relation between structural sites for a specified anion and that anion and for a specified cation and that cation. Note that the same correspondence between anion and cation structural sites is not implied.

A defective compound is then any material in which there is deviation from the conditions specified for the ideal compound, e.g., any compound with impurities is defective by definition.

In Figs. 1(a) and (b), an ideal compound, MgO, crystallizing with a sodium chloride structure (B-6), is represented schematically in two ways. The plane used to represent the three-dimensional crystal is shown in (c). It should be noted in this representation that interstitial positions are not in the plane but a little above or below it, even though they appear as if they occurred in the plane. In (c) it is possible to identify nine cation sites (and cations), sixteen anion sites (and anions), and twelve interstitial positions

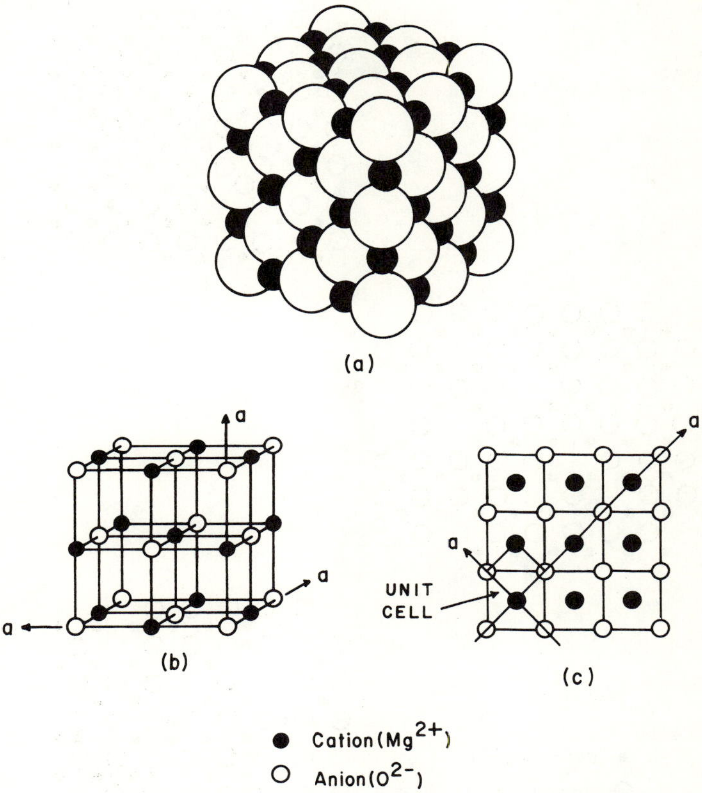

Fig. 1. Hard-sphere model (a), expanded structure model (b), and two-dimensional representation (c), of ideal magnesium oxide (MgO).

that are not occupied. However, in the unit cell, the site ratio* is equal to the anion-to-cation ratio, and each ratio is unity.

In Fig. 2, three hypothetical defective forms of MgO are shown. In (a) it can be seen that a few cations occupy anion sites and, because the anion ratio must be maintained to preserve the net balance of positive and negative charge, an equal number of anions appear on cation sites. The misplaced ions are termed antistructure defects.

In (b) six vacancies have been introduced in the anion structure, and an equal number in the cation structure. Again, to preserve the charge balance

*The site ratio is defined as $= \dfrac{\text{anion sites in unit cell structure}}{\text{cation sites in unit cell structure}}$

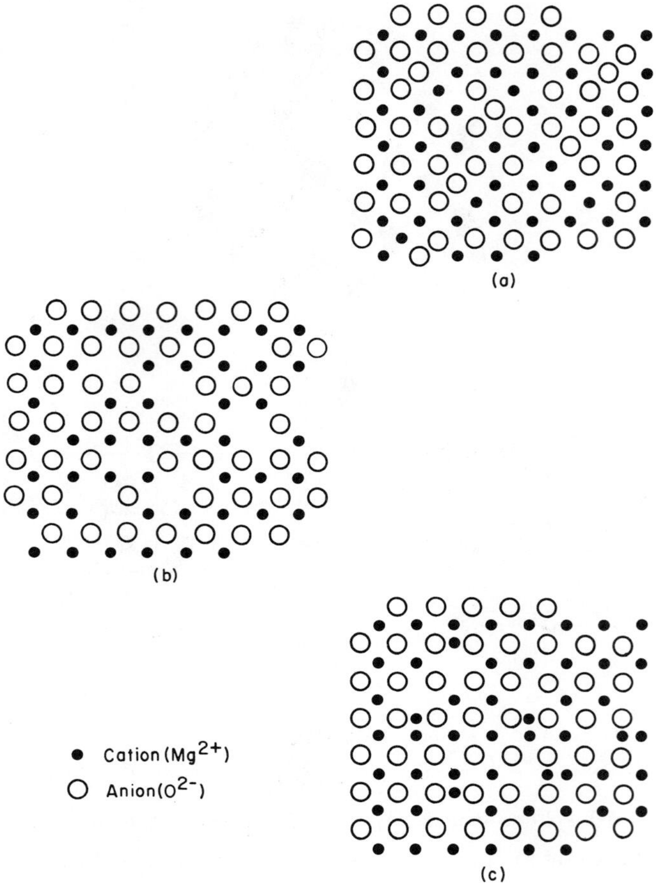

Fig. 2. Schematic representations of three defect types: Antistructure (a), Schottky (b), and Frenkel (c).

between species, equal numbers of each type of vacancy must exist. Defects exhibiting this behavior are termed Schottky defects.

In (c) six cations are present in some of the interstitial positions. If the effective charge that was on the cation is retained on the cation site, the interstitial site is occupied by a neutral atom. The anion-to-cation ratio remains at unity. Defects exhibiting this behavior are termed Frenkel defects.

It is important to note that, in each case, the crystal is "stoichiometric." That is, the ratio of anions to cations is an integer (in these examples, unity).

There are, however, significant changes in the properties of the crystal as shown in Table I.

In Fig. 3, a foreign metal atom is substituted for some Mg, i.e., an atom impurity has been added to the compound. Assuming that the oxidation number of the atom impurity is the same as for a Mg atom on that site, then no vacancies result because charge balance requires an equal number of anions and cations and hence a one-to-one substitution of impurity for Mg. Note that all sites, both anion and cation, are filled; hence, from a structural standpoint, the material is stoichiometric. But if the chemical analysis is reported only in terms of Mg and O, the compound must be considered nonstoichiometric because the ration O/Mg is less than the O-site/Mg-site ratio.

Because confusion may arise as to whether one considers this compound stoichiometric or not, the above case will be referred to as solid solution behavior, and the terms stoichiometric and nonstoichiometric will not be applied in general.

TABLE I

Comparison of Selected Properties between Defect Types

Classification	Ideal	Antistructure	Schottky	Frenkel
Schematic representation	Fig. 1 (c)	Fig. 2 (a)	Fig. 2 (b)	Fig. 2 (c)
Anion to cation ratio	1.000	1.000	1.000	1.000
Compound density	Ideal	Ideal[a]	Less than ideal[a]	Ideal[a]
Diffusion rate	Very low	Very low	Much higher than for "ideal" compound	Diffusion for atom involved in interstitial sites much higher than for "ideal" compounds

[a]It is assumed as a first-order appoximation that the spacing of structural sites remains constant at the "ideal" value. In real crystals, this assumption will not hold. Therefore, the Frenkel case will probably have a density slightly less than the ideal compound.

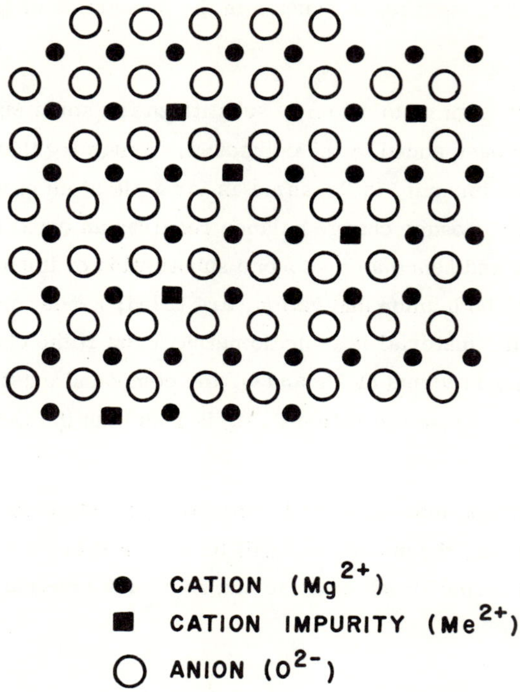

Fig. 3. Solid solution model representation of MgO.

Now consider a situation in which a number of Mg^{2+} ions is removed from the compound. This will leave the material with a net negative charge surplus and with unoccupied cation sites. Suppose that a trivalent cation, such as Fe^{3+}, is added to the material until charge balance occurs again. It is obvious that the imbalance between positive and negative charge will be removed before an Fe^{3+} can be added to every cation site that was emptied. This must be so because each Fe^{3+} carries one more positive charge than each Mg^{2+}. Therefore, when charge balance occurs, there will still be some vacant Mg^{2+} sites (i.e., V_{Mg}). The above description has been traced graphically on Fig. 4. The final structure is shown graphically in Fig. 5(a).

Now on any of the following bases, total anion sites occupied to total cation sites occupied, total O-atoms to total Mg-atoms, or total anions to total cations, the compound is nonstoichiometric, i.e., although these ratios are not equal, each is greater than unity. All of the above ratios are greater

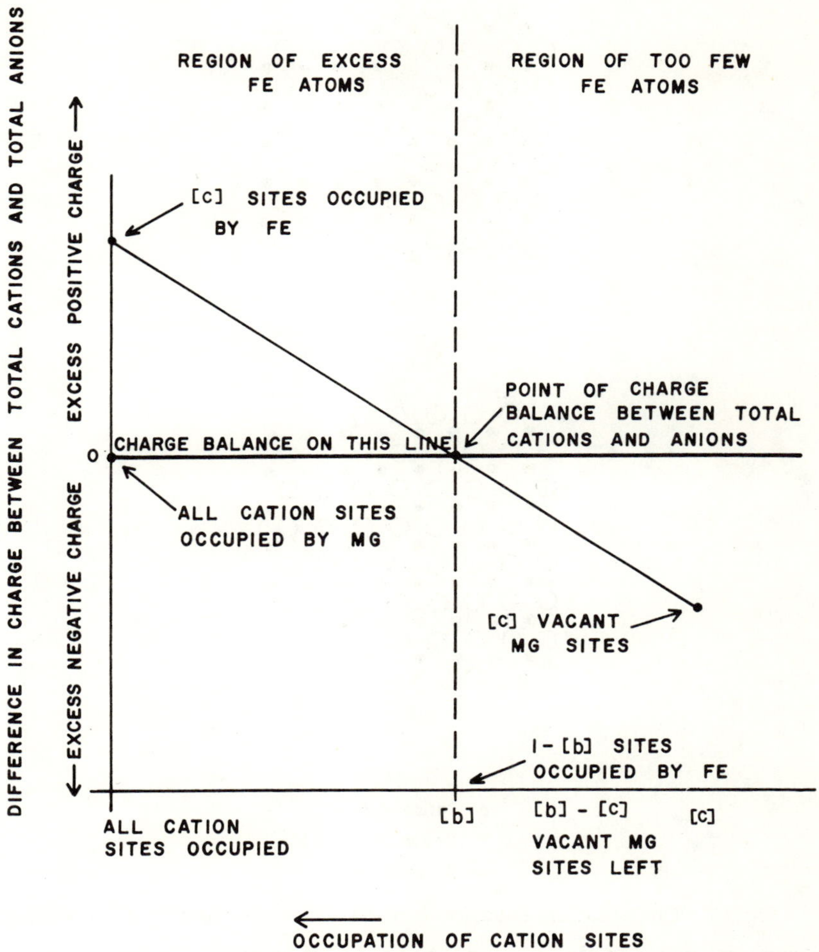

Fig. 4. Charge balance requirements when Fe^{3+} replaces Mg^{2+} in MgO.

than unity because some cation sites are vacant. The ratio of total anion sites to total cation sites, however, is still unity.

Charge balance was brought about in the above case by substituting a cation of greater oxidation number, with cation vacancies appearing as a consequence. The same net result could have been achieved by replacing a given number of Mg^{2+} with Fe^{3+}, then adding anions until the charge imbalance became zero. However, all of the oxygen sites (where the negative

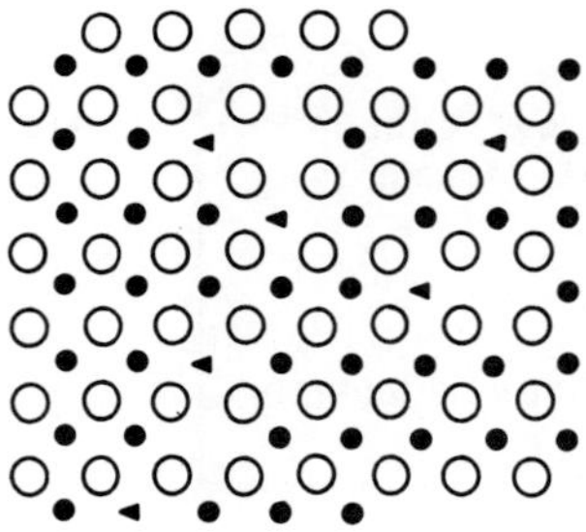

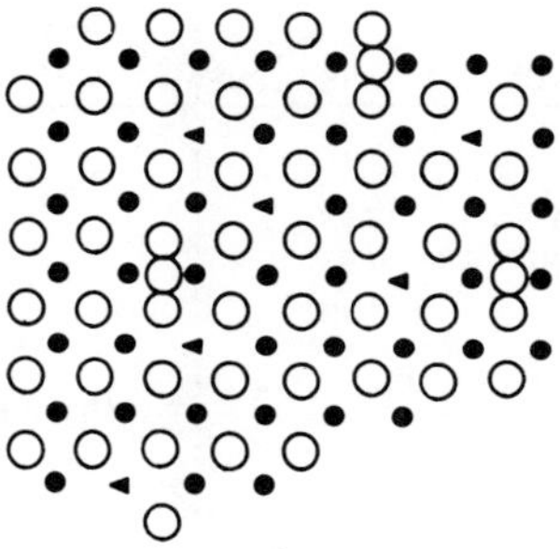

Fig. 5. Charge compensation models for an impurity (M^{3+}) obtained through cation vacancies (a) and through anion interstitials (b).

charges are assumed to reside) were filled; and therefore, additional negative ions could only be accommodated in interstitial vacancies* see Fig. 5(b) This increases the anion-to-cation ratio.

In summary, it has been shown that the anion-cation ratio can be made greater than unity (for an anion-site to cation-site ratio of unity) by either

*If the charge-compensating anions are placed on vacant anion sites at the compound surface, then cation vacancies must be created because anion and cation sites must be added to this structure in the site ratio. That is, the structure cannot be built up by adding only anion or cation sites.

introducing cations with a greater oxidation number than normal for the structural site and creating associated vacancies in the cation structure to allow charge compensation or by introducing anions into interstitial sites (because all structural anion sites are filled). Note that even though the anion-to-cation ratio may be the same for the two cases, if one compound has cation vacancies and the other anion interstitials, their other physical properties will be quite different, as shown in Table II.

Reflection at this point may lead to the thought that anion vacancies and cation interstitials might relate so as to yield an anion-to-cation ratio less than unity. This is of course possible, as shown in Table III.

An additional point is illustrated from the data of Table III. If a cation impurity with an oxidation number less than normal (2+) for the cation site replaces some of the Mg^{2+}, an excess of negative charge arises if all anion sites are left filled; hence, it is expected that anion vacancies should result. If, in addition, some interstitial sites are occupied by cation impurities, then charge balance can be achieved without anion vacancies. It should be noted that the effective charge on the interstitial cation is important only in

TABLE II

Comparison of Selected Properties for Impurity-Compensated Defects V_{Mg} and O_i

	Cation impurity compensated by V_{Mg}	Cation impurity compensated by O_i^{2-}
Anion-to-cation ratio (assumed)	53/50 = 1.06	53/50 = 1.06
No. of vacancies	3	0
No. of interstitials	0	3
Compound density	Less than for ideal structure[a]	Greater than for ideal structure[a]
Expected diffusion rate of impurity cation	High	Low

[a] The atomic weight difference between the cation species will be important at low defect concentrations and must be considered in each case.

TABLE III

Comparison of Selected Properties for Impurity-Compensated Defects V_O and I_i

	Cation impurity compensated by V_O	Cation impurity compensated by I_i^{2+}
Anion-to-cation ratio (assumed)	0.943	0.943
No. of vacancies	3	0
No. of interstitials	0	3
Compound density	Less than for ideal structure[a]	Greater than for ideal structure[a]
Expected diffusion rate of impurity cation	Low	Very low

[a] The atomic weight difference between the cation species will be important at low defect concentrations and must be considered in each case.

that it determines the number of such cations necessary to balance the excess negative charge created by substitution of the normal cations by cation impurities of lower oxidation state.

After examining all of these possibilities, the question naturally arises, in which of the above ways does the compound acquire charge balance? The answer is that the acceptable way is that which results in a crystal with the lower free energy. This solves the problem as far as rhetoric is concerned but yields no method for satisfactory calculation of actual free energy values for comparison and prediction. Therefore, one has to perform experiments on the material to discover how it has adapted itself to its environment. The next section relates how various imperfections lead to different values for the crystal properties and which properties one measures to ascertain the defect type and concentration.

II. GENERAL FORMULATION OF PROPERTY BEHAVIOR

In the previous section the defective states of a compound were related to their formation from the ideal state. This description is sufficient for conceptual purposes, but does not provide any insight into how defective states are prepared and maintained in an experiment. Assuming that the only independent variables used to establish the state of the compound are pressure, temperature and composition, the Gibb's phase rule provides a relation for determining how many of the above variables must be under the control of the experimenter for any given set of conditions. All other quantities (such as resistance, conductivity, density, etc.) measured in the experiment will be dependent on the independent variables.

As an example, consider that two phases, one a crystalline, defective solid and the other a gas, are in equilibrium. It was noted in the first section that for antistructure and Schottky defect situations, only two elements were needed, and the system was binary. In this case, the phase rule requires that two variables must be under the control of the investigator for the other measured variables to be reproducible.

A system of equations relating the concentration of defects in the solid to the independent variables will be developed. This is, of course, the reverse of what is actually done in an experiment, because there we are determining the behavior of certain parameters and deducing the type of defect yielding the behavior. When writing these equations, three requirements must be met:

(1). The site ratio in the ideal structure must be maintained.

(2). The net charge on the structure must be zero, i.e., the sum of all effective positive charges must equal the sum of all negative charges.

(3). Atoms can be neither created nor destroyed.

The idea of an atom and the site it occupies must be tied together by a symbol. Notation of a general nature is illustrated and briefly described in Table IV, see reference A (a) .

The question of how to show charge on an anion or on a cation in a particular site is summarized in Table V. It should be noted that the cation (anion) site in the structure carries the implied, opposite effective charge of the cation (anion); hence, when the cation (anion) site is filled by the proper cation (anion) the unit is regarded as neutral.

As the first example, antistructure defects will be assumed to exist in the compound and the equilibrium equation is written as:

$$Mg_{Mg} + O_O = Mg_O^{\bullet\bullet\bullet\bullet} + O_{Mg}^{''''} \qquad (1)$$

TABLE IV

Nomenclature for Various Defect Types

Mg_{Mg}	=	Mg^{2+} cation on a Mg structural site. No indication of positive charge is given because an ion on this site would normally be 2+.
Mg_i	=	Mg atom (here a neutral atom is meant) on an interstitial site in the structure. As there is only one type of interstitial site in this structure, only one need be identified. If two or more types exist, each has to be identified.
$Mg_i^{\bullet\bullet}$	=	Mg^{2+} cation in an interstitial site. The two dots represent the two positive charges on the cation. Because the interstitial site is considered to be neutral, the oxidation number of any ion on the site must be indicated.
$Mg_O^{\bullet\bullet\bullet\bullet}$	=	Mg^{2+} cation on an O^{2-} structural site. The four dots represent effective positive charges on the cation in this site. The four positive charges arise due to the loss of the two negative charges (equivalent to a gain of two positive charges) resulting from the absence of the oxygen ion and the two additional positive charges of the 2+ cation.
$Fe_{Mg}^{\bullet}$	=	Fe^{3+} cation on a Mg structural site.
Fe_i	=	Neutral Fe atom in an interstitial site.
$Fe_i^{\bullet\bullet}$	=	Fe^{2+} cation in an interstitial site.
$Fe_i^{\bullet\bullet\bullet}$	=	Fe^{3+} cation in an interstitial site.
O_O	=	O^{2-} anion on a structural site assigned to oxygen.
O_i''	=	O^{2-} anion on an interstitial site.
O''''_{Mg}	=	O^{2-} anion on a Mg structural site.

TABLE V

Nomenclature for Charged Defects

Me^{4+} on Mg	=	$Me_{Mg}^{\bullet\bullet}$	A^{4-} on O	=	A_O''
Me^{3+} on Mg	=	$Me_{Mg}^{\bullet}$	A^{3-} on O	=	A_O'
Me^{2+} on Mg	=	Me_{Mg}	$O^{2=}$ on O	=	O_O
Me^{+} on Mg	=	Me_{Mg}'	A^{-} on O	=	$A_O^{\bullet}$
Me^{o} on Mg	=	Me''_{Mg}	A^{o} on O	=	$A_O^{\bullet\bullet}$
A^{1-} on Mg	=	A'''_{Mg}	Me^{1+} on O	=	$Me_O^{\bullet\bullet\bullet}$
A^{2-} on Mg	=	A''''_{Mg}	Me^{2+} on O	=	$Me_O^{\bullet\bullet\bullet\bullet}$

This equation represents an internal equilibrium; i.e., only the solid phase is involved. An equilibrium constant can be formulated* as

$$K_{(1)} = \left[Mg_O^{\cdot\cdot\cdot\cdot}\right] \cdot \left[O_{Mg}^{''''}\right] \tag{2}$$

Pressure does not appear as a variable in this equation. Temperature is a variable because the equilibrium constant is a function of temperature. However, the specific function behavior of temperature is rarely known.

The second case to be considered is that of Schottky behavior as expressed by the following equilibrium relations:

$$0 = V_{Mg} + V_O \tag{3}$$

and

$$K_{(3)} = \left[V_{Mg}\right] \cdot \left[V_O\right] \tag{4}$$

The zero on the left-hand side of the equation may seem confusing; however, this symbolism means that creating Schottky defects in the solid appears to be equivalent to simply expanding the volume of the compound by relocating atoms. The atoms that occupied the now emptied sites effectively have migrated to the surface and extended the structure. Again, the independent variables, temperature and pressure, do not enter explicitly into the equation.

---

*The precise definition of the equilibrium constant should be $K'_{(1)} = \left[(a_{Mg_O^{\cdot\cdot\cdot\cdot}})(a_{O_{Mg}^{''''}})\right] / \left[(a_{Mg_{Mg}})(a_{O_O})\right]$. Where $a_{Mg_O^{\cdot\cdot\cdot\cdot}}$ = activity of Mg on an oxygen site; $a_{O_{Mg}^{''''}}$ = activity of O on a magnesium site; $a_{Mg_{Mg}}$ = activity of Mg on a magnesium site; and a_{O_O} = activity of O on an oxygen site.

However, if it is assumed that the activity (effective concentration) of a species can be approximated satisfactorily by its actual concentration and that the effective concentration of Mg_{Mg} and O_O will be constant in spite of wide variations in the actual concentration of the species $Mg_O^{\cdot\cdot\cdot\cdot}$ and $O_{Mg}^{''''}$, the equation simplifies to:

$$K_{(1)} = \left[Mg_O^{\cdot\cdot\cdot\cdot}\right] \cdot \left[O_{Mg}^{''''}\right]$$

where

$$K_{(1)} = \left[Mg_{Mg}\right]\left[O_O\right] K'_{(1)}$$

The investigation of Frenkel defects follows the same pattern, i.e.,*

$$Mg_{Mg} = Mg_i + V_{Mg} \quad (5)$$

$$K_{(5)} = [Mg_i][V_{Mg}] \quad (6)$$

Again temperature and pressure do not enter these equations explicitly. Consider now the solid solution case:

$$Fe_{(\text{another phase containing } Fe^{2+})} + V_{Mg} = Fe_{Mg} \quad (7)$$

$$K_{(7)} = \frac{[Fe_{Mg}]}{[Fe_{(\text{another phase})}][V_{Mg}]} \quad (8)$$

Again, temperature and pressure are not introduced explicitly. However, there is a new feature; an impurity from another phase is introduced into the compound. It is implied in the equation that the oxidation number of the impurity is the same in the MgO as it is in the other phase. This equation appears to imply further that vacancies must exist for substitution to occur. This is not correct because the full equation should be

$$Fe_{(\text{another phase})} + O_{\text{another phase}} + V_{Mg} + V_O = Fe_{Mg} + O_O \quad (9)$$

where the V_{Mg} and V_O could be surface sites rather than bulk sites in the crystal. The important point is that a one-to-one correspondence exists between new anion and cation sites created to accommodate the material introduced.

Suppose that the effective oxidation number of Fe in the impurity phase is three rather than two and that this oxidation state is preserved in the MgO compound. Then

$$2Fe_{(\text{another phase containing } Fe^{3+})} + 3O_{(\text{another phase})} + 3V_{Mg} + 3V_O$$

$$= 2Fe^{\bullet}_{Mg} + 3O_O + V_{Mg} + 2e' \quad (10)$$

Because charge imbalance would occur in this equation, a charged species must be introduced in the MgO to compensate the unusual charge on the Mg

*It is assumed that the charge on the cation in the cation site (Mg_{Mg}) has remained on the site (V_{Mg}) and the interstitial atom is neutral (Mg_i).

site occupied by the Fe^{3+}. Note that the impurity phase introduces Fe and O in the inverse ratio of the oxidation numbers of Fe and O, i. e., Fe/O = 2/3. In this manner, no extra charge is brought into the MgO compound; the site ratio of the MgO compound is unity, however. Therefore, one Mg site will not be occupied if it is assumed that O goes only into O structural sites. This creates a vacancy, V_{Mg}. In addition, charge balance must be met, but because all structural sites are now accounted for, the charged species cannot change the site ratio; hence, electrons must be used.*

Because oxygen from another phase is a quantity expressed in the equation, we can use the oxygen from any phase, such as a gas, and write Eq. (10) as†

$$2Fe_{(another\ phase)} + \frac{3}{2}O_{2(g)} + 3V_{Mg} + 3V_O$$
$$= 2Fe^{\cdot}_{Mg} + 3O_O + V_{Mg} + 2e' \qquad (11)$$

This equation implies that the solubility (related to the effective concentration of Fe in MgO) is an implicit function of oxygen partial pressure in the ambient atmosphere.

The reduced equilibrium constant for reaction (11) can be written as:

$$K_{(11)} = \frac{\left[Fe^{\cdot}_{Mg}\right]^2 \left[V_{Mg}\right] \left[e'\right]^2}{P_{O_2}^{3/2}} \qquad (12)$$

in which the explicit relation is shown.

If, instead of using cation vacancies to compensate the charge imbalance introduced by Fe^{3+}, interstitial oxygen atoms are introduced; the following equation describes the situation:

$$2Fe_{(another\ phase)} + \frac{3}{2}O_{2(g)} + 2V_{Mg} + 2V_O$$
$$= 2Fe^{\cdot}_{Mg} + 2O_O + O_i + 2e' \qquad (13)$$

---

*The electrons may be trapped on the V_{Mg} yielding either V'_{Mg} or V''_{Mg}. Whether this will occur or not is dependent entirely on free energy considerations and beyond the scope of the present chapter.

† This is because of the following equilibrium: $\frac{1}{2}O_{2(g)} + V_O = O_O$. But $V_O + V_{Mg} = 0$ (Schottky relation); hence, we can also write $\frac{1}{2}O_{2(g)} = O_O + V_{Mg}$.

The equilibrium constant is written as:

$$K_{(13)} = \frac{\left[Fe^{\cdot}_{Mg}\right]^2 \left[O_i\right] \left[e'\right]^2}{P_{O_2}^{3/2}} \tag{14}$$

It can be observed that, at a given temperature and oxygen pressure, the number of charge carriers (e^-) will not be the same for a given solubility of iron when the same number of defects (V_{Mg} or O_i) is introduced because $K_{(11)}$ and $K_{(13)}$ are not equal in general.

When a monovalent cation is substituted for a Mg^{2+} in the structure, the same analysis is used, except that cation interstitials or anion vacancies result and positive holes ($p^{\cdot}$) may be used to provide a charge balance.

The next section will use these general equations and relate them to specific experimental quantities, such as conductivity, density, etc.

III. SPECIFIC FORMULATIONS

A. Electrical Conductivity

An electric current can be maintained in a material by the three carriers; electrons (e'), positive holes ($p^{\cdot}$), and ions. The magnitude of the current is related to the carrier charge, the carrier mobility, and the total number of carriers, by a relation such as:

$$\sigma = \sum_i C_i e_i \overline{V}_i \tag{15}$$

where σ = Total conductivity.

C_i = The number of carriers of type i.

e_i = The effective charge on the type i carrier. Both positive and negative carriers may participate in conduction; however, only the magnitude of e_i is important for total conductivity.

$\overline{V}_i$ = Mean velocity of the type i carrier.

If it is assumed as a first approximation that only a single type of carrier predominates and that the velocity and charge on the carrier are independent of the carrier concentration, then the conductivity will be proportional to the carrier density. This assumption means that the following equations will govern the situation:

$$\sigma = (\text{constant}) \cdot C \tag{16}$$

$$\log \sigma = \log C + \log(\text{constant}) \tag{17}$$

Electrons and positive holes will be the only carriers discussed in this section (ions as carriers will be discussed under diffusion). Equation (12) provides the relation existing between electrons and the independent variables, pressure and temperature, and it provides the tie needed for an evaluation of the functional behavior of conductivity.

The basic steps in relating the conductivity to the independent variables are:

(1). Write down all chemical (sometimes called pseudochemical) equations that relate charge carriers to the independent variables. As an example consider the behavior of singly ionized oxygen vacancies in titania. The equation can be written:

$$O_O = V_O^{\bullet} + e' + \frac{1}{2}O_2(g) \tag{18}$$

(2). Equilibrium constant expressions are formulated from the chemical reactions:

$$K'_{(18)} = \frac{[V_O^{\bullet}][e']}{[O_O]} \cdot P_{O_2}^{1/2} \tag{19}$$

(3). Determine the neutrality condition. This is an equation that includes all charged species and shows that all negative charge is balanced by an equal amount of positive charge because the system must remain neutral. The total neutrality condition assuming only ionized oxygen vacancies could be expressed as:

$$[V_O^{\bullet}] + 2[V_O^{\bullet\bullet}] = [e'] \tag{20}$$

Because it has been assumed that the doubly ionized oxygen vacancy ($V_O^{\bullet\bullet}$) is present in such a small concentration that

$$2[V_O^{\bullet\bullet}] << [V_O^{\bullet}] \tag{21}$$

the neutrality condition reduces to

$$[V_O^{\bullet}] = [e'] \tag{22}$$

That is, the concentration of electrons available for conduction equals the concentration of singly ionized oxygen vacancies.

A set of equations has been obtained that contains the independent variables, P and T, and concentration terms for each of the principal species. When these are combined, the result is

$$[e'] = K_{(18)}^{+\frac{1}{2}} \cdot P_{O_2}^{-\frac{1}{4}} \tag{23}$$

Additional models can be formulated in an analogous manner. The results have been used to prepare Table VI.

In order to obtain Table VI, it was assumed in each model that only a single defect would predominate. In practice, this usually is not the case; therefore, the power laws that are derived may be observed only over narrow pressure ranges.

There are two discernable behavioral patterns evident from the data. Sample conductivity increases with oxygen pressure when ionized titanium vacancies or ionized oxygen interstitials predominate or sample conductivity decreases with oxygen pressure when ionized titanium interstitials or ionized oxygen vacancies predominate. Therefore, if the experimentally observed conductivity increases with oxygen partial pressure, ionized titanium interstitials or ionized oxygen vacancies cannot be of major importance in conduction. If the experimentally observed conductivity decreases with increase in oxygen partial pressure, then ionized titanium vacancies and ionized oxygen interstitials can be neglected. It should be clear that, under the simplifying assumptions for this section, conductivity occurs only when there is an ionized defect in the structure.

It can be seen from Table VI that the following relation is obtained between $[e']$ and P_{O_2}:

$$\log [e'] = -(1/\eta) \log P_{O_2} \tag{24}$$

or

$$\log \sigma = -(1/\eta) \log P_{O_2} + \log (\text{constant}) \tag{25}$$

and for $[P^{\bullet}]$ and P_{O_2}

$$\log [P^{\bullet}] = (1/\eta) \log P_{O_2} \tag{26}$$

or

$$\log \sigma = (1/\eta) \log P_{O_2} + \log (\text{constant}) \tag{27}$$

[illegible] of Electrical Conductivity with Oxygen Pressure for Various Defect Types

Predominate defect type	Formation equation Step 1	Equilibrium constant equation Step 2	Simplified neutrality condition Step 3	P_{O_2} dependency of $[e']$	Conductivity dependency on P_{O_2}	η
Oxygen vacancies						
1. V_O	$O_O = V_O + 1/2\,O_{2(g)}$	$K_1 = P_{O_2}^{1/2} \cdot [V_O]$	None	None	None	
2. $V_O^{\cdot}$	$O_O = V_O^{\cdot} + e' + 1/2\,O_{2(g)}$	$K_2 = P_{O_2}^{1/2} \cdot [V_O^{\cdot}][e']$	$[e'] = [V_O^{\cdot}]$	$[e'] = (K_2)^{1/2} P_{O_2}^{-1/4}$	$\sigma\ \alpha\ P_{O_2}^{-1/4}$ ($\log\sigma\ \alpha\ -1/4 \log P_{O_2}$)	4
3. $V_O^{\cdot\cdot}$	$O_O = V_O^{\cdot\cdot} + 2e' + 1/2\,O_{2(g)}$	$K_3 = P_{O_2}^{1/2} \cdot [V_O^{\cdot\cdot}][e']^2$	$[e'] = 2\,V_O^{\cdot\cdot}$	$[e'] = (2K_3)^{1/3} P_{O_2}^{-1/6}$	$\sigma\ \alpha\ P_{O_2}^{-1/6}$	6
Metal interstitials						
4. Ti_i	$Ti_{Ti} + 2O_O = Ti_i + O_{2(g)}$*	$K_4 = P_{O_2} \cdot [Ti_i]$	None	None	None	
5. $Ti_i^{\cdot}$	$Ti_{Ti} + 2O_O = Ti_i^{\cdot} + e' + O_{2(g)}$	$K_5 = P_{O_2} \cdot [Ti_i^{\cdot}] \cdot [e']$	$[e'] = [Ti_i^{\cdot}]$	$[e'] = (K_5)^{1/2} P_{O_2}^{-1/2}$	$\sigma\ \alpha\ P_{O_2}^{-1/2}$	2
6. $Ti_i^{\cdot\cdot}$	$Ti_{Ti} + 2O_2 = Ti_i^{\cdot\cdot} + 2e' + O_{2(g)}$	$K_6 = P_{O_2} \cdot [Ti_i^{\cdot\cdot}][e']^2$	$[e'] = 2[Ti_i^{\cdot\cdot}]$	$[e'] = (2K_6)^{1/3} P_{O_2}^{-1/3}$	$\sigma\ \alpha\ P_{O_2}^{-1/3}$	3
7. $Ti_i^{\cdot\cdot\cdot}$	$Ti_{Ti} + 2O_O = Ti_i^{\cdot\cdot\cdot} + 3e' + O_{2(g)}$	$K_7 = P_{O_2} \cdot [Ti_i^{\cdot\cdot\cdot}][e']^3$	$[e'] = 3[Ti_i^{\cdot\cdot\cdot}]$	$[e'] = (3K_7)^{1/4} P_{O_2}^{-1/4}$	$\sigma\ \alpha\ P_{O_2}^{-1/4}$	4
8. $Ti_i^{\cdot\cdot\cdot\cdot}$	$Ti_{Ti} + 2O_O = Ti^{\cdot\cdot\cdot\cdot} + 4e' + O_{2(g)}$	$K_8 = P_{O_2} \cdot [Ti_i^{\cdot\cdot\cdot\cdot}][e']^4$	$[e'] = 4[Ti_i^{\cdot\cdot\cdot\cdot}]$	$[e'] = (4K_8)^{1/5} P_{O_2}^{-1/5}$	$\sigma\ \alpha\ P_{O_2}^{-1/5}$	5
Metal vacancies						
9. V_{Ti}	$O_{2(g)} = V_{Ti} + 2O_O$	$K_9 = [V_{Ti}] / P_{O_2}$	None	None	None	
10. V_{Ti}'	$O_{2(g)} = V_{Ti}' + P^{\cdot} + 2O_O$	$K_{10} = \dfrac{[V_{Ti}][P^{\cdot}]}{P_{O_2}}$	$[P^{\cdot}] = [V_{Ti}']$	$[P^{\cdot}] = K_{10}^{1/2} P_{O_2}^{1/2}$	$\sigma\ \alpha\ P_{O_2}^{1/2}$	2
Oxygen interstitials						
11. O_i	$1/2\,O_{2(g)} = O_i$	$K_{11} = [O_i] / P_{O_2}^{1/2}$	None	None	None	
12. O_i'	$1/2\,O_{2(g)} = O_i' + P^{\cdot}$	$K_{12} = \dfrac{[O_i'][P^{\cdot}]}{P_{O_2}^{1/2}}$	$[P^{\cdot}] = [O_i']$	$[P^{\cdot}] = K_{12} P_{O_2}^{1/4}$	$\sigma\ \alpha\ P_{O_2}^{1/4}$	4

*The relation given can be obtained from $Ti_{Ti} + 2O_O = Ti_i + V_{Ti} + 2O_O = Ti_i + V_{Ti} + 2V_O + O_{2(g)}$ but $V_{Ti} + 2V_O = 0$. Hence, $Ti_{Ti} + 2O_O = Ti_i + O_{2(g)}$.

Therefore, when log σ is plotted versus log P_{O_2}, the slope of the data yields directly η. If the value of η is, within experimental error, identical to one of the unique values of η from the models, then identification of the defect has been accomplished. However, because of experimental error and the fact that some models yield the same η value, it may be possible only to determine the behavior pattern and hence eliminate two of the four major types of defect.

The experimental conductivity dependence of TiO_2 on oxygen pressure (see Table VII) has been shown to be related to $(P_{O_2})^{-1/\eta}$ (where η is an integer). From this negative dependency relationship, it can be deduced that TiO_2 does not have a predominate defect involving ionized cation vacancies or ionized anion interstitials. The evidence requires that either anion vacancies or cation interstitials is the predominate type; but one cannot differentiate between them because the value of η cannot be determined precisely enough. In addition, note that both models yield the same value of η (i.e., 6) for a particular state of ionization of each.

TABLE VII

Comparison of Selected Experimental Data with Interpretation

Experimental values of η	Reference	Remarks
5	2	$P_{O_2} <$ 10 mm Hg Interprets data to substantiate cation interstitial model
6	2	$P_{O_2} >$ 10 mm Hg Interprets data to substantiate anion vacancy model
	3	Reinterprets Yahia data to substantiate anion vacancy model with $\eta = 5$ data representing transition from $\eta = 6$ to $\eta = 4$. That is, at high pressures oxygen vacancies doubly ionized, but at the lowest pressures Yahia studied, a mixture of doubly ionized and singly ionized vacancies resulted in a $\eta = 5$ result

In summary, conductivity measurements can be interpreted to separate the possible major defect behavior into two categories, one yielding increased conductivity with increased oxygen pressure and the other yielding decreased conductivity with increased oxygen pressure. If the experimental data are precise enough and the value of η is such that other models do not yield the same value, an identification of the predominate type can be made. Otherwise, an additional experimental method must be introduced to complement the conductivity experiment. Quantitative evaluation of the concentration of a major defect cannot be made by conductivity experiments alone because the carrier mobility and its effective charge will not have been determined.

B. Thermogravimetry

As an oxide builds in or eliminates oxygen atoms in order to adjust its stoichiometry during a change of state, the material loses or gains mass. Thermogravimetric studies of an oxide in selected atmospheres can be subjected to an analysis to aid in the interpretation of the defect type. The absolute concentration of defects cannot be obtained from such studies unless additional methods (such as density determination) are employed.

The basic procedure for investigating possible defect models is:

(1). Write down all chemical equations that relate defects to the independent variables. Suppose the presence of titanium interstitial atoms in defective TiO_2 is suspected. There is a single equation in this case,

$$Ti_{Ti} + 2O_O = Ti_i + O_{2(g)} \tag{28}$$

(2). Write down equilibrium constant expressions. The expression for Eq. (28) is

$$K'_{(28)} = \frac{[Ti_i]}{[Ti_{Ti}][O_O]^2} \cdot P_{O_2} \tag{29}$$

(3). Determine how the weight change, introduced by a changing oxygen content, is related to the defect. In the above case, two oxygen atoms leave the oxide for each interstitial titanium created, i.e.,

$$\text{No. of oxygen atoms eliminated} = 2[Ti_i] \tag{30}$$

Therefore, the weight of oxygen lost is

$$w_{28} = -(\text{No. of oxygens}) \frac{AW_O}{N_o} \quad (31)$$

$$= -2\left[Ti_i\right] \cdot \frac{AW_O}{N_o} \quad (32)$$

where AW_O = atomic weight of oxygen

N_o = Avogadro's number

(4). Combine these expressions to yield the weight dependency. In the above example, the final expression is

$$w_{28} = -2 \frac{AW_O \cdot K_{(28)}}{N_o} \cdot P_{O_2}^{-1} \quad (33)$$

Table VIII includes this result as well as those for other assumptions of defect behavior and the state of oxidation of the defect. Again, in order to simplify the results, it is assumed in each case that only a single defect type predominates.

Equation (33) can be expressed as

$$\log w_{28} = -\log P_{O_2} + \log \left(\frac{-2\, AW_O \cdot K_{(28)}}{N_o}\right) \quad (34)$$

A plot of log (w_{28}) versus log (P_{O_2}) will have a slope that can be related through the results of Table VIII to a particular defect. Therefore, to determine a defect type, the weight loss (or gain) as a function of pressure is determined on a suitable material and the slope of the log (w) versus log (P_{O_2}) curve calculated. This number is compared to the values for the slope given in Table VIII, and a match between a model and the experimental data is sought.

This method fails to provide a unique answer, of course, if the experimental slope is equal to a value that can be obtained from two types of defects, e.g., (-1/2). If the experimental slope is intermediate between two values, then two types of defects of nearly equal importance may be present. In either case, additional experiments will be necessary to interpret the system.

Table IX contains data on TiO_2 obtained from the literature.

TABLE VIII. Comparison of Thermogravimetric Behavior with Oxygen Pressure for Various Defect Types

Predominate defect type	Formation equation Step 1	Equilibrium constant equation Step 2	w dependency on defect Step 3	P_{O_2} dependency of w Step 4	Predicted slope of log w vs log P_{O_2}
Oxygen vacancies	$O_O = V_O + 1/2 O_2(g)$	$K_1 = P_{O_2}^{1/2} [V_O]$	$w_1 = -[V_O] \cdot \frac{AW_O}{N_o}$	$w_1 = -\frac{AW_O \cdot K_1}{N_o} \cdot P_{O_2}^{-1/2}$	-1/2
	$O_O = V_O^{\cdot} + e' + 1/2 O_2(g)$	$K_2 = P_{O_2}^{1/2} \cdot [V_O^{\cdot}][e']$	$w_2 = -[V_O^{\cdot}] \cdot \frac{AW_O}{N_o}$	$w_2 = -\frac{AW_O \cdot K_2^{1/2}}{N_o} \cdot P_{O_2}^{-1/4}$	-1/4
	$O_O = V_O^{\cdot\cdot} + 2e' + 1/2 O_2(g)$	$K_3 = P_{O_2}^{1/2} [V_O^{\cdot\cdot}][e']^2$	$w_3 = -[V_O^{\cdot\cdot}] \cdot \frac{AW_O}{N_o}$	$w_3 = -\frac{AW_O \cdot K_3^{1/3}}{N_o} \cdot P_{O_2}^{-1/6}$	-1/6
Metal interstitials	$Ti_{Ti} + 2O_O = Ti_i + O_2(g)$	$K_4 = P_{O_2} \cdot [Ti_i]$	$w_4 = -2[Ti_i] \cdot \frac{AW_O}{N_o}$ [a]	$w_4 = -2\frac{AW_O \cdot K_4}{N_o} \cdot P_{O_2}^{-1}$	-1
	$Ti_{Ti} + 2O_O = Ti_i^{\cdot} + e' + O_2(g)$	$K_5 = P_{O_2}[Ti_i^{\cdot}][e']$	$w_5 = -2[Ti_i^{\cdot}] \cdot \frac{AW_O}{N_o}$	$w_5 = -2\frac{AW_O \cdot K_5^{1/2}}{N_o} \cdot P_{O_2}^{-1/2}$	-1/2
	$Ti_{Ti} + 2O_O = Ti_i^{\cdot\cdot} + 2e' + O_2(g)$	$K_6 = P_{O_2}[Ti_i^{\cdot\cdot}][e']^2$	$w_6 = -2[Ti_i^{\cdot\cdot}] \cdot \frac{AW_O}{N_o}$	$w_6 = -2\frac{AW_O \cdot K_6^{1/3}}{N_o} \cdot P_{O_2}^{-1/4}$	-1/3
	$Ti_{Ti} + 2O_O = Ti_i^{\cdot\cdot\cdot} + 3e' + O_2(g)$	$K_7 = P_{O_2}[Ti_i^{\cdot\cdot\cdot}][e']^3$	$w_7 = -2[Ti_i^{\cdot\cdot\cdot}] \cdot \frac{AW_O}{N_o}$	$w_7 = -2\frac{AW_O \cdot K_7^{1/4}}{N_o} \cdot P_{O_2}^{-1/4}$	-1/4
	$Ti_{Ti} + 2O_O = Ti_i^{\cdot\cdot\cdot\cdot} + 4e + O_2(g)$	$K_8 = P_{O_2}[Ti_i^{\cdot\cdot\cdot\cdot}][e']^4$	$w_8 = -2[Ti_i^{\cdot\cdot\cdot\cdot}] \cdot \frac{AW_O}{N_o}$	$w_8 = -2\frac{AW_O \cdot K_8^{1/5}}{N_o} \cdot P_{O_2}^{-1/5}$	-1/5
Metal vacancies	$O_2(g) = V_{Ti} + 2O_O$	$K_9 = [V_{Ti}]/P_{O_2}$	$w_9 = 2[V_{Ti}] \cdot \frac{AW_O}{N_o}$ [a]	$w_9 = 2\frac{AW_O \cdot K_9}{N_o} \cdot P_{O_2}$	1
	$O_2(g) = V'_{T_i} + P^{\cdot} + 2O_O$	$K_{10} = [V'_{Ti}][P^{\cdot}]/P_{O_2}$	$w_{10} = 2[V'_{Ti}] \cdot \frac{AW_O}{N_o}$	$w_{10} = 2\frac{AW_O \cdot K_{10}^{1/2}}{N_o} \cdot P_{O_2}^{1/2}$	1/2
	$O_2(g) = V''_{Ti} + 2P^{\cdot} + 2O_O$	$K_{11} = [V''_{Ti}][P^{\cdot}]^2/P_{O_2}$	$w_{11} = 2[V''_{Ti}] \cdot \frac{AW_O}{N_o}$	$w_{11} = 2\frac{AW_O \cdot K_{11}^{1/3}}{N_o} \cdot P_{O_2}^{1/3}$	1/3
	$O_2(g) = V'''_{Ti} + 3P^{\cdot} + 2O_O$	$K_{12} = [V'''_{Ti}][P^{\cdot}]^3/P_{O_2}$	$w_{12} = 2[V'''_{Ti}] \cdot \frac{AW_O}{N_o}$	$w_{12} = 2\frac{AW_O \cdot K_{12}^{1/4}}{N_o} \cdot P_{O_2}^{1/4}$	1/4
	$O_2(g) + V''''_{Ti} + 4P^{\cdot} + 2O_O$	$K_{13} = [V''''_{Ti}][P^{\cdot}]^4/P_{O_2}$	$w_{13} = 2[V''''_{Ti}]\frac{AW_O}{N_o}$	$w_{13} = 2\frac{AW_O \cdot K_{13}^{1/5}}{N_o} \cdot P_{O_2}^{1/5}$	1/5
Oxygen interstitials	$1/2 O_2(g) = O_i$	$K_{14} = [O_i]/P_{O_2}^{1/2}$	$w_{14} = [O_i]\frac{AW_O}{N_o}$	$w_{14} = \frac{AW_O \cdot K_{14}}{N_o} \cdot P_{O_2}^{1/2}$	1/2
	$1/2 O_2(g) = O'_i + P^{\cdot}$	$K_{15} = [O'_i][P^{\cdot}]/P_{O_2}^{1/2}$	$w_{15} = [O'_i]\frac{AW_O}{N_o}$	$w_{15} = \frac{AW_O \cdot K_{15}^{1/2}}{N_o} \cdot P_{O_2}^{1/4}$	1/4
	$1/2 O_2(g) = O''_i + 2P^{\cdot}$	$K_{16} = [O''_i][P^{\cdot}]/P_{O_2}^{1/2}$	$w_{16} = [O''_i]\frac{AW_O}{N_o}$	$w_{16} = \frac{AW_O \cdot K_{16}^{1/3}}{N_o} \cdot P_{O_2}^{1/6}$	1/6

[a] The atomic weight of oxygen is used here because the species involved in the weight change of the sample is still oxygen. Note that two oxygen atoms are eliminated from the structure for each titanium interstitial created, that is, $2[Ti_i] = [O_O]$ eliminated.

TABLE IX

Comparison of Selected Experimental Data with Defect Models

Investigator	Experimental Slope	Type of vacancy prediction
4	-1/6	Unique: double ionized oxygen vacancies
5	-1/6	Unique: double ionized oxygen vacancies

In this instance, a value of the slope was obtained that could only be developed by assuming double ionized oxygen vacancies. Accordingly, the weight loss data yielded a unique answer.

C. Diffusion

The transport of atoms or ions through a solid phase can be expressed by a diffusion equation such as:

$$J_i = D_i \frac{dc_i}{dx} \tag{35}$$

where J_i = the flux of the type i species,
D_i = the diffusion coefficient of the type i species,
c_i = the concentration of the type i species, and
x = the unit of path length.

The total flux of material can be found by summing the J_i terms.

The driving force for the diffusion of the type i species is the concentration of gradient dc_i/dx. The mobility of the diffusing species is a function of the unoccupied sites that are available for it to move into; therefore, the concentration of favorable vacant sites in a material is an important parameter. As a vacant site may represent a defect, the concentration of defects in a material is a parameter contained in D_i. When it can be assumed that D_i is directly related to the concentration of defects of the specific type important in the diffusion process, then the pressure dependency of D_i can be developed.

As already discussed, defective TiO_2 may form via titanium interstitials, and such a mechanism can be written as:

$$Ti_{Ti} + 2O_O = Ti_i + O_2(g) \tag{36}$$

In this model, the $[Ti_i]$ is dependent on the partial pressure of oxygen in the surrounding atmosphere. It is implied in this equation that both the titanium and oxygen structures remain fully occupied, because the equation indicates that a titanium site and two corresponding oxygen sites have been eliminated. Titanium diffusion can only occur then by movement of a titanium atom from one interstitial site to an adjacent vacant interstitial site:

$$K_{(36)} = [Ti_i]\, P_{O_2} \tag{37a}$$

or

$$[Ti] = K_{(36)}\, P_{O_2}^{-1} \tag{37b}$$

The diffusion coefficient D_i is proportional to the concentration of interstitial titanium atoms through the relation

$$D_i = f \cdot D \tag{38}$$

where f = the fraction of sites active in diffusion, and D = the mobility of the diffusing species.

Now,

$$f = F \cdot [Ti_i] \tag{39}$$

where F = proportionality constant related to the total number of interstitial sites. Hence,

$$D_i = F \cdot K_{(36)} \cdot D \cdot P_{O_2}^{-1} \tag{40}$$

or

$$\log(D_i) = -\log P_{O_2} + \log(F \cdot K_{(36)} \cdot D) \tag{41}$$

It is possible that another mechanism for forming interstitial titanium is energetically more favorable, i.e.,

$$Ti_{Ti} = Ti_i + V_{Ti} \tag{42}$$

This mechanism is independent of the ambient oxygen pressure. Diffusion in this mechanism can occur by:

(1). Movements of a titanium interstitial into a neighboring vacant interstitial site

(2). Movement of a titanium interstitial into a neighboring vacant lattice position

(3). Movement of a titanium atom from one lattice position to a neighboring vacant lattice position.

It is observed that diffusion experiments are more complex in their interpretation than previously discussed techniques because the defect important to material transport may be generated via several mechanisms and, the diffusing species is not, in general, the defect itself.

In summary, diffusion experiments should confirm the type of defect proposed to explain the results of other experiments; whereas determining the predominate defect by diffusion experiments alone may be very difficult.

D. Lattice Constant and Density Method

It has been shown in a previous section that a variation of the partial pressure of oxygen in the ambient atmosphere can lead to changes in the weight of a sample. Under these same conditions, a change in sample density will occur also. The variation in density can be measured with sufficient precision so that a model for the predominate defect can be tested. When coupled with a measurement of the x-ray density, one can determine not only the type of defect, but the concentration of the defect as well.

The method can be developed by first calculating the ideal density of the compound assuming occupancy of each crystallographically possible site (and only each such site). As an example, the equations will be formulated assuming defective TiO_2:

$$\rho_{\text{ideal}} = \frac{\eta_{Ti} AW_{Ti}}{V} + \frac{2\eta_{Ti} AW_{O}}{V} \tag{43}$$

where ρ ideal = density of the "ideal" compound,

η_{Ti} = number of cations in the volume V,

AW_{Ti} = atomic weight of titanium, and

AW_{O} = atomic weight of oxygen.

To calculate the variation of density from the ideal value, an equation for a specific defect type is written, such as

$$Ti_{Ti} + 2O_{O} = Ti_{i} + O_2(g) \tag{44}$$

This model assumes that all possible lattice sites are filled and that each interstitial titanium results in a the elimination of two oxygen atoms and hence two oxygen sites, as well as a titanium site. Therefore, the actual density can be written

$$\rho = \rho\,\text{ideal} + \left[Ti_i\right] \frac{AW_{Ti}}{V} \qquad (45)$$

It is assumed that the changes introduced into the volume term are small compared to the changes in the density introduced by the mass term. Such an assumption is usual in this method unless the lattice parameters of the "ideal" compound can be measured experimentally.

The equilibrium constant for Eq. (44) is

$$K_{(44)} = \left[Ti_i\right] P_{O_2}$$

or

$$\left[Ti_i\right] = K_{(44)} P_{O_2}^{-1} \qquad (46)$$

Substituting this into Eq. (45) leads to,

$$\rho = \rho\ \text{ideal} + \left[\frac{AW_{Ti}}{V} K_{(44)}\right] P_{O_2}^{-1} \qquad (47)$$

This model can be tested by plotting log ρ versus log P_{O_2} and determining the slope of the curve.

If the data appear to confirm the assumption of the defect type, then lattice parameters of the samples can be used to calculate the number of interstitial atoms in the following manner. Assuming that the ideal unit cell is fully occupied, a value for the density can be obtained from the unit-cell parameters using the relation

$$\rho\,\underset{\text{x-ray}}{\text{ideal}} = \frac{AW_{Ti} + 2AW_O}{N_o V} \qquad (48)$$

where N_o is Avogadro's number and V is the calculated volume of the unit cell. Therefore, using Eq. (45),

$$\rho - \rho\ \underset{\text{x-ray}}{\text{ideal}} = \left[Ti_i\right] \frac{AW_{Ti}}{N_o V} \qquad (49a)$$

or

$$\left[Ti_i\right] = \frac{N_o V}{AW_{Ti}} \left[\rho - \rho\ \underset{\text{x-ray}}{\text{ideal}}\right] \qquad (49b)$$

This method represents a powerful extension of previous ones because not only can a defect type be deduced, but the actual concentration of the defect can be obtained through equations of the type derived above.

Application of this method to quantitatively determine the extent and type of point defects is very dependent upon the degree of precision of both the lattice parameter and density measurements. The ultimate limit of precision of lattice constant measurements is fixed by spectral purity to about one part in 10^5. To even approach such a degree of precision, meticulous and often sophisticated techniques of x-ray powder diffraction geometry are required, and for these the reader is referred to Klug and Alexander[(6)] and to other standard texts in the field (see Chapter 5).

Ideally, precision of 1 in 10^5 parts is possible in density measurements, but in practice, considerably less precision (1 in 10^4) is sufficient in conjunction with lattice constant measurements (1 in 5×10^4) to determine large concentrations ($>10^{17}$ cc^{-1}) of point defects in crystals.

The application of the method can be readily illustrated by two examples, one using defective TiO and the other defective TiO_2. For the first example, five samples of different composition, all within the range of the homogeneous phase TiO, were prepared by sintering mixtures of Ti and TiO_2 powders at 1400°C, homogenizing at 1000°C, and quenching from that temperature. The lattice constants and linear expansion coefficients of the cubic lattice were determined using the Straumanis asymmetric powder camera and Co-Kα_1 radiation to produce a back reflection line (420) at a $\theta = 74°$. Line shifts due to absorption were avoided by using a cylindrical powder specimen with an average diameter of < 0.2 mm. The linear expansion coefficient was determined over the temperature range 20° - 60°C and the corresponding lattice constants, corrected for refraction, were reduced to the value at 25°C. The densities were obtained using a modification of the method described by Baker and Martin[(7)]. The compositions were determined by chemical analysis. Table X summarizes the x-ray and density data. Values of n' were then calculated from a modified form of Eq. (48).* Examination of Table XI reveals that, as the Ti content is increased, the lattice constant, macroscopic density, and total number of vacant sites also increase. It is also interesting to

*$n' = V\rho N_0/M$ where V is the unit-cell volume, ρ is the experimentally measured density, and M is the formula weight of the "ideal" compound. These values are shown in Table XI.

TABLE X

X-Ray and Density Data in the TiO Phase

x in TiO_x	$a_{25°}$ Å	$\alpha \times 10^6$ Å/°C	ρ_{25} g/cm^3
0.86	4.1850 ± 0.0001	9.6 ± 2	4.998 ± 0.01
0.97	4.1780	5.6	4.953
1.00	4.1766	7.6	4.917
1.06	4.1733	8.3	4.908
1.15	4.1689	6.7	4.854
1.20	4.1661	8.1	4.803

note that, at the stoichiometric composition TiO, the value of n' is 3.38 corresponding to 15.5% vacant sites. Therefore, as pointed out previously, the fact that the stoichiometric ratio of anions to cations is unity in this example does not insure a perfection of the crystal lattice.

The authors conclude that within the homogeneity range of the TiO phase, the various compositions may be regarded as solid solutions of Ti_2O_3 in Ti.

The study of the extent of the TiO_2 phase by Straumanis et al.[(8)] is also interesting for the novel use of the method in determining the stoichiometric ratios of oxygen and titanium. Samples of TiO_2, ranging in color from blue-black to cream, were obtained by heating rutile at 1400°C at increasing equilibrium oxygen pressures.

The Straumanis asymmetric x-ray powder diffraction method was used employing Co-$K\alpha_1$ radiation to provide extremely high θ values of 80° and 83°. Line shifts due to absorption were eliminated as previously. Fluorescence was reduced by covering the film with aluminum foil. The extent of

TABLE XI

Extent of Vacant Sites in the TiO Phase[a]

x in TiO_x	M g/mol	n' TiO_x molecules/u.c.	% Vacancies Ti	O
0.86	61.72 ± 0.06	3.58	10.5	23.03
0.97	63.37	3.43	14.2	16.8
1.00	63.90	3.38	15.5	15.5
1.06	64.83	3.32	17.0	12.0
1.15	66.35	3.19	20.2	8.3
1.20	67.12	3.12	22.0	6.4

[a] n = 4.00.

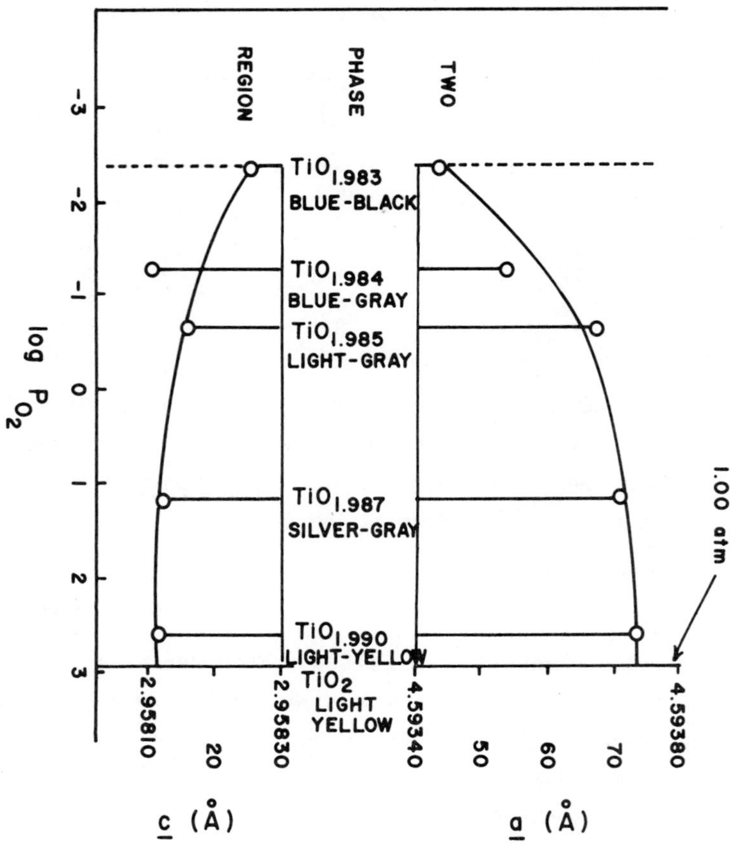

Fig. 6. Variation of tetragonal lattice constants of TiO_x with equilibrium partial pressure of oxygen during heating.

the TiO_2 phase was determined by noting the composition where a new phase appeared as evidenced by new lines in the diffraction photo. The first trace of a new phase appeared for a sample composition of $TiO_{1.983}$ as determined by chemical analysis. The intensities of the new lines increased with decreasing oxygen content into the two phase region. The variation of the tetragonal lattice constants with equilibrium oxygen pressure is shown in Fig. 6. One notes that a decreases and c increases with deepening color and decreasing oxygen content. Accordingly, the amount of oxygen uptake by the TiO_2 phase is dependent upon the partial pressure of the oxygen.

As chemical analysis was not sufficiently accurate to distinguish the compositions of TiO_2 in such a narrow range of homogeneity ($TiO_{1.983}$ - TiO_2), the lattice constant and density methods were employed instead for this purpose. Equation (43) can be rearranged such that, for the crystallographic n, $nM = N_o V\rho$; $N_o = 6.02403 \times 10^{23}$, V was obtained from the lattice constant measurements, and ρ was determined experimentally by the modified Baker and Martin method. If AB_x represents the various compositions, then

$$nM = n(A + xB) = N_o V\rho \tag{50}$$

Assuming n = 2 for "ideal" TiO_2 ($n' = 1.9999 \pm 0.0001$) and further that a loss in B, not detectable by the usual chemical analyses, results, one obtains

$$2A + 2xB = N_o V\rho \quad \text{or} \tag{51a}$$

$$x = (N_o V\rho - 2A) / 2B \tag{51b}$$

The oxygen indices so obtained are those shown in Fig. 6. Similar results are obtained by calculating the loss of oxygen content from the changes in density assuming no change in the volume of the unit cell (62.423 - 62.415 $\pm$ 0.004 $\AA^3$).

The concentration of the vacant oxygen sites per unit cell given by $2(x - 2)$ corresponds to 0.034 for $TiO_{1.983}$, becoming zero at stoichiometric TiO_2. It is suggested that, to compensate for the loss of negative charges, an equivalent amount of Ti^{4+} would be reduced to Ti^{3+}. This is in accord with the darker colors of the Ti-rich compositions since crystals containing Ti^{3+} are usually blue-black.

IV. OTHER TECHNIQUES FOR DETERMINING IMPERFECTIONS

The foregoing dealt with those techniques of determining the nature and amount of imperfections (primarily native imperfections) which have been used extensively and have become so refined as to be considered conventional or classical techniques. One should, of course, also include in this category optical and electrochemical techniques as they have been used for many years, particularly the former in the study of color centers in alkali halides.

As a result of considerable sophistication in analytical instrumentation in the last decade, new techniques, useful in the determination and observation of imperfections in solids, have developed and include among others electron paramagnetic resonance (EPR), electron-nuclear double resonance (ENDOR), field-ion microscopy (FIM), and transmission electron microscopy (TEM).

A. Optical

Changes in optical properties are observed as a consequence of electronic transitions arising from the presence of electrons, holes, and atomic imperfections. The release of electrons or holes trapped at atomic imperfections is often manifested by the appearance of phosphorescence, fluorescence, and photoconductivity. For experimental details and data analysis, the reader is referred to the bibliography under optical effects.

B. Magnetic

The presence of atomic imperfections in solids may give rise to a paramagnetic contribution if the imperfections possess unpaired electrons. The effect is normally small, and static magnetic measurements are usually not sufficiently sensitive to detect perturbations caused by such imperfections. Undoubtedly the most sensitive and powerful tool for detection of changes in magnetic properties of crystals brought about by the presence of imperfections is electron paramagnetic resonance spectroscopy (EPR).

In essence, the spectrometer consists of a microwave source (common frequencies ranging frou about 3 - 40 GHz), a magnetic field serving as the dispersion element, an absorption cell containing the sample (transmission or reflection cavity), and a detector, commonly a silicon crystal diode.

Paramagnetic resonance arises from the precession of the spins of the unpaired electrons of the imperfections (including electrons and holes) about a magnetic field. An absorption spectrum is obtained as a result of quantum transitions occurring between the energy levels. The magnitude of the splitting of the energy levels in the magnetic field is determined by a proportionality constant called the g factor. The magnitude and anisotropy of this splitting factor is useful in ascertaining the nature of imperfections. Additionally, interactions of the magnetic moments of nuclei near the imperfection sites with the electron spin moments and with the magnetic field $\overline{H}$ result in a hyperfine structure (hps) which further permits establishing the precise nature of the imperfections. The measurement of this latter interaction has been further refined by a technique of double resonance called ENDOR. This technique involves the individual determination of the coupling between electron and nuclear spin moments at varying distances. From these data it is possible to plot the wave function of the electron in the vicinity of the imperfection.

Electron paramagnetic resonance has been applied extensively to the study of impurity or defect centers in semiconductors, particularly silicon. A comprehensive review on this subject by Ludwig and Woodbury is recommended. Models for radiation damage centers in Si have been deduced mainly from EPR measurements. Woodbury and Ludwig[(9)] studied transition metal ion impurities in host Si crystals with regard to charge and location employing both EPR and ENDOR techniques. The F centers in alkali halides and in fcc crystals such as MgO have been studied by EPR. The latter system provides a simpler one for studies of F centers and does not require the ENDOR techniques necessary to resolve inhomogeneously broadened lines as occur in the EPR spectra in alkali halides, see Wertz et al.[(10)]

A major drawback of the technique as applied to the detection of imperfections is that not all sites with unpaired spins produce an observable spectrum. Consequently the absence of a resonance effect is not conclusive proof of the absence of such imperfections.

Finally it should be mentioned that another technique, nuclear magnetic resonance (NMR), is also available but is considerably more limited in application than EPR when used independently due to its lower sensitivity in detecting resonances arising from the imperfections themselves. Changes in the line shape and width of NMR spectra have been used to study irradiation damage[(11)]. Also interactions between the nuclear quadrupole moment and electric field gradients arising from the charge distribution in crystals give rise to a resonance effect (NQR) useful in the study of impurity concentrations[(12)] and radiation effects, e.g., neutron irradiated sodium chloride[(13)].

C. Microscopy Techniques

The methods of field ion microscopy (FIM) and transmission electron microscopy (TEM) permit direct observations of atomic imperfections. These techniques are discussed in Chapters 6 and 7.

D. Electrochemical

Electrochemical techniques have been applied in many crystalline systems to determine narrow limits of stoichiometry and to ascertain the nature and extent of atomic imperfections. The scope and diversity of experimental methods in solid state electrochemistry are so far ranging as to be outside the limits of what can be discussed effectively and profitably in this chapter, and accordingly the reader is referred to excellent reviews by Kroger[(14)] and by Raleigh[(15)].

E. Chemical Analysis

Conventional methods of chemical analysis can be employed when the extent of defects is considerable, e.g., large concentrations of atom impurities, large deviations from stoichiometry. However, when concentrations must be known to less than 10^{-7} atom fraction as in semiconductor technology, it is necessary to turn to more sophisticated techniques. The situation is the same in regard to extremely narrow limits of stoichiometric deviations. Accordingly, it is necessary to adopt other techniques, among which are included atomic absorption, colorimetry, fluorimetry, polarography, neutron activation, tracers, vacuum fusion, mass spectrometry, x-ray diffraction, and fluorescence. These analytical methods are briefly reviewed in Chapter 2.

As mentioned previously, analytical techniques based on changes in the physical properties of solids arising from the presence of atomic imperfections can be employed although they have the disadvantage of not always being specific. For example, as mentioned in Section III, A, conductivity measurements alone cannot distinguish the nature of imperfections. The measurements must be combined with other physical or chemical analyses to determine both type and amount of imperfections. Even in those cases where the analytical technique is specific, e.g., optical absorption and paramagnetic resonance, it is seldom possible to determine the overall concentration of atom impurities.

Deviations from stoichiometry of less than 10^{-3} cannot be detected by conventional quantitative analysis, see Section III, D for a specific application of x-ray diffraction in determining a narrow range of stoichiometry. In oxide systems possessing considerable stoichiometric deviation, such as

FeO where O is in excess, chemical analysis can be readily employed to determine directly the concentration of the two oxidation states of the cation. In general, however, the use of chemical analysis in determining the concentration of either the majority or minority cation is seldom strightforward. The reader is referred to a review of such techniques by Kroger[14].

REFERENCES

(1) G. G. Libowitz, The Solid-State Chemistry of Binary Metal Hydrides, Benjamin, New York (1965).

(2) J. Yahia, Phys. Rev., 130, 1711-1719 (1963).

(3) C. J. Kevane, Phys. Rev., 133, A1431-A1436 (1964).

(4) K. S. Förland, Tionde Nordiske Kemistmötef, Stockholm (1959).

(5) P. Kofstad, J. Phys. Chem. Solids, 23, 1579-1586 (1962).

(6) H. P. Klug and L. E. Alexander, X-Ray Diffraction Procedures, Wiley, New York 440-490 (1954).

(7) I. Baker and G. Martin, Ind. and Engr. Chem., Anal. Ed., 15, 277 (1943).

(8) M. E. Straumanis, T. Ejima, and W. J. James, Acta. Cryst., 14, 493-497 (1961).

(9) H. H. Woodbury and G. W. Ludwig, Phys. Rev., 117, 102 (1960).

(10) J. E. Wertz, P. Auzins, R. A. Weeks, and R. H. Silsbee, Phys. Rev., 107, 1535 (1957).

(11) Proceedings of the Symposium on The Chemistry of Extended Defects in Non-Metallic Solids, North Holland, Amsterdam, 1970.

(12) T. P. Das and E. L. Hahn, Solid State Physics, Suppl. 1 (1968).

(13) J. Duchesne, A. Monfils, and J. Depireux, Compt. Rend. Acad. Sci. (Paris), 243, 259 (1956).

(14) F. A. Kröger, The Chemistry of Imperfect Crystals, North Holland, Amsterdam (Wiley, New York) (1964).

(15) D. O. Raleigh, Solid State Chemistry, Pergamon, New York, Vol. 3, p. 83 (1967).

BIBLIOGRAPHY

A. Defects in Solids

(a) W. van Gool, Principles of Defect Chemistry of Crystalline Solids, Academic, New York (1966).

(b) G. G. Libowitz, The Solid-State Chemistry of Binary Metal Hydrides, Benjamin, New York (1965).

(c) F. A. Kröger, The Chemistry of Imperfect Crystals, North Holland, Amsterdam (Wiley, New York) (1964).

(d) T. J. Gray, ed., The Defect Solid State, Interscience, New York (1957).

(e) A. L. G. Rees, Chemistry of the Defect Solid State, Methuen (1954).

(f) Proceedings of the Symposium on The Chemistry of Extended Defects in Non-Metallic Solids, North Holland, Amsterdam, 1970.

B. Symposiums or Meetings Devoted Specifically to Defects

(a) L. Eyring and M. O'Keefe, eds., The Chemistry of Extended Defects in Non-Metallic Solids, London: North Holland, Amsterdam (1970).

(b) H. H. Hausner and M. G. Bowman, eds., Fundamentals of Refractory Compounds, Plenum, New York (1968).

(c) J. B. Wachtman, Jr. and A. D. Franklin, eds., Mass Transport in Oxides, National Bureau of Standards Special Publication 296, 1968.

(d) Calculation of the Properties of Vacancies and Interstitials, National Bureau of Standards Miscellaneous Publication 287, 1966.

(e) L. Mandelcorn, ed., Non-Stoichimetric Compounds, Academic, New York (1964).

(f) Non-Stoichiometric Compounds, Roland Ward, Symposium Chairman, Adv. in Chemistry Series 30, Am. Chemical Soc., 1963.

(g) Symposium on Properties of Non-Stoichiometric Compounds, Trans. British Ceramic Soc., 56, papers 35-39, pp. 553-623 (1957).

(h) Report on the Conference on Defects in Crystalline Solids, Bristol, July, 1954, The Physical Society, 1955.

C. Chapters Devoted to Defects, Occurring in Books on More General Topics

(a) H. Reiss, ed., Progress in Solid State Chemistry, Pergamon, New York, Vol. 1, Chaps. 4 and 5 (1964); Vol. 2, Chaps. 6-8 (1965); Vol. 3, Chaps. 4 and 5 (1966); Vol. 4, Chap. 6 (1967).

1. Electrical Conductivity Experiments

(a) S. P. Mitoff, Progress in Ceramic Science, Pergamon, New York, Vol. 4., pp. 217-264. (1966).

(b) J. A. van Raalte, J. Appl. Phys., 36 (11), 3365-3369 (1965).

(c) S. P. Denker, J. Appl. Phys., 37 (1), 142-149 (1966).

(d) N. M. Tallan and R. W. Vest, J. Am. Cer. Soc., 49 (8), 402-404 (1966).

(e) J. M. Wimmer, L. R. Bidwell, and N. M. Tallan, J. Am. Cer. Soc., 50 (4), 198-201 (1967).

(f) H. C. Graham, N. M. Tallan, and R. Russell, Jr., J. Am. Cer. Soc., 50 (3), 156-163 (1967).

(g) P. Kofstad, J. Phys. Chem. Solids, 23, 1571-1578 (1962).

(h) C. J. Keoane, Phys. Rev., 133 (5A), 1431-1436 (1964).

(i) J. Yahia, Phys. Rev., 130 (5), 1711-1719 (1963).

(j) N. M. Tallan, R. W. Vest, and H. C. Graham, Mat. Sci. Rech., 2, 33-67 (1965).

2. Thermogravimetric Experiments

(a) W. C. Tripp, R. W. Vest, and N. M. Tallan, Vacuum Microbalance Techniques, Plenum, New York, pp. 141-157, (1965).

(b) P. Kofstad, J. Phys. Chem. Solids, 23, 1579-1586 (1962).

(c) W. C. Tripp, R. W. Vest, and H. C. Graham, Vacuum Microbalance Techniques, Plenum, New York, pp. 107-128 (1967).

3. Diffusion

(a) E. A. Secco and W. J. Moore, J. Chem. Phys., 26 (4), 942-948 (1957).

(b) R. W. Redington, Phys. Rev., 87 (6), 1066-1072 (1952).

4. Lattice Constant and Density Method

(a) W. van Gool, J. Mat. Sci., 1, 261-268 (1966).

(b) G. R. Miller and Che-Yu Li, J. Phys. Chem. Solids, 26, 173-177 (1965).

(c) E. R. Jette and F. Foote, J. Chem. Phys., 1, 29-36 (1933).

(d) T. L. Barry and R. Roy, J. Am. Cer. Soc., 50 (2), 105-108 (1967).

(e) I. Estermann, W. J. Leivo, and O. Stern, Phys. Rev., 75 (4), 627-633 (1949).

(f) M. Straumanis, Am. Mineralogist, 38, 662-670 (1953).

(g) G. A. Bell, Australian J. Appl. Sci., 9, 236-244 (1958).

(h) D. A. Hutchison, Phys. Rev., 66 (5 and 6), 144-148 (1944).

(i) C. A. Hutchison and H. L. Johnston, J. Am. Chem. Soc., 62, 3165-3168 (1940).

(j) M. E. Straumanis, T. Ejima, and W. J. James, Acta. Cryst., 14, 493-497 (1961).

(k) M. E. Straumanis, Acta Cryst., 2, 82-84 (1949).

(l) J. Short and R. Roy, J. Phys. Chem., 67, 1860-1861 (1963).

(n) H. P. Klug and L. E. Alexander, X-ray Diffraction Procedures, Wiley, New York, 440-490 (1954).

(o) I. Baker and G. Martin, Ind. and Engr. Chem., Anal. Ed., 15, 277 (1943).

5. Optical Techniques

(a) R. H. Bube, Photoconductivity of Solids, Wiley, New York (1960).

(b) J. H. Schulman and W. D. Compton, Color Centers in Solids, Pergamon, New York (1962).

(c) J. Tauc, Photo and Thermoelectric Effects in Semiconductors, Pergamon, New York (1962)

(d) H. K. Hemsch, Electroluminescence, Pergamon, New York (1962).

(e) D. Curie, Luminescence in Crystals, Wiley, New York (1963).

(f) See Ref. A (C), p. 522 ff.

(g) R. G. Rhodes, Imperfections and Active Centers in Semiconductors, Pergamon, New York (1964).

(h) J. J. Markham, F-Centers in Alkali Halides, Academic, New York and London (1966).

(i) D. L. Greenaway and G. Harbeke, Optical Properties and Bond Structure of Semiconductors, Pergamon, New York (1968).

(j) A. A. Maraduden and G. F. Nardelli, Elementary Excitations in Solids, Pergamon, New York (1969).

6. Magnetic Techniques

(a) Raymond S. Alger, Electron Paramagnetic Resonance, Interscience, New York (1968).

(b) D. J. E. Ingram, Spectroscopy at Radio and Microwave Frequencies, 2nd Ed., Plenum, New York (1967).

(c) S. A. Al 'Tshuler and B. M. Kozyrev, Electron Paramagnetic Resonance, Academic, New York,(1964).

(d) W. Lon., ed., Paramagnetic Resonance, Academic, New York, Vol. II (1963).

(e) G. W. Ludwig and H. H. Woodbury, ESR in Semiconductors in Solid State Physics, Academic, New York, Vol. 13 (1962).

(f) H. H. Woodbury and G. W. Ludwig, Phys. Rev., 117, 102 (1960).

(g) J. E. Wertz, P. Auzins, R. A. Weeks, and R. H. Silsbee, Phys. Rev., 107, 1535 (1957).

(h) G. A. Jeffrey and T. Sokurair, Applications of Nuclear Quadrupole Resonance in Progress in Solid State Chemistry, Pergamon, New York, Vol. I, Ch. 10 (1963).

(i) T. P. Das and E. L. Hahn, Solid State Phys., Suppl. 1 (1968).

(j) J. Duchesne, A. Monfils, and J. Depireux, Compt. Rend. Acad. Sci. (Paris), 243, 259 (1956).

7. Electrochemical

(a) See Ref. A (C), p. 879 ff.

(b) D. O. Raleigh, Solid Sate Chemistry, Pergamon, New York, Vol. 3, p. 83 (1967).

8. Chemical Analysis

(a) L. S. Darken and R. W. Gurry, J. Am. Chem. Soc., 68, 798 (1946).

(b) H. J. Engell, Z. Elecktrochem., 60, 905 (1956).

(c) C. L. Luke and M. E. Campbell, Anal. Chem., 25, 1588 (1953); 27, 1150 (1955); 28, 1340 (1956).

(d) G. W. Chilner and L. J. Slee, Ind. Chemist, 33, 494 (1957).

(e) J. A. Burton, E. O. Koeb, W. P. Slichter, and J. D. Struthers, J. Chem. Phys., 21, 1991 (1953).

(f) G. H. Morrison and J. F. Cosgrove, Anal. Chem., 27, 810 (1955).

(g) N. B. Hannoy and A. J. Ahearn, Anal. Chem., 26, 1056 (1954).

(h) C. D. Thurmond, W. G. Guldner, and A. L. Beach, J. Electrochem. Soc., 103, 603 (1956).

(i) J. A. Lely and F. A. Kröger, Proc. Int. Colloq. on Semiconductors and Phosphors, Garmisch-Partenkirchen, p. 525 (1956).

(j) W. Parrish, Philips Tech. Rev., 17, 209 (1956).

(k) M. E. Straumanis, T. Ejima, and W. J. James, Acta Cryst., 14, 493 (1961).

(l) F. A. Kroger, Luminescence of Solids, Elsevier, Amsterdam, p. 69, (1948).

(m) See Ref. A(C), pp. 25-32.

PART 3

Size, Shape, Strain, and Surface of Powders

Chapter 11

PHYSICAL CHARACTERIZATION

David R. Lankard and Dale E. Niesz

Battelle Memorial Institute
Columbus Laboratories
Columbus, Ohio

I. INTRODUCTION

Almost without exception the methods used to characterize powders provide data in the form of a single parameter which is uniquely associated with the method of measurement, e.g., a diameter, a diagonal, an area, or a volume. A particle diameter or volume measured by one technique may be quite different from that measured by another technique. For this reason, the criteria for selection of a characterization method for a powder should be governed primarily by knowledge of how the data are to be used rather than, time and money considerations. As an obvious and simplified example, data on the specific surface area of a catalyst may be far more relevant regarding the performance of this material in service than data on the mean particle diameter.

The need for powder characterization data may range from quality control in ceramic processing to fundamental research directed at correlating the properties of the powder with the processing behavior or properties of the processed ceramic. Each and every case regarding the selection of a characterization procedure is unique and must be considered on an individual basis. Considerable effort should be devoted to identifying the best characterization method or combination of methods for the case at hand. This chapter briefly reviews some of the standard techniques for determining the physical characteristics of ceramic powders and describes the recent results of research on developing new characterization techniques. Emphasis is placed on describing some of the major merits, limitations, and pitfalls of various analysis techniques and on discussing the effects of various powder characteristics on the processing parameters, microstructural development, and properties of ceramic bodies.

Before attempting a discussion of the characteristics of powders it is helpful to define some terms commonly used in the description of particulate systems.

The working unit of any powder system is the particle, the size and shape of which depends on the manner in which it was formed. An aggregate is an assembly of particles held together by neck areas formed by diffusion. An agglomerate is an assembly of particles and/or aggregates held together by relatively weak cohesive forces. Agglomerates can usually be broken down into ultimate particles through conventional dispersion techniques, whereas aggregates cannot. It is generally acknowledged that the amount of agglomeration and/or aggregation of a powder has an influence not only on the properties unique to the powder but also on the behavior of the powder during processing and on the properties of the finished ceramic. It is, therefore, important to recognize that handling of a powder prior to or during characterization may alter its degree of agglomeration or, less likely, of its degree of aggregation.

The bulk of powder analytical techniques are concerned with the measurement of a particle size and the distribution of this size. It is generally acknowledged that the particle shape is also of importance as regards the response of the powder to processing variables. However, particle shape has been a very elusive factor for the particle analyst to tag quantitatively.

Almost all of the powder characterization techniques in use today are based on principles common to techniques that have been in use for many years. Many of the so-called advances in powder characterization in recent years have simply been the development of more rapid and in some cases more precise means of making the same measurements. There exist today sophisticated instruments and equipment to obtain powder characterization data. However, despite the general usefulness of these data to the ceramic engineer, there are still no established relationships linking raw material characteristics to process variables for virtually all ceramic processing operations.

This situation has led to the claim that new and truly advanced powder characterization techniques or new use of data from existing techniques are required before full use can be made of characterization data as related to ceramic processing controls. A recent study of the strength and pore-size distribution of particle aggregates is an example of an advanced technique which hopefully will provide for more meaningful correlations.

II. PRECHARACTERIZATION PROCEDURES

The usefulness and relevance of any powder characterization data to the user is strongly dependent on the validity of the assumptions that (1) the powder lot from which samples are obtained is actually representative of powder to be analyzed and (2) the sample taken for analysis is representative of the bulk lot. These assumptions may be invalidated by (1) agglomeration due to factors such as dead weight compressive forces and humidity, and (2) particle-size segregation resulting from improper handling. Generally, it is possible to select the powder storage and transport conditions that will minimize the chance of agglomeration during the powder's handling history. The importance of obtaining a representative sample is pointed out by the fact that the characterization study may be run using a few grams of powder that were obtained from a powder lot of a ton or greater.

Recommended procedures for obtaining a representative gross sample are described in the ASTM Manual on Test Sieving Methods[1]. The size of the gross sample is dependent on the size of the powder lot to be evaluated. Once the gross sample has been obtained, it must be further reduced in size to provide a test sample for the characterization method. Depending upon the size of the gross sample, the test sample can be obtained by (1) the cone and quartering method, (2) sample reduction using a chute splitter, or (3) sample reduction using a spinning riffler.

In the cone and quarter process, the material is poured to form a cone which is divided into four equal parts. One quarter of the cone is then isolated and the process is repeated until the desired sample size is obtained. Chute splitters are used to divide a sample into two or more equal fractions, and the process is repeated on one of the isolated fractions. The spinning riffler is a circular disc on which are located a number of compartments. Powder is fed through a funnel located over the compartments as the disc is rotated. Any one of the compartments can be selected to yield the test sample.

The procedures used to obtain test samples have been discussed and assessed as to their relative degree of reliability[2,3]. Of the three sampling techniques just discussed, the spinning riffler appears to be subject to the least amount of operator bias. It cannot be overstressed that efficient sampling technique is of primary importance in an overall powder characterization operation.

Once the powder sample has been obtained, it is recommended that prior to any analyses the sample be examined briefly using an optical or electron microscopy technique. The information that can be gained from this examination includes

1. An indication of the approximate size range of the powder
2. Some idea of the shape of the individual particles
3. An indication of the presence of agglomeration or aggregation and the approximate size of the agglomerates or aggregates
4. The textural and perhaps compositional uniformity of the particles

All of this information will provide guidelines as to the best characterization technique for the powder. Once the most suitable technique has been established, thought can be given to the selection of the instrument or equipment within this class that provides the necessary accuracy, covers the range of values of interest, and is appropriate to the problem.

III. POWDER CHARACTERIZATION PROCEDURES

A wide variety of methods have been used over the years to provide a quantitative measurement of some characteristic of particles in a powder mass. As yet, no formal classification of powder analytical techniques is available. In general, however, the numerous methods of characterization in use today can be grouped into four loosely knit subject areas depending upon the property that is measured as shown in Table I.

In a paper published in 1963[8], 74 different techniques or methods of powder characterization were identified—since then the number has grown even larger. This fact alone points out the importance of providing a full description of the equipment and procedure used when presenting powder characterization data.

A. Particle Size Analysis

By far the largest number of powder analytical techniques fall into the category of particle size analysis. A further classification is possible within this category depending on the manner in which the particle "size" is

TABLE I

Powder Analytical Techniques

A. Particle Size Analysis

A measurement is made of a linear dimension, an area, or a volume of the particles and the distribution of this parameter is determined.

B. Surface Area Measurement

The exposed surface area of a powder sample is determined in terms of the unit surface area per unit weight of material.

C. Permeametry Technique

The resistance that a packed bed of powder offers to fluid flow is measured and is expressed in terms of either particle size or surface area.

D. Powder Strength Analysis

A measurement is made of the strength of particles, aggregates, and agglomerates.

measured as shown in Table II. The subject of particle size measurement has been adequately covered in a number of texts[4-7].

1. Sedimentation Techniques

The usefulness of the sedimentation technique for particle size measurement relies on the fact that the rate of fall of a particle in a fluid medium is proportional to the volume of the particle. The rate of fall (v) of a spherical particle is related to the diameter (d) of the particle by

$$d = \left(\frac{18 \eta v}{(\rho - \rho_f) g} \right)^{1/2} \tag{1}$$

where

ρ is the particle density

ρ_f is the fluid density

g is the gravitational constant

η is the viscosity of the fluid

TABLE II

Methods of Particle Size Analysis

Type	Particle dimension measured
Sedimentation	Stokes diameter; the diameter of a sphere having the same density and the same free falling speed as the particle in a fluid of the same density and viscosity (in the laminar flow region).
Sieving	Sieve diameter; the width of the minimum square aperture through which the particle will pass.
Elutriation	Drag diameter; the diameter of a sphere having the same resistance to motion as the particle in a fluid of the same viscosity and at the same velocity.
Electronic	Volume diameter; the diameter of a sphere having the same volume as the particle.
Microscopic	A physical or statistical diameter or area.

A large number of instruments have utilized this relationship to measure the Stokes' diameter of particles in the range from 0.05 to 1000 μ. The most common range for sedimentation analyses is 9.1 to 100 μ. Many techniques make use of centrifugal effects to permit measurement of particles in the submicron range and to shorten the measurement time while minimizing the effect of Brownian movement, diffusion, and convection upon the results.

It is obvious that the more the particle deviates from sphericity, the greater will be the difference between the volume of the particle and the volume of the equivalent Stokes' diameter sphere. Kaye[9] prefers to look at this situation not as a source of error but as potential information on the shape of the particle. The key to obtaining good analytical data using any sedimentation technique is to obtain complete dispersion of the powder and to adjust the concentration so that particle-particle interference is eliminated. It has been suggested[9] that a good rule is to use a dispersion technique that subjects the powder to a deagglomerating environment similar to what the powder will experience in subsequent processing.

The use of the sedimentation method for particle size analyses of manganese dioxide[10] and alumina[11] have been described recently.

2. Sieving Techniques

One of the most common methods of particle size analysis in the ceramic industry is that of sieving. Many ASTM and industrial specifications are written around particle size distributions obtained using commercially available sieves.

Sieves are available in a coarse series with aperture openings ranging from 4.24 down to 0.25 in. and a fine series with openings ranging from 0.223 down to 0.0015 in. Two types of sieves are used in the U. S.—the Tyler Standard Screen Scale Sieve Series and the U. S. Sieve Series. The only difference in the two series in the fine range of interest for powder measurements is the identification method as shown in Table III. The Tyler sieve numbers are the actual number of meshes per linear inch, while the U. S. sieve numbers refer either to micron designation or an arbitrary number not necessarily related to the mesh count. The width of the aperture opening in successive sieves is different by the factor $2^{1/2}$. This provides a twofold increase in aperture area for each succeeding larger sieve. A

TABLE III

United States and Tyler Sieve Series

Tyler identification, equivalent mesh No.	U. S. identification, mesh No.	Sieve opening	
		mm	inch
3½	3½	5.66	0.223
4	4	4.76	0.187
5	5	4.00	0.157
6	6	3.36	0.132
7	7	2.83	0.111
8	8	2.38	0.0937
9	10	2.00	0.0787
10	12	1.68	0.0661
12	14	1.41	0.0555
14	16	1.19	0.0469
16	18	1.00	0.0394
20	20	0.84	0.0331
24	25	0.71	0.0280
28	30	0.59	0.0232
32	35	0.50	0.0197
35	40	0.42	0.0165
42	45	0.35	0.0138
48	50	0.297	0.0117
60	60	0.250	0.0098
65	70	0.210	0.0083
80	80	0.177	0.0070
100	100	0.149	0.0059
115	120	0.125	0.0049
150	140	0.105	0.0041
170	170	0.088	0.0035
200	200	0.074	0.0029
250	230	0.062	0.0024
270	270	0.053	0.0021
325	325	0.044	0.0017
400	400	0.037	0.0015

complete and comprehensive manual on test sieving methods has been prepared[1] by ASTM Committee E-1.

The relationship that is measured in the sieving technique is the weight of particles in the size interval defined by two different sieve sizes. Recently, the utility of the sieving technique of particle size analysis has been extended through the availability of so-called "micromesh" sieves. These sieves are available in a variety of size ranges extending from 700 to 2 μ within a tolerance of ±2 μ. Specifications on the micromesh sieves have been set by ASTM as shown in Table IV. Micromesh sieves are electroformed from nickel, bonded into a copper support plate for strength, then mounted into a brass frame.

TABLE IV

Precision Electroformed Sieves [a]

Sieve designation nominal size of opening (μm)	Permissible variation of sieve openings (μm)	Openings per linear inch minimum and maximum limits
149	146 to 152	90 to 120
125	122 to 128	110 to 145
105	103 to 107	135 to 175
88	86 to 90	160 to 210
74	72 to 76	190 to 245
63	61 to 65	225 to 290
53	51 to 55	240 to 320
44	42 to 46	260 to 350
37	35 to 39	285 to 400
31	29 to 33	380 to 500
25	23 to 27	450 to 550
20	18 to 22	450 to 550
15	13 to 17	450 to 700
10	9 to 11	450 to 800
5	4 to 6	450 to 1000

[a] From Manual of Test Sieving Methods[1].

Zwicker[12] has reported on an investigation of the use of micromesh sieves for particle size analyses of aluminum hydrates and calcined aluminas. A dry sieving technique was used successfully on sieves of 250-, 149-, 90-, 70-, 55-, 45-, 30-, and 20-μ openings. A wet sieving procedure was developed for micromesh sieves having openings ranging from 45 to 2 μ to obviate the difficulties experienced with sieve "blinding" in the dry technique.

Kaye and Jackson[13] have described a new technique using micromesh sieves called felvation. The new technique greatly minimizes the sieve blinding and wear problems common to conventional means of sieving. The availability and usefulness of micromesh electroformed sieves means that one technique (sieving) can be used as a measure of particle size distribution on powders having a wide range of particle size.

3. Elutriation Techniques

The elutriation process involves the suspension of particles in an upward flowing fluid stream—usually air or water. Elutriation is actually the reverse of sedimentation. The fluid velocity is assumed to be the same as the particle velocity as calculated from the Stokes' equation given previously. This assumption is usually not valid inasmuch as fluid flow near confining surfaces is less than that at points further away resulting in deviations from laminar flow. As is the case with sedimentation techniques, particle agglomeration must be considered, since it affects the elutriation process. In some commercial elutriation units centrifugal forces are used to provide a broader range of particle size measurement.

An advantage of the elutriation process is that it is a continuous operation. As such, the elutriation principle is used in a number of commercial particle classifiers to provide separation and recovery of sized particle fractions.

4. Electronic Techniques

A technique that appears to be gaining in popularity for particle size analyses of ceramic powders is that of electronic sizing. An electrolyte fluid suspension of the powder to be measured moves through a small aperture on each side of which is located an electrode. The concentration of powder is maintained so that the particles go through the aperture substantially one at a time. A particle in the aperture displaces a volume of electrolyte momentarily causing a change in resistance between the electrodes and producing a voltage pulse whose magnitude is in proportion to the volume of the displaced electrolyte. The voltage pulses are electronically amplified and counted.

Allen[14] has discussed the advantages and limitations of electronic techniques and cautions that sizing by this technique is a highly skilled job. Although the precision obtained by this technique appears to be quite good, the absolute values obtained may differ when two different instruments are used. Inasmuch as the volume diameter is measured with this technique, the results obtained may differ considerably from those obtained with other methods of size analysis. Allen[14] concluded that particle size analysis using the electronic technique gives results that are as indicative as those of other techniques and that the data are often far more useful.

The usefulness of electronic techniques in providing rapid and reliable particle size data has been described recently as applied to milled zircon, zirconia, and other ceramic powders[15] and to portland cement[16].

5. Microscopic Techniques

The particle size distribution of ceramic powders can be determined by direct observation and measurement using a microscope equipped with a suitable measuring device. The simplest procedure is the measurement of a preselected linear dimension of the particle using a scale superimposed on the particle image. It is obvious that the choice of the dimension to be measured may be highly subjective, and many techniques have been proposed to overcome this. In some cases a statistical diameter is measured, e.g., Feret's diameter, which is the perpendicular projection onto a fixed direction of the tangent to the extremities of the particle image profile. Particle size data obtained microscopically may be presented either in terms of the number or weight of particles counted.

The normal range of analyses using optical microscopy techniques is from 0.4 to about 100μ (using white light) and using the electron microscope, from 0.001 to about 5μ. Measurements may be made either directly through the microscope, on electronically displayed images, or using photomicrographs. A recommended procedure for particle size analysis by light microscopy is described in ASTM Designation E20-51T (1951).

A basic drawback of the microscopic technique is the time involved in gathering the data. Instruments are available which automate or semiautomate the counting process so that more rapid data acquisition is possible. An example is given in a recent discussion[17] of the measurement of the particle size distribution of ferrite powders using a commercial instrument that permits about 1000 particles to be counted in only 30 min. Other automatic and semiautomatic instruments for microscopic particle size analyses are described in Refs. (5) and (6). An advantage of the microscopic technique is that information on particle shape can be obtained.

B. Surface Area Measurements by Adsorption

One of the most popular techniques used for the determination of the specific surface area of fine powders is that of low temperature nitrogen gas adsorption. What is measured is the amount of nitrogen gas adsorbed by a powder sample at liquid nitrogen temperatures over a range of gas partial pressures. The amount of gas forming a monolayer of the powder surface is usually calculated using the familiar Brunauer, Emmett, and Teller (BET) equation[18]:

$$\frac{P}{[V(P_o - P)]} = \frac{1}{V_m C} + \frac{C-1}{V_m C} \cdot \frac{P}{P_o} \tag{2}$$

where

V is the total volume of gas adsorbed, V_m is the gas adsorbed when the adsorbent surface is covered with a unimolecular layer, P is the equilibrium pressure of the system, P_o is the saturation pressure of the adsorbate gas, C is a constant dependent on the heat of adsorption and heat of liquefaction of the adsorbate gas.

It has been shown that for P/P_o values of 0.05 to 0.30, the plot of $\frac{P}{[V(P_o - P)]}$ versus $\frac{P}{P_o}$ is a straight line having a slope of $\frac{(C-1)}{(V_m C)}$ and an intercept of $\frac{1}{(V_m C)}$ making it possible to calculate both V_m and C.

The specific surface area of the powder (S) is obtained from the relationship:

$$S = \frac{V_m \sigma N}{V_o W} \tag{3}$$

where

σ is the area occupied by one molecule of adsorbate (assumed 16.2 $\mathring{A}^2$ for nitrogen at $-195^\circ C$), N is Avogadro's number, W is the weight of the powder sample, V_o is the ideal gas volume.

Crowl[19] has investigated the precision of surface area measurements on alumina, titania, ZnO, and other powders using the technique of nitrogen gas adsorption. He found a high degree of reproducibility for powders with specific surface areas greater than 1 m^2/g.

Joy[20] has prepared a detailed description of the experimental techniques and equipment that have been used in the determination of specific surface area of powders using nitrogen gas adsorption. Automatic and semiautomatic commercial instruments are now available which greatly reduce the time and manual effort required to obtain a surface area measurement on ceramic powders and are suitable for use with powders having specific surface areas of 0.001 m^2/g and greater.

C. Permeametry Techniques

The use of permeability measurements of fluid flow, through a packed powder bed as a means of characterizing fine powders dates to the 1930's and has as its two main advantages speed of operation and simplicity of equipment. Most permeameters are based on the principles of fluid flow through a packed powder bed as expressed in the Kozeny-Carmen equation[6]:

$$\mu = \frac{\epsilon^3}{(1-\epsilon)^2} \cdot \frac{P}{L} \cdot \frac{1}{B\eta\rho^2 S_w^2} \tag{4}$$

where

ϵ is the porosity of the powder bed, μ is the effective fluid velocity, η is the viscosity of the fluid, L is the length of the powder bed, S_W is the surface area per unit weight of powder, B is a constant, ρ is the density of the powder, P is the fluid driving pressure.

Kay and Jackson[12] have published a detailed discussion of the design and operation of gas flow permeameters. They point out that permeametry is limited to making comparisons of similar powders having different fineness and that reference to a standard powder of known surface area is always involved

D. Powder Strength Analysis

For many applications of powders, the strength of the particles, aggregates, and agglomerates is important. In powder processing, strong, porous aggregates lead to inhomogeneities in powder compacts which result in microstructural inhomogeneities in the sintered or hot-pressed product. Strong or hollow agglomerates formed by spray drying can produce similar inhomogeneities. Many of the powders used in fabricating ceramic bodies are prepared by precipitation or blending followed by calcination. Calcination almost always results in formation of aggregates. The strength of these aggregates depends on the material, as well as the time, temperature, and atmosphere during calcination, and this strength is an important characteristic of a powder with regard to subsequent processing parameters, microstructural development, and properties. Figure 1 shows an example of the microstructural inhomogeneities in a 95% dense alumina body prepared by isostatic pressing and sintering from a powder containing strong porous aggregates. The porous areas are remnants of porous aggregates which were not broken down during pressing. Milling is normally employed in ceramic processing to break down aggregates, and the amount of milling required is quite dependent on the aggregate strength.

A technique for qualitatively evaluating the strength of particles, aggregates, and agglomerates has recently been proposed[22]. The technique is based on interpretation of pressure compaction data plotted as the logarithm of pressure versus relative density. Figure 2 shows data for several materials over a wide range of pressure. The straight-line relationship obtained within selected pressure ranges for compaction data plotted in this manner has been noted by several investigators[23-25]. The use of these data to characterize the strength of the particulates in a powder is based on the assumption that the semilogarithmic relationship originates from the semilogarithmic relation between the mechanical strength of brittle materials and their relative density[26].

The data in Fig. 2 indicate that different mechanisms control compaction in different pressure ranges for each material. Each straight-line portion of the curve for a material in Fig. 2 is believed to indicate a different controlling mechanism. For the low-pressure portion of each curve, particle-particle or aggregate-aggregate friction is believed to be controlling compaction. Alumina II (a dry-milled powder) is believed to be an exception to

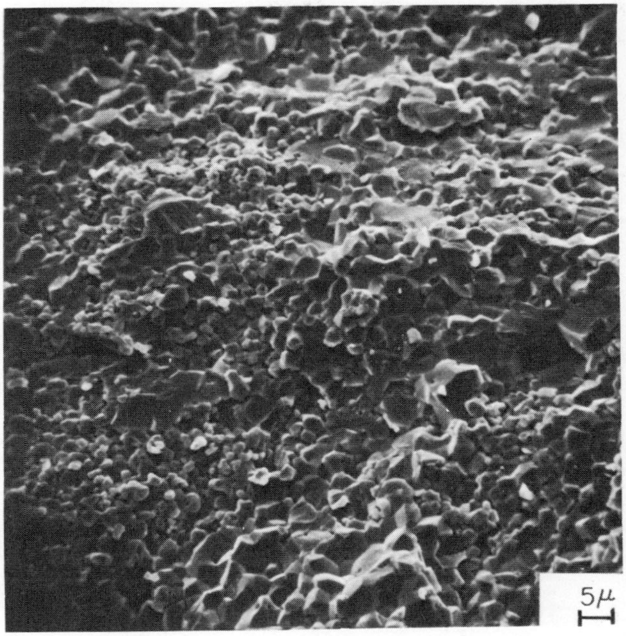

Fig. 1. Fractograph of an alumina specimen isostatically pressed at 100,000 psi and sintered to 95% of theoretical density.

this in that agglomerate strength controls compaction between 1 and 10 psi. Particle-particle friction controls compaction of this material from 10 to at least 100,000 psi. At some pressure the other curves all show a break. This break is indicative of a change in the mechanism controlling compaction. For copper and aluminum, plastic deformation is the controlling mechanism above the break in their curves. The curve for aluminum does not follow a straight line up to 100% relative density. This is probably due to isolation of pores. For graphite (natural Madagascar), the controlling factor is believed to be the rigidity of the thin platelets. Bending of the platelets allows particle rearrangement. At high pressures, the curve for graphite changes slope as particle strength becomes the controlling mechanism.

The curve for Alumina I is typical of a calcined powder. Above the break at about 300 psi, the strength of the porous aggregates controls compaction. When Alumina II is calcined at a series of temperatures between 1800° and 2400°F, a similar break occurs at a pressure which increases

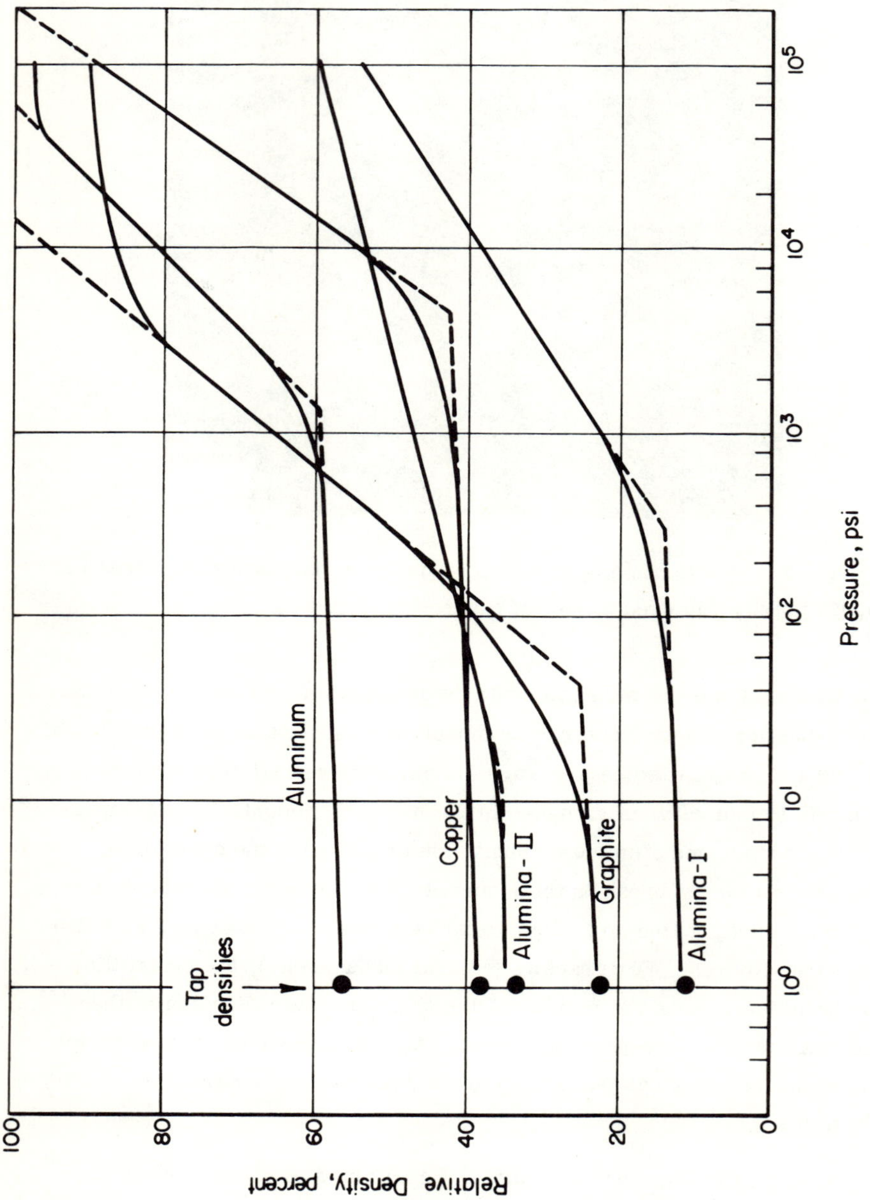

Fig. 2. Compaction behavior of several materials.

with temperature. At 1800°F, it is 60 psi and at 2400°F, it is 2000 psi. For powders containing aggregates, one linear portion of the logarithm of compaction pressure versus relative density curve corresponds to control of compaction by aggregate strength. The pressure at which the extrapolation of this linear portion and the one for the next lower pressure region intersect correlates well with the strength of the aggregates in a powder.

An important aspect in interpreting these data is a definition of the porosity within the aggregates of a powder. These data are also helpful in explaining tap densities of powders. Although there is no well-developed technique for measuring this powder characteristic, a mercury porosimetry technique has been developed(27). Gas absorption techniques might also be useful in characterizing this parameter(28).

IV. GRAPHIC REPRESENTATION OF PARTICLE SIZE DATA

Graphical methods are commonly used to present particle size data in a form that makes it easier both to illustrate the distribution of particle sizes and to compare the differences in particle size distribution of several powders. In these methods particle size (linear dimension, area, or volume) is plotted as a function of the total number, mass, or volume of the particles analyzed. There are two basic ways to present particle size data—as the relative frequency distribution or as the cumulative frequency distribution, both of which are shown in Fig. 3.

The relative frequency distribution portrays the fraction of particles that lie within a narrow size range. These data can be displayed as a smooth curve, as shown, or as a frequency histogram. The cumulative frequency curve is the integral of the relative frequency curve. Any point on the curve defines the cumulative fraction of particles under the stated size. The cumulative frequency curve can also be drawn to show the cumulative fraction of particles over the stated size.

In many instances the treatment of particle size data is rendered more useful if the size distribution of the powder can be expressed as a mathematical function. Three widely used analytical functions are the log-probability

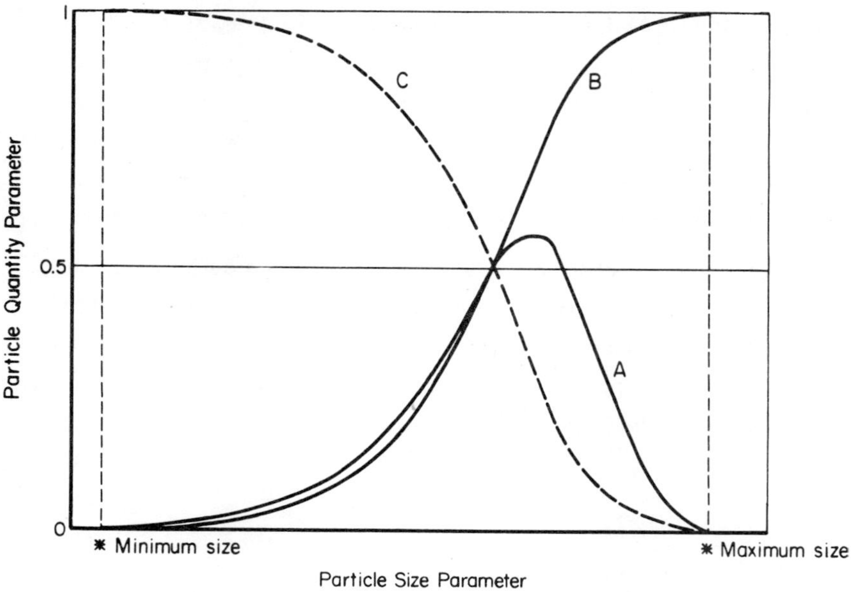

Fig. 3. Examples of graphic representation of particle size distribution: (A). Frequency distribution plot—shows the fraction of particles that lie within a narrow size range relative to the average size of the range. (B). Cumulative distribution plot—expressing cumulative percent of particles under a given size. (C). Cumulative distribution plot—expressing cumulative percent of particles over a given size.

distribution, the Rosin-Rammler distribution, and the exponential distribution. The particle size distribution of a great many powders can be represented by one of these functions. When plotted on the appropriate graph sheets a straight-line relationship results. This considerably facilities interpolation, makes extrapolation more reliable, and means that the particle size distribution can be uniquely defined by only two parameters.

An example[29] of a log-probability plot for an alumina powder is shown in Fig. 4, where particle size is represented on a logarithmic scale and cumulative percent undersize is represented on a probability scale with intervals based on values of the probability interval.

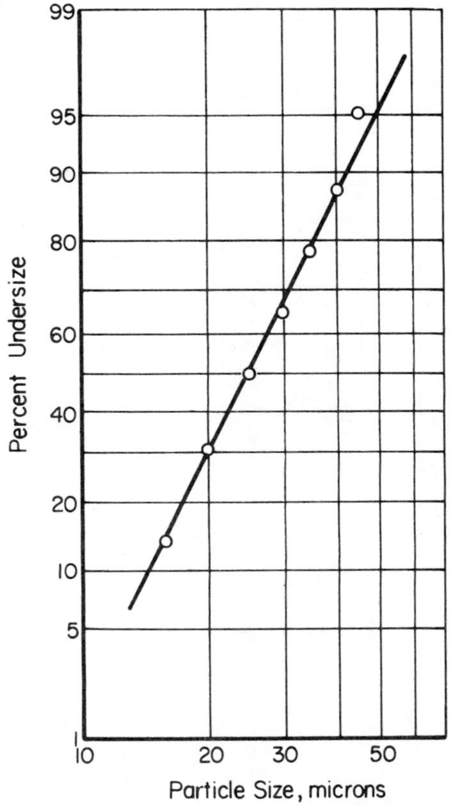

Fig. 4. Log-probability plot of size distribution in powdered alumina[29].

The Rosin-Rammler distribution (as modified by others) plots cumulative percent undersize on a log-log scale and particle size on a log scale as shown in Fig. 5 for raw coal[29].

The exponential distribution is represented as cumulative percent undersize (or oversize) on an exponential scale and particle size on a log scale.

The mechanics of treating particle size data are adequately covered in a number of texts[4-6, 30]. Lewis and Goldman[31] have derived formulas for computing and comparing sample statistics generated by the various particle size measurement methods for both single-component systems and mixtures.

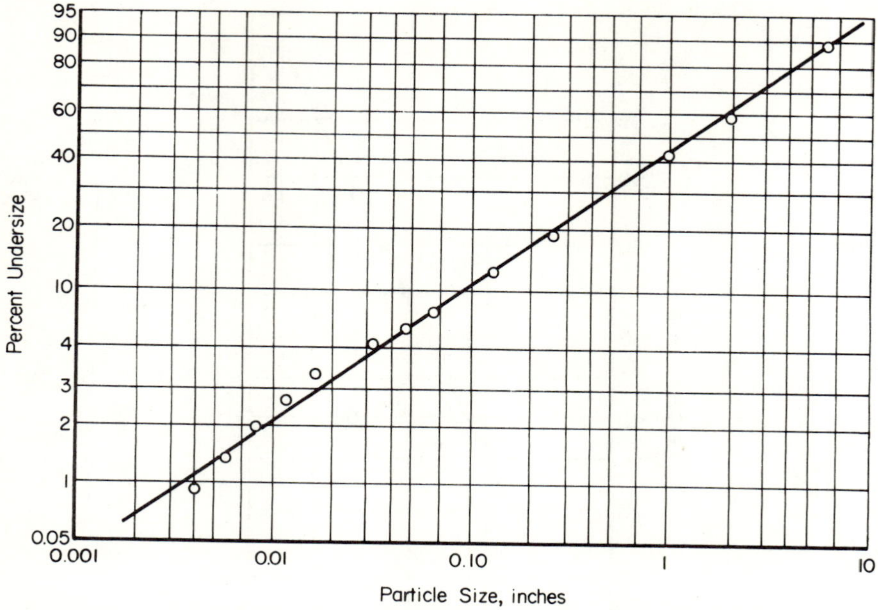

Fig. 5. Size distribution in raw coal plotted according to Rosin-Rammler relation[29].

REFERENCES

(1) Manual of Test Sieving Methods, STP 447, ASTM, Philadlephia, Pennsylvania, 43 pp (1969).

(2) Sampling and Size Analysis, T. Allen, Silicates Industriels, 509-515 (1963).

(3) B. H. Kaye, "An Investigation into the Relative Efficiency of Different Sampling Procedures," Powder Metallurgy, 9, 213-234 (1962).

(4) C. Orr and J. M. Dallavalle, Fine Particle Measurement, Size, Surface, and Pore Volume. Macmillan, New York, 353 pp (1959).

(5) R. R. Irani and C. F. Callis, Particle Size: Measurement, Interpretation and Application, Wiley, New York, 165 pp. (1963).

(6) T. Allen, Particle Size Measurement, Chapman and Hall, London, England, 248 pp (1968).

(7) R. D. Cadle, Particle Size Determination, Interscience, New York, 303 pp. (1955).

(8) "Classification of Methods for Determining Particle Size," Analyst, 88, 156-187 (March, 1963).

(9) B. H. Kaye, "Determining the Characteristics of Fine Powders," Chemical Engineering, 239-246 (November 7, 1966).

(10) A. C. Bryant, et al., "An Appraisal of the Micromerograph for Particle Size Analysis of Manganese Dioxide," Particle Size Analysis, published by The Society for Analytical Chemistry, London, England, 154-165 (1966).

(11) Baker and Bartok, "Particle Size Measurements of Alumina by Centrifuging," Bull. Am. Ceramic Soc., 45(6), 582-585 (1966).

(12) J. D. Zwicker, "Sieve Analysis to Below Two Microns Using Micromesh Sieves," Bull. Am. Ceramic Soc., 45 (5), 716-619 (1966).

(13) B. H. Kaye and M. R. Jackson, "A New Technique for Fractionating Powders," Powder Technol, 1, 43-50 (1967).

(14) T. Allen, "A Critical Evaluation of the Coulter Counter," Particle Size Analysis, published by The Society for Analytical Chemistry, London, England, 11-127 (1966).

(15) L. Wanless, "The Technique of Particle Size Analysis of Zircon Using the Coulter Counter," Trans. Brit. Ceram. Soc., 67 (2), 29-48 (1968).

(16) R. G. Blezard and R. W. Pring, "Electronic Techniques for Particle Size Analysis," Minerals Proc., 16-21 (June, 1969).

(17) Anon, "Semi-Automatic Particle Size Analysis," Ceramic Age, 83 (12), 38-40 (December, 1967).

(18) S. Brunauer, P. H. Emmett, and E. Teller, "Adsorption of Gases in Multimolecular Layers," J. Am. Chem. Soc., 60, 309-319 (1938).

(19) V. T. Crowl, "Surface Area Measurement by Low Temperature Nitrogen Absorption," Particle Size Analysis, The Society for Analytical Chemistry, London, England 288-296 (1966).

(20) A. S. Joy, "Methods and Techniques for the Determination of Specific Surface Area by Gas Adsorption," Vacuum. 3 (3), 254-278 (July, 1953).

(21) B. H. Kaye and M. R. Jackson, "Design Studies of Permeameter for Comparing Fine Powders," Particle Size Analysis, The Society for Analytical Chemistry, London, England, 313-325 (1966).

(22) D. E. Niesz, R. B. Bennett, and M. J. Snyder, Strength Characterization of Powder Aggregates. Paper presented at the 72nd Annual Meeting of the American Ceramic Socity, Philadelphia, Pennsylvania (May, 1970).

(23) Ceramic Fabrication Processes. (W. D. Kingery ed.), Technology Press of Massachusetts Institute of Technology and Wiley, New York (1958). "Pressure Forming of Ceramics," (W. D. Kingery), pp 55-61.

(24) C. A. Bruch, "Problems in Die-Pressing Submicron Size Alumina Powder," Ceramic Age, 83 (10), 44-53 (1967).

(25) V. S. Savin, N. B. Vrhina, and N. A. Fedotov, "Equation of Pressing of Nickel Powders," Poroshkovaya Metallurgiya, (2), 74 11, 14 (February, 1969).

(26) E. Ryshkewitch and W. H. Duckworth, "Discussion of Ryshkewitch Paper," J. Am. Ceram. Soc., "Compression Strength of Porous Sintered Alumina and Zirconia - 9th Communication to Ceramograph," J. Am. Ceram. Soc., 36 (2), 65-68 (1953).

(27) C. Orr, Jr., Application of Mercury Penetration to Materials Analysis, Micrometrics Instrument Corporation, Norcross, Georgia (Undated).

(28) C. G. Shull, "The Determination of Pore Size Distribution from Gas Adsorption Date," J. Am. Chem. Soc., 70, 1405 (1948).

(29) J. B. Austin, "Methods of Representing Distribution of Particle Size," Ind. and Engr. Chem., 11 (6), 334-339 (1939).

(30) G. Herdan, Small Particle Statistics, Elsevier, New York (1960).

(31) H. D. Lewis and A. Goldman, Theoretical Small Particle Statistics: A Summary of Techniques for Data Analysis with Recent Developments in Data Comparison, Notation and Mixture Theory, Los Alamos Scientific Laboratory Report No. LA-3656, 54 pp (November, 1968).

Chapter 12

SMALL ANGLE X-RAY SCATTERING

R. W. Gould

Department of Metallurgical and
Materials Engineering
University of Florida
Gainesville, Florida

I. INTRODUCTION

In a recent article which appeared in Physics Today[1], Andre Guinier summarized small angle scattering by this statement: "In the early days of x-ray diffraction, scattering close to the primary beam was a nuisance, to be disposed of with a large beam trap. Careful study of this scattering has yielded information on features intermediate in scale between that of atomic arrangement in a crystal and that of the images in the electron microscope." As Guinier points out in this article the experimental and theoretical advances in the study of small angle x-ray diffraction started about thirty years ago using focusing crystal monochromators in order to resolve the diffraction pattern close to the primary beam. Guinier[2,3] and others[4] applied the "Form Factor" concept to small angle scattering theory thus providing a statistical measure for the average size and shape of the small heterogeneitics that are responsible for this type of scattering.

The use of SAXS as a tool for materials investigation must compete with electron microscopy. As we know the electron microscope produces highly detailed images which are not available with the SAXS method. Often, however, the very detail in the electron micrograph is a handicap in that it is difficult to interpret. Thus, the statistically averaging properties of the SAXS method become very attractive.

II. GENERAL THEORY

X-ray diffraction is a scattering process, which in the case of crystalline matter, can, for many purposes, be described adequately by Bragg's law, $n\lambda = 2d_{hkl} \sin\theta$. Diffraction effects from "perfect" crystals usually take the form of sharp lines or spots well separated from the direction ($\vec{S}_0$) of the incident x-ray beam. On the other hand, if the material being investigated has a nonperiodic structure or a perturbed lattice, the x-ray scattering pattern is not limited to spots or lines but may contain extended (widely spaced) regions of scattering. It has also been shown that samples containing scattering heterogeneities (having dimensions approximately 10 to 1000 times that of the incident wavelength λ) will give rise to continuous scattering effects in the neighborhood of the main beam. An example of this effect is the halo often seen around the moon which is caused by light being scattered by small particles (water droplets) in our atmosphere. We can see then that x-ray small angle scattering is a somewhat different technique

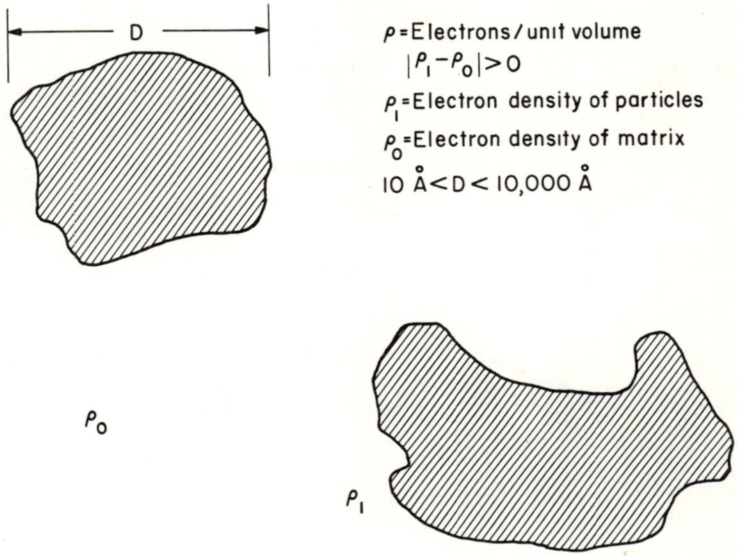

Fig. 1. Scattering of x rays from particles having electron density ρ .

from other x-ray techniques, because its purpose is not generally a study of structure on an atomic scale.

The conditions then for this type of small angle scattering may be summarized as follows: The sample must contain small regions having different scattering ability than their surroundings. Since x rays are scattered by electrons, we say that the heterogeneities must have a different electron density (ρ) than their environs (ρ_0). For x rays normally used in diffraction studies (λ approximately equal to 1 - 2 Å), particles ranging from 10 - 10,000 Å may be investigated (Fig. 1).

Reciprocal Space

X-ray diffraction effects are more easily explained and understood when the reciprocal lattice concept is utilized. The reciprocal lattice of a small crystal is shown in Fig. 2. For discussion of reciprocal lattice concepts the reader is referred to Chapter 5, "X-Ray Diffraction," in this text. In the construction of the reciprocal lattice, each node (hkl) corresponds to a stack of planes having the same (hkl) indices; also, the shape and size of the reciprocal lattice nodes are related to the shape and size of the real crystal. Note in Fig. 2 that the nodes are "football" shaped and rather

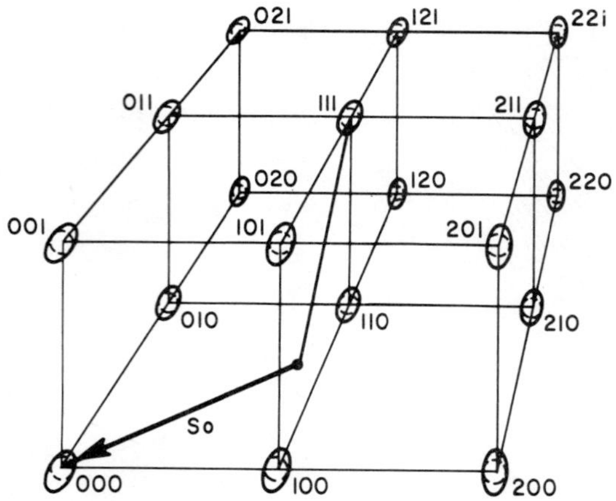

So = Direction of incident x-ray beam. $|So| = \frac{1}{\lambda}$

Fig. 2. A portion of the reciprocal lattice of a small crystal.

large. Large nodes in reciprocal space correspond to small crystals in real space. The intersections of the nodes of the reciprocal lattice with the Ewald sphere (a sphere whose radius, $\vec{S}_o/\lambda$, is directed toward the [000] node of reciprocal space) establish the conditions for a diffraction event to occur. When the reciprocal lattice node contacts the Ewald sphere a vector $\vec{S}$ (which gives the direction of the diffracted beam) may be drawn from the center of the Ewald sphere to the point of contact, and diffraction is said to occur. If the intersection covers a large area, that is, if the node is large, then the diffraction line or spot will be broad. On the other hand, if the reciprocal lattice is that of a "perfect" large crystal, the lattice points in reciprocal space will be very small and their intersection with the Ewald sphere will be pointlike. Note that in the absence of anisotropic lattice strain the reciprocal lattice nodes are all identical in shape, and there is a node at (000) which corresponds to the direction of the main beam. Therefore, there is always a diffraction effect in the direction of the main beam, but in the case of "perfect" large crystals it subtends such a small angle that is is not visible. When, however, the crystals become very small, the size of the node at (000) increases proportionally, and diffraction effects due to size alone extend out into the region around the main beam where they may be detected. Note that the size effects may also be seen at nodes in reciprocal space other than (000), but at these locations the size

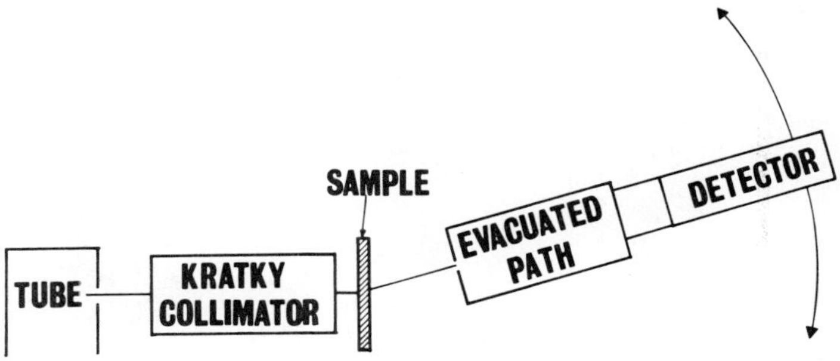

Fig. 3. Components of a Kratky small angle scattering camera.

is not the only cause of node enlargement, i. e., other effects (such as lattice microstrain) can further broaden the strong Bragg peaks. Unfortunately, there are many contributions to the background intensity in all x-ray diffraction measurements, i. e., temperature diffuse scattering, Compton modified scattering, etc. Fortunately, most of these troublesome interferences are minimized in the low angle region.

The observation of low angle x-ray scattering is not generally possible with standard x-ray diffraction equipment since the divergence of the main beam is too great to permit low intensity observations below 3-4°. In order to observe small angle diffraction effects, an effort must be made to severely minimize the divergence* of the main beam. One method of accomplishing this objective has been described by Kratky[5] and may be seen in Figs. 3 and 4. The collimator (Fig. 4) reduces the intensity of the primary line-shaped beam but renders the beam very parallel. The sample is placed immediately in front of the collimator block. The diffracted rays are provided with an evacuated diffraction path to minimize air-scattering. A line-shaped beam is advantageous for intensity considerations, but the scattered intensities must often be corrected as discussed below.

Recently[6,7] a new type of x-ray small angle camera has been marketed. This device uses the multiple reflection technique developed by Bonse and Hart[6] which provides an intense (4×10^6 cps), highly collimated (~4 seconds wide), x-ray beam. Gravatt and Brady[8] have shown that the

*Perpendicular to the plane of the beam for a beam of line-shaped cross section.

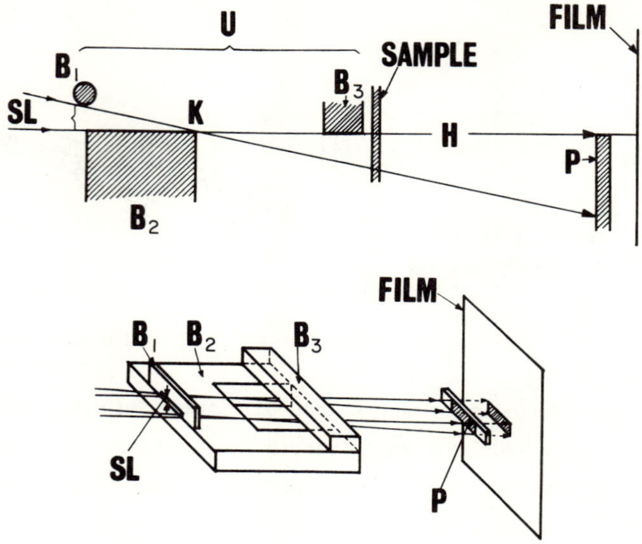

Fig. 4. Collimation system for a Kratky small angle scattering camera.

data obtained from this camera may be satisfactorily treated by standard slit desmearing techniques.

III. MATHEMATICAL RELATIONSHIPS

In the paragraph above, it was mentioned that a line-shaped beam is normally used for intensity reason. It should be mentioned that a correction will generally have to be made in order to compare line-shaped beam scattering data with theory. Correction is necessary because most small angle scattering theory is developed on the basis of a point- or pencil-shaped beam which simplifies much of the theory. Unfortunately, a beam of this shape severely limits the intensities obtainable in small angle scattering.

We would now like to develop some relationships between the scattered intensity I, as a function of $\vec{s}$, where:

$$\vec{s} = (S-S_0)/\lambda \text{ and } \vec{s} = (2\sin\theta)/\lambda = \epsilon/\lambda$$

in order to evaluate our experimental results and extract certain parameters which are characteristic of the scattering particles. It may be shown(2) that, in the vicinity of the (000) region of reciprocal space,

$$I(\vec{s}) = \frac{F^2}{VV_c} \left| \Sigma(s) \right|^2 \tag{1}$$

where F is the structure factor, that is, the scattering ability of the unit cell. Thus, F is related to the electron density in the unit cell; V_c is the volume of the unit cell and V is the irradiated volume; Σ (s) is related to the shape of the node in reciprocal space which in turn is related to the shape of the real particle; Σ (s) (Table I), which is the Fourier transform

TABLE I

Evaluation of Σ(s) for Various Regular Particle Shapes[2]

Spheres

$$I(s) = \left[(\rho - \rho_o)\frac{4}{3}\pi a^3\right]^2 \phi^2(2\pi as)$$

Very Thin Cylinders
of Volume V and Length 2H

$$I(s) = \left[(\rho - \rho_o)V\right]^2 \quad \frac{Si(4\pi sH)}{2\pi sH} - \frac{\sin^2 2\pi sH}{(2\pi sH)^2}$$

Flat Disks
of Radius R and Volume V

$$I(s) = \left[(\rho - \rho_o)V\right.^2 \quad 1 - \frac{1}{2\pi sR} J_1(4\pi sR) \quad \frac{2}{(2\pi sR)^2}$$

$$\phi(2\pi sa) = \left[3 \quad \frac{\sin 2\pi as - 2\pi as \cos 2\pi as}{(2\pi as)^3}\right.$$

$$Si(x) = \int_0^x (\sin t/t)\, dt$$

J_1 = Bessel function of first order

of the form factor of the real particle, may be calculated for many regular particle shapes. For N unit cells of volume V_c, $N=V/V_c$, the intensity:

$$I_N(\vec{s}) = \frac{F^2}{VV_c} \left|\Sigma(\vec{s})\right|^2 \frac{V}{V_c} = \frac{F^2}{V_c^2} \left|\Sigma(\vec{s})\right|^2 \tag{2}$$

Now $(F/V_c)^2 = \rho^2$ where ρ is the electron density, i. e., electrons per unit volume. Therefore:

$$I_N(\vec{s}) = \rho^2 \left|\Sigma(s)\right|^2 \tag{3}$$

The above formula assumes that the heterogeneity, whose electron density is ρ, is embedded in a matrix having and electron density of 0, i. e., a vacuum. In normal situations, the heterogeneity is found in a matrix having a nonzero electron density ρ_o, and therefore (see Fig. 2), the intensity scattered by a single particle, containing N unit cells, can be written as:

$$I(\vec{s}) = (\rho - \rho_o)^2 \left|\Sigma(\vec{s})\right|^2 \tag{4}$$

$$I(s) = (\rho - \rho_o)^2 V^2 \cdot \exp\left[-\frac{4}{5}\pi^2 r^2 s^2\right] \tag{5}$$

A more general form, which expresses the intensity I(s) as a function of the radius of gyration (R) of the particle, is given by

$$I(s) = (\rho - \rho_o)^2 V^2 \exp\left[-\frac{4\pi^2 s^2 R^2}{3}\right] = n^2 \exp -\left(\frac{4\pi^2 \epsilon^2 R^2}{3\lambda^2}\right) \tag{6}$$

where n is the number of effective scattering electrons and $s^2 = (\epsilon/\lambda)^2$. The radii of gyration R for three simple particle shapes are shown in Table II. The Guinier approximation states that the small angle scattering follows a simple exponential law; therefore, if the $\log_e$ of the scattering intensity is plotted as a function of the scattering angle ϵ^2 [(Eq. (6)], the slope will allow R^2 to be obtained directly. If then the particle shape is known, R^2 can be converted into parameters which are characteristic of the particle. For instance, in the case of spheres - once R is known, R may be related to the radius of the sphere by the formula $R = (3/5r)^{1/2}$, where r is the radius of the sphere. Gerold[9] in a review paper has written

TABLE II

Radii of Gyration for Three Simple Particle Shapes[11]

Ellipse with semi-axes a and b
$R^2 = \frac{(a^2 + b^2)}{4}$
Sphere with radius r
$R^2 = \frac{3}{5} r^2$
Ellipsoid with semi-axes a, b, c
$R^2 = \frac{(a^2 + b^2 + c^2)}{5}$

Eq. (4) in a different form [Eq. (6b)] which clearly illustrates the effects of heterogeneity volume fraction (c) and interparticle interference $[\phi(s)]$ upon the scattered intensity I(s):

$$I(s) = c\ V_a \cdot V_p\ (\Delta\rho)^2\ j_\rho(s)\ [1-\phi(s)] \tag{6b}$$

where V_a = atomic volume

V_p = particle volume

$j_\rho(s)$ = scattering function for a single particle using line collimation

It can be shown that interparticle interference has a maximum value at s = 0 and drops to zero for s large.

The exact form of the interparticle interference function is not easy to calculate, and thus dense particle systems will often present a theoretical problem.

A. Limitations of Simple Models

The assumption of identical low density, randomly oriented particles is seldom met in practice. Therefore, small angle scattering results will generally represent average values of particle shape and size. The nature of this averaging process can best be understood by a thorough study of SAXS theory.

B. Mathematical Models

The model that will be used for this discussion of the theory of small angle scattering is described below. Here, it will be assumed that we have identical, randomly oriented particles, where the average volume available for each particle is large compared with the volume of the particle. Note that this is a restrictive model but applicable to some real situations. Under these assumptions there is no interference of scattered x rays between the particles, and the intensities scattered by the various particles simply add. Now, if all the particles have the same orientation then the scattered intensity per particle equals that for a single isolated particle. If the particles are oriented at random, the intensity that is observed is an average intensity scattered by a single particle rotated into all possible orientations. Also, $I(\vec{s})$ is independent of the direction of $\vec{s}$, hence $I(s) = I(\vec{s})$ where $s = |\vec{s}|$. From Eq. (4), it is obvious that if the particle and the matrix do not differ in electron density the intensity near the main beam due to small angle scattering will be zero. In order to evaluate Eq. (4), it is necessary to obtain the average value of $|\Sigma(s)|^2$. This function can be calculated for regularly shaped particles and in Table I, we see the value of $\Sigma(s)$ as a function of the diffraction vector s for spheres, ellipsoids, thin cylinders, and flat disks. Knowing this function and the difference in electron density, we are able to calculate the scattered intensity from particles having this shape. Ideally, we may utilize the scattered intensity and work back to the shape of the particle. Guinier showed that $\Sigma(s)$ may be approximated by an exponential series. This derivation is presented in Chapter 10 of his book. The Guinier approximation, as it is now called, is given in Eq. (5) for spherical particles of radius r.

The above discussion leads us to the point of asking what parameters, other than the radius of gyration, can be obtained from small angle scattering curves (Fig. 5). A list of these parameters is given in Table III.

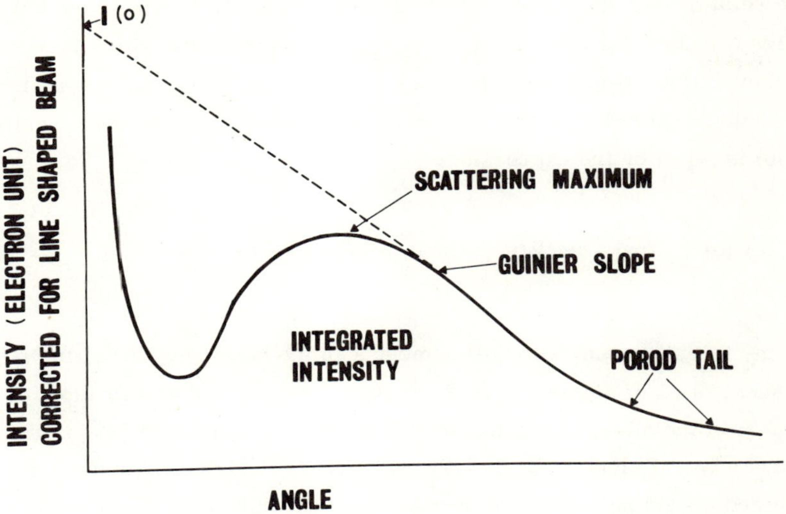

Fig. 5. Typical small angle scattering pattern from spherical particles.

C. Radius of Gyration and Integrated Intensity

We have already discussed the radius of gyration from which the particle shape can often be obtained. This is particularly true if the particles are spherical or nearly spherical, in which case the radius of gyration, R, can be converted directly into a particle radius r. The total volume of the heterogeneity can be obtained by calculating Q(o), the integrated small angle scattering intensity. Gerold has shown for binary alloys that there is a

TABLE III

List of Particle Parameters Obtainable Using the Small Angle Scattering Method

1.	Radius of gyration
2.	Integrated intensity (Q_o)
3.	Particle volume
4.	Number of particles per unit volume
5.	Total surface area of particles
6.	Size distribution of particles

simple relationship between the concentration of atomic specie in the heterogeneities and the total integrated small angle scattered intensity. Hendrix and Borie[10] have discussed, in a very clear presentation, the integrated small angle scattered intensity as it relates to the total scattered intensity. The Q(o) is given by the expression:

$$Q(o) = (m_1 - m_A)(m_A - m_2)\frac{(Z_A - Z_B)^2}{V_a} \tag{7}$$

where m_1 is the concentration of element A in the heterogeneities, m_2 is the concentration of element A in the matrix, m_A is the concentration of element A in the alloy, Z_A is the atomic number of element A, Z_B is the atomic number of element B, and V_a is the atomic volume. The Q(o) can be determined experimentally from absolute measurements of small angle scattering curves. It has been shown that Q(o) is given by:

$$Q(o) = 4\pi \int_0^\infty s^2 I(s)\, ds \tag{8}$$

when a point-shaped beam is used. Here, as before, $s = (2 \sin\theta / \lambda = \epsilon / \lambda$ where ϵ is the diffraction angle and λ is the wavelength of radiation. For spherical particles and/or for random distribution of particle orientations, I(s) is independent of the direction of s. In order to solve this equation for the case of a line-shaped beam, a transformation must be made. Guinier and Fournet show that under this transformation Eq. (8) above becomes:

$$Q(o) = 4\pi \int_0^\infty s \int_0^\infty I\ (s^2 + t^2)^{1/2}\ dt\, ds \tag{9}$$

Thus, it is possible to calculate from the small angle scattering curve, using a line-shaped beam, the quantity Q(o) which may be directly related to the compositions and the total volume of the heterogeneities in binary systems. It is well to mention that several computer programs exist[11, 12] for slit-shape corrections, i. e., conversion of slit data to point beam data.

D. Volume per Particle (V_p) and Number of Particles per Unit Volume $N_v(R)$

By extrapolating a small angle scattering curve uncorrected or slit-shaped beam effects to $\epsilon = 0$, one can obtain the intensity $J^*(o)$. It can easily be shown[9] that:

$$V_p = \frac{R_g J^*(o)}{3\pi \int sJ^*(s)\, ds} \tag{9b}$$

where $J^*(o)$ is a theoretical intensity function (at $\epsilon = 0$) for a slit-shaped beam. The asterisk signifies that interparticle interference has been neglected. In general, however, $J^*(o)$ is difficult to obtain accurately.

If the volume fraction of particles, c, is known then the number of particles per unit volume $N_v(R)$ can easily be calculated provided a particle shape can be assumed. For spherical particles, all of radius R, we have:

$$C = 4/3\,\pi \sum_i N_v(R)\, R^3$$

The volume fraction c can be obtained by first calculating the composition of the zones using the integrated intensity Q(o). Bonfiglioli and Guinier[13] have illustrated this procedure for calculating c for various Al-Zn alloys.

E. Surface Area of Particles

Porod has shown that the intensity in the tail of the scattering curve can be related to the total surface area of the scattering particles. The tail of the scattering curve (Fig. 5), or Porod region as it is called, is defined as that part of the curve for which in I(S) plotted against $\ln \epsilon^2$ results in a straight line of slope -3 for a line-shaped x-ray beam, or -4 for a point-shaped beam. An alternate method to establish the Porod region is to plot $I(s)s^3$ versus s^3 and look for the straight portion of this curve. It should be mentioned that in order for Porod's law to be valid there must be a sharp interface[14, 15] between the particle and the matrix. If I_o is the intensity of the main beam then:

$$\frac{I(s)}{I_o} = \frac{K(\rho - \rho_o)^2}{s^3}\, S_T \tag{10}$$

for a line-shaped beam. Here K is a constant determined by experimental conditions, S_T is the total surface area, and s is the usual diffraction vector.

For spherical particles the total surface area S_T may be calculated using a relationship introduced by Guinier[3] and used by Gerold[9] and others[16]:

$$S_T = \frac{4cV(1-c)\ \mathrm{Lim}\left[h^3\ J(h)\right]}{\int_0^\infty hJ(h)dh}$$

where c = volume fraction of particles, V = irradiated volume, and $h = 2\pi s = \dfrac{4\pi \sin\theta}{\lambda}$.

As we have seen the Guinier approximation, for a system containing spherical particles, will provide a radius of the particles. It has also been shown by Guinier that the maximum in the scattering may also be used to determine a particle radius and, in fact, it is even possible using the Porod approximation to determine another type of particle radius based on the total surface area of the particles. We will show in the next section how these radii compare and how they have been used to estimate the distribution of particle sizes in certain systems.

F. Particle Shape

The Porod region may be used to estimate particle shape. Schiller and Mering et al.[17] have shown [Fig. 6(a)] how rounded particles produce a curve that oscillates before becoming asymptotic. In contrast sharp-edged particles [Fig. 6(b)] show little oscillation. Gerold[9] has presented a somewhat quantitative particle shape factor W_1 which is normalized to unity for spherical particles.

$$W_1 = V_p/V_s$$

V_p was given by Eq. (9b) and V_s is the volume of a spherical particle whose radius of gyration, as obtained from the Guinier approximation, is R_g. Thus,

$$V_s = \frac{4\pi}{3}\left(\frac{5R_g^2}{3}\right)^{3/2} = \frac{4\pi R_s^3}{3}$$

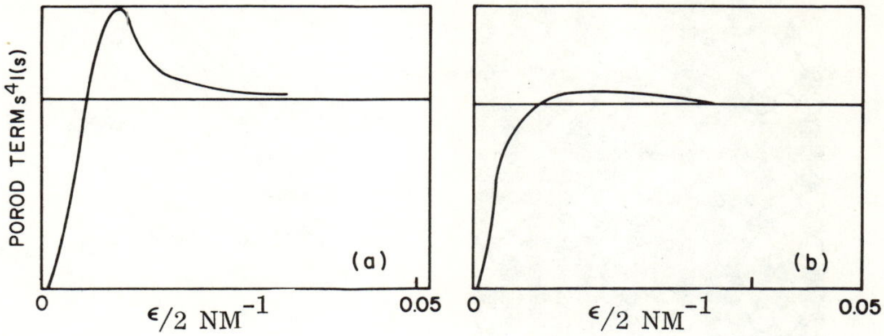

Fig. 6. Rounded particles produce a curve that oscillates (a) before becoming asymptotic. Sharp edged particles (b) show little oscillation. From Schiller and Mering (17).

Gerold presents another shape factor W_2 (see Ref. (9), page 287 and 305 for more details).

IV. APPLICATIONS

A. Age Hardening Studies

Small angle x-ray scattering has been used extensively in the study of age hardening alloys in which heterogeneities take the form of Guinier-Preston (GP) zones. In an age hardening alloy such as Al-Zn the alloy is heat treated in the solid solution region and quickly quenched at room temperature. Aging at room terperature produces and increase in yield strength and hardness. This strengthening is associated with the growth of Zn-rich clusters or the GP zones in the quenched solid solution. The rate of growth of these particles* is strongly affected by the quenching temperature and the aging temperature because age hardening is a diffusion process and the quenched-in vacancy concentration is a major rate-controlling factor. Since the atomic radii of Al and Zn atoms are nearly the same, the heterogeneities which form in the Al-Zn system are spherical during most of the growth process. The condition of this alloy during heat treatment is shown schematically in Fig. 7 where we see the solid solution containing small clusters

*Similar measurements have been made on nonmetallics such as LiF containing small concentrations of Ni^{2+} (18).

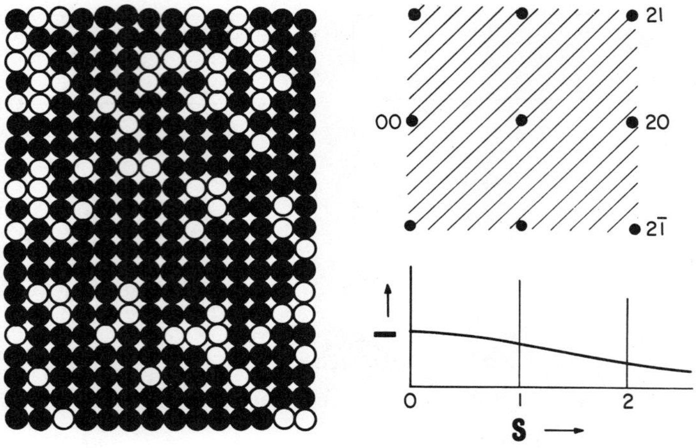

Fig. 7. Alloy of atoms A and B at a temperature above the solvus.

of Zn atoms. When this alloy is quenched, the clusters of Zn atoms called GP zones result (Fig. 8). These figures also show a portion of the reciprocal lattice and the x-ray scattering intensity as a function of angle. In Fig. 8, the slope of the intensity surrounding the (000) node is small. After quenching and growth, however, the slope of the intensity around the (000) node increases sharply. Note that there is an intensity shoulder around each Bragg peak which is also caused by the GP zones in the solid solution. Further aging produces larger zones (Fig. 9) and also an increase in the small

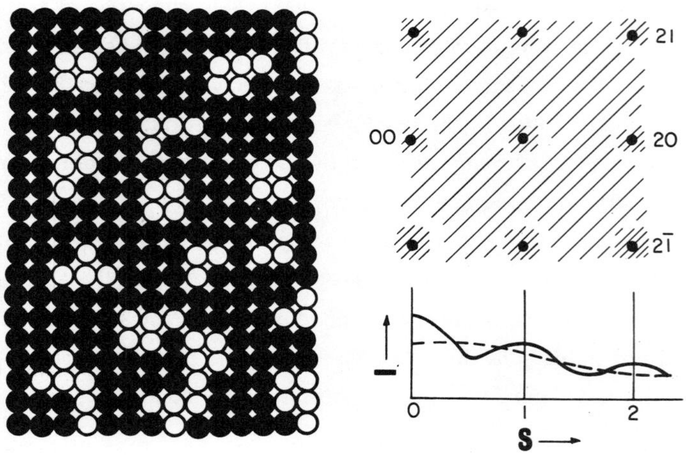

Fig. 8. Guinier-Preston zones in a binary alloy.

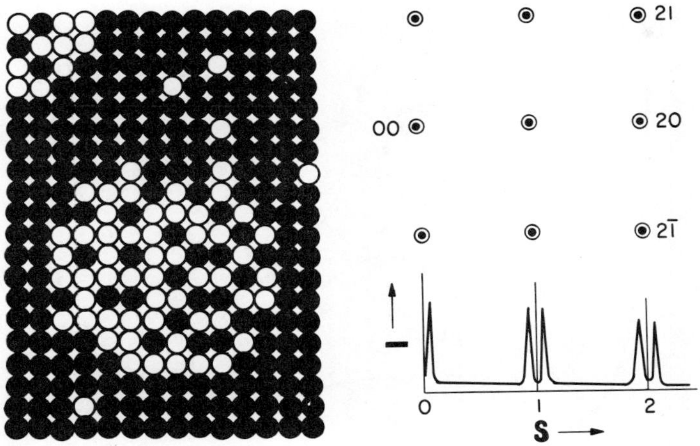

Fig. 9. Large Guinier-Preston zones in a binary alloy.

angle scattering intensity. In alloys such as Al-Ag, and Al-Zn-Ag, the atoms within the zone tend to order. In this case the diffuse x-ray scattering takes on an additional attribute, namely, a small peak between the first Bragg peak and (000) as shown in Fig. 10. Small angle scattering has been used to measure the kinetics of the growth of these GP zones in many alloy systems. Figure 11 shows a study of this type in which small angle scattering curves are plotted as a function of time at temperature. From these curves, one can obtain the parameters listed in Table III and study them as

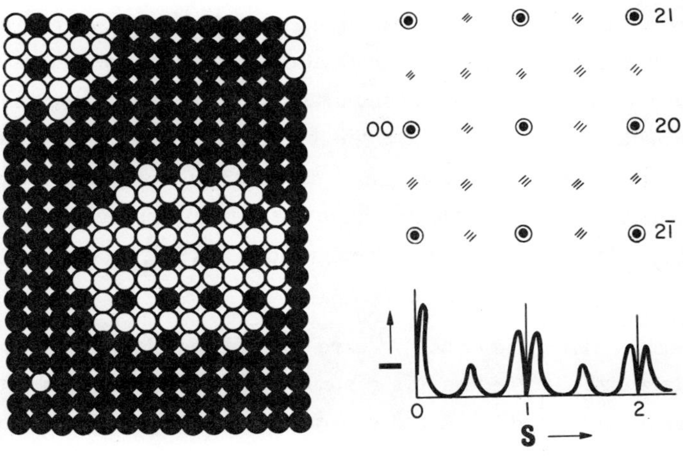

Fig. 10. Ordered Guinier-Preston zones in a binary alloy.

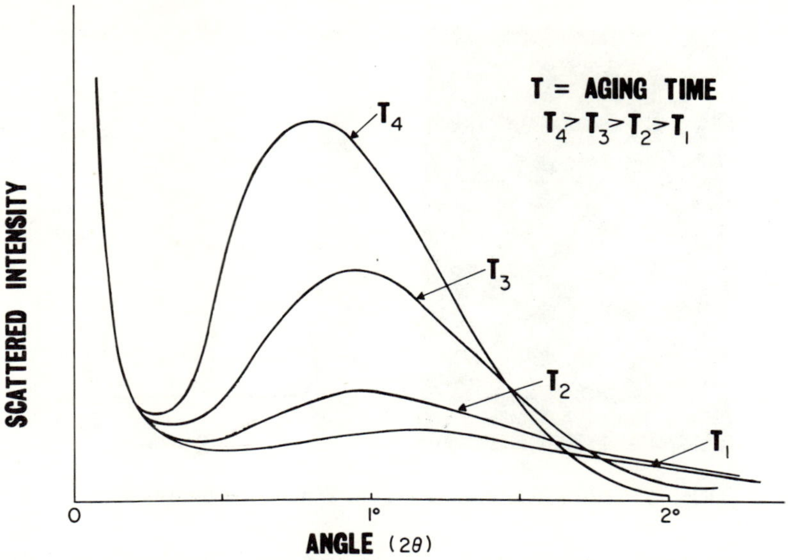

Fig. 11. Kinetics of particle growth as shown by a change in small angle scattering curves.

a function of time and various other physical properties. Since this is a thermally activated process, its rate may be increased or decreased by a small change in aging temperature.

B. Particle Size Distribution

Baur[19] has compared the small angle scattering method with transmission electron microscopy for the evaluation of particle radii. His results show that the two methods give comparable results, and furthermore, the three radii that he measures from the small angle scattering patterns might be used to approximate a particle size distribution curve (Fig. 12). Note that the radius R_I was determined from the angular position of the scattering maxima, R_{II} from the Guinier approximation, and R_{III} from the Porod region of the small angle scattering curve.

Recently, Harkness et al.[20] have shown that it is possible to determine the particle size distribution of a system of spherical particles if the Guinier sphere radius (R_s) and the Porod radius (R_p) are known. This method assumes a log normal size distribution (although a normal distribution with R_s

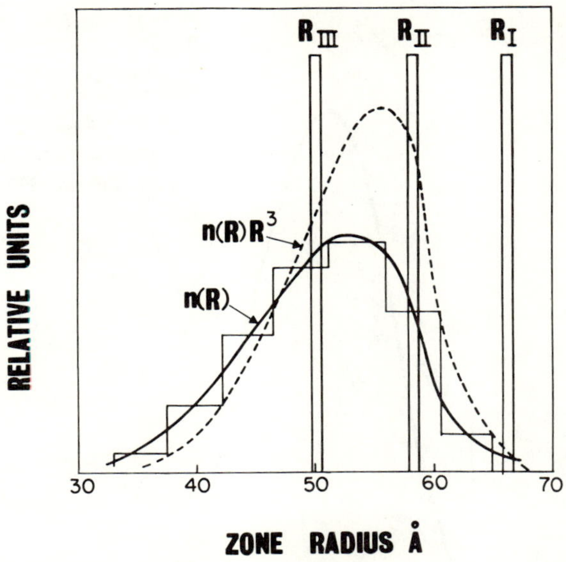

Fig. 12. Relationship between the radii, R_I, R_{II}, R_{III}, obtained from small angle scattering to size distribution curve from transmission electron microscopy.

and R_p. The relationship between a measured (by transmission electron microscopy) distribution and the x-ray results are shown in Fig. 13.

Schmidt and his students[(21-23)] have developed particle size distribution methods which require no prior assumptions concerning the nature of the particle size distribution function. In general, his procedures like the ones described in the preceding paragraphs work best with spherical particles having a simple particle size distribution.

C. Small Angle Scattering Studies in Nonmetallic Materials

The small angle x-ray scattering method is gaining wide acceptance for investigation of nonmetallic materials. In this section, we will briefly discuss a few of these applications.

Arnott[(24)] has measured the particle sizes and size distribution of clay minerals (Kaolinite, Montmorillonite, and Illites). These results are based upon the early theoretical work of Schull and Roess[(25)].

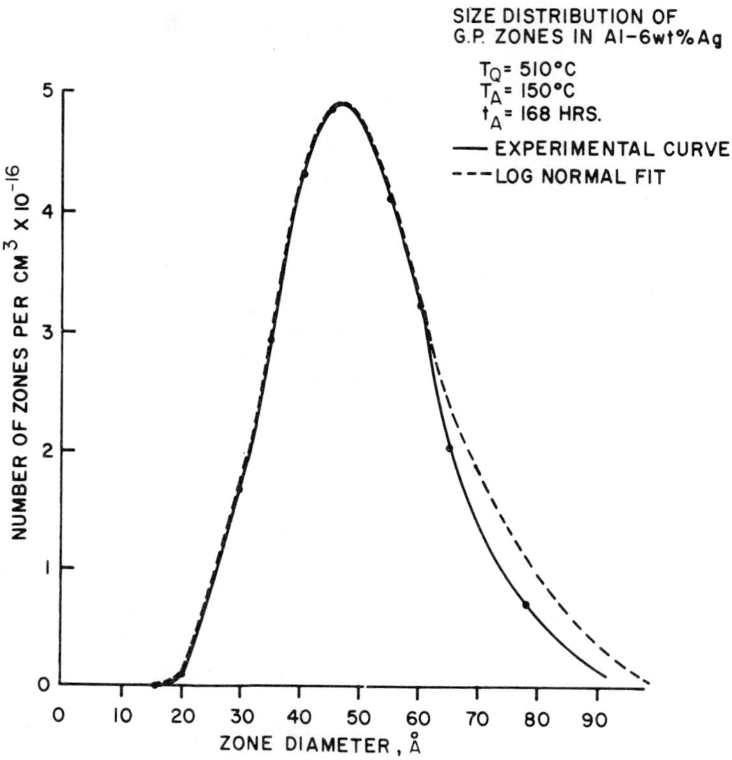

Fig. 13. Size distribution of Guinier-Preston zones in Al-Zn determined by SAXS after Harkness et al. (20).

More recently, Watanabe[26] has studied allophone, hydrated halloysite, and kaolinite using Jellinek et al. [27] methods together with Porod's theory[14]. Watanabe calculated the radius of gyration $<\bar{R}g>$, coherency distance 1c, and has investigated the Porod region of the SAXS curve for these materials. Veen and Wiersma[28] have used the SAXS method to study textured periodicity in Gibbsite.

Other materials that have been investigated by this method include TiO_2 emulsions[29], Quartz $\alpha \rightarrow \beta$ transition[30], polymers[31,32], and alumina catalyst surfaces[33].

Surface area measurement by SAXS can quantitatively reveal the presence of microporosity. Schiller[17,34] et al, have studied the shape of the micropores in graphite. Tchovbar-Vallet[35] has investigated porosity in concentrated systems of polyflurotetra ethylene as well as MgO[15]. Brumberger[16] has recently reviewed the subject of specific surface area determinater by SAXS.

The SAXS from glasses and other amorphous solids was the subject of a recent symposium[36] held at the University of Missouri. The small angle method is being applied to phase separation in glasses[37-40] radiation damage[41, 42] and to amorphous semiconducting glasses[42].

Biological application of SAXS have been made in studies of bone structures[44], RNA[45, 46], and proteins[47] to mention only a few.

V. SUMMARY AND CONCLUSION

The small angle x-ray method has been shown to have a wide application in the field of materials science. It has been used extensively for the study of age hardening in metallurgical systems, and is currently being used in the study of devitrification of glass-ceramics, catalysts, polymers, bone, etc. While the method is simple and easy to apply, one must exercise reasonable care in interpretation, particularly with regard to the limitations imposed by the chosen model. With proper care, one can obtain easily many parameters which are not available by other competing methods.

REFERENCES

(1) A. Guinier, Physics Today, 22 #11 , 25 (November, 1969).

(2) A. Guinier, X-Ray Diffraction, Freeman, 1963.

(3) A. Guinier and Fournet, Small Angle X-Ray Scattering, Wiley, New York, 1955.

(4) R. Hoseman and S. N. Bagchi, Direct Analysis of Diffraction by Matter, North Holland, 1962.

(5) O. Kratky, Zeitschrift fur Electric Chemie, 58, 49, (1954).

(6) U. Bonse and M. Hart, Small Angle X-Ray Scattering, Gordon & Breach, 1965.

(7) D. M. Koffman, Advances in X-Ray Analysis, Vol. 11, p. 332, 1967.

(8) C. C. Gravatt and G. W. Brady, "Slit Smearing Effects in the Hart-Bonse Small Angle X-Ray Diffractometer," Applied Crystallography (1969).

(9) V. K. Gerold, Small Angle X-Ray Scattering, Gordon and Breach, 1965.

(10) R. W. Hendrix and B. Borie, Small Angle X-Ray Scattering, Gordon and Breach, 1965.

(11) P. Schmidt, "Collimation Corrections in Small Angle X-Ray Scatterings," Acta Crystallographica, 19, 938 (1965).

(12) R. W. Hendrix and P. Schmidt, Acta Physica Austriaca, 26, 97 (1967)

(13) A. Bonfiglioli and A. Guinier, Acta Metallurgica, 14, 1213 (October, 1966).

(14) G. Porod, Z. Kolloid, 125, 83 (1951).

(15) Denise Tchoubar-Vollat, C. R. Acad Sc. Paris, 263, 1030 (November, 1966).

(16) H. Brumberger, Determination of Specific Surfaces by Small Angle X-Ray Scattering Methods X-Ray and Electron Methods and Analysis, p. 76, Plenum, New York, 1968.

(17) C. Schiller, J. Mering, et al., C. R. Acad. Sc. Paris, p. 309-312 (January, 1967).

(18) A. M. Levelut, M. Lambert, and A. Guinier, Acta Crystallographica, A24 4 , 459 (1968).

(19) R. Baur and V. Gerold, Acta Met., 12, 1449 (1964).

(20) S. D. Harkness, R. W. Gould, and J. J. Hren, Phil. Mag., 19 157 , 115, (January, 1969).

(21) P. W. Schmidt, G. G. Weil, and O. L. Brill, X-Ray and Electron Methods of Analysis, p. 86, Plenum, New York, 1968.

(22) O. L. Brill and P. Schmidt, Applied Physics, Vol. 39, #5, P. 2274, 1968.

(23) J. H. Letcher and P. W. Schmidt, Applied Physics, 37 2 , 649 (1966).

(24) R. J. Arnott, Am. Mineralogist, 50, 1563 (October, 1965).

(25) C. G. Schull and L. C. Roess, Applied Physics, 18, 295.

(26) T. Watanabe, Am. Mineralogist, 53, 1015 (1968).

(27) M. H. Jellinek, et al., Ind. Eng. Chem., 18, 172 (1946).

(28) A. W. L. Veer and J. Wiersma, Am. Mineralogist, 53, 1033 (1968).

(29) O. Kratky and H. Leopold, Colloid and Interface Sc., 22 #3 , (September, 1966).

(30) H. Brumberger, Physical Review Letters, 22 #11 (1969).

(31) R. G. Kirste, Small Angle X-Ray Scattering, Gordon and Breach, Science, p. 33, (1967).

(32) W. W. Beenan, Small Angle X-Ray Scattering, Gordon and Breach, Science, p. 197, (1967).

(33) G. A. Somorjai, X-Ray and Electron Methods of Analysis, p. 101, Plenum, New York, 1968.

(34) C. Schiller and J. Mering, C. R. Acad. Sc. Paris, 264, 247-250 (January, 1967).

(35) D. Tchoubar-Vallet, C. R. Acad. Sc. Paris, 263, 981-984 (October, 1966).

(36) Conference on Small Angle X-Ray Scattering in Glasses and High Temperature Materials, Sept. 9-13, 1969, Univ. of Missouri.

(37) W. Vogel, Microstructure of Glasses Resulting from Sequential Phase Separation, Conference on Small Angle X-Ray Scattering in Glasses and High Temperature Materials, Sept. 9-13, 1969, Univ. of Missouri (Dept. of Ceramic Eng. and Dept. of Physics), Rolla, Mo.

(38) R. K. MacCrone, M. Tomozawa, and H. Herman, Small Angle Scattering in Vitreous Systems, Univ. of Missouri, Rolla, Mo., 1969.

(39) M. Tomozawa, H. Herman, R. K. MacCrone, A Study of Phase Separation in Na_2O-$S1O_2$ Glass Using Small Angle X-Ray Scattering, Univ. of Missouri, Rolla, Mo., 1969.

(40) P. W. Schmidt and S. M. Ohlberg, Electron Density Differences between Droplets and Matrix in Na_2O-SrO-$S1O_2$ Glass, Univ. of Missouri, Rolla, Mo., 1969.

(41) H. D. Bale, Small Angle X-Ray Scattering from Neutron Irradiated Vitreous Silica, Univ. of Missouri, Rolla, Mo., 1969.

(42) G. Den Outen, "Small Angle Scattering of X-Rays by Neutron Irradiated Germanium," Applied Physics, 39 #10 , 4509-4517 (1968).

(43) L. L. Hench and S. R. Bates, Small Angle Scattering in Semiconducting Glasses, Univ. of Missouri, Rolla, Mo., 1969.

(44) E. D. Eanes and A. S. Posner, X-Ray Scattering Study of Bone Materials, p. 493, Gordon & Breach, 1965.

(45) J. A. Lake and W. W. Beeman, "Yeast Transfer TNA: A Small Angle X-Ray Study," Science, 156, #3780 , 1371-1373 (June, 1967).

(46) J. W. Anderegg, X-Ray Scattering from Small RNA Viruses in Solution, p. 243, Gordon & Breach, 1965.

(47) W. W. Beeman, Some Experimental Procedures and Recent Results in Liquids, Solutions and Globular Proteins, p. 197, Gordon & Breach, 1965.

Chapter 13

X-RAY LINE PROFILE ANALYSIS

R. W. Gould

Department of Metallurgical and Materials Engineering
University of Florida, Gainesville, Florida

I. INTRODUCTION

The term "particle size" has been used frequently throughout this book. In most instances, the reader clearly understands what the author means by size, i.e., the measurable outside dimensions of the particle. Even in these cases, however, we must mentally disregard shape when we speak of "particle size" because "size" implies that a single parameter exists which can adequately describe the particle. This would be true in the case of spherical particles, but it is not true for the more general shapes. The situation becomes even more confused when we consider the "particle sizes" measured by x-ray diffraction. With x rays, the size measured may not even come close to the outside physical dimensions of the particle. Often, in fact, x-ray "sizes" are several times smaller than sizes measured by transmission electron microscopy. Must of the confusion arises from a lack of understanding concerning what is being measured by the x-ray diffraction technique. In

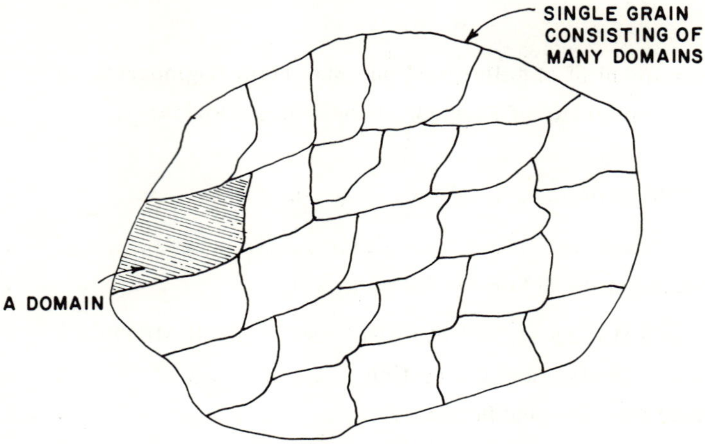

Fig. 1. Schematic illustration of a domain structure.

general, the x-ray technique does not see the particle as a unit, but rather as a collection of smaller units. These smaller units are called domains, coherently diffracting domains or crystallites. Figure 1 is a schematic illustration of a domain structure. The details of a domain wall structure will not be discussed in this article, we are only concerned here with the domain, i.e., nearly perfect regions of the particle which are separated from each other by narrow boundaries consisting of more or less imperfect crystalline material[(1)]. These domains are, in general, slightly misoriented with respect to one another. Thus, Bragg's law is not satisfied for each domain at exactly the same angle θ .

We saw earilier, in the chapter on basic x-ray diffraction, how small crystals give rise to enlarged nodes in the reciprocal lattice and thus, to large diffraction spots. Therefore, if we have a particle which consists of many small misoriented perfect particles, we can expect that the diffraction lines will be broadened "enlarged." Such small perfect regions diffract x rays like small single crystals and thus are referred to as coherently diffracting domains.

Size is not the only factor that can produce diffraction line broadening. We will show in this chapter that lattice strain can also cause the line shapes to change. Because the higher order lines (small d-spacings) are more sensitive to strain than the low angle reflections, we will be able to separate strain effects from "particle size" effects. In addition, there are other factors which can lead to shape changes and/or peak shifts of the diffraction profile. These factors are related to imperfections and irregularities in the stacking sequences of various atomic planes in the lattice. We will not cover the theoretical details of stacking fault and twin fault broadening, but we will present the relationships that can be used to measure and interpret these effects.

In order to remain consistent with the existing literature, the term "particle size" will be used throughout this chapter. The reader is encouraged to remember that we are, however, referring to a size, measured in a particular direction [hkl], of many coherently diffracting domains.

This chapter can only serve to introduce the reader to the subject of x-ray line profile analysis. For this reason, there are several excellent review articles listed in the appendix of this chapter for those interested in pursuing the subject in greater detail.

II. THEORY

Figure 2(a) shows a diffractometer tracing of the (004) line of cold worked aluminum. Comparing this with Fig. 2(b), the same sample annealed at 500°C, we can see dramatic differences in the two line shapes. The most obvious difference between these two lines is the width or narrowness of the annealed peak compared with the cold worked profile. As early as 1918 Scherrer[2] showed that the angular width in radians of a line at a point equal to one-half of its maximum intensity is related to the "particle size." This relationship is given in Eq. (1):

$$L = Nd = \frac{0.9\,\lambda}{(\Delta 2\theta)_c \cos\theta_o} \tag{1}$$

where L = the particle size (crystal thickness)

d = the spacing of the reflecting planes

$(\Delta 2\theta)_c$ = the angular width of the line at half-maximum intensity corrected for instrumental broadening

θ_o = the Bragg angle of reflection

λ = the wavelength of the radiation

N = the number of planes of spacing d

The derivation of this equation is shown clearly by Guinier[3], but the apparent size determined by this technique has no rigorous meaning. If

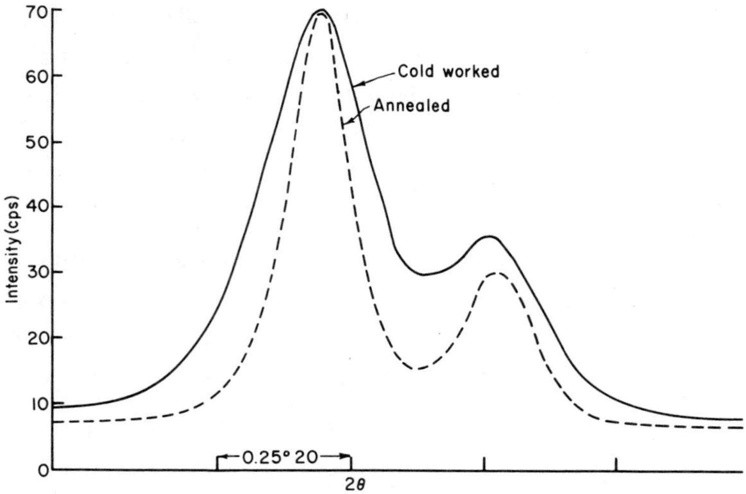

Fig. 2. (a) Diffractometer profile of (004) line of cold worked Al; and (b) diffractometer profile of (004) line of Al annealed at 500°C.

instead of using a breadth at half-maximum intensity, one uses the integral width, then the Scherrer size can be shown to be equal to the volumetrically averaged diameter of the particles in the system. The integral width of a line is simply defined as the area under the peak divided by the peak maximum intensity as shown in the Eq. (20):

$$\beta = \frac{\int_{\infty}^{\infty} y(2\theta)\, d\theta}{y_{max}} \tag{2}$$

The equation relating the integral width with the particle size is similar to Eq. (1) as shown by:

$$L = \frac{\lambda}{\left(\beta_{\substack{\text{integral} \\ \text{corrected}}}\right) \cos\theta}$$

The integral line width or half-maximum line widths must be corrected for instrumental effects before they can be used. This is due to the fact that even a large grained sample gives rise to a diffraction peak having a finite width. This width is due to the instrument itself, i.e., the slits, x-ray source size, etc. In order to eliminate this effect, the instrumental portion of the broadened line width must be removed. This subject is discussed in detail by Taylor[4] (see especially Chapter 14). As a first approximation, if

$\Delta 2\theta'$ is the one-half maximum breadth of the well-annealed large particle sample then:

$$(\Delta 2\theta)_{corrected} = \Delta 2\theta - \Delta 2\theta' \tag{3}$$

and if β' is the instrumental integral breadth then:

$$\beta^{corrected}_{integral} = \left[(\beta_{integral})^2 - (\beta'_{integral})^2 \right]^{1/2} \tag{4}$$

if the peak has a Gaussian shape.

Limitations of the Scherrer Technique

Even though the Scherrer technique provides a volume averaged particle diameter, it cannot distinguish size broadening from other causes of line broadening. It has been shown that lattice strain and errors in the stacking faults give rise to line broadening. Stacking faults may be of several types, i.e., deformation or twin. These are shown in Fig. 3. It is possible that a broadened line may derive all of its excess breadth from such defects, thus the Scherrer "size" in these cases would be meaningless. In order to extract size, strain, and faulting information from broadened diffraction line profile, a different approach must be taken. One such approach was discussed originally by Warren[5] who presented a technique for separating size, strain, and faulting. This material is covered thoroughly in his new

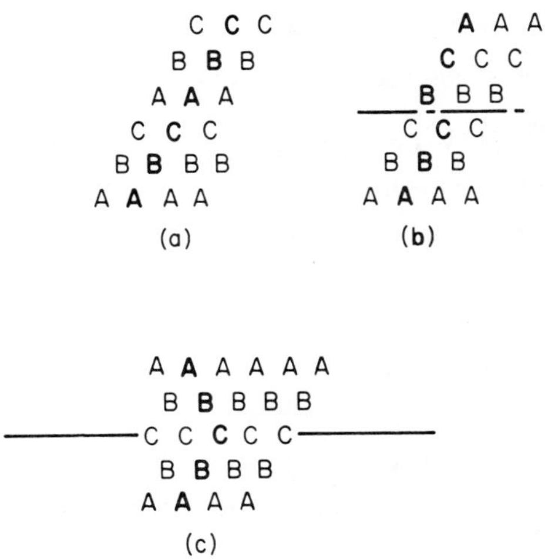

Fig. 3. Stacking irregularities in fcc crystals. (a) Correct; (b) deformation fault; (c) twin fault.

text and elsewhere. Thus, we will present here only the highlights of this method.

III. LINE PROFILE ANALYSIS (WARREN-AVERBACH METHOD)

A. Data Collection and Correction

Various authors have discussed the subject of proper data correction for line profile studies. Here, we will emphasize a few of the more important factors. In general, line profile analysis studies are done using a well-aligned diffractometer adapted with a primary beam or a diffracted beam monochromator. Beta filtered radiation used with pulse height selection will also provide good results if sample fluorescence is not a problem. There is no general rule with regard to resolution. Often with very broad peaks, one does not use the highest possible resolution. This is primarily due to the fact that the high resolution system in this case would not be necessary and would sacrifice a great deal of intensity.

Many of the L. P. A. problems arise from inaccuracies in measured peak profiles[6]. There are two regions of the peak which are most sensitive to inaccuracy; the long tails of the peak and the valleys separating the K_{α_1} and K_{α_2} peaks. A general tendency is to read the backgrounds of the broadened peaks too high due to tail overlap. Wagner[7] discusses this matter in his excellent review article. Warren[8] suggests using the background of the annealed sample as the true background as the peak tails for sharp peak do not contribute much to the background between the peaks.

Data can be obtained by slow continuous scanning or by point counting. If the diffraction laboratory is equipped with a step scanner this method will provide greater precision than the continuous scanning techniques. The data collection increments need not be closely spaced in the long peak tails, but must be closely spaced where the intensity is changing rapidly. It generally requires 80 - 250 intensity readings per peak depending on the peak shape. As shown by Young et al.[6] the number of steps should be such that there are not significant features on a scale smaller than the step size and the actual profile can be satisfactorily approximated by a series of straight lines.

If data is taken with K_α radiation, one must separate out K_{α_1} from K_{α_2}[7, 8]. Several methods exist for accomplishing this[9, 10], but they do not always work without difficulties. With a high resolution diffractometer, partially resolved $K\alpha_1$ and $K\alpha_2$ peaks can cause problems when a Rachinger separation is being made by a computer. Data must be collected carefully, particularly near the $K\alpha_1$-$K\alpha_2$ valley to avoid problems. When possible, it is often advisable to use a lower resolution slit system to avoid partially resolved α_1, α_2 peaks. Partially resolved α_1, α_2 peaks can often be separated better by graphical methods rather than using a computer technique.

B. Small Domains and Strain

Figure 4 is a schematic illustration of a small domain. It was shown by Bertaut[11] that only the "thickness" of the domain perpendicular to the diffracting planes can give rise to line broadening. Obviously the "thickness" is an average thickness since in general the domains have an irregular shape. In Fig. 4, we see the "column of unit cells" approximation for a domain. The thickness perpendicular to (00ℓ) may be described as being the average column length perpendicular to the (00ℓ) plane. The mathematics of line profile analysis is formulated using this column of cells approximation. The

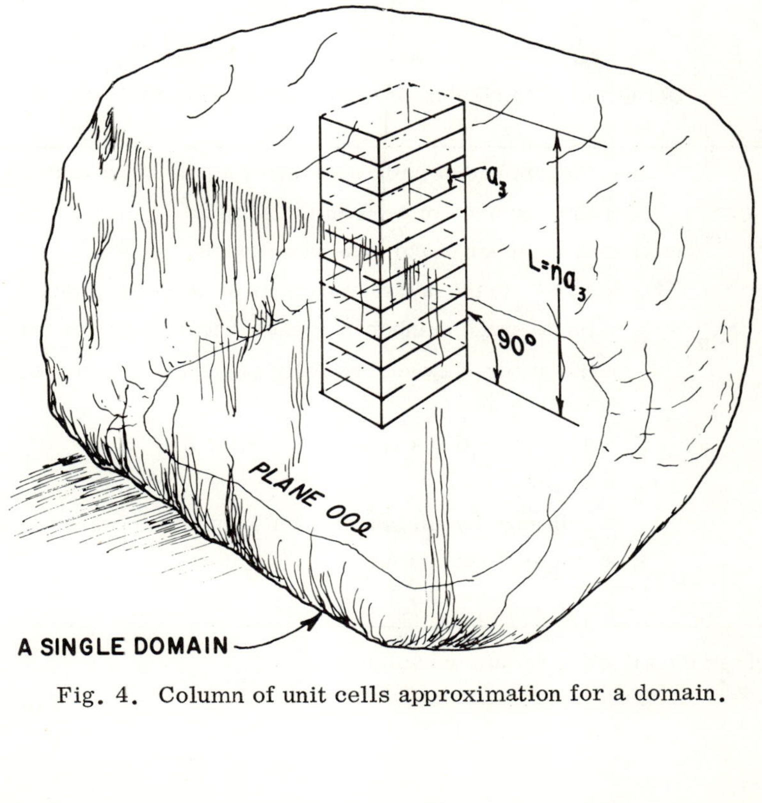

Fig. 4. Column of unit cells approximation for a domain.

procedure described by Warren[8] involves writing an expression for the total diffracted power in a reflection as a function of the average number of cells per column perpendicular to the reflecting plane and the average strain or displacement between cells in a column perpendicular to the reflecting plane. In this section, we will assume that no other sources of broadening are present, i.e., stacking irregularities, etc.

For mathematical convenience, all reflections can be considered to be reflections of the type (00ℓ). This is accomplished by a simple transformation of crystal axes. It is a theoretical consideration only and does not affect the generality of application. No attempts will be made here to derive the various relationships that are used to obtain size and strain data. As far as size and strain determinations are concerned, the method to be described is applicable to any crystal system. However, when faulting irregularities are being calculated the crystal system is very important.

Table I defines a set of terms which will be used in the discussion that follows. Some of these terms are shown in Fig. 4. These terms are based upon the work of Warren[8].

TABLE I

Definition of Variables Used in Line Profile Analysis

n	=	the number of cells separating a pair of cells in a column
N	=	total number of cells in the sample[a]
$N(\text{col})$	=	the number of columns in a sample
N_3	=	N/N_{col}; the average number of cells per column.
N_n/N_{col}	=	the average number of (n) separated pairs per column
Z_n	=	the strain between two cells separated by a distance of small n cells
K	=	a constant that varies slowly with θ
h_3	=	$2a_3 \sin\theta / \lambda$
a_3	=	cell edge length in a particular crystallographic direction
L	=	column length $= na_3$

[a] Sample is the effective irradiated volume.

The expression which gives the total diffracted power $P'(2\theta)$ in any powder diffraction line is given by Eq. (5).

$$P'(2\theta) = \frac{KNF^2}{\sin^2\theta} \sum_{n=-\infty}^{n=+\infty} \frac{N_n}{N_3} \langle \exp 2\pi i \ell Zn \rangle \quad \exp 2\pi i n h_3 \qquad (5)$$

Expanding the exponentials in Eq. (5) using Eulers formula and letting $Z_{-n} = -Z_n$ as shown by Warren[8], we can show that Eq. (5) has the form of a Fourier series. Allow:

$$A_n = \frac{N_n}{N_3} \langle \cos 2\pi \ell Zn \rangle \qquad (6)$$

and

$$B_n = \frac{-N_n}{N_3} \langle \sin 2\pi \ell Zn \rangle \qquad (7)$$

and we have:

$$P'(2\theta) = \frac{KNF^2}{\sin^2\theta} \sum_{n=-\infty}^{\infty} \left\{ A_n \cos 2\pi n h_3 + B_n \sin 2\pi n h_3 \right\}$$

where

$$h_3 = \frac{2 \sin\theta}{\lambda |b_3|} = \frac{2a_3 \sin\theta}{\lambda}$$

Our peak shape can therefore be represented by the sum of a series of sine and cosine terms. Note, if the peak shape is symmetrical the odd sine coefficients B_n must be zero. It is now possible to show how the coefficients may be used to obtain information relating to the domain size, strain, and faulting probabilities. We will not discuss the method of obtaining the Fourier coefficients from the peak shape. Many computer programs are available which rapidly perform this function as well as several other data clearing operations.

C. Fourier Coefficients and Separation of Size and Strain

Examination of Eq. (6) shows that A_n has some unique properties. Note that A_n is the product of two terms: N_n divided by N_3 and $\langle \cos 2\pi \ell Zn \rangle$. Note also that the first of these terms contains only size information while the second contains only strain or distortion (Z_n) information. Thus, A_n is written as:

$$A_n(\ell) = A_n^S A_n^D \qquad (8)$$

where A_n^S is the size term and A_n^D is the strain or distortion term. In Eq. (6), A_n approaches a value of unity as the small n index approaches zero. And A_o, itself, has physical significance. When a curve such as a diffraction tion peak is expanded into a Fourier series, the A_o coefficient is proportional to the total area under the curve.

Note that the distortion term A_n^D contains ℓ, the order of the reflection. Thus, the distortion term A_n^D, as well as A_n itself, must be a function of the order ℓ (angle 2θ) while the size term A_n^S is not. In general*, therefore, it will be necessary to have data from several orders, i.e., 002, 004, 111, 222 of the same reflection in order to separate size and strain. If, in A_n^D, we expand the cosine term and discard terms higher than the order two, we obtain:

$$< \cos 2\pi \ell Z_n > = 1 - 2\pi^2 \ell^2 < Z_n^2 > \tag{9}$$

taking logarithms of both sides of the Eq. (8):

$$\ln A_n(\ell) = \ln A_n^S + \ln\left(1 - 2\pi^2 \ell^2 < Z_n^2 >\right)$$

Since for small x the log (1 - x) = -x, we have:

$$\ln A_n(\ell) = \ln A_n^S - 2\pi^2 \ell^2 < Z_n^2 > \tag{10}$$

Equation (10), which is valid only for small values of the product ℓZ_n is the basis for the separation of domain size and strain. In order to illustrate its use, let us look at Table II which contains a set of computer calculated Fourier coefficients for two pure aluminum diffraction peaks. Peak A is from well-annealed aluminum while Peak B is from aluminum in the cold worked state. In this particular case, the cold work was in the form of low cycle fatigue. The coefficients shown in Table II must now be manipulated in order to remove all effects of instrumental broadening from the cold worked profile. This procedure is known as Stokes[13] correction. If H_r and H_i are the cosine and sine coefficients of the cold worked peak and G_r and G_i are the

---

*In some cases only one order can be measured, but Rothman and Cohen[12] have shown that a satisfactory analysis of size and strain can still be made.

TABLE II

Computer Calculated Cosine Coefficients for Cold Worked and Annealed (004) Peaks of Al

Cold worked		Annealed	
An (C. W.)	L = na	L = na	An (annealed)
1.000000	0.0	0.0	1.000000
		6.49	0.998980
0.999487	12.11	12.99	0.995929
		19.48	0.990902
0.997947	24.21	25.98	0.983969
		32.47	0.975236
0.995388	36.32	38.97	0.964830
		45.46	0.952897
0.991823	48.42	51.96	0.939597
0.987269	60.53	58.45	0.925091
		64.95	0.090556
0.981751	72.63	71.44	0.893163
		77.94	0.876077
0.975295	84.74	84.43	0.858450
		90.93	0.840424
0.967940	96.84	97.42	0.822123
		103.92	0.803652
0.959728	108.95	110.41	0.785097
		116.90	0.766525
0.950710	121.05	123.40	0.747987
		129.30	0.729513
0.940943	133.16	136.39	0.711119
0.930495	145.26	142.88	0.692811
		149.38	0.674583

corresponding coefficients of the annealed peak, then the corrected cosine and sine coefficients A_n and B_n are given by:

$$A_n = \frac{H_r(n)G_r(n) + H_i(n)G_i(n)}{G_r^2(n) + G_i^2(n)} \tag{11}$$

$$B_n = \frac{H_i(n)G_r(n) - H_r(n)G_i(n)}{G_r^2(n) + G_i^2(n)} \tag{12}$$

Table III lists the computer calculated Stokes corrected coefficients. These corrected coefficients must now be plotted in a special way in order

TABLE III

Computer Calculated Stokes Corrected
Coefficient for the (004) Peak of Al

An	L	An	L
1.000000	0.0	0.937898	71.94
0.999497	6.49	0.929006	84.43
0.997985	12.99	0.919916	90.93
0.995496	19.48	0.910699	97.42
0.992051	25.98	0.901411	103.92
0.987705	32.47	0.892097	110.41
0.982515	38.97	0.882790	116.90
0.976549	45.46	0.873509	123.40
0.969886	51.96	0.864260	129.89
0.962599	58.45	0.855035	136.39
0.954780	64.94	0.845818	142.88
0.946518	71.44	0.836585	149.38

to separate domain size broadening from strain broadening. For cubic crystals, it is convenient to rewrite Eq. (10) in the form:

$$\ell n\, A_L(h_\circ) = \ell n\, A_L^S - \frac{2\pi^2 <\epsilon_L^2> L^2\, h_\circ^2}{a^2} \qquad (13)$$

where $A =$ the lattice parameter

$h_\circ^2 = h^2 + k^2 + 1^2$

$L = Na_3$

$\Delta L = a_3 Z_n$

$\frac{\Delta L}{L} = \epsilon_L$, the strain component along the a_3 direction averaged over a distance L

$\frac{Z_n}{n} = \epsilon_L$

If we plot the Stokes corrected cosine coefficients $\ell n\, A_L(h_\circ^2)$, as a function of $h_\circ^2$ for several orders of a reflection, we will obtain results similar to Fig. 5. In Fig. 5, we have plotted $\ell n\, A_L$ versus $h_\circ^2$ for several values of L, the column length or averaging distance. Differentiation of both sides of Eq. (13) with repsect to $h_\circ^2$ shows that the slope of these lines in Fig. 5 will be equal to $-2\pi^2 <\epsilon_L 2> L^2 / a^2$ from which ϵ_L^2 can be obtained since L and a are known. Also, we see that the intercept of each line with the ordinate ($h_\circ^2 = 0$) will give A_L^S, a size coefficient. It should be noted that we used

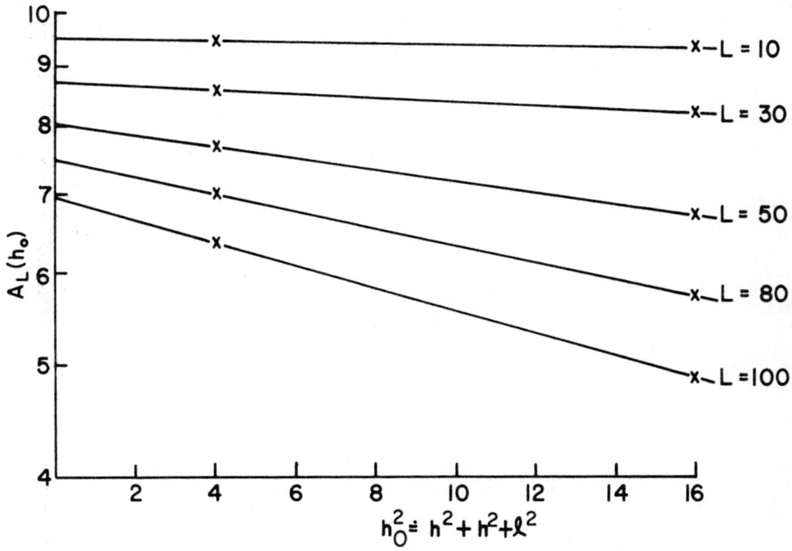

Fig. 5. Stokes corrected cosine coefficient $A_L(h_o)$ plotted against h_o^2.

several orders of a single reflection, i.e., 111, 222 or 200, 400 to draw Fig. 5. If the sample had been elastically isotropic, such as tungsten, we could have used all the reflections and still obtained a straight line. The size coefficients A_L^S must now be treated to obtain the effective domain size. It can be shown that the negative initial slope of the plot of A_L^S versus L will give the average column length N_3. Thus,

$$d\,A_L^S/dL = -1/N_3 \tag{14}$$

In Fig. 6, the values of A_L^S obtained from Fig. 5 are plotted versus $L = Na_3$. Note that the data points form a curve for small values of L. This curvature is called the "hook effect" and is discussed by Warren[(8)] and Wagner[(7)]. Warren[(8)] shows that it is not theoretically possible to have a concave down portion of this curve. The downward concavity generally comes from intensity errors in the long tails of the peaks. A non-exact correction for this effect is to neglect the curvature and extrapolate small L values from the curve in Fig. 6 and renormalize all values so that A_L^S will equal unity when $L = 0$. When this is done, the intercept on the abcissa, $L = na$, will equal the effective domain size D_{eff} according to Eq. (14). Rothman and Cohen[(12)] have outlined a unique method for removing the hook effect.

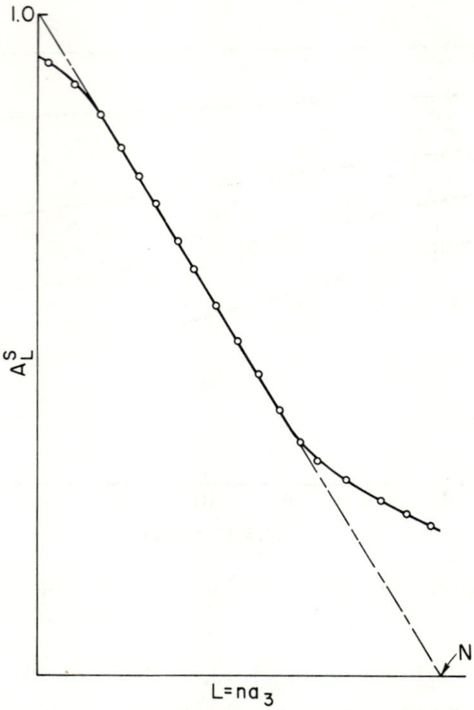

Fig. 6. Plotting the size coefficients A_L^S versus $L = na_3$.

We have now determined an effective domain size and strain averaged over a range of averaging distances L. In general, $<\epsilon^2>^{1/2}_{50 Å}$, the root mean squared microstrain averaged over 50 Å, has been used by many authors as a characteristic strain value. In this section, we assume that size and strains were the only causes of line broadening. In the next section, we will see how stacking irregularities can also contribute to line broadening. The contribution of these other effects will be found in the A_L^S coefficients, thus the size as determined by the above method must be considered as an "effective size" (D_{eff}) until all contributions due to stacking irregularities have been separated from the pure size broadening. We will discuss this matter in the next section.

D. Line Broadening and Stacking Irregularities

In a perfect crystal the stacking of atomic planes is very regular. When such a crystal is mechanically deformed, stacking irregularities can develop.

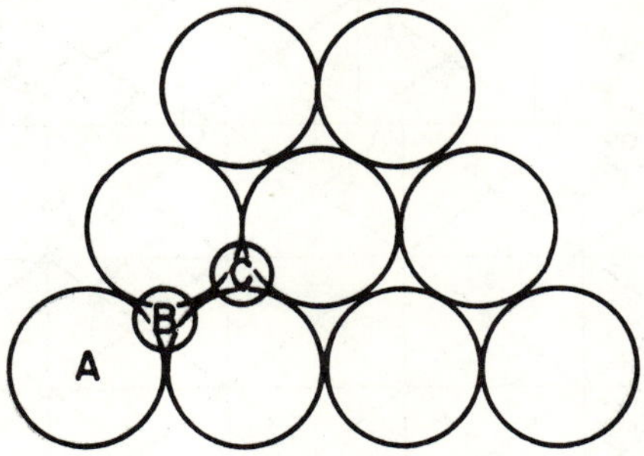

Fig. 7. The (111) layer in the fcc system showing various stacking positions.

In this section, we will examine how these stacking faults, as they are called, affect the diffraction profile and conversely what information concerning faults can be obtained by careful line profile analysis. In order to present the subject most easily, we will confine our discussion to the fcc crystal system, realizing, however, that Warren[8] and others have also analyzed the hcp and bcc systems. In the fcc system, faulting occurs in the stacking sequences of the (111) slip planes. Patterson[14] showed that in fcc system deformation faulting should produce symmetrical line broadening and peak shifts, while twin or growth faults should produce assymmetric peaks. We will first discuss the case of deformation faulting.

E. Deformation Faults on the fcc System

The (111) layer in the fcc crystal system is shown in Fig. 7. Note that the next layer can stack in either a B position or in a C position. In fcc metals, the correct stacking sequence is ABC, ABC, etc. As a result of deformation, a certain number of the (111) planes are displaced from their proper stacking sequence, i.e., a jog can occur in the sequence as shown in Fig. 3. The probability that such faults will occur is given by α , the deformation fault probability. The average number of layers between deformation* faults is $1/\alpha$, thus $\alpha = 0.02$ means that on the average there is one faulted layer for every fifty perfect layers. As Patterson[14] points out, we can expect deformation faults to produce two effects: peak shifting and

* β, the twin fault probability, is defined in a similar fashion.

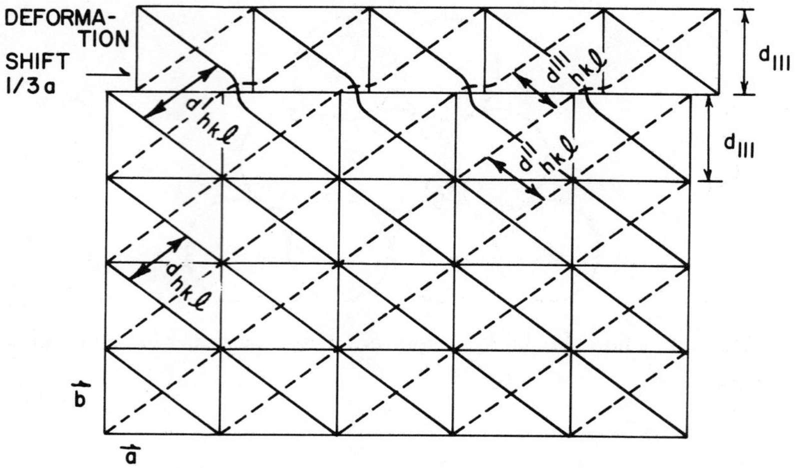

Fig. 8. Schematic illustration showing why peak shifting can occur as a result of deformation faulting.

peak broadening. Figure 8 illustrates schematically why peak shifting should occur. As a result of a slip on the (111) plane, the d_{hkl} spacing increases to d'_{hkl} while d''_{hkl} decreases to d'''_{hkl} and d (111) remains uneffected, thus if d values of various planes change as a result of faulting, we should expect to observe some diffraction peaks shifting. In fcc materials it can be shown that the (111) peak shifts to higher angle while the (200) shifts to lower angles. Thus, the peak shift can be measured between a pair of peaks on the same pattern (Fig. 9). Equation (15) shows the relationship between the change in peak separation for the (111) - (200) line pairs and the deformation fault parameter α. Warren[8] has very clearly covered the theory that predicts the magnitude of the peak position shifting that will be produced. For details on the derivation of Eq. (14), the reader is referred to his text.

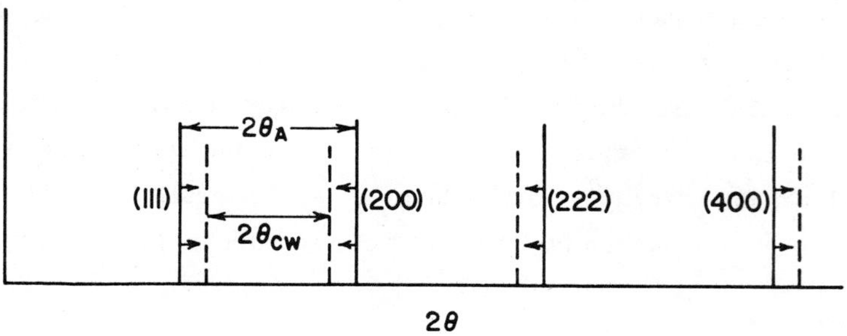

Fig. 9. Peak shifting in fcc as a result of deformation faulting.

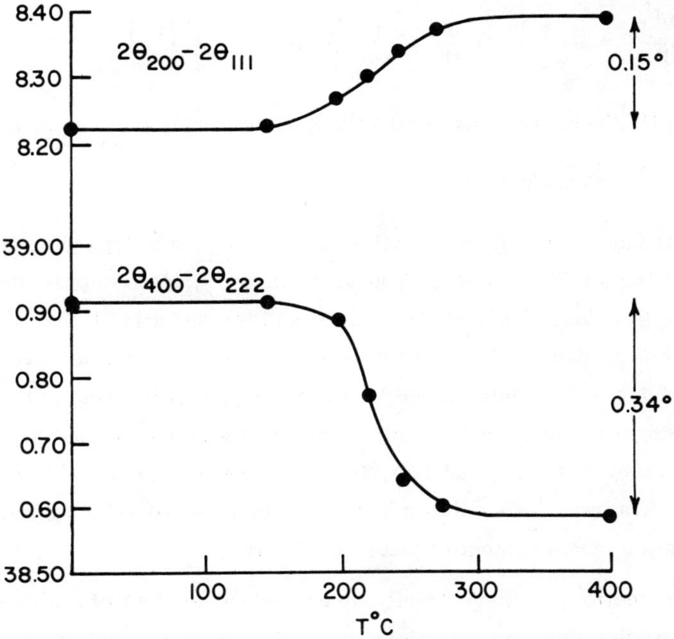

Fig. 10. Peak shifting as a result of cold working in α Brass (Warren [5]).

$$\Delta(2\theta_{200} - 2\theta_{111}) = \frac{-90\sqrt{3}\ \alpha}{\pi^2} \left(\frac{\tan\theta_{200}}{2} + \frac{\tan\theta_{111}}{4} \right) \tag{15}$$

Equation (16) represents a similar result for the (222) - (400) pair:

$$\Delta(2\theta_{400} - 2\theta_{111}) = \frac{90\sqrt{3}}{\pi^2} \left(\frac{\tan\theta_{400}}{4} + \frac{\tan\theta_{222}}{8} \right) \tag{16}$$

Figure 10 shows some typical results for cold worked subsequently annealed alpha brass. Note the change in peak separation as a result of heat treatment.

F. Twin Fault Probability in fcc

When only particle size broadening is present, α , $\beta = 0$ and we obtain the particle size using Eq. (14) as discussed in the previous paragraph. However, when α , β are greater than zero, what we obtain from a plot such as Fig. 6, is D_{eff} not D. It can be shown(8) that the following equation holds when α , β are greater than zero.

$$-\frac{dA_L^S}{dL} = \frac{1}{D_{eff}} = \frac{1}{D} + \frac{1.5\alpha + \beta}{ah_o(u+b)} \sum_b |L_o| \qquad (17)$$

Table IV shows the calculated values of the term:

$$\sum_b |L_o| / h_o(u+b)$$

for several face centered cubic reflections; $1/D_{eff}$ will depend upon which pair of hkl lines is being used. Note that since D_{eff} does depend upon hkl, we can use this dependence to separate D and the quantity $(1.5\alpha + \beta)$ from D_{eff}. The derivative of A_L^S with respect to L is measured for two pairs of lines and $1.5\alpha + \beta$ is determined by solving Eq. (17) for both pairs of lines and obtaining a simultaneous solution. Since α was determined by peak shifting as shown by Eqs. (15) and (16), it is possible to separate β from the sum $1.5\alpha + \beta$. Warren[8] also shows how β can be determined independently from a simple measurement of peak asymmetry.

In this section, we have briefly discussed the subject of faulting and its effect on the line profile. Relations similar to those described above also exist for the bcc and hcp cases[8]. It should also be mentioned that double layer stacking irregularities also exist. These faults have a probability α''.

TABLE IV

Values of the Term $\sum_b |L_o| / h_o(u+b)$

for Faulted fcc Metals

hkl	$\sum_b \frac{\|L_o\|}{h_o(u+b)}$
111	$\sqrt{3/4}$
200	1
220	$\sqrt{2/2}$
311	$3/2\sqrt{11}$
222	$\sqrt{3/4}$
400	1

The deformation fault parameter α is related to α' (single layer stacking fault) and α'' by the relationship $\alpha = \alpha' + \alpha''$. α'' is not generally separated from α' in most cases.

G. Applications of Line Profile Analysis

Mikkola and Cohen[15] have reviewed the application of a line profile technique to a wide variety of metallic systems. They showed, for example, how elastic anisotropy can be determined by measuring the ratio of D_{eff} for two crystallographic directions. Microstrain ratios for two crystallographic directions were shown to have the same variation with direction as the reciprocal of Young's modulus. The variation of α and β with alloy composition is also discussed. More recently, Lewis[16] has applied the line profile analysis technique to nonmetallic materials. Table V lists some crystallite

TABLE V

Crystallite Sizes and Microstrains in Ball-Milled LiF (16)

Treatment	[hkl]	Warren-Averbach
		D_{eff} (Å)
1 h	100	690
Perspex	111	900
1 h	100	420
Agate	111	510
1 h	100	300
Steel	111	410
		$<\epsilon_L^2>^{1/2}$
1 h	100	0.00066
Perspex	111	0.00039
1 h	100	0.0014
Agate	111	0.00087
1 h	100	0.0023
Steel	111	0.0016

[a] Averaged over effective crystallite sizes.

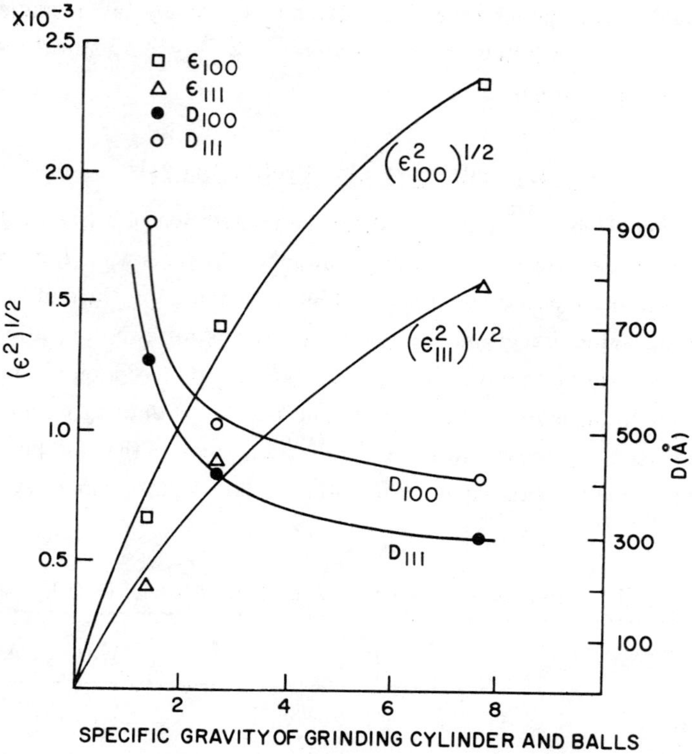

Fig. 11. Microstrain and crystallite size in LiF as a function of the density of grinding balls and cylinders Lewis[16].

sizes and microstrains for ball-milled LiF measured by the Warren-Averbach technique. Figure 11 shows the effect of the density of the grinding cylinder and balls upon microstrain and crystallite size. Cutter and McPherson[17] have also applied the line profile analysis technique to an abraded single and polycrystalline MgO and compared their results with other workers as shown on Table VI. Figure 12 shows how LPA measures the effect of annealing on the abraded MgO. Lewis and Wheeler [18] have also applied the integral breadth technique[7] to the analysis of temperature effects on microstrains and crystallite growth in annealed alumina.

IV. SUMMARY

The line profile analysis technique has found a wide acceptance in the field of materials for the measurements of crystallite size and microstrain.

TABLE VI

Particle Sizes and Strains in Single and Polycrystalline MgO (17)

Material	Treatment	Warren-Averbach ϵ_{rms} x 10^3	Warren-Averbach D_{eff} (Å)	Reference
MgO	Abraded single crystal	1.5	200	
MgO	Abraded polycrystal	1.05	245	
MgO	Abraded polycrystal	—	—	
MgO	Abraded polycrystal	—	—	
MgO	Ball-milled	—	—	Lewis and Lindley (1964)
MgO	Irradiated single crystal	2.5		Walker and Hickman (1965)
MgO	Irradiated single crystal	2.2	500	Walker and Hickman (1965)
MgO	Explosive shock, single crystal	0.55	160	Klein and Rudman (1966)
MgO	Explosive chock, powder	2.30	170	Keckel and Youngblood (1968)
W	Filed	2.0	210	Halder and Wagner (1966)
Ta	Filed	2.2	170	Schoening (1956)
Cu	Filed	—	—	Agnihotri and Mitra (1963)
Mo	Filed	2.5	240	Agnihotri (1963)
Mo	Filed	2.7	500	Williamson and Smallman (1954)
Al	Filed	0.7	400	Wagner and Aqua (1963)

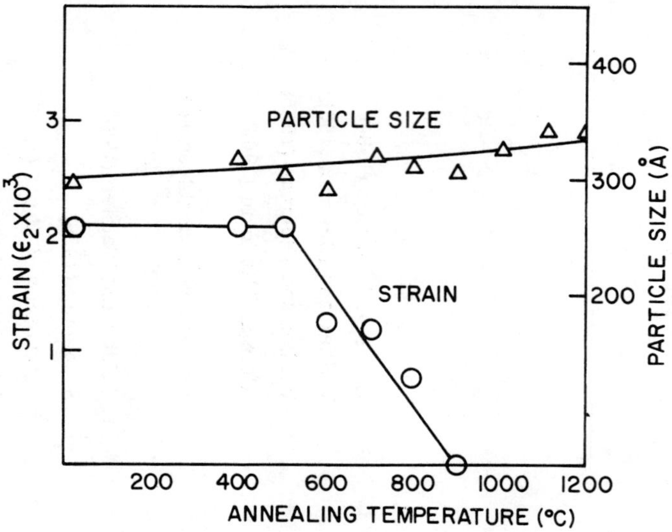

Fig. 12. The effect of annealing on abraded MgO as shown by x-ray line profile analysis.

It has widely supplanted the simple Scherrer technique since it provides so much more additional information and because of computer techniques requiring only a slightly greater amount of time. Various authors have related the many parameters obtainable by this technique, i.e., $<\epsilon_L^2>^{1/2}$, D, D_{eff}, α, α', α'', and β, to changing physical properties of a variety of materials. It is anticipated that this technique will find many new applications in the field of ceramic science and engineering.

REFERENCES

(1) P. L. G. M. LaFleur, Acta Crystallographica, A 25, Part 6, p. 634 (1969).

(2) P. Sherrer, Nachr. Göttinger Gesellschaft., 98 (1918), Zsigmondy's Kollodchemie, 3rd Ed., p. 394.

(3) A. Guinier, X-Ray Diffraction, Freeman, 1963.

(4) A. Taylor, X-Ray Metallography, Wiley, New York, 1961.

(5) B. E. Warren, Progress in Metal Physics, Vol. 8, Pergamon, New York, 1959.

(6) R. A. Young, R. J. Gerdes, and A. J. C. Wilson, Acta Crystallographica, 22, Part 2, (February, 1967).

(7) C. N. J. Wagner, Local Atomic Arrangements Studies by X-Ray Diffraction, Met. Soc. Conf. of AIME, Vol. 36, Gordon & Breach, 1966.

(8) B. E. Warren, X-Ray Diffraction, Addison Wesley, New York, 1969.

(9) W. A. Rachinger, J. Sci. Inst., 25, 254 (1948).

(10) J. W. M. Dumond, Rev. Mod. Physics, 5, 30 (1933).

(11) F. C. R. Bertaut, Acad. Sci., Paris 228, 187 (1949).

(12) R. L. Rothman and J. B. Cohen, Adv. X-Ray Analysis, 12, 1969 Plenum, New York, p. 208.

(13) A. R. Stokes, Proc. Phys. Soc., London 61, 382 (1948).

(14) M. S. Patterson, J. Appl. Physics, 23, 805 (1952).

(15) D. E. Mikkola and J. B. Cohen, Local Atomic Arrangements Studied by X-Ray Diffraction, Met. Soc. Conf. of AIME, Vol. 36, Gordon & Breach, 1966.

(16) D. Lewis and D. O. Northwood, Brit, J. App. Phys., 2 [2], 1969.

(17). I. Cutter and R. McPherson, Phil. Mag., 20, No. 165, 489 (September, 1969).

(18) D. Lewis and E. J. Wheller, J. Materials Sci., 4, 481-684 (1969).

Chapter 14

SCANNING ELECTRON MICROSCOPY

S. R. Bates

Department of Metallurgical and Materials Engineering
University of Florida, Gainesville, Florida

I. INTRODUCTION

The scanning electron microscopy (SEM) is the latest member of the electron microscope family and offers the advantages of large depth of focus and easy sample preparation. As indicated by Tufts in Chapter 6, SEM offers many advantages over conventional replication techniques often used by electron microscopists. This chapter will discuss the application of the new SEM techniques to the analysis of materials.

Much current scientific effort is centered around the determination of the effect of microstructure on the overall behavior of materials. Detailed studies of surface structure are becoming increasingly more important. For this reason, high resolution and large depth of field is needed in order to determine the shape and size of highly irregular submicron (>200 Å) surface structures.

Modern optical microscopes have limited diffraction resolution of the order of a few thousand (1000-5000 Å) angstroms, but their depth of field at their resolution limit is very restricted. Using electrons, diffraction effects do not become important until a resolution limit of fractions of an angstrom has been reached.

II. HISTORY OF SEM

In 1935 Knoll[(1)] proposed that an electron microscope could be made that used a focused electron beam which was scanned across a specimen with the emitted current being monitored as a function of beam position. In 1938 v. Ardenne[(2)] constructed the first scanning electron microscope; v. Ardenne original specimens were thin sections and the transmitted current was used to obtain the micrographs. Although the original instrument did not compare favorably to conventional transmission microscopes, v. Ardenne did suggest improvements that could be made and suggested using the secondary electron current to examine opaque samples. Zworykin et al. [(3)] incorporated v. Ardenne's and other ideas into an SEM which used the secondary electron emission for image formation. In their arrangement the secondary electrons were accelerated onto a phosphor screen held at 10 kV and the light emission used in conjunction with a photomultiplier. Their instrument has a resolution of approximately 500 Å.

Following World War II much of the development was carried out in France and England. Many of the major improvements came from the English group working at Cambridge. While all of the Cambridge developments are too numerous to name, some of the more important involved Oatley and Everhart's[(4)] reports of the ability to detect changes in surface voltages and Everhart and Thornley's[(5)] development of a collector system using a scintillator-photomultiplier combination which is the system most commonly used at the present time.

III. BASIC SYSTEM AND FUNDAMENTAL INTERACTIONS

The SEM can be divided into three basic units:

1. The electron-optical column together with the appropriate electronics.

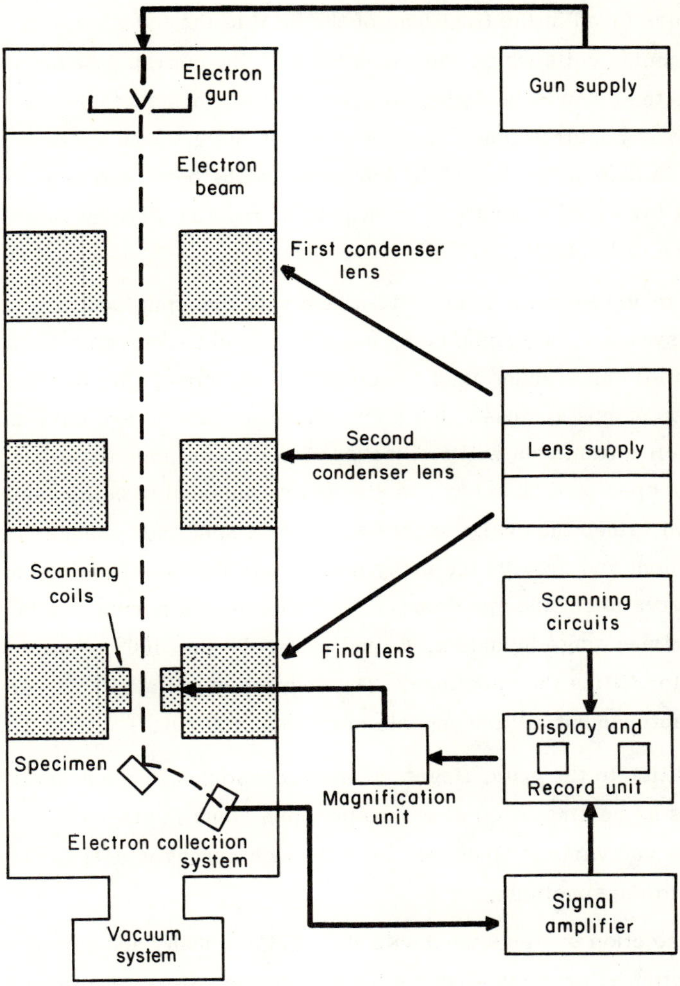

Fig. 1. A schematic diagram of the scanning electron microscope.

2. The vacuum system which will include the specimen chamber and stage
3. The signal detection and display systems.

A simplified diagram of an SEM is shown in Fig. 1.

The design of the electron column varies with the different manufacturers of the equipment, but most of the instruments feature magnetic lenses with their number varying from two to four. The primary function of the lenses is to demagnify the electron beam and beam diameters of approximately 150-200 Å are now commonly attainable with commercial instruments. One other feature found in the final lens of the SEM is the incorporation of two sets of scanning coils which when coupled with the appropriate scan generator cause the beam to be deflected over the specimen surface in a raster-like pattern. Features found in other electron optical lens systems such as apertures to help define the angle subtended by the beam and provisions for correcting for small amounts of astigmatism are also present in the electron column of the SEM.

The vacuum systems commonly in use with the equipment are usually two pump systems, one pump being used for the electron column and the other used for the evacuation of the specimen chamber. In most instances oil diffusion pumps are used, but some oil-free systems are currently in use and their advantages will be discussed in the section on specific applications. The specimen chamber and stages are included with the vacuum system section, since the design must be such that specimen changes are rapid and convenient and provide the many movements that are required for detailed observation of the specimens. The movements normally attainable in the specimen stage include x, y, and z translation, 360° rotation, and provision for tilting the specimen. The degrees of tilt available vary with the manufacturer but all are capable of at least 60° tilt.

In addition to the basic stages, many accessories are now available for such things as heating or cooling the specimen, tensile stages for deformation studies and various electrical lead-through arrangements for applying potentials to the specimen.

The detection systems used with the SEM are many and varied since the interaction of primary electron beam with the specimen produces a variety of effects such as secondary electron emission, reflected or back-scattered electron currents, beam induced conduction, x-ray production, and in some cases, cathodoluminescence. All of the signals can be detected, amplified, and used to control the brightness of a cathode ray tube. The deflection of the electron beam in the CRT is controlled by the same scan generator which determines the position of the electron beam on the sample so there is a point-to-point correspondence of the display on the CRT and the beam position on the sample. In general, two types of CRT are used, one having a long-persistence phosphor for visual observation and the other a short-persistence phosphor for photographing. Magnifications are also controlled by the scan generator in a unique fashion. The signal fed to the deflection coils of the CRT is such that raster size remains constant and, at the same time, the signal fed to the scanning coils is variable. By decreasing the scan area on the specimen surface and maintaining a constant display area on the CRT a magnification is attained. For example, the raster on the specimen surface is $2\mu \times 2\mu$ and this is displayed on the CRT with a raster of 10 cm x 10 cm; the linear magnification is 50,000X. One generally has a wide range of magnifications available from 10-20X to 150,000-200,000X, but in general the upper limit of useful magnification is presently about 60,000-70,000X.

Before proceeding further, it will be useful to introduce the terminology that is commonly used to describe the modes of operation of the SEM. While there has been no formal standardization of notation adopted, the terminology used in the remainder of this chapter will be that set forth by Thornton[(6)].

1. If the current of emitted electrons is being exploited to obtain information, the SEM is being operated in the emission mode. This notation is used whether secondary electrons or reflected primary electrons are being used and independently of the type of detector used.

2. When the examination involves the use of any electric current created by the primary beam, it is said that the conductive mode of operation is being used.

3. When the light emitted from the specimen under electron beam excitation is being used to diagnose specimen properties, the SEM is being used in the luminescent mode, regardless of the exact mechanism whereby light is created.

The electron beams commonly used in the SEM's are accelerated through a voltage in the range of 1 to 50 kV. When the accelerated electron beam strikes the sample many types of interactions can occur and some of the more important are shown in Fig. 2. The initial type of interaction is shown in Fig. 2(a), in which there is no exchange of energy between the incident electron and atoms of the surface of the sample, only a change in momentum. The next step is represented by Fig. 2(b), where the incident electron is scattered from the sample surface after a single inelastic collision. Instead of being scattered out of the sample, the incident electron can be deflected down into the specimen as is illustrated in Fig. 2(c). The electrons ejected in the manner shown in Fig. 2(c) can themselves produce further ionization

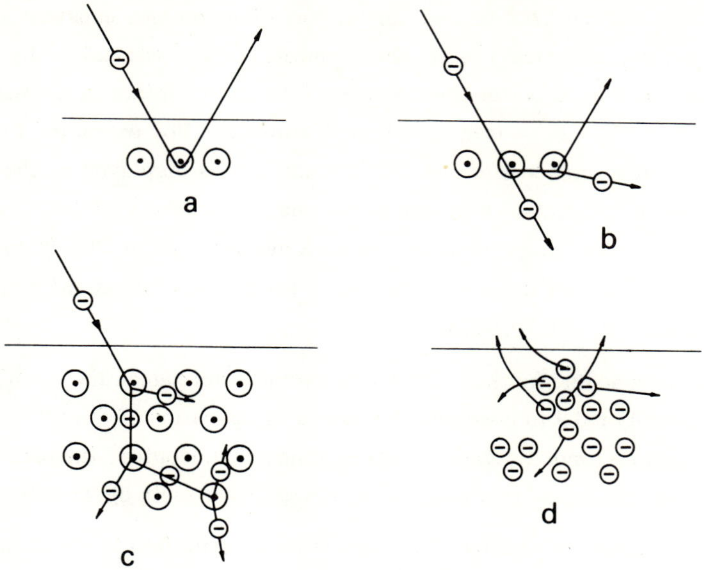

Fig. 2. Interaction of the electron beam with the specimen surface: (a) no energy loss; (b) inelastic scattering out of specimen; (c) incident electron deflected into specimen; and (d) secondary electron formed near specimen surface.

until their kinetic energy is reduced to approximately 3 to 8 eV. The electrons shown in Fig. 2(d) can be considered to represent the secondary electrons which are formed near the surface of the material and will diffuse toward the surface and, if they have sufficient kinetic energy, can penetrate the surface potential of the material. Fig. 2(a) schematically represents what is thought of as back-scattered or primary reflected electrons. While rather arbitrary, the criteria for distinguishing between secondary and back-scattered electrons is as follows: electrons having a kinetic energy less than 50 eV are considered as secondary electrons and those with energies greater than 50 eV are back-scattered electrons.

IV. SIZE OF THE ELECTRON BEAM

The electron optics found in commercial SEM's are very similar to those found in many conventional transmission electron microscopes. Due to the availability of numerous comprehensive discussions of the factors and problems associated with electromagnetic lens and electron guns, they will not be discussed in this chapter. It is sufficient to note that for practical reasons electromagnetic lenses are used almost exclusively, since the coils can be mounted outside the vacuum system and thus the lens surfaces can be easily cleaned. If it is assumed that such factors as astigmatism in the final lens, misalignment of the electron column, stray fields, and mechanical vibrations have been minimized, the diameter of the electron beam can be given by:

$$d^2 = d_o^2 + d_s^2 + d_c^2 + d_f^2 \qquad (1)$$

where d_o = diameter of the Gaussian image, d_c = diameter of the disc of confusion due to spherical aberration, d_f = disc of confusion due to diffraction, and d_c = diameter of disc of confusion due to chromatic aberration.

V. ELECTRON GUNS

All of the current microprobes and SEM's use a hot-filament electron gun as a source of electrons. The filament wire is about 100 μ in diameter, and the effective area at the tip from which electrons are emitted is about 100 x 150 μ. The filament is biased positive with respect to the cathode shield by the order of 100-1000 V to improve stability and to concentrate the emergent electrons into a nearly parallel beam. The closer the filament tip is to the cathode shield, the greater is the bias voltage required, and the greater is the beam current.

Spherical aberration arises because the focal length for electrons passing near the center of the lens is different from that for electrons passing near the periphery of the lens. The disc of confusion diameter d_s can be expressed as

$$d_s = \frac{1}{2} C_s \alpha^2 \qquad (2)$$

where C_s = spherical aberration coefficient and α = half-aperture angle.

Chromatic aberration arises because there is a spread of electron energies in the primary electron beam with the net effect that the lens is unable to focus electrons of different energies to the same point. The disc of confusion due to chromatic aberration can be expressed as:

$$d_c = C_c \frac{\Delta V}{V} \tag{3}$$

where C_c = chromatic aberration coefficient and $\Delta V/V$ = energy spread of the electron beam.

The disc of confusion due to diffraction can generally be considered as negligible due to the high voltages used but it can be expressed as:

$$d_f = 1.22\ \lambda/\alpha \tag{4}$$

where λ = electron wavelength.

Pease and Nixon[7] demonstrated that by differentiating Eq. (1) with respect to α and setting the resulting equation equal to zero, an optimum semi-aperture angle which can be combined with the theoretical current density from the source gives an equation relating to electron beam diameter as follows:

$$d_{min} = 1.29 C_s^{1/4} \lambda^{3/4} (7.92 \frac{iT}{J} \times 10^9 + 1)^{3/8} \tag{5}$$

where i = beam current, J = cathode emission current density, T = absolute temperature, and λ = electron wavelength.

Using the values C_s = 1 cm, C_c = 0.8 cm and i = 10^{-12} A, d_{min} = 50 Å at 15 kV according to the calculations of Pease and Nixon. This is an indication of the importance of low beam currents, but this requirement conflicts with needs of low noise requirements which are reduced with increasing beam currents. This equation also points out the importance of ultra stable power supplies to minimize C_s. It should also be noted that the possibility of high resolution electron microprobe analysis is limited by the fact that the incident beam current varies with the 8/3 power of the beam diameter, while x-ray emission is linearly dependent on beam current. In addition to the problems of an enlarged x-ray source size due to scattering effects, there is the problem of low x-ray intensities due to the lower beam currents used in the SEM.

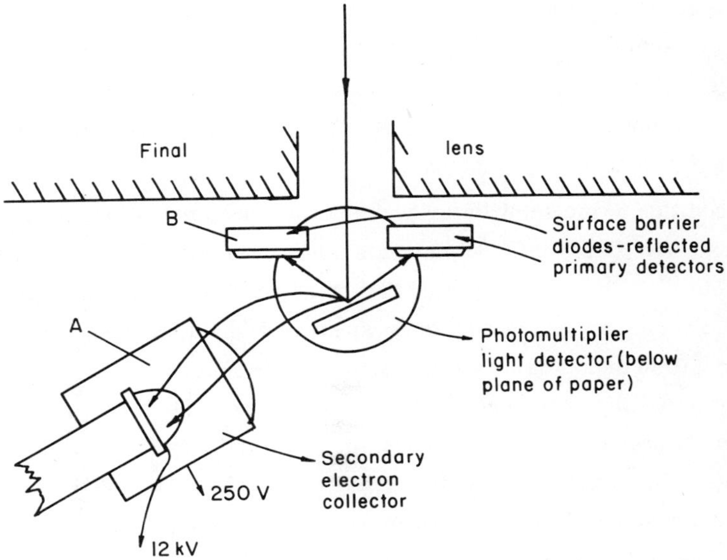

Fig. 3. Various electron detection systems currently in use.

VI. DETECTION SYSTEMS

Owing to the low beam currents involved, it is important that efficient, low noise, detectors be used. A summary of some of the detection systems currently in use are shown in Fig. 3.

The micrographs that are most commonly associated with the SEM are due primarily to secondary electron emission. The detection system that is in the widest use at the present time was designed initially by Everhart and Thornley[5]. The secondary electrons emitted from the specimen surface are attracted to a Faraday cup which normally has a positively charged wire mesh of +200 to 250 V. Some of the electrons penetrate through the wire mesh and are accelerated toward the scintillator. The scintillator is usually a plastic light guide empregnated with scintillating material which is coated with a layer of aluminum approximately 500 Å thick. The layer of aluminum is held at 10-12 kV and, therefore, each electron which reaches the scintillating material has a kinetic energy in excess of 10 kV and is of sufficient energy to excite the material. Many of the resulting photons of light travel through the plastic light guide to a photomultiplier. The output from the photomultiplier is used to modulate the brightness of a CRT.

This same detection system may be used for back-scattered electrons simply by reversing the polarity of this potential applied to the Faraday cup. Since the high energy primary have sufficiently high energies not to be deflected by a low grid potential, the net effect of a -250 V potential will be to exclude the secondary electrons.

In addition to the scintillator-photomultiplier combination, a solid state back-scattered electron detection system has been described by Kimoto and Hashimoto[8]. They demonstrated that by having two detectors located 180° apart and adding or subtracting the respective signals, micrographs based on composition or topography could be obtained. Thornton[6] reported the use of pairs of surface barrier counters made of n-type Si having a depletion layer width at zero bias of approximately 10 μ. In Thorton's arrangement the primary beam passes through a hole in the center of the detector and in this case the detectors are perpendicular to the primary electron beam. The reflected electrons that reach the detectors lose their energy by creating electron-hole pairs and thereby cause a current to flow in an external circuit. These signals are further amplified and either added or subtracted and used to modulate the brightness of a CRT.

When a specimen is struck by an incident electron beam, electrons which are not back-scattered or secondarily emitted must be taken to ground potential in order to minimize charge collection on the sample surface. This current has come to be known as neutrality or specimen absorbed current and can often be used to obtain information about the sample. Heinrich has shown that the specimen current may contain the following components:

$$i_s = i_B + i_n + i_{nn} + i_{nSe} - i_{Se} + i_{SeT} + i_{SeSe}$$

where

i_B = beam current incident upon the sample

i_n = back-scattered current

i_{nn} = back-scattered current first back-scattered by the sample then by the lens and collected by the sample

i_{nSe} = secondaries collected by the sample ejected by primary back-scattered electrons striking the lens

i_{Se} = true secondary electron current

i_{SeT} = secondaries ejected by the sample but recollected by the sample

i_{SeSe} = secondaries collected by the sample emitted from the lens struck by secondaries from the sample.

Colby et al.[9] have published a method for evaluating each of the contributions. Due to the higher beam currents (10^{-6} A) normally used in electron microprobe analysis, specimen current imaging has been used rather extensively. Additional information can be obtained from beam induced currents concerning the phenomenon of charge collection in the semiconduction field. The interested reader is referred to Chapter 9 of the book Scanning Electron Microscopy by P. R. Thornton[6] for additional information.

The luminescent mode of operation has not been as widely used in the SEM as the other modes of operation. The types of detection systems have largely been improvised in the laboratories to suit the specific needs of the user. Excitation of materials, such as phosphors, by an electron beam has long been utilized to convert electrical signals into visible light. By proper selection of a detector various wavelengths may be measured (infrared, ultraviolet, or visible). Thus a wide range of processes may be examined. Optical microscopists have used luminescence processes for many years, but now the SEM should provide additional insight since it is not limited by diffraction effects.

VII. IMAGE CONTRAST AND RESOLUTION

The factors which have the most significance, to the average user of an SEM for the interpretation of an SEM image, are: surface topography, mean atomic number and magnetic and electric fields in or near the sample surface. The relative importance of each of these factors is dependent upon the type of detection system used. Considering first the back-scattered electrons, Thornton[6] has indicated that we need to know the dependence of the following quantities on the primary beam energy:

1. The total back-scattered coefficient, η; i.e., the total flow of back-scattered electrons per incident electron summed over all angles.

2. The angular distribution of the back-scattered electrons.

3. The mean energy of the reflection electrons.

4. The most probable energy of the reflected electrons.

Values of η are tabulated in Table I and two trends are apparent: one is a gradual increase in η with increasing atomic number and a small variation of η, with increasing atomic number. Since this is the most important and commonly encountered effect of the back-scattered electrons, the other three factors will not be discussed.

TABLE I

Values of Back-Scattering Coefficient η

Material	Atomic number	Beam energy (keV)		
		5	10	30
Carbon	6	0.085	0.072	0.06
Aluminum	13	0.186	0.177	0.155
Silicon	14	0.197	0.186	0.162
Titanium	22	0.270	0.268	0.254
Chromium	24	0.285	0.283	0.270
Iron	26	0.300	0.296	0.288
Nickel	28	0.333	0.323	0.308
Copper	29	0.352	0.339	0.319
Zinc	30	0.352	0.342	0.330
Germanium	32	0.362	0.349	0.334
Molybdenum	42	0.367	0.381	0.385
Silver	47	0.418	0.420	0.420
Tungsten	74	0.472	0.483	0.501
Platinum	78	0.486	0.503	0.516
Gold	79	0.489	0.501	0.521
Uranium	92	0.495	0.513	0.534

In order to consider the various factors affecting secondary electron emission it is convenient to subdivide them into two groups. The first group contains specimens having high electrical conductivity and the second group consists of semiconductors, some conductors, and insulators.

According to Thornton[5] the following factors should be considered:

1. The secondary electron yield as a function of the primary beam energy.

2. Energy distribution of the emitted electrons.

3. The angular distribution of the emitted electrons.

4. The effect of varying the angle of incidence of the primary beam.

5. The temperature dependence of secondary electron emission.

6. The relationship between the secondary emission yield and other properties such as the work function, intensity, and atomic number.

The secondary electron yield initially increases with increased primary energy and subsequently decreases at higher energies. The subsequent decrease in secondary electron emission at higher energies is attributed to the increased depth at which the secondary electrons are formed.

Most of the secondary electrons have energies in the range of 0-30 eV with an average value of approximately 5 eV.

The secondary electron emission has been shown to depend upon the angle of incidence of the primary beams:

$$\sigma(\theta) = \sigma(o) \exp d_s \alpha (1 - \cos \theta)$$

where σ = secondary electron emission coefficient, θ = angle between incident beam and specimen normal, d_s = mean depth of escape of secondary electrons, and α = coefficient of absorption of the secondary electrons.

An additional topographic effect can enter into a yield of secondary electrons as a function of increasing angle. This can be an increased yield of secondary electrons at spikes or edges of specimens due to the increased probability of electrons escaping from sharp edges. A schematic representation of this can be seen in Fig. 4.

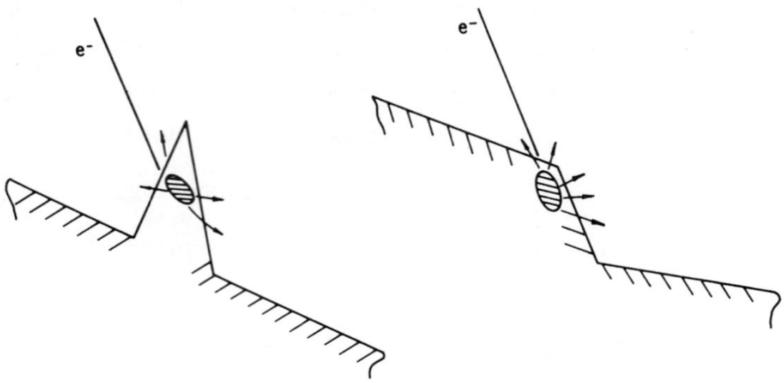

Fig. 4. Effect of sharp surface irregularities on electron emission.

In general, secondary electron emission is rather insensitive to changes in specimen temperature, work functions, and atomic number.

VIII. SOME APPLICATIONS OF SCANNING ELECTRON MICROSCOPY

The current materials science literature abounds with scanning electron micrographs, and this chapter will not attempt to review this literature. However, several micrographs are shown in Figs. 5, 6, and 7 in order to illustrate the application of the scanning electron microscopy technique to several highly specialized ceramic research areas. Figure 5, for example, is a cross section of rat femur containing a ceramic implant. The bone (B), ceramic (C) interface, is clearly discernible as is the prominent porosity of the bone material.

Surface flatness is often characterized by optical microscopy. An optically flat dielectric substrate is shown in Fig. 6. This 2000X scanning electron micrograph reveals a highly irregular surface consisting of many small hillocks.

The science of sintering has benefited significantly from the introduction of the scanning electron microscope. Incompletely sintered bodies are characterized by highly complex, irregular surfaces which require great depth of field for visual interpretation. Figure 7 illustrates the effect of sintering upon previously irregular UO_2 particles. This 20,000X micrograph illustrates neck growth and surface faceting, features which are impossible to observe by any other technique.

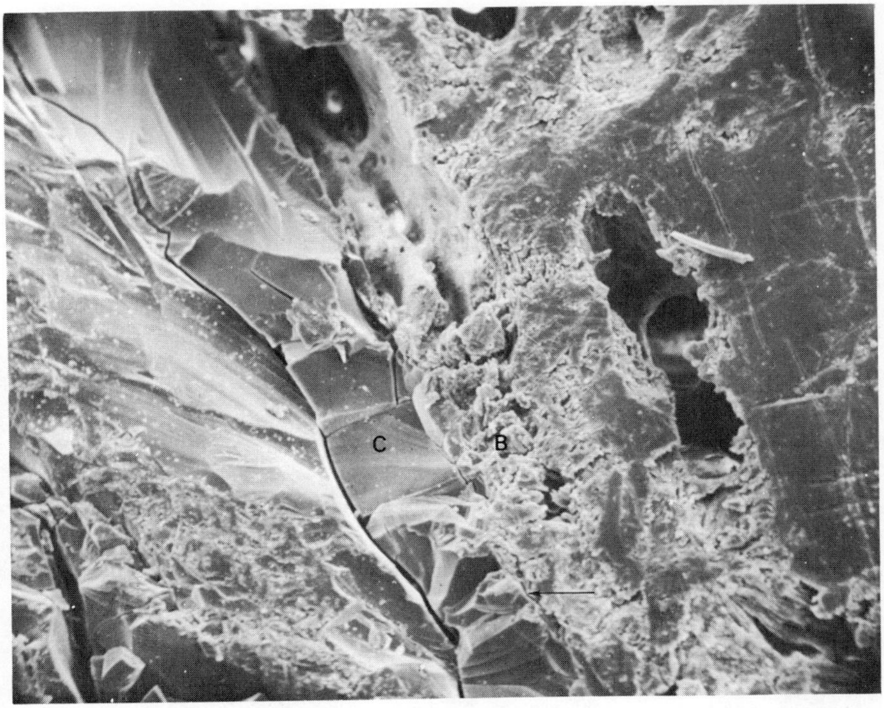

Fig. 5. Cross section of a rat femur (B) containing a ceramic implant (C).

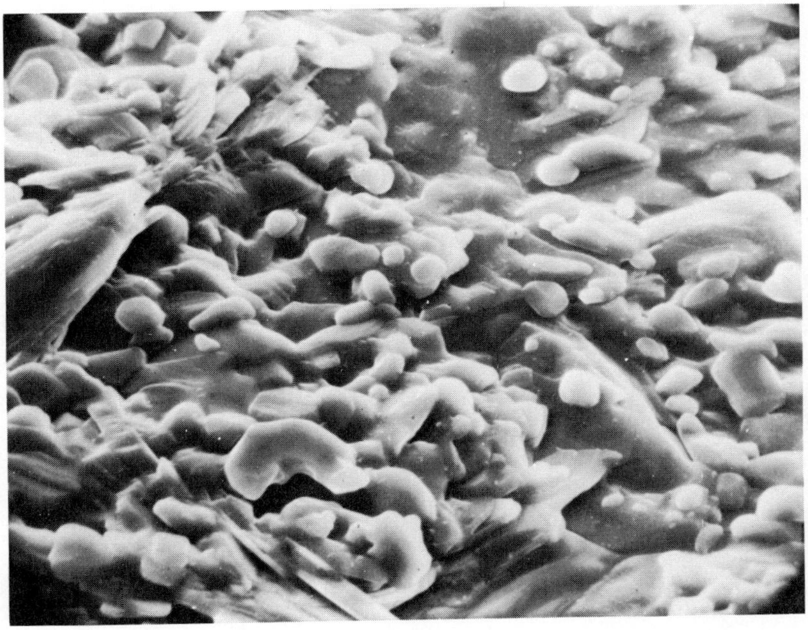

Fig. 6. Optically flat dielectric substrate (2000X).

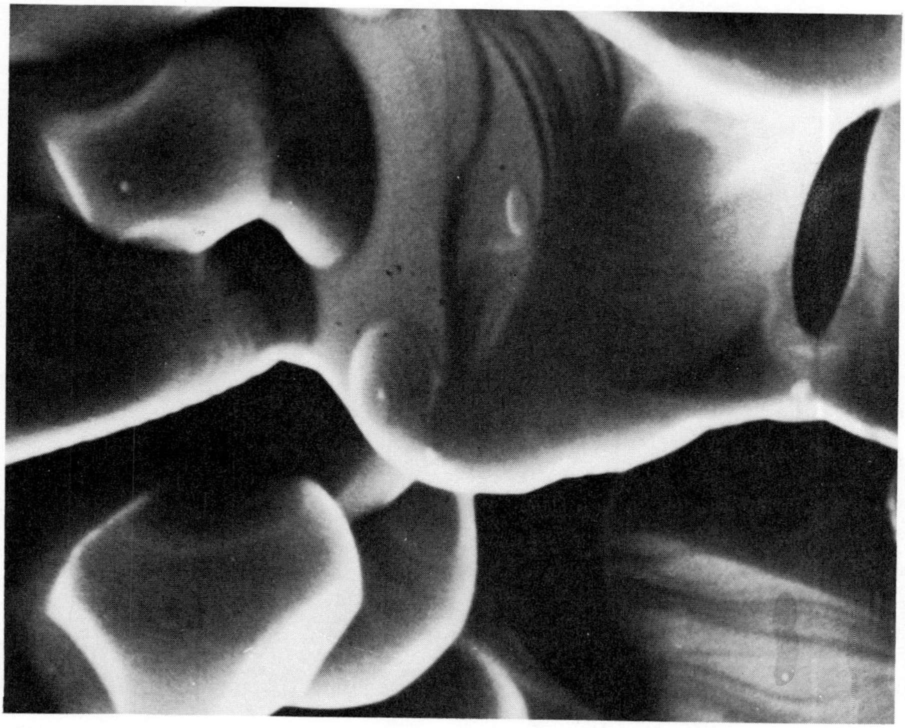

Fig. 7. Sintered UO_2 (1700°C - 2 hr) particles at 20,000X.

REFERENCES

1. M. Knoll, Z. Tech. Phys., 16, 367 (1935).
2. M. von Ardenne, Z. Phys., 109, 553 (1938).
3. V. K. Zworykin, J. Hillier, and Snyder, ASTM Bull., 117, 15 (1942).
4. C. W. Oathly and T. E. Everhart, J. Electron., 2, 568 (1957).
5. T. E. Everhart and R. F. M. Thornley, J. Sci. Inst., 37, 246 (1960).
6. P. R. Thornton, Scanning Electron Microscopy, Chapman and Hall, 1968.
7. R. F. W. Pease and W. C. Nixon, J. Sci. Inst., 42, 81 (1965).
8. Kimoto and Hashimoto, The Electron Microprobe, Wiley, New York, p. 480, (1966).
9. J. W. Colby, W. N. Wise, and D. Conley, Adv. X-ray Analysis, Plenum, New York, p. 10 (1967).
10. H. E. Bishop, Ph.D. Dissertation, Cambridge, 1966.

Chapter 15

LIGHT SCATTERING

John H. Boughton
602 Marsh Road
Wilmington, Delaware

I. INTRODUCTION

Light scattering is a powerful and versatile tool for characterization of ceramic particles. There is a growing appreciation for it in laboratories, but little industrial use has been made of it. As some of the instrumental problems are solved, we may expect to see a rapid expansion of light scattering into routine industrial quality control and specifications.

The study of small particle systems by light scattering has two significant advantages over most other techniques; the system under study can remain essentially undisturbed, and we may study systems which are not conveniently at hand (such as interstellar dust). Additionally, one can determine sizes and study size distributions in the $0.5 - 1.5\,\mu m$ diameter range, which are virtually inaccessible by any other technique except transmission electron microscopy; a tedious, slow and expensive route.

The intention of this chapter is not to present a detailed story of light scattering nor to develop the theories rigorously but rather to stimulate interest and open the door to a fascinating area of science still largely unexploited. Several excellent books have been written reviewing and developing the theory of light scattering; of particular note are van de Hulst's[1] 1957 classic and a more recent excellent monograph on scattering by Kerker[2]. These contain sufficiently extensive bibliographies for a review of the activity in any section of light scattering. The study of optics has undergone a renaissance during the years since World War II. Sparked by such things as the availability of large capacity digital computers and the advent of lasers, interest and development of optical light scattering intensified. The range of applications is rapidly expanding.

Light is scattered when it passes through a region where refractive index changes. Usually, we think of this in terms of solid particles or liquid droplets in air, but density fluctuations in liquids or in gas-mixture systems may also contribute.

Examples of scattered light in nature are all around us and have been recognized as such, or guessed at, for centuries. A classic example is the blue color of the sky. If it were not for the scattering of sunlight by the molecules of the air, the sky would appear black; da Vinci appears to have had an idea of this reason as far back as the early sixteenth century. The extent of visibility through the atmosphere is inversely proportional to the amount of scattering from solid and liquid particles in the air. Turbidity of liquids and opacity of solids are due to the same general type of scattering, but usually more intense because of the greater concentration of scattering particles. Rainbows, halos around the sun and moon, and the colors of sunset are other natural examples of light scattering.

The real, systematic study of light scattering began with Tyndall[3] in 1869 followed by the tremendous contributions of Lord Rayleigh during the next forty years. The field grew during this time and later, particularly after the realization that the scattering is a function of the relationship between the particle size and the wavelength of the incident radiation rather than wavelength alone. The whole electromagnetic radiation spectrum, of which the visible section is only a small fraction, follows the same scattering laws. Thus, x rays scattered by crystal lattices, visible light scattered by dust or moisture particles, microwaves scattered by raindrops, or radio waves scattered by satellites, all scatter in the same general way.

Obvious applications in the area of larger particles are the meteorologists' use of radar backscatter to locate and identify rain, snow, hail, etc.

Astronomers use the technique to study interstellar dust, which, at least until now, was not accessible to more direct means of analysis.

Our discussion will principally concern the visible part of the spectrum; 4000—8000 Å or 0.4 - 0.8 μm.

What is light scattering and how is it caused? As mentioned earlier, any change in the refractive index of the medium through which the light passes causes scattering. This may be regarded in a very simple sense as the result of the interaction between the electric field of the light wave and the oscillating dipoles of the atoms making up the scattering object (Rayleigh scattering).

RAYLEIGH SCATTERING

Scattering by particles which are small with respect to the wavelength of the scattered light may be treated simply with Rayleigh scattering theory. This theory is generally applicable when the particles of interest have a radius of about 5% or less of the wavelength of the scattered light. Detailed discussions and development of this theory may be found in many different sources[1,2,4,5] and in the additional references they contain. For our purposes, we will discuss the assumptions and arguments only enough to get a physical feel for the theory's uses.

Centuries ago it was observed that very small particles scattered blue light strongly (blue sky) and Tyndall showed the light scattered at $90°$ to be linearly polarized. Further theoretical development by Lord Rayleigh soon gave the physical explanation. If the particle is homogenous, isotropic, and small compared to the wavelength of light, the oscillating electric field of the light wave will induce a synchronously oscillating polarization of the particle and the energy emitted by this dipole is the scattered energy. From this consideration, it has been shown that the $90°$ scattering must be linearly polarized.

To be experimentally accessible, we need a measure of intensity as a function of other physical parameters.

If the incident light is unpolarized, it may be resolved into two linearly polarized components; one polarized parallel and one polarized perpendicular to the scattering plane. (The scattering plane contains the light source, the scatterer, and the observer.) These are incoherent. The intensity may be written as:

$$I_u = \frac{I_1 + I_2}{2} = \frac{8\pi^4 a^6}{r^2 \lambda^4} \left(\frac{n^2 - 1}{n^2 + 2}\right)^2 (1 + \cos^2\theta) \tag{1}$$

where I_u = unpolarized intensity

I_1 and I_2 = parallel and perpendicular components, respectively

a = particle radius

r = distance from particle to observer

λ = wavelength of light

n = refractive index

θ = angle of observation, from the incident direction to the scattered direction

A typical example is shown in Fig. 1.

From this statement, the inverse dependence of the scattering on the fourth power of the wavelength is apparent. Further, the degree of polarization for unpolarized incident light is

$$P = (1 - \cos^2\theta)/(1 + \cos^2\theta) \tag{2}$$

The polarization ratio, $P_u(\theta) = \cos^2\theta$ (3)

is defined as the ratio of the intensities of the horizontal to vertical components when the scattering plane is horizontal.

The total scattering from a random sample of like particles is the sum of the scatter from each particle. If we have N particles of volume v per unit volume, Eq. (1) may be rewritten as

$$NI_u = \frac{9\pi^2 N v^2}{2r^2 \lambda^4} \left(\frac{n^2-1}{n^2+1}\right)^2 (1 + \cos^2\theta) \tag{4}$$

This emphasizes the dependence of scattering on the particle volume; the larger the particle volume, the more intense the scattering.

Other useful measurable quantities are available as well. The scattering cross section may be written as

$$C_{SCA} = \frac{128\pi^5 a^6}{3\lambda^4} \left(\frac{n^2-1}{n^2+2}\right)^2 = \frac{24\pi^3 v^2}{\lambda^4} \left(\frac{n^2-1}{n^2+2}\right)^2 \tag{5}$$

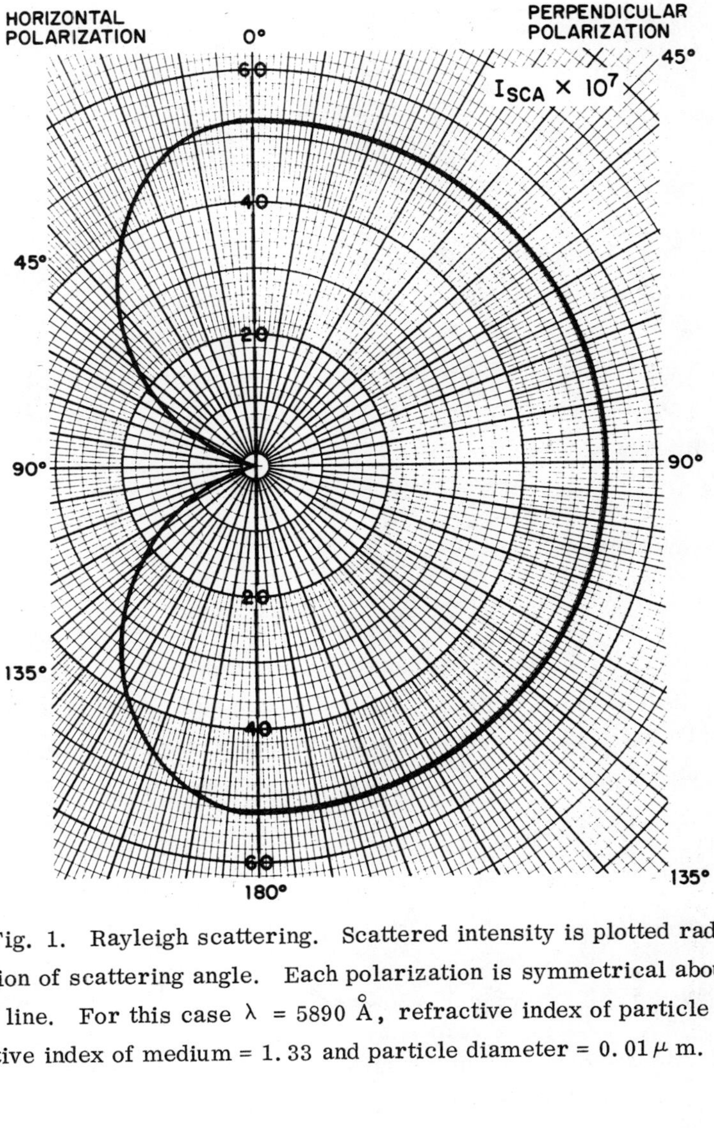

Fig. 1. Rayleigh scattering. Scattered intensity is plotted radially as a function of scattering angle. Each polarization is symmetrical about the 0-180° line. For this case λ = 5890 Å, refractive index of particle = 1.55, refractive index of medium = 1.33 and particle diameter = 0.01 μm.

The scattering efficiency factor is the scattering cross section divided by the geometrical cross section, or πa^2 for a sphere,

$$Q_{SCA} = \frac{128\pi^4 a^4}{3\lambda^4} \left(\frac{n^2-1}{n^2+2}\right)^2 \tag{6}$$

Using a size parameter common in the literature,

$$\alpha = 2\pi a/\lambda \tag{7}$$

Eq. (6) becomes

$$Q_{SCA} = \frac{8\alpha^4}{3}\left(\frac{n^2 - 1}{n^2 + 2}\right)^2 \tag{8}$$

As long as the total incident intensity is 1, the expressions for C_{SCA} and Q_{SCA} work for any combination of polarized and unpolarized incident radiation.

The transmission of a beam through a dispersion of equal sized Rayleigh scatterers can be related to the scattering cross section. The attentuation due to scattering is

$$-dI/dx = NC_{SCA} I \tag{9}$$

Then, the light transmitted is

$$T = I_{\ell}/I_0 = \exp(-NC_{SCA}\ell) \tag{10}$$

where ℓ is the path length. The quantity called "turbidity," , is defined by $T = \text{esp}(-\Upsilon\ell)$. Thus, $= NC_{SCA}$. For Rayleigh scattering, turbidity is related to the intensity of scattering at any angle. Comgining Eq. (5) and (1), we can readily derive

$$= \frac{16\pi r^2 NI_u}{3(1 + \cos^2\theta)} \tag{11}$$

The Rayleigh ratio,

$$R\theta = r^2 NI_u = \frac{3}{16\pi}(1 + \cos^2\theta) \tag{12}$$

is the energy scattered by a unit volume of dispersion per steradian in the direction θ, when the dispersion is illuminated with unit intensity of unpolarized light. This ratio is used quite commonly in particle size measurements and will be discussed again later.

III. GENERAL LIGHT SCATTERING THEORY

For particles too large to be handled by the Rayleigh theory, and too small to be treated directly by the geometrical ray theory, the general theory of scattering by a sphere must be used. This is commonly called the Mie theory[6]. Recently, Kerker[2] questioned this name as the best choice and suggested that the names of at least Debye and Lorenz should also be connected with it. By whatever name, it is a rigorous scattering theory for spheres of arbitrary size.

Again, we shall avoid detailed development of the theory, for the equations mean little unless considerable computer time is available for the student to devote to their analyses. Many thorough treatments are available, for example[1-3, 6, 7]. Effectively, Mie theory defines the intensity of the scattered field at any angle as

$$I_\theta = \frac{\lambda^2}{8\pi^2 r^2} (i_1 + i_2) \tag{13}$$

Again, we measure the scattering angle from the incident to the observed direction. The scattering plane contains the light source, scatterer, and observer and assumes the scatterer is an optically isotropic sphere with a constant refractive index. This time, we require the incident beam to be linearly polarized and essentially monochromatic. The terms i_1 and i_2 are defined in terms of the coefficients of the electric and magnetic wave components, a_n and b_n. The result is a complicated expression involving α and n.

Recall that Rayleigh scattering is calculated from the synchronously oscillating electric dipole induced in the scatterer by the incident ray. The Mie theory proceeds from this concept and further considers the contributions from magnetic dipoles, electric and magnetic quadrupoles, octupoles and higher order multipoles.

Further, the intensities of scattering with crossed polarizations are zero.

Actual calculations can be done for spheres, but not for arbitrary particle shapes. This is because the symmetry of the sphere permits us to determine the component of amplitude of each order in any direction by multiplying the a's and b's of that order by the corresponding direction function. Then, squaring the sum of the amplitudes of all orders gives the intensity. An arbitrary particle shape would not permit us to calculate these coefficients which are free of direction. Most real systems only approximate this requirement: but many are close enough for practicality.

For practical purposes, we are concerned with the scattering cross sections and efficiencies we may observe and calculate. Those materials which absorb light may be characterized by a complex refractive index which is treated the same way as a real one. Therefore, with Mie scattering, as well as other types, the total cross section, or total energy abstracted from the beam, is due to the sum of the scattering and absorption cross sections:

$$C_{ext} = C_{SCA} + C_{ABS} \tag{14}$$

This evaluation was done by Mie[6] and resulted in

$$C_{SCA} = (\lambda^2/2\pi) \sum_{n=1}^{\infty} (2n+1) \left\{|a_n|^2 + |b_n|^2\right\} \tag{15}$$

where a_n and b_n are functions of particle size and refractive index, and

$$C_{ext} = (\lambda^2/2\pi) \sum_{n=1}^{\infty} (2n+1) \left\{\mathrm{Re}(a_n + b_n)\right\} \tag{16}$$

when Re is the real part of the complex amplitude function.

As with Rayleigh scattering and efficiency, dividing C_{SCA} and C_{ext} by the geometrical cross sections gives the efficiency factors for scattering and extinction, respectively:

$$Q_{SCA} = (2/\alpha^2) \sum_{n=1}^{\infty} (2n+1)\left\{|a_n|^2 + |b_n|^2\right\} \tag{17}$$

$$Q_{ext} = (2/\alpha^2) \sum_{n=1}^{\infty} (2n+1)\left\{\mathrm{Re}\ (a_n + b_n)\right\} \tag{18}$$

While the Mie theory is able to rigorously handle scattering from spheres, as well as systems of isotropic spheres of uniform size and refractive index, widespread practical use of it awaited the advent of high speed digital computers to handle the cumbersome mathematics.

We have now surveyed methods of treating very small particles by the simple Rayleigh approach for $\alpha / \lambda \leq 0.05$ and the more general Mie approach for any size scatterer. The Mie theory is most frequently used when particle size is of the order of the wavelength of light. Mie expressions reduce to the corresponding Rayleigh expressions with very small particles and approach geometric-ray theory as the particle radius becomes very large.

We will not discuss geometric optics here because it is seldom of interest with respect to ceramic powders and the field is covered very well elsewhere. See Longhurst, for example[8].

Extensive tables of scattering functions have been computed, and a comprehensive bibliography of these may be found in Kerker[2], pp. 78-82.

Typical plots of intensity versus scattering angle are shown in Figs. 2 and 3. The plots shown in Figs. 2(a) and (b) show the effects on scattering of polarization and size. The right side of each graph shows the change in the scattered intensity of the perpendicularly polarized light with angle while the left side shows the effect on a parallel polarized beam. This sequence in Fig. 2 clearly shows the development of the forward scattering lobe as the particle diameter increases. Figures 2 and 3 are for the reasonable and realistic case of quartz suspended in water and illuminated by the sodium D line.

Figures 3(a) and (b) are plotted as semi-log plots to show the overwhelming forward bias of the scattered intensity and also to point out that fine structure exists at these larger sizes which may need to be considered in data handling of particle systems of similar size. It might be interesting for the reader to convert one of these to a polar plot like Figs. 2(a) and (b) to get a better feel for the scale.

Complete programs for calculation of various Mie scattering functions are available from the National Bureau of Standards and other sources for little or no cost. They cover a wide range of particle size and relative refractive index and save the experimentalist a tremendous amount of time. However, while every advantage must be taken of time-saving aids, one must not lose sight of the basic assumptions made and arrive at conclusions which the data do not support.

We mentioned earlier that the advent of lasers and high-speed computers had revived interest in light scattering.

The laser simplifies the experimental work to a considerable degree by providing a very intense light source which has a very narrow wavelength spread: effectively monochromatic. Even low-powered lasers, now available at quite reasonable prices, work well in scattering experiments. One needn't use a laser, however, to do excellent work, but their use is becoming routine.

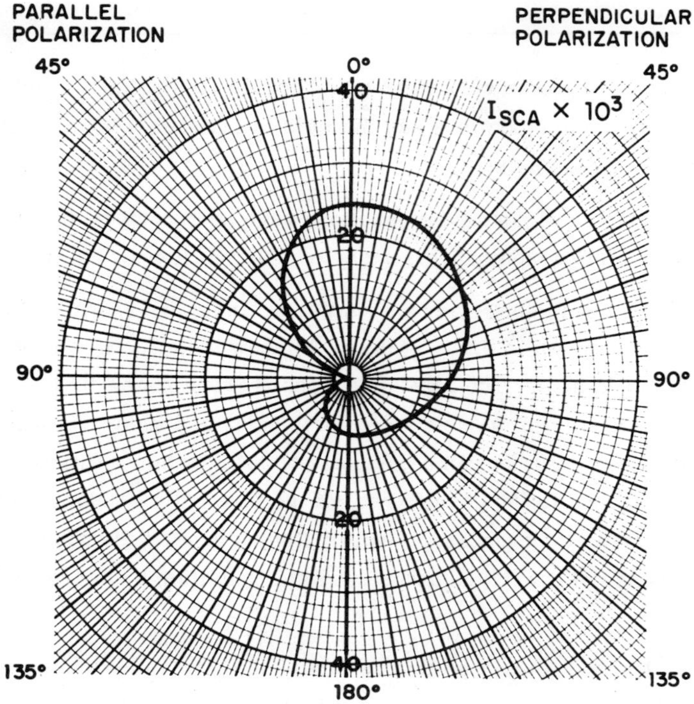
PARALLEL
POLARIZATION
PERPENDICULAR
POLARIZATION
45°
0°
45°
40
$I_{SCA} \times 10^3$
20
90°
90°
20
135°
40
180°
135°

(a)

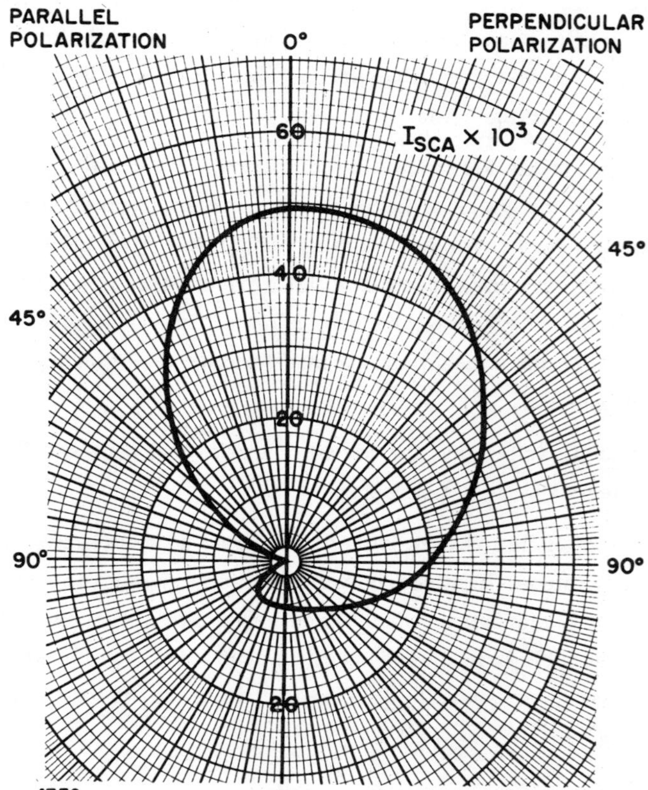
PARALLEL
POLARIZATION
PERPENDICULAR
POLARIZATION
0°
60
$I_{SCA} \times 10^3$
45°
40
45°
20
90°
90°
20
135°
180°
135°

(b)

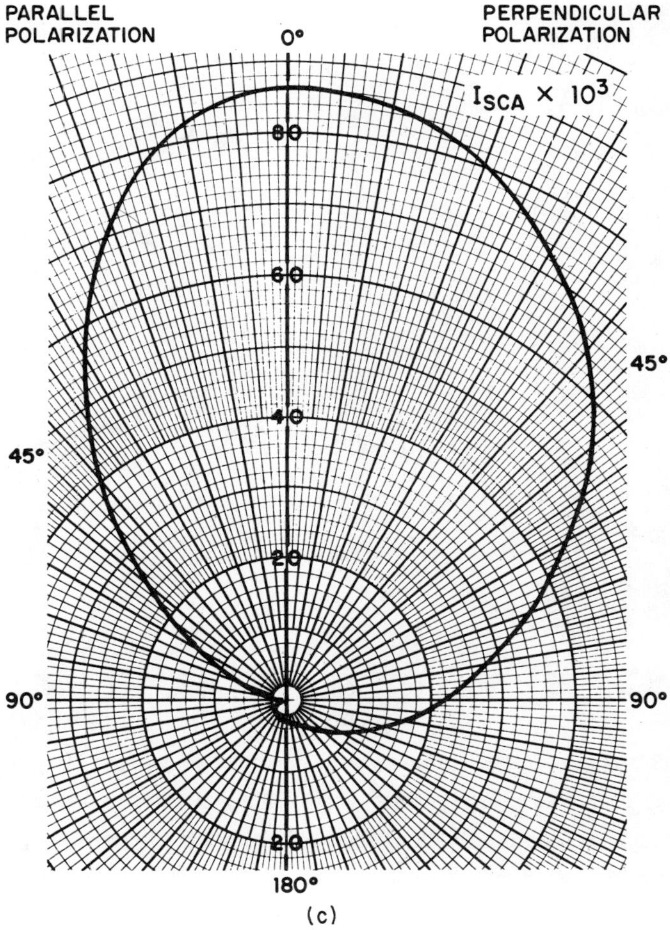

(c)

Fig. 2. (a) Scattered intensity as a function of angle. Scattered intensity is plotted radially as a function of scattering angle. Each polarization is symmetrical about the 0-180° line. For these cases, λ - 5890 Å, refractive index of particle = 1.55, refractive index of medium = 1.33. Particle diameter = 0.16 μ m. (b) Scattered intensity as a function of angle. Particle diameter = 0.20 μ m. (c) Scattered intensity as a function of angle. Particle diameter = 0.24.

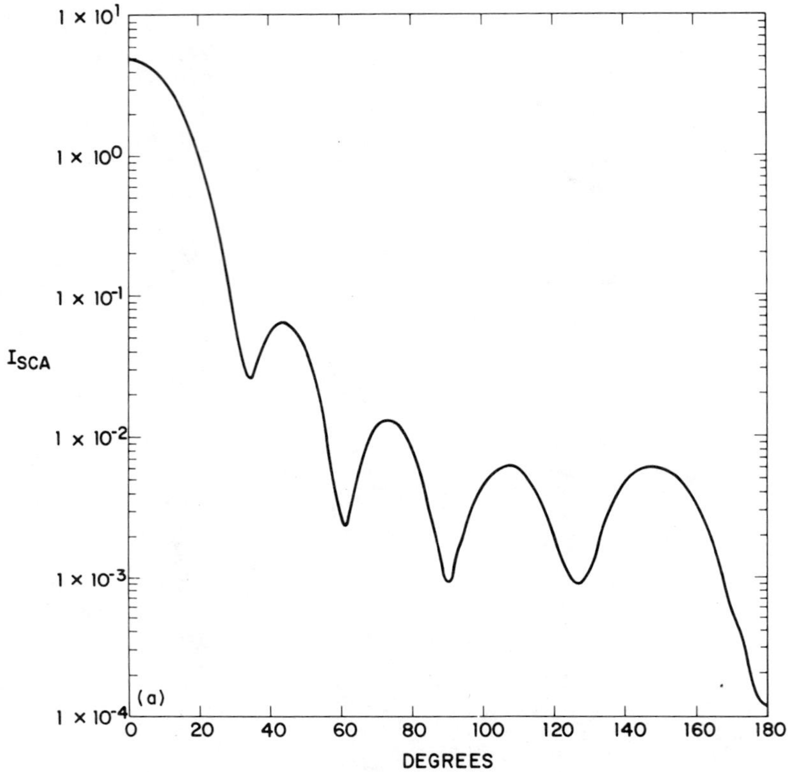

Fig. 3. (a) Scattered intensity as a function of angle. Perpendicular polarization, diameter = 0.98 μ m. λ = 5890 Å, refractive index of the particle = 1.55, refractive index of the medium = 1.33. Parallel polarization, diameter = 0.98 μ m.

The Mie theory provides an exact description of the light scattering for spheres of any size, but the treatment of large particles can be greatly simplified by diffraction theory. Generally, particles in the range $40 < \alpha < 400$ may be treated this way if there is a reasonable difference in refractive index between the particles and the medium. The lower limit of size for which the method is good depends on the materials. For opaque or very high refractive index materials, $\alpha \sim 10$ may be satisfactory, but this range is a reasonably safe guide.

To most easily understand diffraction, or the "bending" of light passing near an edge into the "shadow" zone, consider the effect of placing a pinhole in the path of a collimated, monochromatic light beam in front of a screen.

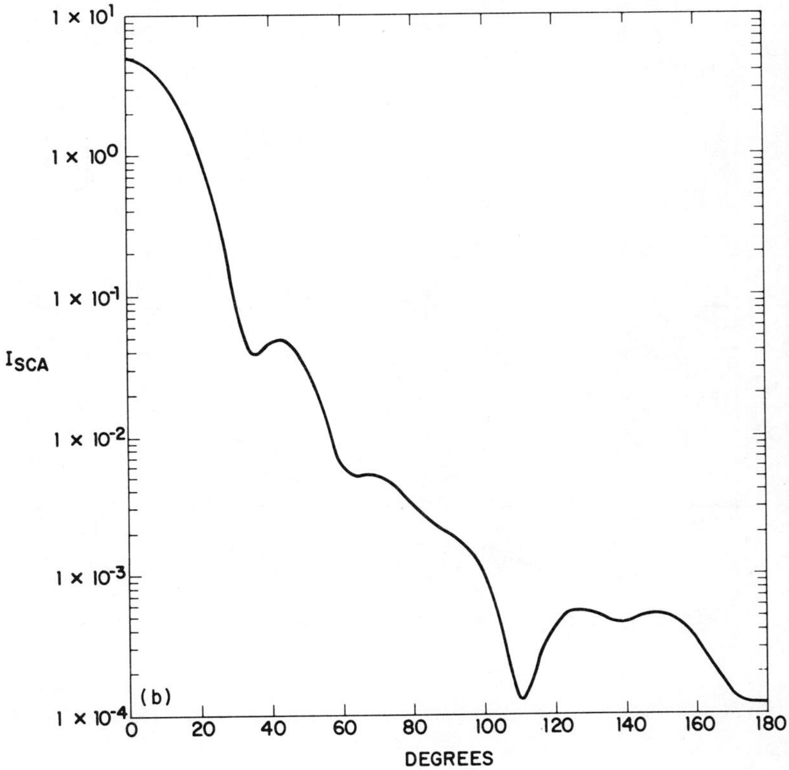

Fig. 3. (b) Scattered intensity as a function of angle. Parallel polarization, diameter = 0.98 μm. Perpendicular polarization, diameter = 0.98 μm. λ = 5890 Å, refractive index of the particle = 1.55, refractive index of the medium = 1.33.

The light passing through the pinhole is spread as shown in Fig. 4. From this graph, we see the diffraction pattern as a bright center spot surrounded by light and dark rings of rapidly diminishing intensity.

The first dark band location is given in radians by the expression,

$$\theta_1 = 1.220 \frac{\lambda}{D} \tag{19}$$

where D is the aperture diameter. This diffraction ring also limits the resolution of optical instruments such as the telescope and microscope. (Two patterns are resolved when the central maximum of one falls on the first dark ring of the other.)

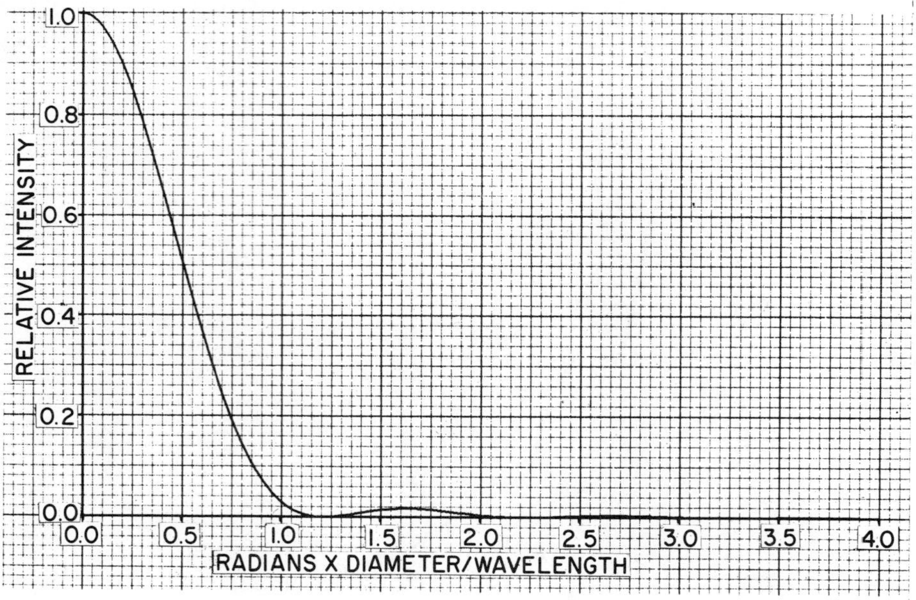

Fig. 4. Relative intensity in airy diffraction pattern.

A telescope with a 5-cm diameter objective lens and 50-cm focal length would have a central disc of:

$$\theta = \left[1.220\ (5.6\ \mathrm{X}\ 10^{-5})\right] / 5 = 1.366\ \mathrm{X}\ 10^{-5}\ \text{rad.}$$

This half-angle, multiplied by the focal length, gives the radius of the spot, 50 cm X (1.366 X 10^{-5}) rad = 68.3 X 10^{-5} cm radius, or a diameter of 0.001366 cm. (We assumed the effective wavelength of white light to be 5600 Å.) This disc would represent the central part of the diffraction pattern for a point source such as a distant star.

The same principles apply to the resolution limits of a microscope. Here, the object is close to the objective lens and this subtends a large angle, 2i, at the object plane. In high magnifying power microscopes, the space between the object and the objective lens is oil filled, and Abbe found the minimum distance of separation to be given by

$$s = \frac{1.22\ \lambda}{2n \sin i} \tag{20}$$

where n is the refractive index of the oil.

The product, n sin i, is called the "numerical aperture" and is characteristic of a given objective. Good microscopes commonly have numerical aperture values of 1.25 — 1.3 with at least one reported as high as 1.6. Thus, the best microscopes, in light of effective wavelength of 5600 Å, resolved

$$s = \frac{(5.6 \times 10^{-5})}{2\ (1.3)} = 2.2 \times 10^{-5}\ \text{cm} = 0.22\ \mu\text{m}$$

Now consider, instead of the pinhole in a screen, an opaque spherical particle in the collimated light beam. Light passing close to the particle will be diffracted. The distribution of intensity with angle will be identical with that of a circular aperture of the same diameter. The total amount of light diffracted by the particle is the same as the amount of light which strikes the particle. Of course, superimposed on this diffraction pattern is the remainder of the beam which passes too far from the particle to be diffracted. With particles which are not opaque, refraction also occurs. But, for particles large enough for us to consider in this part of the discussion, the diffracted light is scattered into a small angular zone, whereas, the refracted light is spread over a large angle, provided that the refractive index of the particle is not too close to that of the medium. Therefore, the refracted light can be ignored in the region where the diffracted light is intense.

If the particles are monodisperse, randomly arranged, and in dilute suspension, the individual diffraction patterns add, and the result is a more intense pattern of the same distribution as that of a single particle. (See Fig. 4.) Thus, we could easily calculate the particle size.

In practice, however, we encounter a range of particle shapes and diameters. Many kinds of particles are close enough to spheres for this treatment. The distribution of sizes results in a diffraction pattern which is still the sum of the patterns of the individual particles, but the bands are now smeared out. This pattern gives average particle size, and in many cases, size distribution may be determined.

III. CURRENT APPLICATIONS OF LIGHT SCATTERING

This section, on applications, is intended to survey the kinds of work being done currently by light scattering. We will not discuss each in detail but try to develop a feel for the versatility and usefulness light scattering already has. These are only typical applications, and we expect to see new ones developed constantly.

Probably, the most active field in light scattering has been in small particle size and size distribution determination. A variety of facets of light scattering theory have been explored to find the best experimental approach for various small particle problems.

A typical example of particle size determination by light scattering measured by transmission is given by Maron et al.[9] using turbidimetry. For particles too large to be handled by Rayleigh's law, the specific turbidity expression is used.

This expression may be easily derived from basic relations,

$$\text{specific turbidity} = \frac{\tau}{c} = \frac{C_{SCA}}{M} \tag{21}$$

where M = mass of the particle = $v\rho_2$

ρ_2 = the density of the scattering spheres

v = volume of the particle = $\frac{4}{3}\pi a^3$

c = concentration in grams of solute (or suspended material) per milliliter of solution (or slurry)

$$\tau/c = \frac{C_{SCA}}{v\rho_2} = \frac{3\,C_{SCA}}{4\pi a^3 \rho_2} \tag{21a}$$

Remembering that $Q_{SCA} = \frac{C_{SCA}}{\pi a^2}$

$$\tau/c = \frac{3\,Q_{SCA}}{4a\rho_2} \tag{21b}$$

Since $\alpha = \frac{2\pi a}{\lambda}$

$$\frac{\tau}{c} = \frac{3\pi Q_{SCA}}{2\alpha\lambda\rho_2} \tag{21c}$$

Experimental measurements made at several wavelengths have shown considerable internal consistency, and this method seems to be a quick and accurate way to determine particle sizes in the 2 μ m and lower region. Further details of experimental procedures and equipment may be found in [10,11].

For particles small enough to treat by Rayleigh scattering, simplifications can be made to further ease the determination.

The light transmitted may be measured directly as we discussed earlier by Eq. (11), or using the Rayleigh ratio, by Eq. (12);

$$R_\theta = r^2 N I_\mu = \left(\frac{3}{16\pi}\right)\tau(1 + \cos^2\theta) \tag{12}$$

τ, the turbidity, is NC_{SCA}. As before,

$$C_{SCA} = \frac{24\pi^3 v^2}{\lambda^4}\left(\frac{n^2-1}{n^2+2}\right)^2 \tag{5}$$

For convenience, these may be transformed in several very useful ways. Turbidity may be related to the volume fraction of the scattering material by:

$$\frac{\tau}{\phi} = 24\pi^3\left(\frac{n^2-1}{n^2+2}\right)^2 \frac{v}{\lambda^4} \tag{22}$$

Then the weight fraction may be obtained from

$$\frac{\tau}{W'} = 24\pi^3\left(\frac{n^2-1}{n^2+2}\right)^2\left(\frac{\rho_{12}}{\rho_2}\right)\frac{v}{\lambda^4} \tag{23}$$

ρ_{12} is the density of the solution or slurry. Heller[12] showed that molecular weight could be substituted for particle size in this very simple way,

$$\frac{\tau}{c} = \frac{24\pi^3}{N}\left(\frac{n^2-1}{n^2+2}\right)^2 \frac{M}{\rho_2^2 \lambda^4} \tag{24}$$

where N is Avogadro's number and M is the weight average molecular weight.

To avoid multiple scattering, contact of the light with more than one scatterer before observation, the measurements should be made at a few different low concentrations and extrapolated to infinite dilution.

In spite of the fact that these measurements may be made quickly and easily and the calculations are short and straightforward, this technique is not in wide use. However, it appears that it is ready for rapid development; for example, in systems needing rapid, frequent routine analyses. It should readily adapt to full automation.

Nonturbidimetric scattering work is based on the Rayleigh ratio at 90° for small particles and relates very closely to the results obtained by turbidimetry. For larger particles, in the Mie region, the measurements are made to exploit the nonsymmetrical distributions of scattered intensity as a function of angle. (See Figs. 2 and 3, for example.)

A typical experimental apparatus for light scattering work is shown in Fig. 5. A number of clever devices have been reported such as those of Tabata and Mori[13], Kunugi and Jinno[14], and Ericson, et al.[15]; all of them depend primarily on variations in angular intensities.

This approach has been widely used in laboratories to measure particle sizes and sometimes size distributions in flowing streams of liquid and

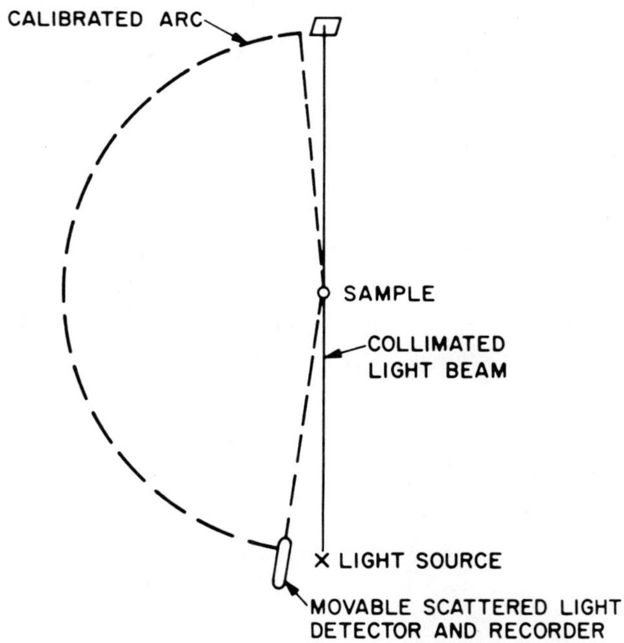

Fig. 5. Typical experimental apparatus to measure scattered light. Appropriate slits and filters are not shown.

solid particles[16-33], including a holographic technique[34], combustion and flames[35-36], turbulent flows[37-44], kinetics[45-46], plasmas[47], liquid structure[48-50], polymerization mechanisms, and molecular weights[51-53], etc.

Basically the same technique using x rays scattered at small angles has proved useful for determining small grain sizes in metals[54] and oxides[55]. Small-angle visible light scattering has been used to study crystal imperfections[56] and as a rapid particle sizing tool[57].

Light scattering theory was developed in such a way that the scattering could be predicted from a known size distribution of spheres. But, we normally measure the scattered light and try to deduce the system which caused it. There is no direct way to do this, but we can frequently get useful approximations by matching experimental data with theoretical values for assumed size distributions or by other indirect methods.

However, for reasonably precise measurements of particles in the range of 0.5-2.0 μ m, light scattering offers power and versatility not yet fully appreciated and applied. It is especially useful for flowing systems, very hot or cold systems like plasmas, and remote systems such as interplanetary dust which are hard to work with any other way.

ACKNOWLEDGMENTS

I am very grateful to Mr. W. D. Ross of Du Pont, Wilmington, Delaware, for the Mie theory calculations and his many thoughtful and practical suggestions for improvements in the text. I also thank Mr. E. D. Lewis of Du Pont for his helpful criticism and support of this project.

REFERENCES

The references listed here are generally articles of broad coverage, of significance in their fields and containing pertinent bibliography. Through them the interested reader will be able to follow the development of the area of his interest.

(1) H. C. van de Hulst, Light Scattering by Small Particles, Wiley, New York, 1957.

(2) M. Kerker, The Scattering of Light: and Other Electromagnetic Radiation. Academic, New York, 1969.

(3) J. Tyndall, Phil. Mag., 37, 384; and 38, 156 (1869).

(4) J. Stratton, Electromagnetic Theory. McGraw-Hill, New York, 1941.

(5) J. R. Hodkinson, in Aerosol Science, Chapter X (C. N. Davies, ed.), Academic, New York, 1966.

(6) G. Mie, Ann. Physik, 25, 377, (1908).

(7) D. Deirmendjian, Electromagnetic Scattering on Spherical Polydispersions, American Elsevier, New York, 1969.

(8) R. S. Longhurst, Geometrical and Physical Optics, 2nd ed., Wiley, New York, 1967.

(9) S. H. Maron, P. E. Pierce, and I. N. Ulevitch, J. Colloid Sci., 18, 470 (1963).

(10) E. J. Meehan and J. K. Miller, J. Phys. Chem., 72, 1523 (1968).

(11) A. G. Walton and T. Hlabse, Talanta, 10, 601 (1963).

(12) W. Heller, Rev. Mod. Phys., 31, 1072 (1959).

(13) N. Tabata and M. Mori, Mitsubishi Denki Laboratory Reports, 10, 15-37 (1969).

(14) M. Kunugi and H. Jinno, 11th Intern. Symposium Comb., 257-264 (1966).

(15) W. D. Ericson, G. C. Williams, and H. C. Hottel, Comb. Flame, 8, 127-132 (1964).

(16) F. D. Bryant and P. Latimer, J. Colloid Interface Sci., 30, 291-304 (1969).

(17) F. T. Gucker and R. L. Rowell, Disc. Faraday Soc., 30, 185-191 (1960).

(18) P. A. Hiltner and I. M. Krieger, J. Phys. Chem., 73, 2386-2389 (1969).

(19) J. R. Hodkinson, Appl. Optics, 5, 839-843 (1966).

(20) A. C. Holland and J. S. Draper, Appl. Optics, 6, 511-518 (1967).

(21) M. Kerker, E. Matijevic, W. F. Espenscheid, W. A. Farone, and S. Kitani, J. Colloid Sci., 19, 213-222 (1964).

(22) M. Kerker, W. A. Farone, L. B. Smith, and E. Matijevic, J. Colloid Sci., 19, 193 (1964).

(23) S. Kielich, J. Colloid Interface Sci., 28, 214-218 (1968).

(24) P. J. Marteney, NASA-CR-211. Experimental Investigation of the Opacity of Small Particles, 1965.

(25) E. Matijevic, M. Kerker, and K. F. Schulz, Disc. Faraday Soc., 30, 178-184 (1960).

(26) G. D. Parfitt and J. A. Wood, Kolloid-Zeitschrift and Zeitschrift für Polymere, 229, 55-60 (1969).

(27) S. A. Schleusener, Rev. Sci. Instr., 39, 1916 (1968).

(28) S. A. Schleusener and A. A. Read, Rev. Sci. Instr., 38, 1152 (1967).

(29) A. F. Stevenson, W. Heller, and M. L. Wallach, J. Chem. Phys., 34, 1789 (1961).

(30) T. P. Wallace and J. P. Kratohvil, J. Polymer Sci., B5, 1139-1144 (1967) and C25, 89-98 (1968).

(31) T. P. Wallace and J. P. Kratohvil, Appl. Optics, 8, 824-826 (1969).

(32) M. L. Wallach, W. Heller, and A. F. Stevenson, J. Chem. Phys., 34, 1796 (1961).

(33) M. Yajnick, J. Witeczek, and W. Heller, J. Am. Chem. Soc., Div. Polym. Chem., Preprints, 8(1), 280-283 (1967).

(34) M. E. Fourney, J. H. Matkin, and A. P. Waggoner, Rev. Sci. Instr., 40, 205-213 (1969).

(35) W. E. Kaskan, Comb. Flame, 5, 93-98 (1961).

(36) D. C. Vogel, R. R. Circle, and R. S. Powell, Rev. Sci. Instr. 38, 499-501 (1967).

(37) H. A. Becker, H. C. Hottel, and G. C. Williams, 16th Can. Chem. Eng. Conf., Windsor, Ont., 1966.

(38) E. Rolf, J. K. Silk, S. Booth, K. Meister, and R. M. Young, "Laser Doppler Velocity Instrument," NASA Contractor Report, NASA CR-1199, December, 1968.

(39) G. C. Williams, H. C. Hottel, and R. N. Gunmitz, 12th Symp (Intern.) Combustion, Poitier, France, 1968. Pub. Comb. Inst., Pittsburgh, Penna., pg. 1081, 1969.

(40) H. A. Becker, H. C. Hottel, and G. C. Williams, J. Fluid Mech., 30, 259-84 (1967).

(41) D. E. Christiansen, I. and E. C. Fundamentals, 8(2), 263-271 (1969).

(42) P. diPorto, B. Crosignani, and M. Bertolotti, J. Appl. Phys., 40, 5083-5087 (1969).

(43) J. O. Nye and R. S. Brodkey, Rev. Sci. Instr., 38, 26-28 (1967).

(44) L. H. Tanner, J. Sci. Instr., 43, 125 (1966).

(45) H. Otouma and H. Ukihashi, Bull. Chem. Soc. Japan, 39, 1634-1637 (1965).

(46) H. Schönert, Ber. Bunsenges, Physik Chem., 73, 366-368 (1969).

(47) S. E. Schwarz, J. Appl. Phys., 36, 1836-1841 (1965).

(48) L. Blum and Z. W. Salzburg, J. Chem. Phys., 48, 2292-2309 (1968) and 50, 1654-1660 (1969).

(49) L. Blum, J. Chem. Phys., 50, 17 (1969).

(50) W. H. Nichols, C. R. Kunsitis-Swyt, and S. P. Singal, J. Chem. Phys., 51, 5659-62 (1969).

(51) P. J. Flory, Principles of Polymer Chemistry, Chap. VII, Cornell Univ. Press, Ithaca, New York, 1953.

(52) S. Krause, J. Phys. Chem., 65, 1618 (1965).

(53) S. N. Timasheff, J. Colloid Interface Sci., 21, 489-497 (1966).

(54) R. W. Gould, Advances in X-Ray Analysis, 10, 185 (1967).

(55) R. W. Heckel and J. L. Youngblood, J. Am. Ceram. Soc., 51, 398-401 (1968).

(56) C. A. Arguello, D. L. Rousseau, R. Luzzi, and S. P. S. Porto, Materials Resch. Bull., 4, 463-476 (1969).

(57) P. J. Livesey and F. W. Billmeyer, Jr., J. Colloid Interface Sci., 30, 447-472 (1969).

Chapter 16

CHARACTERIZATION OF POWDER SURFACES

L. L. Hench

University of Florida, Gainesville, Florida

I. INTRODUCTION

Analyzing the surface character of powders is perhaps one of the most important and most difficult procedures in ceramics. The importance of a powder's surface is that it can control the extent of agglomeration or aggregation of the particles and the strength of the interparticle bond. As discussed in Chapters 1 and 19, the state of powder aggregation or agglomeration strongly influences the morphology developed in sintering, particularly the tendency towards an "end-point density."[1] Recent theoretical sintering studies[2] have proven that the initial number and distribution of interparticle contacts establishes the entire course of morphological changes occurring in sintering and hot pressing steps. Alterations in the interparticle topology during forming processes also are directly related to the number and strength of interparticle bonds. In addition, the reactivity of powders is largely related to surface structure, and therefore sintering kinetics, catalysis, and the ion exchange behavior of ceramics are functions of the particle surfaces. Therefore, it would seem that surface characterization should be a well established procedure in the manufacture of many ceramic products.

However, the understanding of powder surfaces is perhaps the least developed of any field of science and technology. Consequently, specific powder surface characterization methods are not well established at all. Correlations of powders with well-characterized surfaces and various ceramic processing methods are practically nonexistent. There is good reason for the lack of development in the field of powder surface characterization. The scale of observation required for surface structure analysis is beyond the resolution limits of existing instrumental techniques.

II. DIRECT MEANS OF CHARACTERIZATION

Direct transmission electron microscopy (TEM) comes the closest to analyzing powder surface structures. But even this method cannot detect atomistic surface irregularities, amorphous surface films several atom layers thick, and of critical importance, TEM analysis requires vacuum measurements. Insitu analysis of surface hydrosphere structures, which exist for most ceramic powders under standard humidity conditions, is therefore impossible.

Sampling problems in TEM are obviously formidable, and analysis of the agglomeration or aggregation state of the powder is extraordinarily laborious due to particle overlap problems. In order to distinguish whether a group of particles (a clump) are sintered together (aggregate) or bonded by physisorbed layers, van der Waals bonds, or electrostatic charges (agglomerates), it is necessary to obtain a three-dimensional view of the interparticle contacts. This requires the use of a goniometer stage and the tilting of every pair of particles in the clump to insure that particle overlap is not responsible for the apparent bonding. There are several difficulties associated with such an analysis. It is restricted to clump sizes of 10 - 15 particles in order to obtain transmission images. If the number of particles per clump is too large, the sample will appear opaque to the electron beam. Also, if two or more phases are present as particles it is necessary to use quantitative stereological analysis of the TEM images to determine volume fraction distributions per clump, surface areas, mean free paths, etc. (see Chapter 18). Such analysis must use the mathematical corrections for overlap errors[3] and also must apply thickness error corrections[4]. As recently shown[4], if sampling thickness effects are ignored errors of as much as 300% can be introduced. (See Chapter 19 for details of these and other possible errors in quantitative analysis of TEM and SEM images).

Direct TEM analysis of particle surface layers is not particularly encouraging for several reasons. The resolution of selected area diffraction is insufficient to isolate surface layers of 5 - 20 Å thick. The disordered and potentially amorphous structure of the surface layers also is unfavorable for diffraction analysis.

Other direct means of surface characterization have been developed and successfully applied to single-crystal surfaces as discussed in Chapter 20. These techniques include low energy electron diffraction (LEED)[5], Auger spectroscopy[6], field emission microscopy (FEM)[7], field ion emission microscopy (FIM)[8], and scanning electron diffraction (SED)[9]. Each of these tools share the difficulty with TEM that a vacuum is required. The experimental difficulties of applying these methods to the analysis of powders has not been overcome at the present time. In fact, quantitative data interpretation of well-characterized, clean, single-crystal surfaces is still in its adolescence[8, 10].

III. INDIRECT MEANS OF CHARACTERIZATION

The above discussion emphasizes the need for indirect methods of analysis if the characterization of powder surfaces is to be accomplished. A variety of such tools have been developed by physical chemists over the years and the various methods generally fall into two categories: (1) interaction of the powder with a liquid, and (2) interaction of the powder with a gas. Different techniques arise by analyzing different types of interactions in each of the above cases and by varying the nature of liquid or gas employed. All such indirect methods have in common the requirement that experimental results must be interpreted in terms of a surface structural model. Although the experimental methods in themselves may be reasonably straightforward and reproducible, it is in the interpretative models that difficulties and controversy arises.

The reason for ambiguity in interpretation is simple and implied in the previous section. These is at present no valid, direct means of characterizing most ceramic powders which can be used as standards in evaluating the models for indirect characterization. This difficulty is beginning to be overcome, as shown by recent gas absorption studies on directly characterized carbon[11] and silica powders[12]. Once agreement as to model interpretations for these and several oxide powders has been achieved, then indirect means of characterizing ceramic surface structures may become standard tools. Granted the enormity of the need for powder surface characterization and assuming future development of standardization procedures, let us examine the type of information that may be potentially achieved by indirect methods.

IV. SURFACE FORCES

A. Types of Forces

Surface phenomena are a consequence of asymmetrical or unbalanced forces between atoms and molecules at the interface of the particle. Consequently, in order to develop methods of surface characterization it is necessary to understand the various types of surface forces and their range of spatial influence.

Surface forces are electronic in origin and may be either attractive or repulsive. As discussed in a variety of sources[13], the electrostatic potential energy of interaction between two ions of unlike charge, $-Z_1e_1$ and $+Z_2e_2$, separated by a distance (x) is

$$\epsilon_1(x) = -Z_1e_1Z_2e_2/x \tag{1}$$

The potential energy between a surface ion of charge Ze and a dipole of moment μ can also be shown[13] to be

$$\epsilon_2(x) = -Ze\mu/2x^2 \tag{2}$$

Interaction between a surface dipole and a dipole at distance x away involves an interaction energy[13] of

$$\epsilon_3(x) = -\mu^2/4x^4 \tag{3}$$

However, thermal agitation tends to randomize the motion of the dipoles in each of the above cases which increases the distance dependence. For example, the temperature-dependent orientation energy[13] of interaction for polar molecules at a surface is

$$\epsilon_4(x) = -2\mu^4/3x^6\ kT \tag{4}$$

Induced dipole-quadrupole interactions give rise to an energy term related to the inverse eighth power of distance and dipole-octapole or quadrupole-quadrupole interactions give a term involving x^{-10}.

A general attraction at a particle interface will also arise due to London-van der Waals dispersion forces[14] regardless of how dissimilar the particles and environmental molecules may be chemically. London dispersion forces arise from the interaction of fluctuating electron dipoles within a material with induced electron-nuclei dipoles in the atoms or molecules comprising a neighboring substance. It can be shown[13, 14] that the mutual attraction energy due to dispersion forces of two particles of polarizabilities α_1 and α_2 and average vibrational energies of $h\upsilon_1$ and $h\upsilon_2$ is:

$$\epsilon_5(x) = -\frac{3}{2}\,\frac{\alpha_1\alpha_2}{x^6\left(\frac{1}{h\upsilon_1}+\frac{1}{h\upsilon_2}\right)} \tag{5}$$

Potential covalent contributions to surface bonding may also exist. The calculations of mutual sharing of electrons between nuclei show that the distance dependence, x, is on the order of the radii of the ions or atoms involved in the bond. The covalent bond energy is expressed qualitatively as:

$$\epsilon_6(x) = -\frac{C_6 \kappa}{x^n} \tag{6}$$

where κ is the exchange energy due to a positive quantum electron interaction and C_6 and n are constants dependent on the number of electrons, nucleus charge, and symmetry of atom arrangements in the covalent bond. For a detailed analysis of covalent bonding the reader is referred to Slater[15].

Each of the interactions in Eqs. (1) - (6), with the exception of the coulomb force between isolated charges, are short range and thereby contribute to the system's energy only over a few angstroms distance. These short-range forces constitute the binding that holds the solid together and, being additive, give rise to the total potential energy of solid, as discussed in the next section. It is the extension of these surface forces to atoms or molecules in the proximity of a particle that results in monolayer chemisorption or physisorption at the particle surface.

Although the distinction between chemisorption and physical adsorption is somewhat arbitrary, it is generally considered that chemisorbed species form a chemical bond nearly as strong as the substrate. Chemisorbed surface interactions with ceramics therefore will normally involve ionic-covalent interaction energies ϵ_1 and ϵ_6. On the other hand, more weakly bonded, physical adsorption species on ceramic particles will usually involve interaction energies ϵ_2, ϵ_3, ϵ_4, ϵ_5. Because of the secondary or induced nature of these surface forces, physisorption can also occur on top of a chemisorbed layer.

B. Long-Range Interactions

By the very specific nature of the surface interactions involved, chemisorption is usually restricted to a monolayer of coverage or less with the extent of the surface forces limited to several angstroms. However, there is considerable evidence that appreciably long-range forces are also present at the surface of particles. For example, platelike Fe_2O_3 or WO_3 hydrates arrange themselves in horizontal layers upon settling with as much as 8000-Å thick layers of water between the platelets[16]. Brownian motion of the platelets also suggests long-range attractive forces[17]. Clay minerals also exhibit the well-known property of arranging in highly ordered "card pack" formations with large layers of water between the particles[18].

Three major contributions to long-range forces have been established. They are the primarily electrostatic interactions of electrical double layer systems, dispersion force interactions, and dipole-dipole interactions. Calculations of the extended dispersion or van der Waals attractions between particles of specific geometries have been made. To illustrate, in the case of two spheres of radius r, the energy of van der Waals attraction will be

$$\epsilon'_5 = \frac{rH}{12\,x} \tag{7}$$

when the distance of separation of the spheres, x, is less than r. H in Eq. (7) is known as the Hamaker constant and is equal to $\frac{3}{4}\pi^2 n^2 h \nu_0 \alpha^2$, where n is the number of atoms per cubic centimeter and the other terms are as defined in Eq. (5). Since H is on the order of 10^{-13} erg, the energies so calculated from Eq. (7) show that the interaction energy from this source is comparable to that of thermal motion when the particles are separated by distances of the order of their dimensions. Equation (7) also shows that the dependence of ϵ'_5 is inverse first power; consequently, two particles need only to come within a distance of separation of approximately 2r in order to be bonded by dispersion forces. Although only weakly bonded agglomerates arise from such interactions, additional shorter range interactions become more favorable as the particles are brought closer together by the dispersion interaction.

Direct measurement of ϵ_5 has been made between macroscopic plates of glass[19], quartz[20], and mica[21] using delicate balance-type arrangements. Forces of the order of 10^{-2} dynes/cm^2 were obtained at distances of 1μ m or so. Experimental verification of the relative magnitude of the dispersion force contribution to the surface reactivity of ceramic powders has not been obtained to the author's knowledge.

Another potentially significant contribution to the long-range interaction between ceramic particles is due to the extension of surface charge induced dipoles throughout many layers of an adsorbed species[13]. Extending the interaction energy ϵ_2 of a polarizable species adsorbed on charged surface sites to the ith layer leads to a locus of ϵ_2 values represented by an exponential relationship

$$\epsilon_2(x) = \epsilon_{2(o)} e^{-ax} \tag{8}$$

where x represents the distance from the surface and a is given by

$$a = -\left(\frac{1}{d_o}\right) \ln(\alpha/d^3) \tag{9}$$

where d_o is the atomic diameter of the molecule of polarizability α at a distance of separation from the surface of d.

Adsorption of inert gases on alumina and graphite can be interpreted to be due to such a polarization of the absorbate by strong electrostatic surface fields[22]. Long-range orientational interactions in multilayer absorbed water films on ceramic powders in air or in suspension may also be interpreted in terms of such an induced dipole orientational model. It is the transmission of such long-range interactions between particles that produce agglomerates.

C. Surface Charges in Liquids

In liquid suspension many ceramic powders develop an electrical double layer in response to surface charges on the particles. The electrical double layer is usually negatively charged and repulsive toward other particles. However, dispersion and dipole attractive forces will be arithmetically additive to the repulsive force of the double layer in the total response of the interparticle system.

Let us first consider the Guoy-Chapman model for the diffuse double layer at a charged planar surface[13, 23]. The electrical potential at the surface is taken to be ψ_o; it decreases into the solution and at any point is a value $\psi_{(x)}$. Any ion in the electric field will thus have a potential energy $Ze\psi$, where Z is the ion valence and e is the electron charge. Consequently, the probability of finding an ion at distance x from the surface, $n^{\pm}$, in a total concentration of n_o, is proportional to the Boltzmann factor; i.e.,

$$n^{+} = n_o e^{-Ze\psi/kT} \tag{10}$$

$$n^- = n_o e^{+Ze\psi/kT} \tag{11}$$

By using suitable boundary conditions and integrating the ion distribution as a function x, it is possible to achieve the functional decrease in ψ from the surface value ψ_o. For the simplest case, when ψ_o is less than kT, the result is

$$\psi = \psi_o e^{-kx} \tag{12}$$

where k is defined as $[(4\pi e^2/DkT)\ \Sigma_i n_i Z_i^2]^{1/2}$, D is the dielectric constant of the medium containing n_i ions of valence Z_i. At 25°C, kT has a value of 25.7 mV, so for singly charged ions at room temperature, Eq. (12) will hold for $\psi_o < 25$ mV.

The exponential dependence of ψ on x is shown in Fig. 1 for several concentrations of the solution medium. The quantity 1/k is the distance at which the potential has reached $1/e^{th}$ of its value at the surface. Since this distance coincides with the center of gravity of the space charge, the plane at x = 1/k

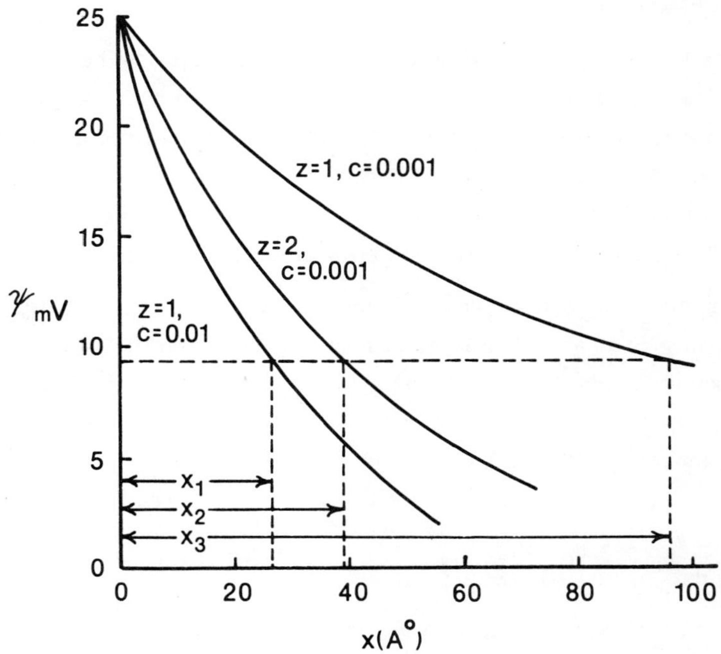

Fig. 1. Change in powder surface potential ψ with distance from the surface, x, for various electrolyte concentrations c and valences Z.

is taken as the thickness of the diffuse double layer. Figure 1 shows that the double layer thickness decreases as the ionic concentration increases. This is because the probability of finding a charge compensating ion close to the surface increases proportionally with ionic concentration. A comparison of two different valences of ions in the solution at equivalent concentrations is also shown in Fig. 1. As expected from Eq. (12), the thickness of the double layer decreases as the charge per ion increases.

The total surface charge Σ can be obtained from the gradient of the potential with respect to x as

$$\Sigma = -(D/4\pi)(d\psi/dx)_{x=o} \tag{13}$$

which leads to the relation between the total surface charge and the surface potential as

$$\Sigma_G = Dk\psi_o/4\pi \tag{14}$$

At the other limiting condition when ψ_o is much larger than kT, the distance dependence of ψ reduces to

$$\psi = (4kT/Ze)\,e^{-kx} \tag{15}$$

as long as $x >> 1/k$. In this case the potential falls away in the manner of Eq. (12) but only at some distance from the surface and with an apparent ψ_o. For singly charged ions at 25°C the apparent ψ_o is fixed at approximately 100 mV. The breakdown of the Guoy-Chapman theory close to the surface at high ψ_o values is a result of neglecting ionic diameters in the model. The Stern layer modification eliminates this difficulty[24].

Stern suggested that the double layer should consist of two parts. Close to the surface is an absorbed layer of ions forming a compact double layer of a thickness d reducing ψ_o to a value of ψ_d. The Guoy diffuse layer then extends outward from ψ_d, as shown in Fig. 2. The potential gradient in the compact layer will thus be

$$\frac{d\psi_s}{dx} = \frac{\psi_o - \psi_d}{d} \tag{16}$$

and, therefore, the charge density for the compact layer will be

$$\Sigma_S = (D'/4\pi)(\psi_o - \psi_d) \tag{17}$$

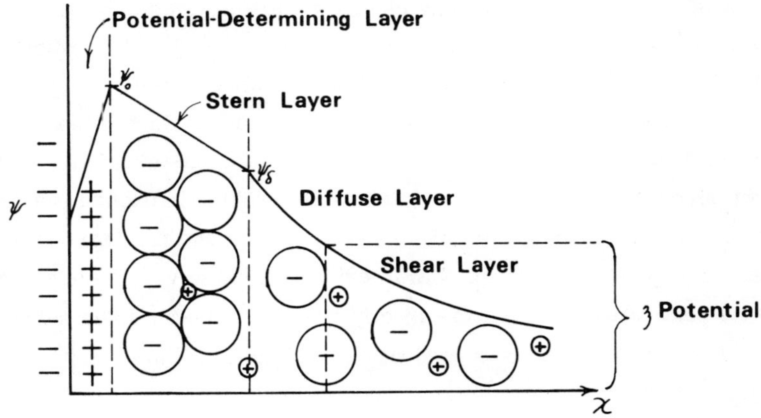

Fig. 2. Surface charge model for a powder in an electrolytic solution.

D' is a local dielectric constant which may differ from that of the bulk solvent. $D'/4\pi d$ represents the capacitance of the compact Stern layer and is experimentally accessible by electrocapillarity measurements and other techniques to be discussed.

The total surface charge density will be $\Sigma_S + \Sigma_G$ and the total electrical capacitance will be given by the series addition

$$\frac{1}{C} = \frac{1}{C_S} + \frac{1}{C_G} \tag{18}$$

In concentrated solutions C_G becomes large so the surface capacitance is almost entirely due to the compact Stern layer, i.e., $C \cong C_S$.

An additional variable that influences the surface charge is the adsorption of charged ions at the surface of the particle at the concentration proportional to their concentration in solution. Such ions drastically alter the intrinsic surface potential ψ_i to a new value ψ_o and can even change the sign of the surface charge. Since ψ_o is thus a function of the concentration of such ions in solution, they are termed the potential determining ions. The Stern and diffuse layers are preceded by a potential determining layer of ions as shown in Fig. 2. Examples of potential determining ions includes Ag^+ ions on silver iodide, Cl^- ions on gold sols, H^+ ions on clays, and H^+ ions on proteins and biological macromolecules. A schematic of the ion distributions at the surface of a clay particle that are responsible for the three charge layers is also shown in Fig. 2.

Because of the relatively strong electrostatic and van der Waals bonding of the ions in the Stern layer to the particle surface, they will be relatively immobile if a force is applied to the particle. Motion of the particle in response to the force consequently will occur at a shear plane at some distance in the diffuse layer. The surface potential that exists at this shear plane is termed the zeta potential. Various electrokinetic phenomena can be used to obtain a measurement of the zeta potential and therefore it is a parameter for surface structure characterization. However, as described above, the zeta potential is only indirectly related to the intrinsic surface charges of the particle and is dependent on the liquid medium employed in the measurement.

V. SURFACE ENERGY

The various ionic and atomic bonding forces described in the previous section are generally uniformly distributed within the bulk of a solid and give rise to a lattice energy which can be fairly accurately calculated and measured[(25)]. However, at the surface of a solid the local atomic arrangements are not maintained, coordination shells are unfilled, and unsatisfied bonds are present. Consequently, the solid surface energy can be defined as the energy required to restore the bulk lattice configuration at the surface of a perfect crystal. It is a potential energy which is often expended in the chemisorption and physisorbtion of ions in the medium within which the particle is suspended. For this reason the measurement of surface energies can be an important characterization step for many ceramic powders or prodcuts.

As suggested by Duga[(26)], the various techniques for measuring surface energies can be divided into two categories: (1) mechanical and (2) thermodynamic. The concept applied in the mechanical methods is that new surfaces are created during the fracture of a brittle solid, as most ceramics are at room temperature[(27)]. The surface energy of the solid thus is simply the energy required to effect fracture divided by the area of the newly formed surfaces. The thermodynamic methods involve the interaction of the solid surface with a liquid of known properties. Measurements of the solid-liquid contact angle or the heats evolved during immersion of dissolution can be used to calculate the surface energy of the solid.

TABLE I

Experimental Techniques for Determining Surface Energy[a]

Mechanical methods	Thermodynamic methods	Miscellaneous
Crack Propagation	Contact angle	X-ray lattice
Strain energy release	Heat of immersion	Parameter shift
Crushing	Heat of solution	
Drilling	Dissolution and Ostwald ripening	
Bent wafer	Equilibrium shapes	
Fiber stress		

[a] From J. J. Duga (26).

Various techniques in each of these categories are summarized in Table I adapted from Duga[(26)]. The crack propagation methods, when applied to completely brittle single crystals in high vacuum, yields the most reliable surface energy values for ceramics available[(28)]. Highly reliable data on the change in surface energy due to adsorbed films are also achieved when the mechanical method is applied in a controlled environment[(29)]. Accurate surface energy measurements of powders have not been achieved using mechanical methods.

Thermodynamic measurements appear to offer the widest application for all types of solids, including powders over wide temperature ranges.. However, the proper choice of a series of liquids is difficult. Impurities localized at particle surfaces or adsorbed films must be removed in order for the measurements to represent equilibrium conditions. Diffusion of the solid or liquid across the interface, or chemical reactions at the interface invalidate the thermodynamic analysis.

A brief review of several of the most reliable methods of surface energy measurement follows:

The Griffith criterion[(30)] for crack propagation is the basis of later developments of mechanical techniques for surface energy measurement. The Griffith equation

$$\sigma = \left(\frac{4E\ \gamma}{\pi c} \right)^{1/2} \tag{19}$$

stems from a balance between the rate of decrease in strain energy associated with the growth of the crack under stress and the rate of increase of surface energy due to the two new surfaces being created at the crack. The parameters in Eq. (19) are: σ = minimum applied stress required for crack growth; E = Young's modulus; γ = surface energy; and c = length of ideally grown crack.

An excellent experimental application of the Griffith concept to the measurement of surface energy was advanced by Gilman[31]. A schematic of the cantilever beam method is shown in Fig. 3 where various terms and dimensions are defined. The Gilman analysis of the relationship between the force (F) required to propagate the crack and the surface energy of the solid (γ) is given in Eq. (20):

$$\gamma = \frac{6F^2L^2}{Ew^2t^3}\left[1 + \frac{\alpha E}{4G}(t/L)^2\right] \tag{20}$$

This equation also includes a second term developed by Westwood and Hitch[32] which accounts for contributions from end effects and shearing forces; G is the shear modulus, and α is a parameter determined by the boundary conditions in the vicinity of the crack tip. One value of the Westwood and Hitch modification is that measurements made as a function of t/L enable α to be calculated and therefore errors in γ due to plastic relaxation at the crack tip can be eliminated. Table II gives surface energy values for a number of ceramic crystals obtained by this method.

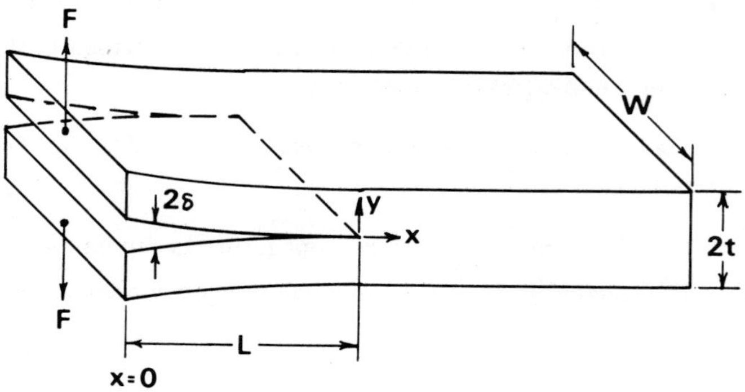

Fig. 3. Double cantilever method for determining surface energies.

TABLE II

Surface Energies of Selected Ceramic Single Crystals Using Double Cantilever Method

Crystal	Plane	Temp °C	Surface energy ergs/cm^2
Al_2O_3	$(10\bar{1}1)$	25	6000
Al_2O_3	$(10\bar{1}0)$	25	7300
$CaCO_3$	$(10\bar{1}0)$	-196	230
CaF_2	(111)	25	355
	(111)	-196	450
	(111)	-273	542
	(110)	-273	1081
LiF	(100)	25	563
	(100)	-196	340
MgO	(100)	25	1700
NaCl	(100)	-196	280
Si	(111)	-196	1240
SiO_2 (quartz)	$(11\bar{2}0)$	25	760
SiO_2 (quartz)	$(10\bar{1}1)$	25	450
SiO_2 (quartz)	$(10\bar{1}0)$	25	1030

Of the thermodynamic methods, the critical angle of wetting technique yields perhaps the most accurate and most easily interpreted values of the surface energy. However, this technique is also restricted to single crystals of dense bodies and is therefore more appropriately discussed in Chapter 20.

The heat of immersion method is particularly suited for evaluating the surface energy of powders. The technique involves the calorimetric measurement of the heat evolved when a higher energy solid-vapor surface is replaced by a lower energy solid-liquid surface. By knowing the surface area involved (A), the change in energy per unit area of surface can be determined, i.e.,

$$q_{im} = A(H_{SV} - H_{LV}) \tag{21}$$

For details of the technique and applications of this method to a number of solids, the reader is referred to Zettlemoyer[33, 34].

Heat of solution measurements are a similar calorimetric technique for evaluating the surface energy of powders. The method involves the measurement of the small differences between the heat of solution of coarse powders and of finely divided powders. By measuring the heat capacities of the fine versus coarse material down to low temperatures, the heat capacity difference, ΔC_p, can be used to calculate the temperature dependence of the surface energy. The surface entropy and free energy also can be obtained from the calculation as:

$$\overline{H}^s{}_T = H^s{}_{298°K} + 1/A\int_{298°K}^{T} \Delta C_p dT \quad (22)$$

$$\overline{S}^s = 1/A\int_{o}^{T} \Delta C_p d \ln T \quad (23)$$

$$\overline{G}^s = H^s - T\overline{S}^s \quad (24)$$

The bars on the quantities above reflect the fact that surface energy measurements on powders represent average energy values of the local surfaces exposed on the particles. The variations in surface energy associated with different crystallographic planes[35], as well as edges and corners, dislocations, and other imperfections are averaged into one representative quantity. The influence of such averaging can be important for, as Bruce has shown[36], the surface energy of an fcc structure such as MgO can vary as much as a factor of 1.225 from the minimal energy (110) plane.

Examples of the application of the heat of solution method to ceramic materials include a series of studies conducted on MgO. Giauque and Archibald[37] prepared coarse and fine MgO by calcination from the hydroxide. A difference in heat of solution of 888 cal/mole was determined. A surface area measurement of 86 m^2/g was made by Jura and Garland for fine MgO powder prepared in a similar manner. On combination of the two results, a surface energy of 1090 ergs/cm^2 for MgO is obtained[13]. By employing Eqs. (22) - (24) and temperature-dependent specific heat data, they concluded that the average surface free energy of MgO decreased from 1040 ergs/cm^2 at 0°K to 1000 ergs/cm^2 at 298°K.

For purposes of characterization, it may often be desirable to know the surface energy of a material at a specific temperature in either the solid or liquid state. Bruce[38] has developed a procedure for estimating the surface energy temperature dependence of ceramics if experimental values are unavailable. The surface energy at 0°K can be estimated from the relation:

$$\gamma_s^{o} = H_s b/ZcN \tag{25}$$

Between 0°K and the melting point T_m:

$$\gamma_s^{T} = \gamma_s^{o} - \left(\frac{\gamma_s^{o}}{T_c}\right) T \tag{26}$$

At T_m,

$$\gamma_s^{T_m} - \gamma_L^{T_m} = \Delta\gamma = \eta H_f/AN \tag{27}$$

and between T_m and the critical temperature, T_c:

$$\gamma_L^{T} = k\,(1 - T/T_c)^n \tag{28}$$

The terms in Eqs. (25) - (28) are:

H_s = molar heat of sublimation (ergs/mole)

N = Avogadros number

b = # bonds/cm^2 of the lattice plane considered

c = # bonds/molecule

A = surface area of a molecule

η = ratio of free bonds to normal coordinate bonds

n = adjustable exponent, about 1.2

H_f = latent heat of fusion

$$k = \frac{H_s b}{2Nc}\;\frac{(1 - T_m/T_c) - (2\eta c/bA)\,H_f/H_s}{(1 - T_m/T_c)^n}$$

If the critical temperature is unknown, Bruce has shown that the following ratios of critical temperature to boiling point are approximately equal for several classes of oxides: $T_c/T_b \cong 1.52$ for MO, $T_c/T_b \cong 1.50$ for MO_2, and $T_c/T_b \cong 1.40$ for M_2O_3.

Application of Bruce's method to MgO yields the following results which are in reasonable agreement with experimental evidence:

$$\gamma_o = 1300 \text{ ergs/cm}^2$$

$$\gamma_s^{T_m} = 1300 - 0.238T$$

$$\gamma_s^{T_m} = 1300 - 0.238(3075) = 565 \text{ ergs/cm}^2$$

$$\Delta\gamma = 270 \text{ ergs/cm}^2$$

$$\gamma_L^{T_m} = 295 \text{ ergs/cm}^2$$

$$\gamma_L^{T} = 2(295)\left(1 - \frac{T}{5470}\right) = 590 - 0.108T$$

Several other techniques have been applied to determine the surface energy of powders including Ostwald ripening of dissolving powders(39), equilibrium shapes of crystals(40), and expansion of lattice parameters from x-ray diffraction(41). The latter method appears to hold the most promise, although strain contributions to the lattice parameters must be eliminated (see Chapter 13). The reader is referred to Bickerman(42) for a critical discussion of these as well as each of the above methods for measuring surface energy.

VI. SURFACE STRUCTURE

As pointed out earlier, structural features at an atomistic level can appreciably affect local surface charges and the local surface energy of a particle. Such features include vacancies, interstitials, impurity ions, dislocations, subgrain boundaries, twin boundaries, surface ledges, and steps. Stranski(43) catalogs several dozen types of surface energy perturbing features and estimates their energy on a sodium chloride crystal structure. Observation and analysis of most of these contributions to the surface energy is possible only with field ion microscopy (Chapter 7) or transmission electron microscopy (Chapters 6 and 7), if then. For most ceramic powders such analyses are presently impossible. Even if examination were possible, sampling problems for powders with real size distributions are nearly insurmountable.

For example, consider the surface energy of a subdivided 1-g cube of MgO. In addition to the 1000 ergs/cm planar surface energy, determined in a manner cited above[44], there will be an edge energy which can be estimated as 1×10^{-5} ergs/cm. The change in total surface energy and edge energy (ergs/g) of the cube as it is subdivided is shown in Table III. Small particles in the 100-Å range exhibit a very large surface and edge energy. Consequently, if the size distribution of a 0.1-μm commercial MgO powder is represented as 70%, 0.1-μ particles; 10%, 1.0-μm particles; and 20%, 0.01-μm particles, the distribution of the surface energy would be 25.8% from the 0.1-μm particles, 74.1% from the 0.01-μm particles, and 0.1% from the 1.0-μm particles. Because of the disproportionate percentage of surface energy for the small particles, it is essential that the low side of the size distribution be well characterized. However, it is these particles that will be most easily agglomerated and most difficult to isolate into small enough clumps for direct size analysis by TEM and most difficult to disperse for indirect size analysis.

Another important factor is that the density of crystal defects will also vary with the particle size. It is to be expected that edge steps, kinks, ledges, etc., will be much more prominent on very small particles. Consequently, the edge energy contribution listed in Table III probably becomes underestimated by as much as a factor of 2 - 10 for 0.01-μ particles. For these reasons, a complete surface energy characterization of a ceramic powder should involve an energy distribution analysis including surface, edge, and local defect contributions. The fraction of the energy contribution "tied up" in agglomerates would also need to be included. This level of direct characterization lies many years in the future and may seldom be practical for commercial powders. However, as shown in a later section, indirect adsorption techniques can be used to obtain a good approximation to the site energy distribution of ceramic powders.

An additional complication is the well-documented decrease in surface energy associated with adsorbed gas films[29]. Reduction in surface energies by as much as 5 have been reported when adsorbed water layers are present on ceramics. Such adsorption effects on surface energy may or may not be dependent on the particle size of the powders. If they are, the energy distribution curves for the powders will be shifted. This type of information is presently not available but is critical if powder characterizations are to be related to processing and end use behavior.

TABLE III

Surface Energy of a Subdivided 1-g Cube of MgO

Number of particles	Single particle weight g	Single particle volume cc	Side length cm	Total area cm^2	Total edge length cm	Surface energy ergs/g	Edge energy ergs/g
1	1.0	0.27	0.65	2.5	7.8	2.5×10^{3}	7.8×10^{-5}
2702	3.7×10^{-3}	1×10^{-3}	0.1	28	550	2.8×10^{4}	5.5×10^{-3}
2.7×10^{5}	3.7×10^{-6}	1×10^{-6}	0.01	280	5.5×10^{4}	2.8×10^{5}	0.55
2.7×10^{8}	3.7×10^{-9}	1×10^{-9}	0.0001	2.8×10^{3}	5.5×10^{6}	2.8×10^{6}	55
2.7×10^{11}	3.7×10^{-12}	1×10^{-12}	10^{-4} (1μ)	2.8×10^{4}	5.5×10^{8}	2.8×10^{7} (0.76 cal)	5.5×10^{3}
2.7×10^{14}	3.7×10^{-15}	1×10^{-15}	10^{-5} (.1μ)	2.8×10^{5}	5.5×10^{10}	2.8×10^{8} (6.6 cal)	5.5×10^{5} (0.13 cal)
2.7×10^{17}	3.7×10^{-18}	1×10^{-18}	10^{-6} (100 Å)	2.8×10^{6}	5.5×10^{12}	2.8×10^{9} (66 cal)	5.5×10^{7} (1.3 cal)

VII. POWDER-LIQUID INTERFACE TECHNIQUES

A variety of methods have been developed to measure the electrical behavior of powder surfaces in a powder-liquid system. The methods involve the relative motion between a charged surface and the bulk solution at the plane of the zeta potential. If an electric field is applied, the particles move in response to the applied force and their velocity is related to the charge on the particles. Or, if the charged particles are mechanically made to move, an electrical potential will be induced proportional to the particle charge density. The different methods are summarized in Table IV, after Adamson[13].

VIII. ELECTROPHORESIS

A potential gradient X is established in a solution containing the charged particles, and the resulting velocity v of the particles is measured by a microscope or a moving boundary apparatus. The force exerted on the particle surface, per cm^2, is ΣX and at steady state is balanced by the viscous drag $\eta v/\tau$. τ is the thickness of the diffuse double layer from the shear plane out, Fig. 2, and η is the viscosity of the solution.

Although the total charge of the particle is not known, Σ will be related to the zeta potential at the shear plane as

$$\Sigma = D\xi / 4\pi\tau \tag{29}$$

Equation (29) is the equation for a parallel plate condenser with a thickness τ and dielectric constant D.

Equating the force applied with the viscous drag, and substituting Eq. (29) yields the desired expression for the particle velocity:

$$v = \xi DX / 4\pi\eta \tag{30}$$

TABLE IV

Powder-Liquid
Electrocharacterization Methods

	Nature of solid surface	
Potential	Stationary	Moving
Applied	Electroosmosis	Electrophoresis
Induced	Streaming potential	Sedimentation potential

Consequently, a measure of the velocity per unit field gives the zeta potential (ξ).

The reader is referred to several sources for a discussion of a more detailed theory[45, 46] and the complicating effects of electrophoretic retardation and relaxation contributions.

IX. SEDIMENTATION POTENTIAL

An applied mechanical force such as developed by a centrifuge will cause the particles to move relative to the solution, inducing a potential E. If the mass of dispersed substance per cc, C_m, is rotated with an angular velocity w, the sedimentation potential between two points R_2 and R_1 along the axis of rotation will be given by[13]

$$E = (C_m/k)(D\xi/6\pi\eta)(w^2/R)(R_2^2 - R_1^2) \tag{31}$$

X. ELECTROOSMOSIS

This method involves measuring the velocity of a liquid moving past an immobile charged surface. Particles possessing a certain zeta potential are compacted into a cylinder with pores of a radius r. When the cylinder is located as a plug in a solution bath (see Fig. 4) and a potential X is applied, the mobile diffuse double layer will move through the pores. The volume

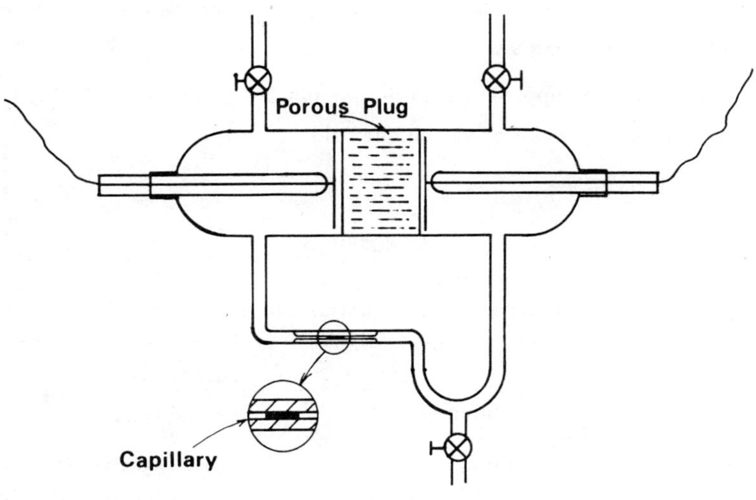

Fig. 4. Electroosmosis method for analyzing the zeta potential of a powder.

flow V, in cc/sec, is related to the size of the pore, r, and the linear velocity of the liquid in the pores, v, as:

$$V = \pi r^2 v \tag{33}$$

and by applying Eq. (30)

$$V = r^2 \xi DX/4\eta \tag{32}$$

Observation of the liquid transport can be by the motion of an air bubble in the capillary providing the return path as shown in Fig. 4. For example, if ψ_o is 100 mV, a field of 1500 V/cm will produce a velocity of 1 cm/sec in water where D = 80 at 25°C.

XI. STREAMING POTENTIAL

When the solution is caused to flow through the pores past the charged surface, a potential E is induced. The relationship between the applied pressure P, the ξ potential, and the induced field is[13]

$$E = \xi PD/4\pi\eta k \tag{34}$$

where k is the specific conductance of the solution.

XII. POWDER-GAS INTERFACE TECHNIQUES

Ceramic powders are in equilibrium with a gaseous environment throughout most processing steps. Consequently, understanding the powder-gas interface is a vital characterization requirement. As discussed earlier, direct means of studying this interface are not available because of the 4-20 Å thickness involved. For this reason, a variety of indirect methods have been developed which employ variations of the gaseous atmosphere itself as an analytical tool. The major analytical variables are the types of gases used (molecular size, polar or nonpolar), temperature of the powder, and partial pressure of the gas. Most techniques involve measuring the volume of gas sorbed onto the powder surface as a function of the above variables [13, 47, 48]. It is also possible to combine the fundamental data on variations in the volume of sorbed gases with other physical property measurements such as dielectric losses[49, 50], capacitance[49], conductance[51], NMR[52]* or IR[53]†. Such combined experiments offer the most powerful approach to powder surface characterization available today.

---

*Nuclear magnetic resonance.

†Infrared analysis.

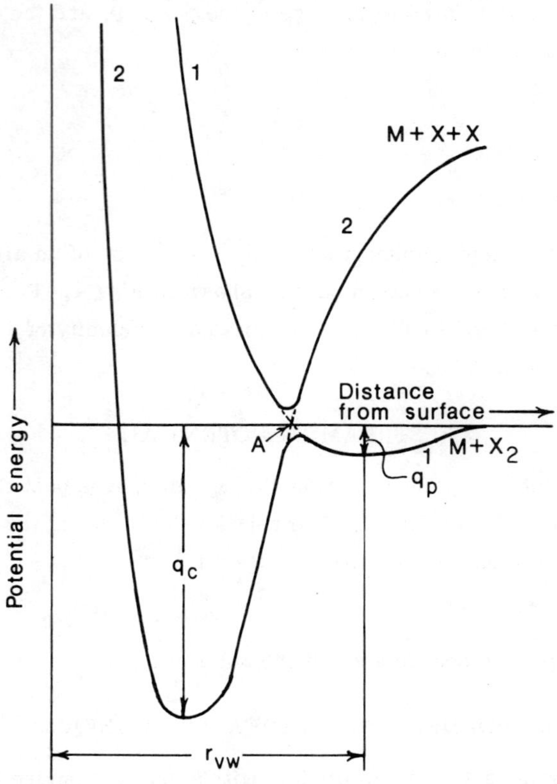

Fig. 5. Energy model for chemisorption q_c and physisorption q_p of a gas X_2 on a surface(54).

XIII. TYPES OF ADSORPTION

Gas-surface interactions can be divdied into two somewhat arbitrary categories: physical adsorption (physisorbtion) and chemical adsorption (chemisorbtion). A physisorbed gas is bonded to the surface primarily through general van der Waals-London forces with a bond strength of approximately 7 kcal/mole or less. Chemisorbed species on ceramics, however, will have developed a strong chemical bond with the surface of 10-40 kcal/mole or more. The origin of the differences in bond energy between the two types of adsorption can be seen from Fig. 5.

Figure 5, from Hayward and Tropnell(54), represents the variation in potential energy of a gas specie as it approaches a planar surface. As the molecule X_2 approaches the surface, the van der Waals attraction of physical adsorption will lower the energy, curve 1, to a minimum, q_p, established at a distance r_{vw}. The value of r_{vw} will be the sum of the van der Waals radii of the surface atom and the adsorbate molecule. At distances less than r_{vw}, quantum repulsion is experienced by the adsorbate molecule and the potential energy rises. There is no activation barrier for physisorbtion; therefore, it is very rapid (except when transport is limited as in a porous substance). It is reversible and the gas may be many layers thick because of the nonspecific nature of the forces involved.

It is also possible that the gaseous specie X_2 will dissociate on the surface, M, by forming a chemical bond. This process is depicted in Fig. 5 by curve 2. The bond strength formed on chemisorbtion will be q_c, which is much larger than the physisorbed minimum, q_p, and therefore chemisorption will be a stable process. The potential curves cross and become rounded at the crossing region because of a wave mechanical mixing of electron energy states. The adsorbate will follow the lowest energy curve and thus can cross from the physisorbed to the chemisorbed state. However, the potential changes involved may be such that an appreciable energy barrier E exists between the two states. If this is the case, the transition time from physisorbtion to chemisorption will be very long and may not be statistically significant for a powder, as long as total processing times are short. However, the transition may take place in calcinated powders within the period of hours to days and produce drastic aging effects on surface reactivity[55, 56]. Table V, after Adamson[13], summarizes the adsorption spectra in terms of surface bond energy Q, transition time τ, and surface concentration r. τ will also be affected by the particle size of the powder and the partial pressure of the gas. For example, the author[57] has shown that MgO prepared by calcination from $Mg(OH)_2$ will hydrate at room temperature within 30 hr at 85% relative humidity, whereas the surface is nearly unchanged after 10 days at 25% relative humidity. Even if the chemisorption rate is unimportant at room temperature, at a higher temperature it may dominate.

As the powder temperature is increased to T_c for the adsorbate, the physical adsorption becomes progressively less, curve 1 in Fig. 6, because of the low adsorption energy. The chemisorption rate may still be appreciable, curve 2 in Fig. 6, so a transition will occur following a path dependent on the equilibration time. Such a transition may occur for adsorbed water

TABLE V

Gas Adsorption Spectrum

τ sec (25°C)	Q kcal/mole	r moles/cm^2	
4×10^{-11}	3.5	10^{-12}	Physisorption
4×10^{-7}	9.0	10^{-8}	
100	20.0		Chemisorption
10^{17}	40.0		

on the surface of ceramic powders during drying, mixing, and grinding steps or in the early stages of sintering.

XIV. LANGMUIR MODEL

The approach that can be used to characterize the fraction of surface covered by a chemisorbed layer was first developed by Langmuir[58]. The rate of condensation on the surface is proportional to the number of unoccupied surface sites S_o and the gas pressure P, i.e.,

$$R_c = k_c P S_o \quad (35)$$

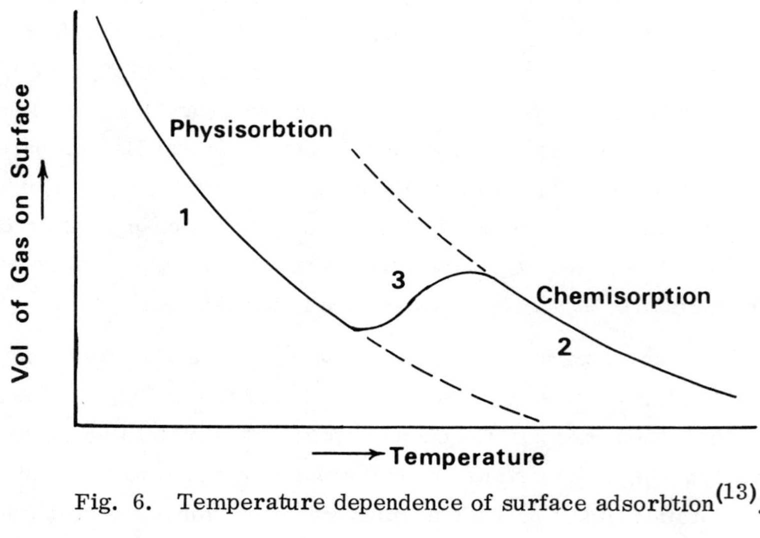

Fig. 6. Temperature dependence of surface adsorbtion[13].

Evaporation of the gas from the surface is proportional only to the number of occupied surface sites S_1 of a total number S, or

$$R_e = k_e S_1 \tag{36}$$

At equilibrium, R_c and R_e can be equated and upon rearrangement yields,

$$\frac{S_1}{S} = \frac{k_c P}{k_e + k_c} \tag{37}$$

S_1/S is the fraction of surface covered, θ, and the ratio k_c/k_e is known as the Langmuir constant b. Thus, Eq. (37) can be written as

$$\theta = bP/(1 + bP) \tag{38}$$

which is the Langmuir adsorption isotherm. Alternatively, if θ is replaced by v/v_m, where v is the volume of adsorbed gas at P_i and v_m is the volume, cc/g at STP, adsorbed at the monolayer point, the Langmuir equation reads as

$$v = v_m bP/(1 + bP) \tag{39}$$

The temperature dependence of adsorption enters in Eq. (39) through the Langmuir constant b. The rate of evaporation, k_e, is the reciprocal of the adsorption time, or

$$k_e = (1/\tau_o)e^{-Q/RT} \tag{40}$$

The rate of molecules impinging on the site area σ_o will describe the adsorption rate. From gas kinetic theory, k_c will then be

$$k_c = N\sigma_o/(2\pi MRT)^{1/2} \tag{41}$$

The ratio of Eqs. (41) and (40) yields b, which is most handily written as a product of a frequency factor b_o, where

$$b_o = N\sigma_o\tau_o/(2\pi MRT)^{1/2} \tag{42}$$

with thermal activation; i.e.,

$$b = b_o e^{Q/RT} \tag{43}$$

A plot of Langmuir isotherms at several temperatures is shown in Fig. 7.

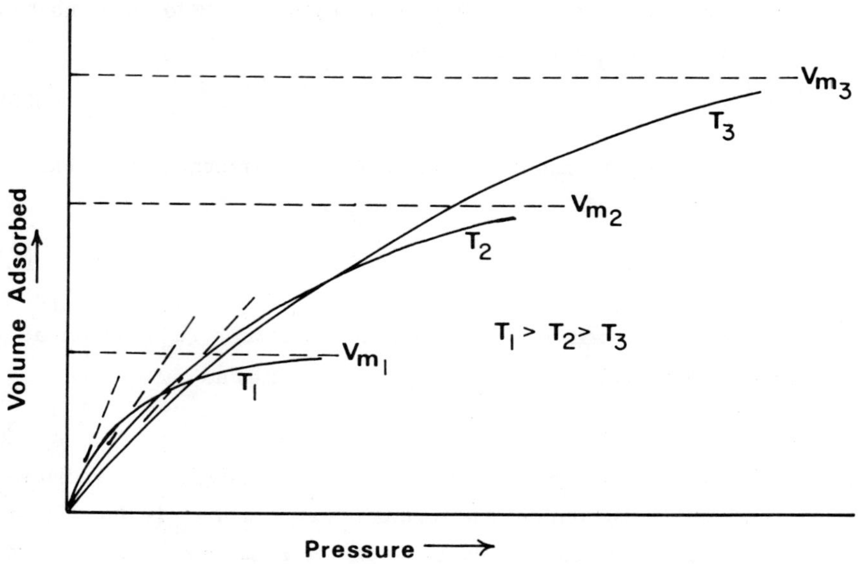

Fig. 7. Typical Langmuir isotherms.

For a more formal statistical mechanical derivation of the monolayer adsorption isotherm, the reader is referred to Fowler and Guggenheim[59] or Adamson[13]. The major difference in the final equation is that the kinetic Langmuir model assumes equal internal partition functions for the adsorbed and desorbed states. This assumption is not valid because the gas species loses rotational degrees of freedom on adsorption.

Although chemisorption does reasonably occur as a monolayer phenomena, experimental results often do not follow the Langmuir relation. This is because of surface heterogeneity and lateral interaction effects and the lack of accounting for entropy changes between states, mentioned above. Detailed discussions of these modifications are given in Adamson[13] and Bickerman[42]. The Langmuir adsorption model is also appropriate for describing submonolayer physical adsorption and has, in fact, been successfully applied to a large number of materials[60]. Data are usually plotted as P/v versus P according to the linear form of Eq. (39),

$$P/v = 1/bv_m + P/v_m \tag{44}$$

The constants b and v_m can be calculated from the best linear fit. Assuming a value for σ_o, usually taken from the liquid density of the adsorbate, the value of v_m can be related to the surface area of the solid, S, as

$$v_m = Sv_o/N\sigma_o \tag{45}$$

where v_o is the STP number 22,400 cc/mole. Values of σ_o in Å^2/molecule normally used are: N_2 (16.2); O_2 (14.1); A (13.8); Kr (19.5); and n-C_4H_{10} (18.1).

XV. MULTILAYER MODELS

A major difficulty of the Langmuir model is that it is inappropriate for describing multilayer formation so characteristic of physical adsorption. Brunauer[61] has considered that there are five principal types of adsorption isotherms, as illustrated in Fig. 8. Type I is the Langmuir type, Type II is that usually observed for multilayer physisorbtion, Type III occurs rarely when the heat of adsorbtion is less than the heat of liquefaction of the adsorbate, and Types IV and V reflect capillary condensation phenomena. The adsorbate condensation causes the adsorbed volume to level off prior to the saturation pressure and often results in hysteresis effects.

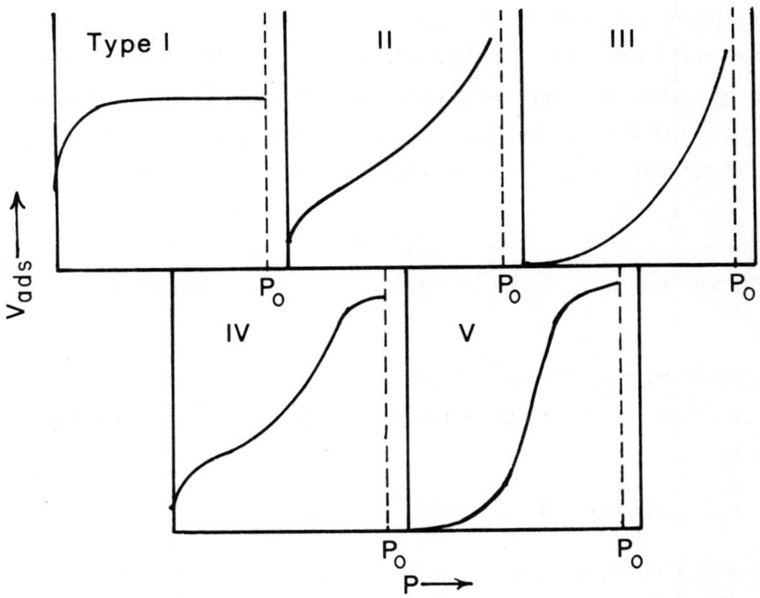

Fig. 8. Five types of adsorption isotherms[61].

The Brunaeur, Emmett, and Teller analysis of the Type II isotherm[62] for N_2 adsorbtion at 77°K is the basis for the now classic method of surface area analysis of powders. Use of the BET method was discussed in Chapter 12. Although the BET equation covers Types I, II, and III isotherms, it should be kept in mind that the fit is usually valid only over the range of $P/P_o = 0.05$ to 0.3.

Because of the relatively gross assumptions involved in the BET derivation, a large literature of criticism, modifications, and rebuttals has evolved[47]. Most changes have involved the introduction of another variable parameter accounting for lateral interactions within the molecules of the multilayer.

A number of other models for multilayer adsorption have also been developed, including ones based on two-dimensional film pressures[63], a two-dimensional equation of state employing the ideal gas law[64] or the van der Waals equation[65], and a condensed film model[66]. None of these appear to offer significant improvement over the BET results for surface area analysis, however.

Another approach to multilayer adsorbtion which leads to structural interpretation of the adsorbed surface film is called the potential theory[61]. The model is basically that of a liquid film condensed on the surface of the particles by a van der Waals-dispersion force potential field. If one assumes that the liquid has negligible compressibility, the thickness of the film, x, is related to the adsorbed volume v as:

$$v = Sxv_o/V_1 \tag{46}$$

where S is surface area, v_o is 22,400 cc/mole, V_1 is the molar volume of the liquid.

The potential ϵ_x that will produce condensation at a distance x from the surface is related to the gas pressure in equilibrium with the adsorbed film, P, by

$$\epsilon_x = RT \ln P_o/P \tag{47}$$

As discussed earlier, the dispersion potential function of a planar surface can be approximated by an inverse cube power of the distance, or

$$\epsilon(x) = \epsilon_o/(a+x)^3 \tag{48}$$

where a is the distance of the order of a molecular radius. Upon combination of Eqs. (47) and (48) one obtains

$$RT \ln (P_o/P) = \epsilon_o / (a + x)^3 \tag{49}$$

which on solving for x and substituting into Eq. (46) yields

$$v = -\alpha + \beta \left[\ln P_o/P\right]^{-1/3} \tag{50}$$

with $\alpha = av_oS/V_1$ and $\beta = (v_oS/V_1)(\epsilon_o/RT)^{1/3}$.

Equation (50) fits Types II adsorption isotherms over a wide range of $P/P_o = 0.1$ to 0.8. Equation (50) has the further advantage of yielding surface area measurements by plotting v versus $[\ln P_o/P]^{-1/3}$, which are decidedly less dependent on the adsorbate used than the BET method[13]. A modification is known as the Frenkel-Halsey-Hill[67] equation. Surface heterogeneity effects have also been included by Halsey[68].

When either the surface potential at a distance x from the surface or the density of the gas film at x is plotted as a function of the volume of gas adsorbed at thickness, the plot is termed a characteristic curve[69]. An example of such a characteristic curve for the adsorption of SO_2 on silica gel is shown in Fig. 9. Two features of the curve are especially significant. First, an extensive surface volume of adsorbed gas is required to decrease the surface potential from a value of 4 kcal/mole to zero. The thickness of the adsorbed layer could be calculated using Eq. (46) to be a value of approximately 10 Å. Secondly, the characteristic curve is temperature independent; i.e., the adsorbate remains gaseous and is merely compressed in the potential field. At a sufficiently lower temperature the adsorbate pressure will reach the vapor pressure of liquid, P_o, and condensation will occur. Equation (50) will describe the adsorption isotherm of the solid.

Another approach that has been taken to describe the theoretical variation of ϵ with x is that of surface induced polarization of nonpolar adsorbate molecules[70]. The surface polarization model described earlier, Eq. (8), when combined with Eqs. (46) and (47) results in the polarization adsorption isotherm. This approach has been suggested as the explanation for the adsorption of gases on alumina[71] and carbon surfaces[13]. A significant consequence of the polarization models is that it predicts strong orientational effects within multilayers and a perturbation from the normal liquid structure which may be expected from contact angle phenomena[13].

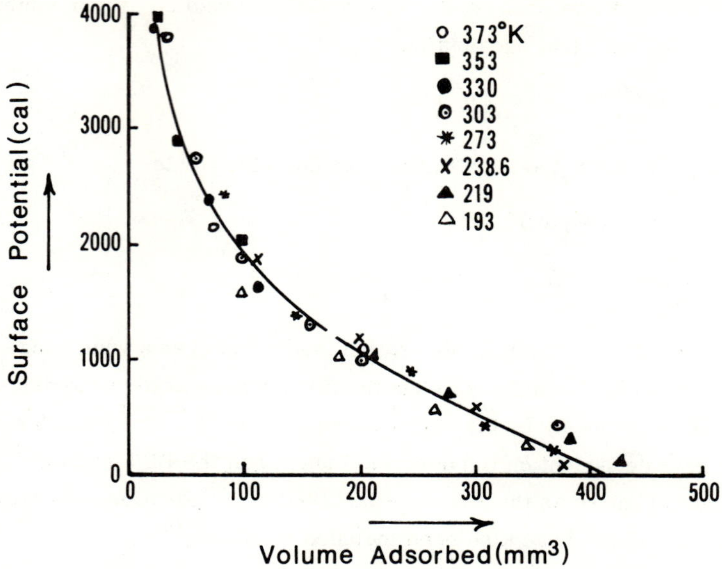

Fig. 9. Characteristic curve for the adsorption of SO_2 on silica gel[69].

Analysis of the thickness of adsorbed surface layers is most easily achieved by employing the t-curve, discussed recently in detail by Broekhoff and Linsen[72]. Shull[73] was the first to show for a number of nonporous solids that a straight line results when the ratio of the adsorbed volume V_a to the volume of a unimolecular layer, V_m, is plotted as a function of the relative pressure. Consequently, if the thickness of the layer is known at one pressure, the linear relationship permits the thickness of the adsorbed layer to be calculated for the entire range of pressures. Broekhoff and Linsen[72] have proposed the use of a statistical thickness t based on the assumption that the adsorbed liquid layers have the same density as condensed capillary liquid. The defining equation for t is

$$t = (MV_{sp}/22{,}400) \cdot (V_a/S) \times 10^4 \text{ Å} \tag{51}$$

where V_a is the adsorbed volume of the adsorbate in ml gas STP/g of adsorbent, V_{sp} is the specific volume of the adsorbate in ml/g, M is the molecular weight of the adsorbate, and S is the specific surface area in m^2/g adsorbent.

For nitrogen, Eq. (51) reduces to

$$t = 15.47\,(V_a/S) \text{ Å} \tag{52}$$

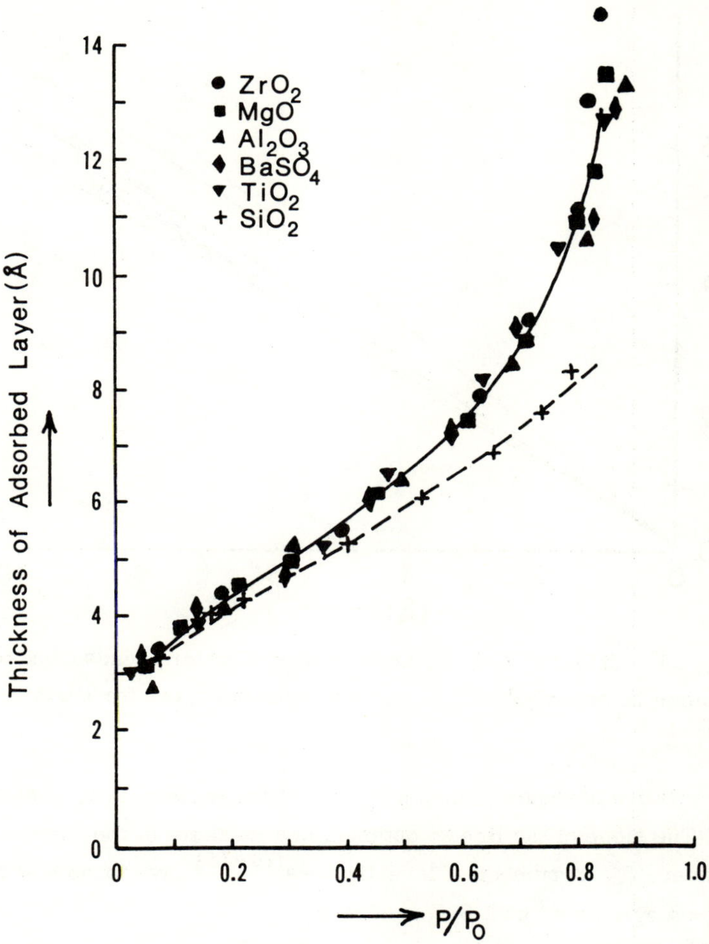

Fig. 10. The t-curves of various ceramic powders as a function of relative pressure of N_2, 77°K[74].

By substituting S, obtained from a BET analysis, and V_a for a given relative pressure, P/P_o, into Eq. (52), t can be calculated as a function of P/P_o for a powder. The t vs. P/P_o curve for N_2 adsorbtion on a number of nonporous powders, e.g., TiO_2, $BaSO_4$, ZrO_2, Al_2O_3[74] all show a common t-curve (Fig. 10). However, a porous SiO_2 gel (Aerosil)[75] shows a marked departure from the common t-curve. The relative pressure at which departures occur can be used to evaluate the powder porosity.

It is also apparent that a series of t-values calculated from Eq. (52) will produce a straight line when plotted against V_a, the total volume adsorbed at

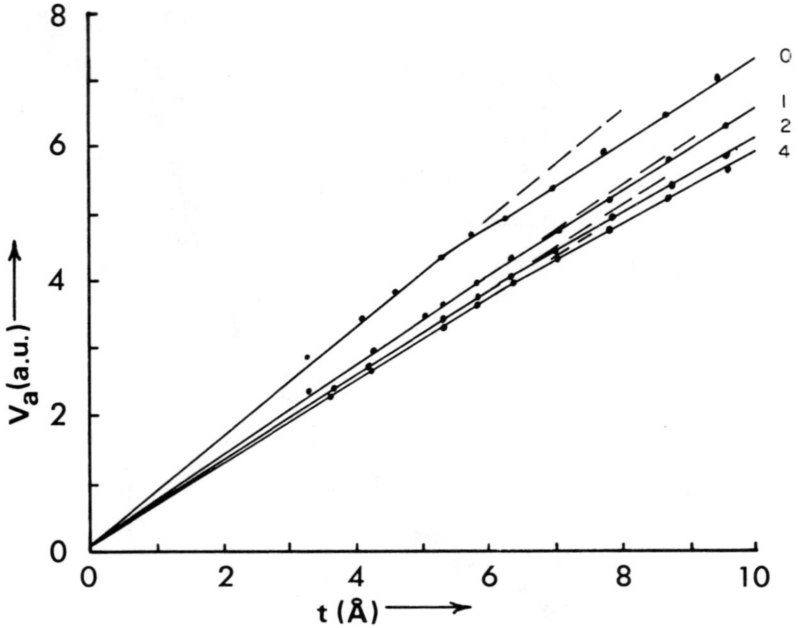

Fig. 11. Effect of 0, 1, 2, and 4 layers of water preadsorbed on the total volume N_2 adsorbed, 78°K, versus layer thickness for anatase TiO_2 powder(72).

the given relative pressure. Such a V_a - t plot for anatase TiO_2 is shown in Fig. 11. The slope of the line so obtained is a measure of the surface area of the powder, S_t. A number of investigators(76-78) have found that S_t values were in good agreement with S_{BET} values.

Broekhoff and Linsen(72) have provided a handy tabulation of the functional dependence of t on P/P_o for nonporous samples. Their table is readily used in a preliminary analysis of absorption effects.

Deviations from the calculated linear t vs. V_a plots provide an important indication of the powder's pore structure. No deviation implies no surface pores. If the curve bends upward at a given pressure (see Fig. 12), capillary condensation is occurring in pores of a certain shape and dimension. A downward bend to the curve is indicative of slit-shaped pores or large holes being filled. A number of detailed interpretations of pore shapes and dimensions have been made using such V_a - t plots(76-78).

Broekhoff and Linsen(72) have recently reviewed the detailed calcuations of adsorption and desorption from various shaped pores that provide the

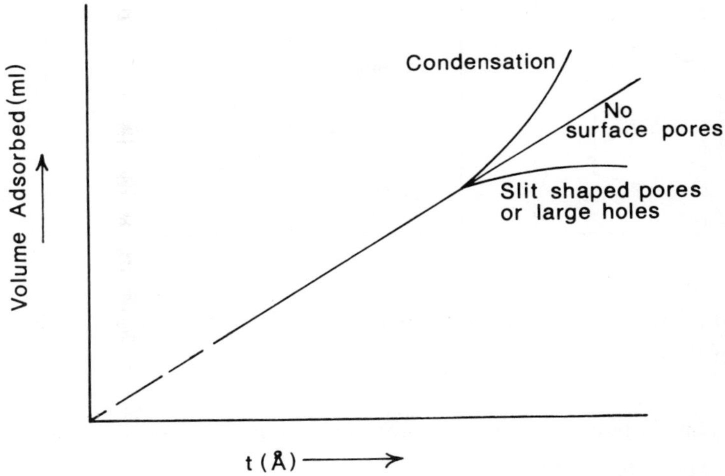

Fig. 12. Three fundamental shapes of the total volume adsorbed versus layer thickness plot.

quantitative evaluation of the adsorption behavior of porous powders. A basic concept in the analysis is that the thermodynamic potential of the adsorbed phase of thickness t is determined by the distance of the interface between the adsorbate in a condensed state and the vapor in a gaseous state; i.e.,

$$\mu_a = \mu_L - F(t) \tag{53}$$

where μ_L is the thermodynamic potantial of the liquid in the bulk state at the same temperature and F (t) is a function which can be determined from a t-curve of multimolecular adsorption. The analytical method is too involved to be presented here and the reader is referred to the original reference. The approach has been successfully applied to calculating the pore size distributions in powders of alumina[79], zirconia[8], porous glasses[81], graphite oxides[82], a nickel-silica catalyst[75, 83], MgO[84], various SiO_2 gels[85], sol-gel ThO_2[86], and sol gel (U, Th) oxides[87].

The adsorption isotherms of the above powders also exhibit an interesting and useful hysteresis behavior. Three of the types of hysteresis loops categorized by de Boer[64] are also of particular importance in interpreting sorption isotherms. They are designated Types A, B, and E and are shown in Fig. 13. Type A is characteristic of a powder with particles that contain cylindrical pores open at each end. Tubular pores with widened sections and

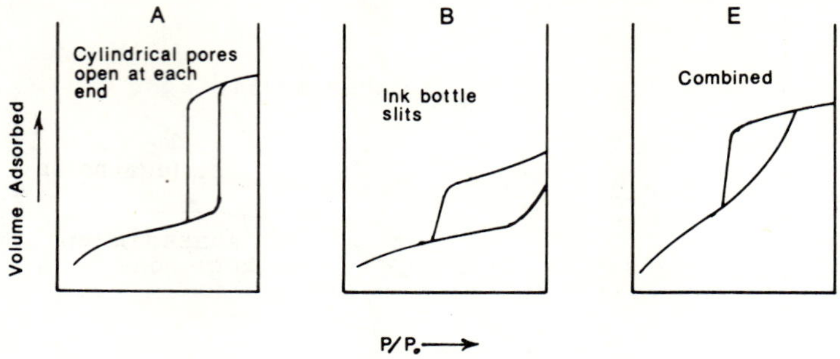

Fig. 13. Three types of adsorption hysteresis loops according to de Boers' classification(64).

ink-bottle shaped pores with narrow necks and small bodies will result in distorted Type A hysteresis.

Ink-bottle pores with very wide bodies and narrow necks or slit-shaped pores open at all sides give rise to B-type hysteresis. The E-type loop is intermediate to A and B and is due to a distribution of pores with narrow necks of approximately uniform diameters, but with wide bodies of varying diameters. Separation of these structural types is obviously difficult and should be accompanied by TEM analysis of the powders.

XVI. APPLICATIONS

Although obtaining fundamental information of the surface behavior of powders is fraught with the assumptions cited above, comparative studies utilizing this technique can be especially revealing. Consider the following examples among many others cited in recent volumes on this subject (13, 34, 47, 48, 50).

The V_a versus t N_2 adsorption curve for anatase TiO_2 shown in Fig. 11 was measured as a function of pre-adsorbed layers of water. The change in the powder surface to that of a hydrosphere does not change the nature of the multilayer N_2 adsorption. Since the slope of the curves decreases, the adsorbed water acts to decrease the surface area of the powder, presumably due to blocking N_2 from micropores(75, 88). However, a change in the density of the adsorbed N_2 layer due to the H_2O substrate would produce a similar effect. A density distribution analysis discussed earlier could be used to distinguish between these two possibilities. At a particular thickness

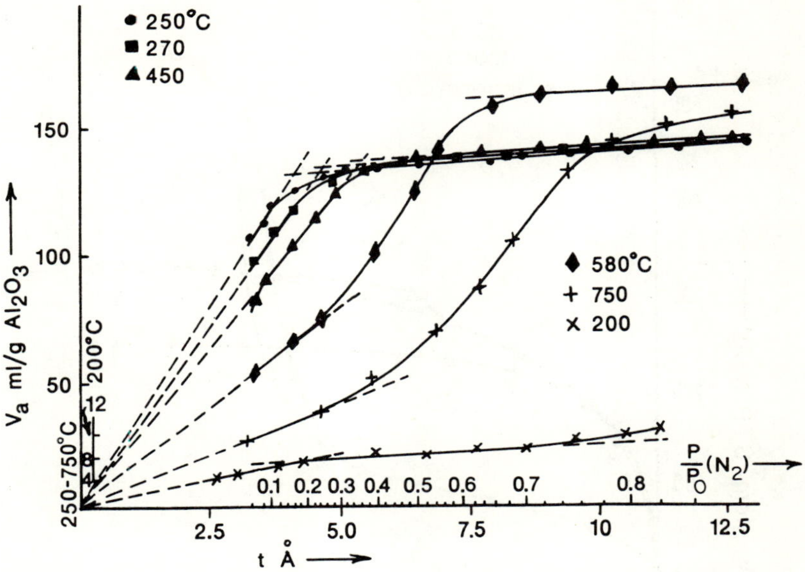

Fig. 14. Influence of heating Bayerite Al_2O_3 powder on the volume of adsorbed N_2 versus thickness dependence of the gas layer[89].

of the N_2 layer the curves depart downward from linearity, indicating capillary condensation of the N_2 which indicates micropores in the powder. The increasing thickness of N_2 required for condensation is further indication that the pre-adsorbed water has filled micropores.

The variations in surface area and pore structure of Al_2O_3 powders produced by thermal decomposition of bayerite[89] are shown in Fig. 14. As temperature is increased above the decomposition temperature, > 230°C, the surface area (slope of the V_a - t plot) rapidly increases and then decreases as sintering of the powders occurs. The t plots show an appreciable amount of micropores giving rise to high surface areas, 480 m^2/g. After sintering occurs, the pore structure changes as shown by the change in slope of the curve.

Characterization of porous silica gels has also been accomplished using N_2 adsorption analysis[85]. For example, Fig. 15 illustrates that the hysteresis loops for SiO_2 gels change from Type E to Type A as the pH of precipitations of the gel is changed from 4.2 to 6.6. This is due to a large

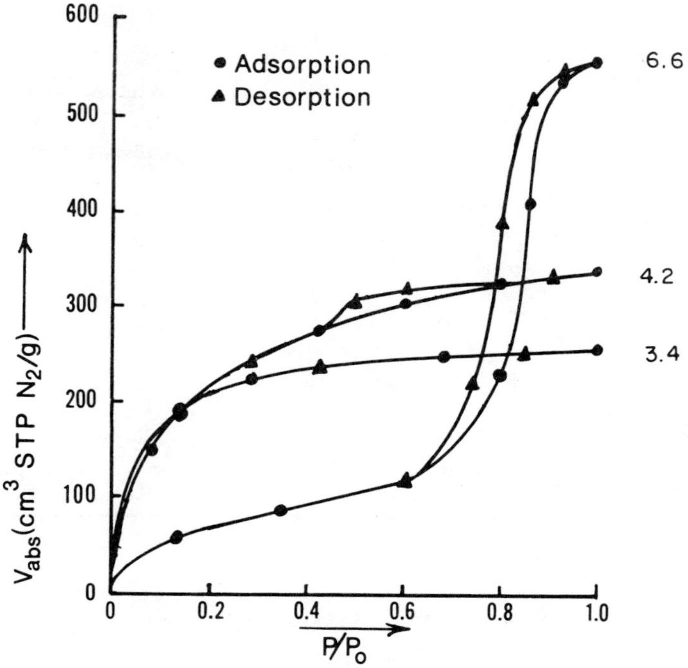

Fig. 15. Effect of precipitation pH on the N_2 adsorption isotherms, -196°C, of silica powders[85].

variation in the pore size distribution (Fig. 16) associated with the kinetics of precipitation.

Redistributions of MgO pore sizes associated with varying the thermal decomposition of magnesites have been calculated using adsorption isotherm analysis[84] and are illustrated in Fig. 17. Another particularly dramatic example of the use of adsorbtion methods is the characterization of sintered (U, Th) oxide powders produced by sintering gels at different temperatures[87]. The change in the isotherms with increasing sintering temperature is shown in Fig. 18.

XVII. DIELECTRIC TECHNIQUES

Dielectric measurements offer a potentially powerful technique to use in conjunction with adsorption isotherms. Measurements of conductance and capacitance can be made on the same sample as the adsorption measurements[50]. By being able to vary the measuring frequency and applied voltage

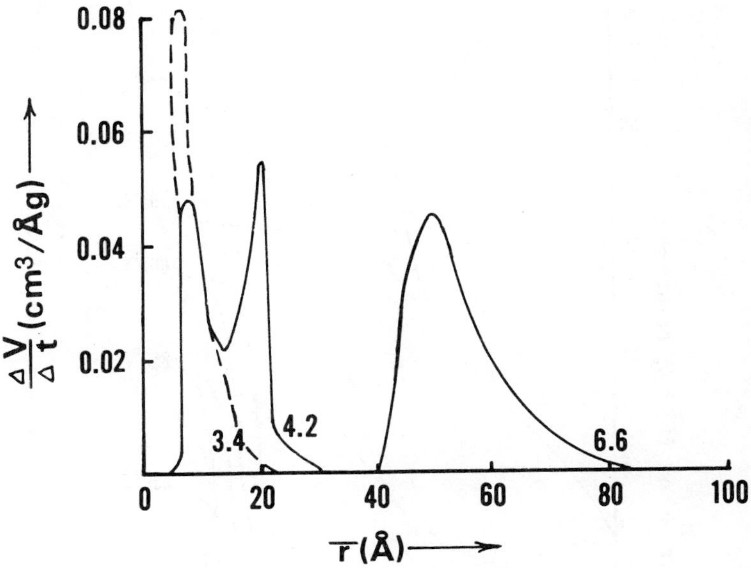

Fig. 16. Variations in pore size distributions of SiO_2 powders resulting from differences in precipitation pH's[85].

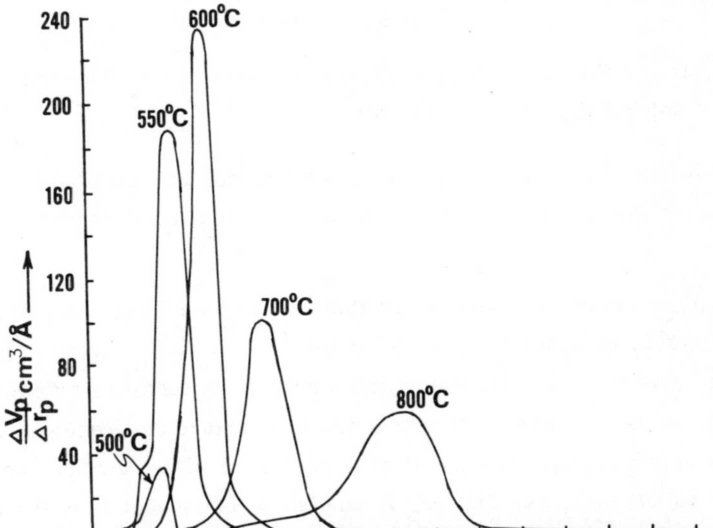

Fig. 17. Change in pore size distributions due to heating MgO powder to various temperatures[84].

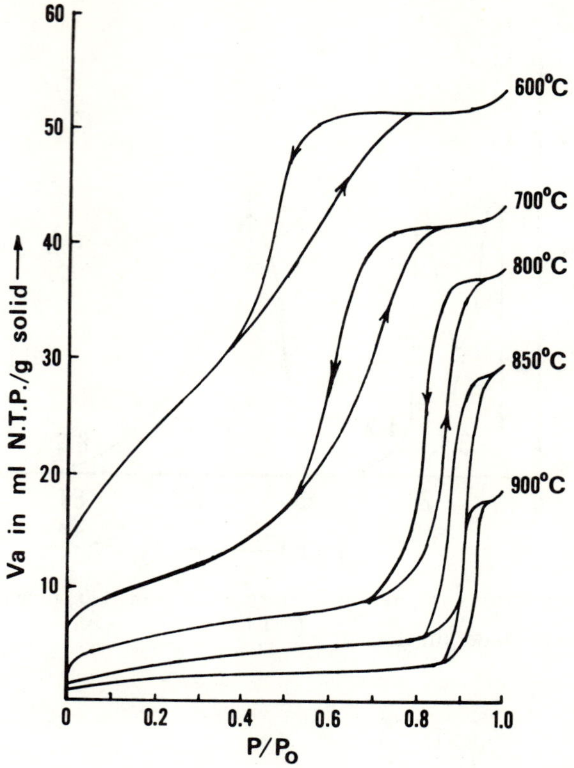

Fig. 18. Influence of sintering temperature on the N_2 adsorption isotherms of sol-gel (U, Th) oxide containing 11% U [87].

under conditions of constant temperature and constant amount of adsorbate, dielectric analysis can be employed to probe the structure of the adsorbed layer.

Several series of dielectric adsorbtion studies have been made on γ - alumina powders using water as the adsorbate [50, 90, 91]. The results obtained show that there is a frequency and temperature sensitive maximum in the dielectric loss ϵ'' at a particular amount of surface coverage θ. The charging constant ϵ' also exhibits breaks in plots of ϵ' vs. amounts adsorbed at similar values of θ, see Fig. 19. A consistent interpretation of the various studies does not seem to have been reached by the authors cited. However, the most plausible explanation appears to be one associating the appearance of loss peaks with capillary condensation producing an isolated high conductivity phase, thus satisfying the Maxwell-Wagner-Sillars [92] requirements for losses in a heterogeneous dielectric.

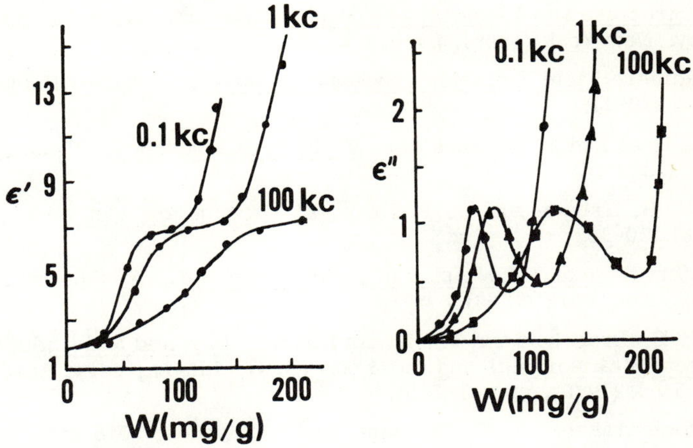

Fig. 19. Dependence of the dielectric charging constant ε ' and loss constant ε '' of porous γ alumina on the adsorption of H_2O [(50)].

XVIII. CONCLUSIONS

The various direct and indirect surface characterization methods discussed provide a large armamentarium for the physical ceramist. However, there have been few efforts to employ a variety of methods on a series of systematically varied ceramic powders. When such studies are conducted and interpreted in terms of the several theoretical surface models, it will be possible to develop a structural theory of ceramic surfaces. Such a theory can be used as a basis for understanding the changes in behavior associated with the handling and wet or dry processing of ceramic powders.

REFERENCES

(1) R. L. Coble, Ceramic Microstructures (J. Pask and R. Fulrath, eds.), Wiley, New York (1969), p. 658.

(2) R. T. DeHoff and R. N. Rhines, "Topology of Sintering," Proceedings of Powder Metallurgy Conference, Stuttgart (1969), p. 1.

(3) A. N. Holmes, Petrographic Methods and Calculations, Thomas Marby, London (1927).

(4) C. V. Gokularathnam, S. W. Freiman, R. T. DeHoff, and L. L. Hench, J. Am. Ceram. Soc., 52, 327-331 (1969).

(5) The Structure and Chemistry of Solid Surfaces (G. A. Somorjai, ed.), Wiley, New York (1969), papers 1-57, 19, 21, 23, 28, 30, 31, 33-35, 37, 50, 63, 69 70-73, 77, 78.

(6) The Structure and Chemistry of Solid Surfaces (G. A. Somorjai, ed.), Wiley, New York (1969), papers 13, 29.

(7) R. Gomer, Field Emission and Field Ionization, Harvard University Press (1961).

(8) J. J. Hren and S. Ranganathan, Field Ion Microscopy, Plenum, New York (1968).

(9) C. W. B. Grigson and D. B. Dove, J. Vacuum Sci. and Tech., 3 3 (1966) 120-132.

(10) The Structure and Chemistry of Solid Surfaces (G. A. Somorjai, ed.), Wiley, New York (1969).

(11) J. C. P. Broekhoff and R. H. van Dongen, Physical and Chemical Aspects of Adsorbents and Catalysts (B. G. Linsen, ed.), Academic, New York (1970), pp. 63-146).

(12) G. Hochstrasses, J. F. Antonini, and I. Peyches, "MS and ESR Studies of Dangling Bonds and Adsorbed Ions on the Pristine Surface of Silica," in The Structure and Chemistry of Solid Surfaces (G. A. Somorjai, ed.), Wiley, New York (1969), pp. 37-41.

(13) A. W. Adamson, Physical Chemistry of Surfaces, 2nd Ed., Interscience, New York (1967).

(14) E. M. Lifshitz, Soviet Physics, 2 [1] (1956) 73-83.

(15) J. C. Slater, Quantum Theory of Atomic Structure, Vol. 1, McGraw-Hill (1960), p. 166.

(16) S. Levine, Trans. Faraday Soc., 42B, 102 (1946).

(17) A. Hachisu and K. Furusawa, Sci. Light (Tokyo), 12 [1] (1963) 157.

(18) Ceramic Fabrication Processes (W. D. Kingery, ed.), Wiley, Cambridge, Mass., The Technology Press (1958), Chapters 1 and 2.

(19) W. Black, J. G. V. de Jongh, J. Th. G. Overbeck, and M. J. Sparnaay, Trans. Faraday Soc., 56 (1960) 1597.

(20) I. I. Abrekossova and B. V. Denjaguin, Phys. Chem. Solids, 5 (1958) 1.

(21) D. Tabor and R. H. S. Winterton, Proc. Roy. Soc., A312 (1969) 435.

(22) S. W. Benson and J. W. King, Jr., Science, 150 (1965) 1710.

(23) G. Guoy, Ann. Phys., 7 [9] (1917) 129.

(24) O. Stern, Z. Elektrochem, 30 (1924) 508.

(25) J. C. Slater, Insulators, Semiconductors and Metals: Quantum Theory of Molecules and Solids, Vol. 3, McGraw-Hill, New York (1967), p. 206.

(26) J. J. Duga, "Surface Energy of Ceramic Materials," Defense Ceramic Information Center Report 69-2, Battelle Memorial Institute, Columbus Laboratories (1969).

(27) R. J. Stokes, "Mechanical Behavior of Polycrystalline Ceramics," in Ceramic Microstructures (R. Fulrath and J. Pask, eds.), Wiley, New York (1968), p. 379.

(28) S. M. Weiderhorn, "Fracture of Ceramics," in Mechanical and Thermal Properties of Ceramics (J. B. Wachtman, Jr., ed.), NBS Publication 303 (1969), p. 217.

(29) D. J. Bryant, L. H. Taylor, and P. L. Gutchall, "Cleavage Studies of Lamellar Solids in Various Gas Environments," in Trans. Tenth National Vacuum Symposium (1963), pp. 21-26.

(30) A. A. Griffith, Phil. Trans. Roy Soc. (London), A221 (1920) 163.

(31) J. J. Gilman, J. Appl. Phys., 31 (1960) 2208.

(32) A. R. C. Westwood and T. T. Hitch, J. Appl. Phys., 34 (1963) 3085.

(33) A. C. Zettlemoyer, "Hydrophobic Surfaces," in Hydrophobic Surfaces (F. M. Fowkes, ed.), Academic, New York (1969), pp. 1-28.

(34) Z. C. Zettlemoyer and K. S. Narayon, "Heats of Immersion and the Vapor/Solid Interface," in The Solid-Gas Interface (E. A. Flood, ed.), Vol. I, pp. 145-175, M. Dekker, New York (1966).

(35) W. L. Winterbottom, "Crystallographic Anisotropy in the Surface Energy of Solids," in Surfaces and Interfaces (J. J. Burke, N. L. Reed, and V. Weiss, eds.), Vol. 1, Syracuse University Press (1967), pp. 133-169.

(36) R. H. Bruce, "Aspects of Surface Energy of Ceramics: I. Calculation of Surface Free Energies," in Science of Ceramics, Vol. 2 (G. H. Stewart, ed.), Academic, London (1965), p. 359.

(37) W. F. Giauque and R. C. Archibald, J. Am. Chem. Soc., 59 (1937) 561.

(38) R. H. Bruce, "Aspects of Surface Energy of Ceramics: II. Calculation of the Average Surface Free Energy and the Effect of Some Variables," in Science of Ceramics, Vol. 2 (G. H. Stewart, ed.), Academic, London (1965), p. 368,

(39) J. P. Borel, Compt. Rend., 257 (1963), 1847.

(40) B. E. Sundquist, Acta Met., 12 [67] (1964), 585.

(41) M. M. Nicholson, Proc. Roy. Soc., A228 (1955), 490.

(42) J. J. Bickerman, Physical Surfaces, Academic, New York (1970).

(43) I. N. Stranski, Z. Physik Chem., 136 (1928), 259.

(44) A. W. Adamson, Physical Chemistry of Surfaces, 2nd Ed., Interscience, New York (1967), p. 278.

(45) H. R. Kruyt, Colloid Science, Elsevier, New York (1952).

(46) J. Th. Overbeek, Advances in Colloid Science, Vol. III, Interscience, New York (1950), p. 97f.

(47) Physical and Chemical Aspects of Adsorbents and Catalysts (B. G. Linsen, ed.), Academic, New York (1970).

(48) Fundamentals of Gas-Surface Interactions (H. Saltsburg, J. N. Smith, Jt., and M. Rogers, eds.), Academic, New York (1967).

(49) N. K. Nair and J. M. Thorp, Trans. Faraday Soc., 61, Part 5 (1965) 962-989.

(50) R. L. McIntosh, Dielectric Behavior of Physically Adsorbed Gases, M. Dekker, New York (1966).

(51) A. J. Mountvala, G. Y. Onoda, E. J. Onesto and A. G. Pincus, "Characterization of Ceramic Particles: Analysis of Alumina by Dielectric Measurements," ONR Report, Contract No. N00014-68-C-0305, Nov. 1969.

(52) W. S. Brey, Jr. and K. D. Lawson, J. Phys. Chem., 68 (1964), 1474.

(53) M. R. Basila, T. R. Kantner, and K. H. Rhee, J. Phys. Chem., 68 (1964), 3197.

(54) D. O. Hayward and B. M. W. Trapnell, Chemisorption, Butterworths, London (1964).

(55) S. J. Gregg, The Surface Chemistry of Solids, Reinhold Publishing Corp., New York (1961), pp. 300-328.

(56) L. L. Hench, "Sintering and Reactions of MgO and Cr_2O_3," Dissertation Abstracts, University of Michigan Microfilm (1964).

(57) L. L. Hench, "Dielectric Relaxation in Materials Analysis," Society of Aerospace Materials and Process Engineers, Proceedings of the 14th Annual Symposium, Nov. 1968. U. of F. EIES Reprint.

(58) I. Langmuir, J. Am. Chem. Soc., 40 (1918), 1361.

(59) R. H. Fowler and E. A. Guggenheim, Statistical Thermodynamics, Cambridge Univ. Press, Cambridge, England (1952).

(60) D. M. Young and A. D. Growell, Physical Adsorption of Gases, Butterworths, London (1962).

(61) S. Brunauer, The Adsorption of Gases and Vapors, Vol. 1, Princeton Univ. Press, Princeton, New Jersey (1945).

(62) S. Brunauer, P. J. Emmett, and E. Teller, J. Am. Chem. Soc., 60 (1938), 309.

(63) S. J. Gregg, J. Chem. Soc. (1942) 696.

(64) J. H. de Boer, The Dynamical Character of Adsorption, Clarendon, Oxford (1953).

(65) S. Ross and J. P. Olivier, On Physical Adsorption, Interscience, New York (1964).

(66) W. D. Harkins and G. Jura,(J. Am. Chem. Soc., 66 (1944), 1366.

(67) T. L. Hill, Adv. Catalysis, 4 (1952), 211.

(68) G. D. Halsey, Jr., J. Am. Chem. Soc., 73 (1951), 2693.

(69) J. McGavack, Jr. and W. A. Patrick, J. Am. Ceram. Soc., 42 (1920), 946.

(70) J. H. de Boer and C. Zwikker, Z. Physik Chem., B3 (1929), 407.

(71) S. W. Benson and J. W. King, Jr., Science, 150 (1965), 1710.

(72) J. C. P. Broekhoff and B. G. Linsen, "Studies on Pore Systems in Adsorbents and Catalysts," in Physical and Chemical Aspects of Adsorbents and Catalysts (B. G. Linsen, ed.), Academic, New York (1970), pp. 1-63.

(73) C. G. Shull, J. Am. Chem. Soc., 70 (1948), 1405.

(74) J. H. de Boer, B. G. Linsen, Th. van der Plas, and G. J. Zondervan, J. Catalysis, 4 (1965), 649.

(75) J. H. de Boer, B. G. Linsen, J. C. P. Broekhoff, and Th. J. Osinga, J. Catalysis, 11 (1968), 46.

(76) B. C. Lippens, B. G. Linsen, and J. H. de Boer, J. Catalysis, 3 (1964), 32.

(77) J. H. de Boer and B. D. Lippens, J. Catalysis, 3 (1964), 38, 44.

(78) W. F. N. M. de Vleerschauwer, "Active Magnesia," Thesis, Delft University of Technology, The Netherlands (1967).

(79) B. C. Lippens, "Structure and Texture of Aluminas," Thesis, Delft University of Technology, The Netherlands (1961).

(80) (a) J. H. de Boer, B. G. Linsen, and Th. J. Osinga, J. Catalysis, 4 (1965) 643.
(b) H. Th. Rijnten, "Formation, Preparation and Properties of Hydrous Zirconia," in Physical and Chemical Aspects of Adsorbents and Catalysts (B. G. Linsen, ed.), Academic, New York (1970), pp. 315-371.

(81) P. H. Emmett and M. Cines, J. Phys. Colloid Chem., 51 (1947), 1260.

(82) A. B. L. van Dorn, "Grafietoxyde," Thesis, Delft University of Technology, The Netherlands (1957).

(83) B. G. Linsen, "The Texture of Nickel-Silica Catalysts," Thesis, Delft University of Technology, The Netherlands.

(84) W. F. N. M. de Vleerschauwer, "Active Magnesia," in Physical and Chemical Aspects of Adsorbents and Catalysts (B. G. Linsen, ed.), Academic, New York (1970), pp. 265-316.

(85) C. Okkerse, "Porous Silica," in Physical and Chemical Aspects of Adsorbents and Catalysts (B. G. Linsen, ed.), Academic, New York (1970), pp. 213-265.

(86) E. G. P. Cornelissens, in Physical and Chemical Aspects of Adsorbents and Catalysts (B. G. Linsen, ed.), Academic Press, New York (1970), p. 420.

(87) M. E. A. Hermans, "Hydrolytic Phenomena in U(VI)-Preparation," in Physical and Chemical Aspects of Adsorbents and Catalysts (B. G. Linsen, ed.), Academic, New York (1970), pp. 373-423.

(88) F. E. Karasz, W. M. Champion, and G. D. Halsey, Jr., J. Phys. Chem., 60 (1956), 376.

(89) B. C. Lippens and J. H. de Boer, J. Catalysis, 4 (1965), 319.

(90) M. G. Baldwin and J. C. Morrow, J. Chem. Phys., 36 (1962), 1591.

(91) G. Ebert, Kolloid Z., 184 (1962), 148.

(92) L. K. H. von Beek, in Progress in Dielectrics, Vol. 7, CRC Press (1966), p. 69.

PART 4

Microstructure

Chapter 17

ELECTRON MICROPROBE

Gordon Lewis

Graduate Center for Materials Research
and Department of Ceramic Engineering
University of Missouri at Rolla
Rolla, Missouri

I. INTRODUCTION

The electron microprobe offers a unique tool to the ceramic industry(1) in that the microprobe allows an essentially nondestructive chemical analysis of an extremely minute volume of material (a lower volume limit is usually assumed to be several cubic microns). In many investigations it may be of great advantage to investigate the elemental makeup of a small inclusion or precipitate and compare it to the surrounding matrix. It has been shown, however, that it is often difficult to find a material for a standard that is homogeneous on the micron scale[2]. All elements[3, 4], except hydrogen and helium, can be displayed as an "elemental concentration map" on an oscilloscope with the variations in brightness on the cathode ray tube yielding a rough indication of concentration[5, 6].

Although microprobe analysis has found its primary application in metallurgical research, its range of application is steadily growing as evidenced by the following selected list:

1. In metallurgical research — studies of phase systems[7-9], both for elemental makeup as well as for solid solution, diffusion[10, 11], analysis of inclusions[12-14], and segregation[15].
2. In nonmetallic research — studies of semiconductors[16-20], both in active and inactive states, meteorites[21], mineral specimens[22, 27], glasses[28-32], phase studies[33, 34], and nonstoichiometry[35].
3. In medical, dental, and biochemical research — studies of the elemental makeup, and the elemental distribution in bone[36], teeth[37], and tissue[38, 39].
4. In production — characterization of electrical contact surface[40], glasses[29], other ceramic objects[41], (i. e., refractory brick, catalysts, insulators, contamination spots on thin film resistors, etc.)

II. ELECTRON MICROPROBE ANALYZER

The electron microprobe[42] can be broken down, for purpose of discussion into units representing four older and simpler instruments, all housed in a common vacuum chamber:

1. Conventional x-ray tube
2. X-ray emission spectrograph
3. Light microscope
4. Electron microscope

In a conventional x-ray tube (as used in an x-ray diffraction unit or in an x-ray fluorescence unit) the characteristic radiation[43-46] from a single-element target is excited by electron bombardment. The same excitation principle is used in the electron microprobe with the sample of unknown chemical composition functioning as the target. Therefore, at an appropriate electron energy, each element in the sample will produce characteristic x-radiation from which it can be identified and a quantitative estimate of its concentration can be obtained. Electromagnetic lenses are incorporated in the electron beam path (hence the relation to the electron microscope). The electron lens system allows the relatively large electron beam from the electron gun (approximately 100 to 200μ) to be defocused into a beam of approximately 1μ diameter at the sample surface[47, 48]. An electron optics system is shown schematically in Fig. 1.

The inclusion of deflection plates in the electron optical system, see Fig. 1, makes it possible to display on a cathode ray tube various output data, such as sample current, back-scattered electron current, or x-ray detector output, from a selected area of the sample. The same sweep voltages are applied to both the deflection plates in the microprobe and the deflection plates in the oscilloscope. The intensity of the electron beam in the oscilloscope is modulated by the chosen output mode thereby yielding a "T. V. -like display" of the sample surface which can be directly related to chemical composition. Figure 2(a) shows a back-scattered electron display from a Ti-V-O sample. The V distribution in Fig. 2(b), the Ti distribution in Fig. 2(c).

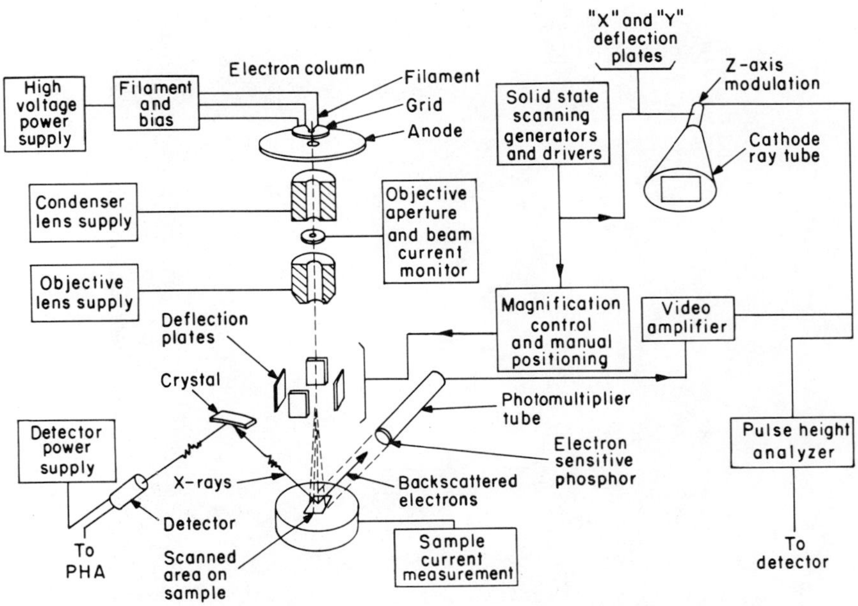

Fig. 1. Basic electron optics system with beam scanning

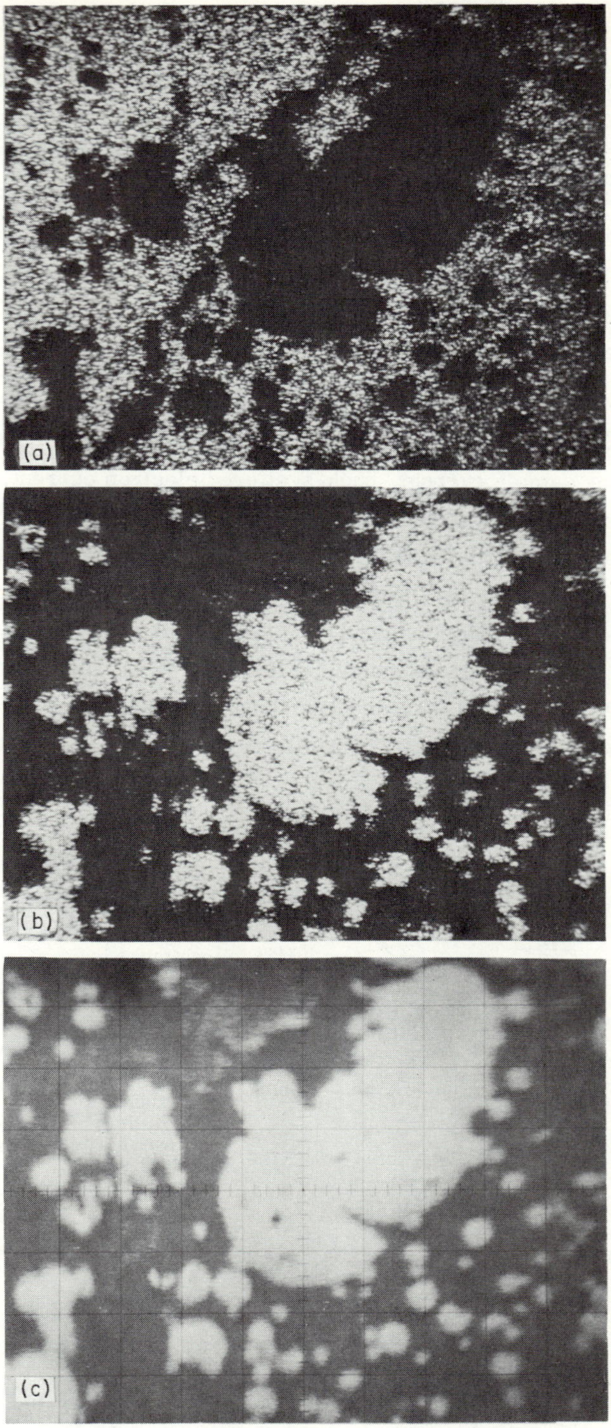

Fig. 2. Oscilloscope displays from beam scanner: (a) V-rich particles in Ti-rich matrix shown by back-scattered electrons, (b) V K_α x rays indicate V-rich areas, and (c) Ti K_α x rays indicate Ti-rich areas.

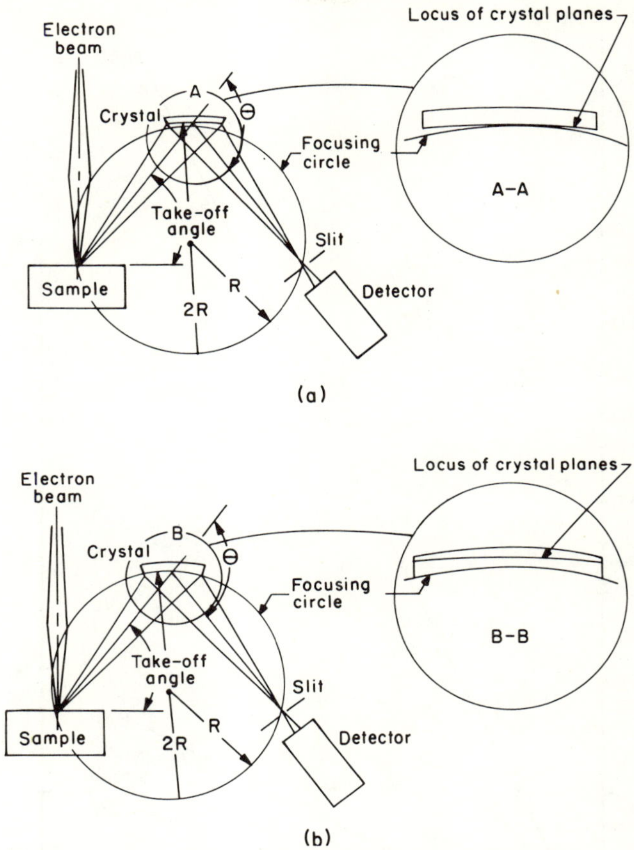

Fig. 3. X-ray optics for Johann and Johansson crystals: (a) Johann geometry and (b) Johansson geometry.

The x-ray photons produced in the target (unknown sample) are propagated in all directions. The x-ray spectrometer (two different crystal geometry designs are shown schematically in Fig. 3) is used to detect those photons emitted in a small solid angle corresponding to the projected area of the diffracting crystals, as indicated in Fig. 3. It should be noted that the solid angle does not change as a function of diffraction angle. When Bragg's law[43-49] is satisfied for a particular wavelength, diffraction of the incident radiation of that wavelength will occur.

The selection of a given set of planes in the diffracting crystal fixes, according to Bragg's law, the maximum wavelength that can be diffracted by the crystal. Because of design considerations in the mechanism whereby

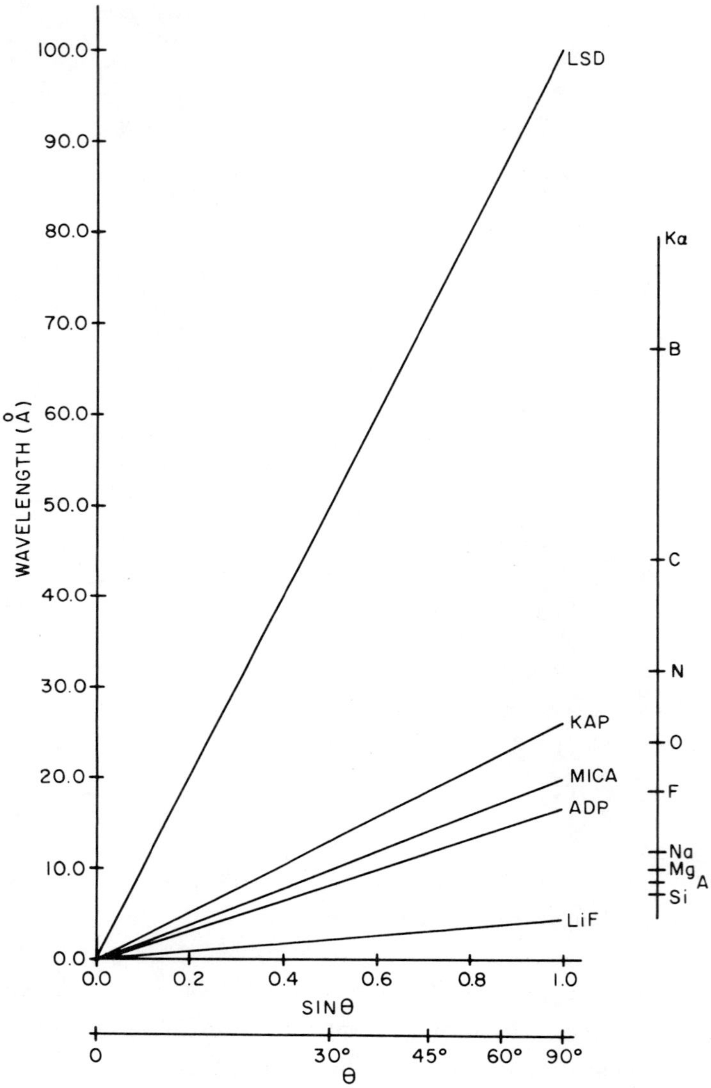

Fig. 4. Diffraction wavelength ranges for selected crystals. Data from White et al.[50].

the correct geometry is maintained between the x-ray source (on the sample surface), the diffracting crystal, and the detector slit, the maximum wavelength diffracted by a given set of planes is less than the 2d value. The relationship between the diffracted wavelength and a scale on the drive mechanism is shown for five common crystals in Fig. 4. The specific wavelengths of selected K_α lines are shown on the right-hand margin of Fig. 4.

It can be seen from this diagram that, although a lead stearate crystal can diffract all wavelengths shown, the limitations of the scanning mechanism may require that other crystals be used to cover selected portions of the spectrum.

The crowding of the characteristic wavelengths in the short wavelength region of the spectrum also requires that a crystal with greater dispersion be selected. Dispersion is a function(51) of both the interplanar spacings of the crystal (expressed in units of 2d) and the diffraction angle. In order to provide the maximum dispersion, it is necessary to select crystals with the smaller interplanar spacings. It can be observed from the figure that a crystal, such as LiF, spreads a very small portion of the spectrum over the total scanner range, whereas lead stearate covers the whole spectrum over the same scanner range. Thus the slopes of the various crystal curves are inversely proportional to the dispersion.

When a planar crystal is used, the diffracted beam continues to diverge after diffraction, and only a small portion of the diffracted beam is directed into the detector. It can be shown(52, 53) that if a crystal is given an appropriate radius of curvature in the plane of the focusing circle,* the x-rays will appear to have been specularly reflected from the crystal surface and focusing of the diffracted beam on the detector slit will occur.

There are two crystal curvature relationships. In Johann geometry(47-48, 54), the crystal is bent to conform to a radius of curvature of twice the radius of the focusing circle (see Fig. 3). The crystal will then be tangent to the focusing circle only at the center of the crystal and, therefore, the diffracted beam will not be perfectly focused on the detector slit. In Johansson geometry(47-48, 55) (see Fig. 3), the crystal is bent to the same radius of curvature as in Johann geometry (that is, twice the radius of curvature of the focusing circle). After bending, however, the crystal face is ground to conform to a radius of curvature of that of the focusing circle

---

*The focusing circle is defined by three points: (1) the intersection of the electron beam with the sample surface, (2) the center of the crystal face, and (3) the center of the detector slit or detector window.

(hence the crystal face is tangent to the focusing circle along the entire crystal length). The diffracted beam should then be perfectly focused on the detector slit and some gain in resolution and intensity[56] may be realized for a given crystal.

Gas-filled detectors are used in the proportional range[57-61] because pulse height discrimination[62-64] can be used to improve the signal-to-noise ratio by rejecting background pulses of unusually high or low energy. A second advantage is realized if a high-order reflection from a higher energy radiation is present in the same spectral region. Pulse height discrimination can be used to reject these pulses because they will occur in a different portion of the energy spectrum even though they have the correct effective wavelength for diffraction.

This relationship is shown in Fig. 5. For each element in the first-order column, the ordinate gives both the energy and the wavelength of the characteristic x ray. For the same element, its position in the higher order columns gives its effective wavelength only, because the energy of the photon still corresponds to the first-order energy. As an example, suppose an analysis for phosphorus were to be performed. Looking in the higher order columns, and the L spectra column as well, it can be seen that the second-order K_α from Sc, the third-order K_α from Mn, and the first-order L_α from Zr occur at about the same effective wavelength. The Sc and Mn K-radiations both have significantly higher energy photons than P; therefore, with a properly adjusted pulse height analyzer system, P can be analyzed in the presence of Sc and Mn. The case for Zr is different, because the L-radiation from Zr not only has the same wavelength as P, but the Zr line is a first-order line, therefore, it has the same energy. Therefore Zr L_α and P K_α cannot be separated by pulse height analysis. To identify such an interference, the K-radiation from Zr could be investigated first.

III. SPECIMEN PREPARATION

The preparation of a specimen for electron microprobe analysis must be considered an art[23, 40, 65] as opposed to a science. The main point to remember is that the x rays are generated in a volume of material that penetrates only about 2 to 10μ below the surface[66-68] of the specimen. Great care must be taken to ensure that this material represents the bulk

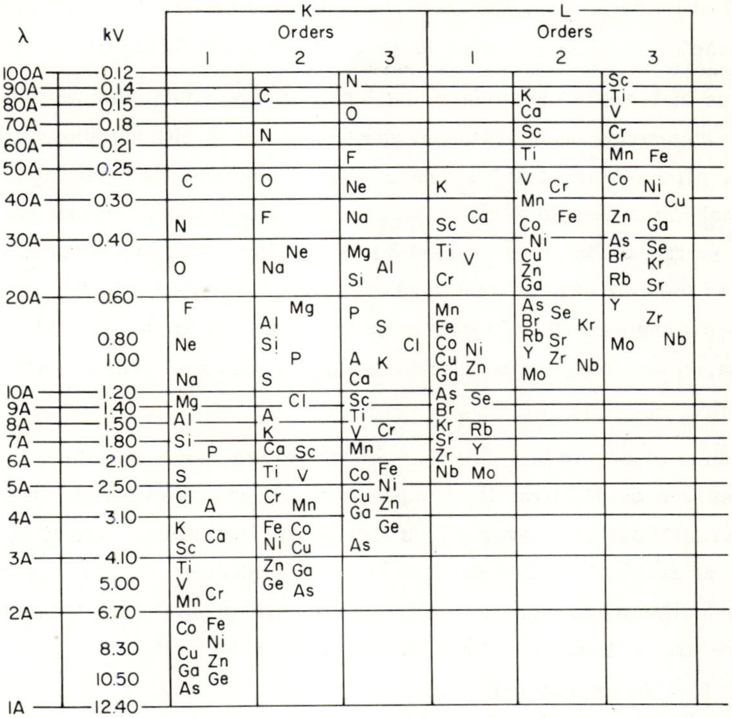

Fig. 5. Emission line wavelength as a function of diffraction order.

phase to be analyzed, both chemically and with regard to any defects or damage introduced in the grinding-polishing operations. Therefore, any treatment of the sample surface which alters the chemical composition of this region, that is, degradation of the material by grinding procedures or leaching with various fluids either deliberately by etching[69] or accidentally in polishing, will lead to erroneous results. Smearing of one phase onto another can be included in this class also.

The surface topography of the sample can become an important consideration, especially near phase boundaries. The roughness that can be tolerated must be defined in terms of the spatial resolution desired in the electron beam, the changes in take-off angle that can be tolerated by the x-ray optics, and the changes in absorption in the specimen. In analysis, the diameter of the electron beam is not nearly as important as the effective diameter from which x rays are emitted[70-71]. Features in the order of magnitude of the diameter of the beam can, therefore, be tolerated because the effective excited diameter of x-ray emission is approximately two to ten times larger.

Gross features in the surface, such as pores or voids, can be readily shown by comparing back-scattered electron displays with sample current displays. The detector for the back-scattered electrons is at an angle to the normal to the sample surface; therefore, surface imperfections can be oriented so that a shadow is present on the display. The shadow arises because of the loss of back-scattered electrons to the defect walls. Figure 6(a) shows a back-scattered electron picture of a copper/silver grid. It can be compared with the sample current display of the same grid as shown in Fig. 6(b).

Because electrons are used to excite the characteristic radiation, a path must be provided from the sample to ground if space charge buildup and erratic results are to be avoided. If the sample is metallic, mounting it in copper- (or graphite-) filled Bakelite* provides adequate electrical contact. If nonconductive mountings are used, then the silver paint used for printed circuit repair or Aquadag(72) can be used to provide a conduction path from the sample to the sample holder.

When nonconductive minerals or inorganic samples are investigated, a conductive layer of carbon or a metal, such as aluminum, copper, etc., must be plated over the sample surface. A considerable amount of information is available in the literature(40, 65) regarding successful methods. Each experiment must be approached as a unique case, and parameters must be established by trial and error.

It appears that, in some cases, a metal layer can also serve as a heat sink(73-74) and hence reduce errors generated through change of the sample composition by heating. It must be appreciated that the power densities can be quite high in a finely focused electron beam.† Two major effects can

*A typical product is Diall 1610 - Conductive Molding Compound, sold by Mesa Products - Plastics Div., 12270 Nebraska Ave., Los Angeles, Calif. 90025.

†Consider a 0.1-μA beam of 25-kV electrons focused to a 1-μ diameter at the sample surface. This represents a power density of

$$\frac{1\times10^{-7}\ \text{A} \times 2.5\times10^{4}\ \text{V}}{\pi/4\ (10^{-4})^2\ \text{cm}^2} = 3.2 \times 10^{5}\ \text{W/cm}^2$$

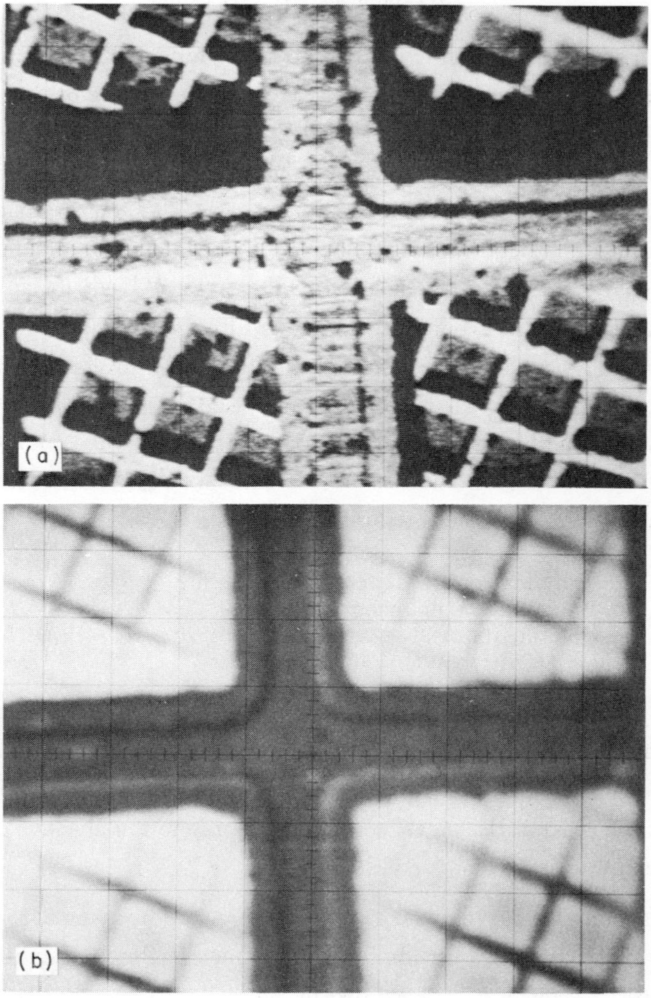

Fig. 6. Surface topography effects shown by back-scattered electrons: (a) back-scattered electron display and (b) sample current display.

result from overheating: (1) change of composition by vaporization of a constituent and (2) change in composition by a constituent diffusing into or out of the hot region depending on its relative solubility[31]. These effects will be most pronounced in materials that are relatively poor thermal conductors.

IV. QUANTITATIVE ANALYSIS

The application of the microprobe as a tool for micro-quantitative chemical analysis is probably the most outstanding feature of this instrument. The possibility of singling out a 10-μ particle (smaller in special cases where only rough data are sought or special techniques[75-76] can be used) for analysis in situ is indeed exciting. This asset would seem to make the instrument unique enough; however, an even further advantage is realized in chemical analysis through the deduction by Castaing[77-79] that the intensity of a characteristic line from a given element in a specimen of unknown composition is, in a first approximation, directly related to the intensity of the same line from the pure element. This observation and its subsequent experimental verification in selected systems have opened the door to a vast number of hitherto experimentally impossible investigations. It must be remembered, however, that the simple relation:

$$C_A = \frac{\text{intensity of characteristic line for A in unknown}}{\text{intensity of characteristic line for A in elemental sample}} = I_A/I_{(A)}$$

where C_A = the weight concentration of element A, is only valid in a first approximation[80]. Strictly speaking, this expression is correct for the radiation generated in the sample before any of the radiation is absorbed, and hence can predict the measured intensity only in the absence of significant absorption and secondary fluorescence.

As a schematic aid in understanding the overall process of x-ray excitation and transmission, consider Fig. 7. Electrons impact on the sample surface with an energy corresponding to the accelerating voltage in the electron source. If this energy is to be converted into x-ray photons, each electron must cause a series of ionizations in the solid sample, each of which in turn yields an x-ray photon in the process which returns the ion to its ground state. The total number of possible ionizations is simply equal to the total electron energy divided by the energy lost in each ionizing collision. However, there are many processes by which an electron can spend its energy and each process can be rated according to its effective "cross section" (or its relative probability of occurrence). The process of x-ray photon production is a low cross-section process, therefore, only a small percentage[47-48] of the electron energy is converted into x-ray photons. The

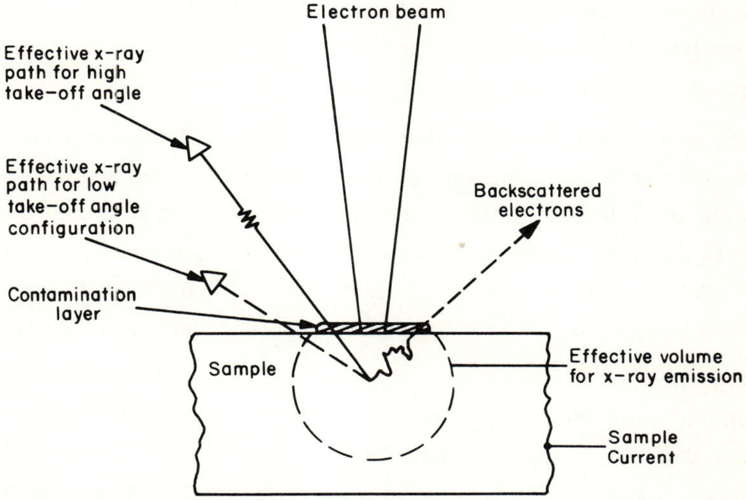

Fig. 7. Simplified representation of electron interaction with sample.

remainder can appear as heat or light (that is, the energy goes into the production of radiation in some other portion of the spectrum) or part may be carried back out of the sample by electrons suffering collisions which scatter them out of the sample.

In general then, the electrons will diffuse away from the surface along some path[66-68], leaving a trail of excited atoms behind them. Some electrons may be deflected back out of the sample at some point in their lifetime. Their energy may have been reduced by ionizing collisions, or they may carry nearly their full energy back out of the sample.

It is obvious that, in order to calculate the number of x-ray photons generated in the sample, many processes have to be considered and each must be given its proper weight. It has been observed that the "cross sections" for many processes depend on the electron environment existing around the atom. More exactly, the energy levels of the atom are perturbed by its neighbors, and the probability of removing an electron from a certain level or of its exchange between levels depends on its neighbors and their states. Attempts to provide models for this effect, the so-called "atomic number effect,"[81-83] have not progressed to the point of complete understanding, and most correlations for this effect are more or less empirical.

It has been observed that, in samples where the elements differ by only a few atomic numbers, such effects can generally be ignored[84]. When the elements differ widely in atomic number, such as iron in uranium, a significant deviation of the observed intensity from that expected on the basis of the simple linear relationship will be found[7].

X-ray photons will be propagated in all directions; those properly oriented will travel a path through the sample and exit from the surface. The sample has a "cross section" for absorbing the photon and, therefore, some of the photons, even though on the exit path, will be reabsorbed in the sample. This leads to an absorption effect[77, 85-89], and the effective calculated concentration must be increased to account for photon absorption.

The photons which are absorbed can lead to other ionization processes in the sample which yield x-ray photons in a longer wavelength portion of the spectrum, that is a fluorescent radiation appears[77, 90-94]. This does not affect the exciting radiation, except as it reduces its intensity via absorption. However, the transition that yields the fluorescent radiation may be used to characterize another element and hence would have to be separated from the radiation of this wavelength produced by electron excitation. This is the fluorescence effect, and it is observed that it can be considered to be atomic number dependent. That is, some elements will have a relationship such that one will cause fluorescence in the other, whereas other pairs or combinations will not. Fluorescence may be caused by any radiation of the correct wavelength, regardless of the transition involved. Therefore, any characteristic line used for analysis must be investigated to see if it has a contribution from fluorescence or if it causes fluorescence in some other element in the sample.

In summary, the concentration of a given element in a specimen can be shown to be related to the factors mentioned above by the following relation:

$$C_A = \frac{I_A}{I_{(A)}} \cdot \{\text{Abs}\} \cdot \{\text{ATN}\} \cdot \{\text{Fluor.}\}$$

where Abs = Correction factor for absorption of line in unknown compared to absorption of line in standard.

ATN = Correction factor for average atomic number of sample compared to average atomic number of standard.

Fluor. = Correction factor for enhancement of the observed line by fluorescence.

A discussion of the details of each correction is beyond the scope of this Chapter and, therefore, the literature cited should be examined for detailed derivations and applications of each factor.

The precision that can be expected in a given analysis is related mainly to the counting statistics and to such time-dependent factors as drift in the electronic components of the instrument[93, 95-96], contamination of the specimen surface[97-98], and evaporation or migration of a constituent. The contamination of the sample surface by carbon deposited through an interaction of the electron beam and the residual gases in the vacuum system leads to an absorption of electrons (of some importance) and an absorption of x-rays (of minor importance). The contamination can build up quite rapidly at high beam currents; however, each case must be considered independently (such factors as the vacuum level achieved and contaminants are important) if a correction is to be applied.

The level of accuracy that can be achieved in a given analysis appears to depend mainly on the homogeneity of the unknown sample (and standard) and also on the relative concentration of the analyzed element in the standard sample and in the unknown sample. For example, when elemental standards are used, the accuracy of the determination tends to decrease as the concentration of the analyzed element decreases in the unknown. Even though corrections exist for this effect, it is doubtful, with the present state of the art, that accuracies better than ±5% of the amount present can be obtained in a routine manner.

By selection of the data sources (for mass absorption coefficients, fluorescent yields, sigma factors, alpha factors, etc.) which give the best calculated concentrations on samples whose composition is known (from wet chemistry or other methods), a level of ±1% of the amount present is possible. However, this method appears to be equivalent to determining a calibration curve over the range of concentration of the unknown element.

Perhaps, as better data for the various factors becomes available, and as the models for the correction terms are refined, it will be possible to obtain 1% accuracy on a routine basis.

If a sample is not homogeneous at the micron level (few materials appear to be completely homogeneous at this level)[1], the electron microprobe results can be influenced by the nonhomogeneous regions. The x rays are emitted over an approximately hemispherical region* which has a radius of several microns. If a concentration gradient exists for the element being analyzed, i.e., a grain boundary or any other type of steep concentration gradient, then investigations too close to the discontinuity may lead to results either far too high or far too low[93,95].

Unless the sample is quite refractory, the analysis may not truly be nondestructive. As indicated previously, the amount of heat generated in the sample can be quite high, and decomposition, phase changes, melting, diffusion, etc., can easily be induced.

In this vein, the analysis of glass samples has been particularly troublesome[31]. Because the analysis time must be chosen in order to achieve sufficient statistical precision, and the count rates are low for the light elements, long counting times cannot be avoided and high sample currents are required.

V. SPECIAL TECHNIQUES

The analysis of the intensity relations does not constitute the only profitable area in microprobe analysis.

For analytical purposes, the wavelength of the characteristic radiation of an element can generally be considered as fixed. This is a satisfactory assumption as long as the energy levels involved in the transition do not participate in the chemical bond. However, when elements in the first period are investigated, and also when the transitions involve L and M levels in heavy elements, the emission wavelength can be changed through the influence of the atom's neighbors. The wavelength shift has been correlated with several parameters[100-106], the coordination number, the electron transfer in the bonding process, and the small chemical variations involved in nonstoichiometry in some compounds. Much of this work is relatively recent, and theories explaining such behavior are very tentative.

The most promising aspect of this work for quantitative analysis is the observation that wavelength shifts can be correlated to very small compositional changes[42, 108], much below the level of change that can be observed by changes in the line intensity.

*The shape of the region is related to the sample density and the electron accelerating voltage and can be far from hemispherical in some cases[68].

Trace analysis can be considered under special techniques because the most sophisticated method[109] uses elaborate auxiliary equipment. The limit on trace analysis is provided by the smallest signal-to-noise ratio that can be resolved. Ultimately, the question that must be answered is how long can experimental variables be held reasonably constant. The more variables that can be freed of time variation, the more sensitive the method can be made. Analyses of iron and nickel in a uranium matrix in the 10-40 ppm range have been reported. Short term stability (in the order of 10 sec) is easily achieved, and, therefore, a complete single analysis is made in this short time interval. By accumulating the information from a series of the short runs for periods as long as 24 hr, the desired level of precision can be reached.

Fluorescence (cathodoluminescence)[110-113] in the optical spectrum is often observed under the electron beam in the microprobe. This often aids the operator in locating and focusing the beam. This effect is being used to study carrier lifetimes in semiconductors and other related effects. Because this effect appears to be very sensitive to small concentration changes, it could probably be used in trace analysis and studies involving small concentration gradients.

Studies involving sample current changes have indicated several interesting applications. One application[114] leads to a quantitative analysis scheme similar to that which can be obtained from the x-ray emission lines. The other[115] is used to study electronic microcircuits in the "active" and "inactive" states; that is, when current is being passed or is not being passed in the circuit. This has indicated immediately the trouble spots in various circuits and has indicated where the emission analysis should begin.

BIBLIOGRAPHY

A. General

A comprehensive bibliography on electron probe microanalysis and related subjects has been prepared (the last supplement dated January, 1965) by Dr. K. F. J. Heinrich. Copies of the bibliography can be obtained by writing to Dr. Heinrich, Spectrochemical Analysis Section, Analytical Chemistry Division, National Bureau of Standards, Washington, D. C. 20234.

Reference books for microanalysis can be classified under three categories:

B. Electron Probe Microanalysis and Related Subjects from Books

a. L. S. Birks, Electron Probe Microanalysis, Chemical Analysis, Vol XVII, Interscience, New York (1963).

b. H. A. Elion, Instrument and Chemical Analysis Aspects of Electron Microanalysis and Macroanalysis, Progress in Nuclear Energy Series IX, Analytical Chemistry, Vol 5, Pergamon, New York (1966).

c. R. Theisen, Quantitative Electron Microprobe Analysis, Springer-Verlag, New York (1965).

C. Electron Probe Microanalysis from Conferences and Symposia

a. Advances in X-ray Analysis, Vol. 1 (1957) to Volume 14, Plenum, New York, 1970.

b. ASTM Special Technical Publication No. 317, Symposium on Advances in Electron Metallography and Electron Probe Microanalysis (1962).

c. ASTM Special Technical Publication No. 349, Symposium on X-ray and Electron Probe Analysis (1964).

d. T. D. McKinley, K. F. J. Heinrich, and D. B. Whittry, eds., The Electron Microprobe, Proceedings of a Symposium, (1964), Wiley, New York (1966).

e. A. Engstrom, V. Cosslett, and H. Pattee, eds., X-ray Microscopy and X-ray Microanalysis, Proceedings of the Second International Symposium (Stockholm, 1960), Elsevier, Amsterdam, (1960).

f. X-ray Optics and X-ray Microanalysis, Proceedings of the Third International Symposium (Stanford, 1962), Academic, New York (1963).

g. K. F. J. Heinrich, ed., Quantitative Electron Probe Microanalysis, NBS Special Pub. 298, (1968).

D. X-Ray Theory and X-Ray Spectrochemical Analysis from Books

a. L. S. Birks, X-ray Spectrochemical Analysis, Chemical Analysis, Vol XI, Interscience, New York (1959).

b. M. A. Blokhin, Methods of X-ray Spectroscopic Research, Pergamon, New York (1965).

c. H. A. Liebhafsky, H. G. Pfeitter, E. H. Winslow, and P. D. Zemany, X-ray Absorption and Emission in Analytical Chemistry, Wiley, New York (1960).

REFERENCES

(1) C. W. Mead and M. E. Morse, First National Conference on Electron Probe Microanalysis (1966) Paper No. 25.

(2) H. Yakowitz, et al., Adv. X-ray Analysis, 9, 289 (1966).

(3) B. L. Henke, Adv. X-ray Analysis, 7, 460 (1964).

(4) D. W. Fischer and W. L. Baun, Adv. X-ray Analysis, 5, 460 (1964).

(5) J. B. Wittry and R. Fitzgerald, Adv. X-ray Analysis, 5, 538 (1962).

(6) K. F. J. Heinrich, Adv. X-ray Analysis, 6, 291 (1963).

(7) D. B. Wittry, Adv. X-ray Analysis, 3, 197 (1960).

(8) J. W. Colby, Adv. X-ray Analysis, 8, 352 (1965).

(9) P. Pietrokowsky and J. R. Maticich, C-f, p. 591.

(10) A. V. Manzione, F. J. Huegel, and D. E. Fornwalt, Adv. X-ray Analysis, 7, 353 (1964).

(11) L. S. Birks and R. E. Seebold, J. Nucl. Mater., 3, 260, 249 (1961).

(12) J. Philibert, J. Inst. Metals, 90, 241 (1962).

(13) D. A. Melford, J. Inst. Metals, 90, 217 (1962).

(14) B. R. Banerjee and W. D. Bingle, C-d, p. 653.

(15) D. A. Melford, C-f, p. 577.

(16) T. E. Everhart, C-d, p. 665.

(17) P. Lublin and W. J. Sutkowski, C-d, p. 677.

(18) D. F. Kyser and D. B. Wittry, C-d, p. 691.

(19) J. N. Ramsey and P. Weinstein, C-d, p. 715.

(20) C. C. Nealy, et al., C-d, p. 748.

(21) C. A. Anderson, K. Keil, and B. Mason, Science, 146, 256 (1964).

(22) A. K. Baird and D. H. Zenger, Adv. X-ray Analysis, 9, 487 (1966).

(23) A. K. Temple, et al., C-d, p. 784.

(24) J. V. Smith, J. of Geology 73, 830 (1965).

(25) J. V. Smith, J. of Geology 74, 1 (1966).

(26) J. V. Smith and P. H. Ribbe, J. of Geology, 74, 197 (1966).

(27) P. H. Ribbe and J. V. Smith, J. of Geology, 74, 217 (1966).

(28) R. V. Adams, H. Rawson, D. G. Fisher, and P. Worthington, Glass Technology, 7, 98 (1966).

(29) M. P. Borom and J. A. Pask, J. Amer. Cer. Soc., 49, 1 (1966).

(30) J. R. Ryan, E. Rah, and C. Clark, Cer. Bull., 45, 260 (1965).

(31) R. V. Adams, et al., Glass Technology, 7, 99 (1966).

(32) A. K. Varshneya, A. R. Cooper, and J. McCable, Appl. Phys., 37, 2199 (1966).

(33) J. R. Shappiro, Adv. X-ray Analysis, 9, 304 (1966).

(34) D. E. Fornwalt, et al., C-d, p. 581.

(35) J. E. Holliday, Adv. X-ray Analysis, 9, 365 (1966).

(36) R. C. Mellors, et al., C-d, p. 834.

(37) K. C. Hoerman, et al., Norelco Reporter XIV, 69 (1967).

(38) J. V. P. Long and H. Rockert, C-f, p. 513.

(39) A. J. Tousimis, C-f, p. 539.

(40) H. Schreiber, Adv. X-ray Analysis 8, 363 (1965).

(41) D. H. Speidel, Adv. X-ray Analysis 9, 259 (1966).

(42) D. B. Wittry, Adv. X-ray Analysis 3, 185 (1960).

(43) A. H. Compton, and S. K. Allison, X-rays in Theory and Practice, 2nd ed. D. Van Nostrand Co., Inc., (1953).

(44) M. J. Buerger, X-ray Crystallography, Wiley, New York, (1942).

(45) H. P. Klug and L. E. Alexander, X-ray Diffraction Procedures, Wiley, New York, (1954).

(46) B. D. Cullity, Elements of X-ray Diffraction, Addison-Wesley Publishing Company, Inc. (1956).

(47) L. S. Birks, B-a.

(48) H. A. Elion, B-b.

(49) W. L. Bragg, ed., The Crystalline State, Bell, London (1933).

(50) E. B. White, et al., X-ray Wavelengths, 2nd ed. Special Publication No. 3-64, Mineral Industries Experiment Station, The Pennsylvania State University, University Park.

(51) A simple picture of dispersion relations can be obtained by inspection of the equation $n\lambda = 2d \sin\theta$ where n = order of diffraction, λ = diffracted wavelength, d = interplanar spacing of diffracting crystal, and θ = diffraction angle. Differentiating and assuming d is a constant yields $n d\lambda = 2d \cos\theta \, d\theta$ or, in dispersion form $\frac{d\theta}{d\lambda}$ = dispersion = $n/2d \cos\theta$. In summary, as d increases, dispersion decreases; as θ increases, dispersion increases; as n increases, dispersion increases.

(52) M. deBroglie and F. A. Lindemann, Compte Rendue, 158, 944 (1914).

(53) J. W. M. Dumond and H. A. Kirkpatrick, Rev. Sci. Inst., 1, 88 (1930).

(54) H. H. Johann, Z. Phys., 69, 185 (1931).

(55) T. Johansson, Naturwissenschaffen, 20, 758 (1932).

(56) M. P. Johnson, et al., Adv. X-ray Analysis, 8, 259 (1964).

(57) W. R. Kiley, Norelco Reporter VII (6), 143 (1960).

(58) D. H. Wilkinson, Ionization Chambers and Counters, Cambridge University Press (1950).

(59) P. H. Dowling, C. F. Hendee, T. R. Kohler, and W. Parrish, Norelco Reporter IV(2), 23, (1957).

(60) K. F. J. Heinrich, D. Vieth, and H. Yakowitz, X-ray Analysis, 9, 208 (1966).

(61) S. L. Bender and E. J. Rapperport, C-d, p. 405.

(62) D. C. Miller, Norelco Reporter IV(2) 37 (1957).

(63) K. F. J. Heinrich, Adv. X-ray Analysis 4, 370 (1961).

(64) R. M. Dolby, Proc. Phys. Soc., 73, 81 (1959).

(65) J. V. P. Long, C-f, p. 279.

(66) G. D. Archard, J. Appl. Phys., 32, 1505 (1961).

(67) G. D. Archard and T. Mulvey, Brit, J. Appl. Phys., 14, 626 (1963).

(68) P. Duncumb and P. K. Shields, Brit. J. Appl. Phys., 14, 617 (1963).

(69) M. Feller-Kniepmier and H. H. Uhlig, Geochim. Cosmochim. Acta, 21, 257 (1961).

(70) D. B. Wittry, Jour. Appl. Phys., 29, 1543 (1958).

(71) C. A. Andersen, C-d, p. 58.

(72) Aquadag is a commercial preparation of colloidal graphite which can be obtained from Acheson Colloids Co., Port Huron, Mich.

(73) G. S. Almasi, et al., Jour. Appl. Phys., 36, 1848 (1965).

(74) B. Gale and K. F. Hale, Brit. J. Appl. Phys., 12, 115 (1961).

(75) E. J. Rapperport, Contribution No. 38, First National Conference on Electron Probe Microanalysis, College Park, Md. (1966).

(76) L. S. Birks, D-a, p. 115.

(77) R. Castaing, Thesis, University of Paris, O. N. E. R. A. Publ. 55, (1951).

(78) R. Castaing, Adv. X-ray Analysis, 4, 351 (1961).

(79) R. Castaing, Adv. Electronics and Electron Physics, 13, p. 317, Academic, New York (1960).

(80) J. D. Brown, Adv. X-ray Analysis, 7, 340 (1964).

(81) G. D. Archard and T. Mulvey, C-f, p. 393.

(82) D. M. Poole and P. M. Thomas, C-f, p. 411.

(83) D. M. Poole and P. M. Thomas, C-f, p. 269.

(84) L. S. Birks, B-a, p. 109.

(85) R. Castaing and J. Descamps, J. Phys. Radium, 16, 304 (1955).

(86) L. S. Birks, J. Appl. Phys., 33, 233 (1962).

(87) M. Green, C-f, p. 361.

(88) J. Philibert, C-f, p. 379.

(89) P. Duncumb and P. K. Shields, C-d, p. 284.

(90) D. B. Wittry, Ph. D. Thesis (1957), California Institute of Technology, Pasadena, California.

(91) L. S. Birks, J. Appl. Phys., 32, 387 (1961).

(92) P. Duncumb and P. K. Shields, C-f, p. 329.

(93) R. R. Dils, L. Zeitz, and R. A. Huggins, C-f, p. 341.

(94) L. S. Birks, et al., C-d, p. 199.

(95) S. J. B. Reed and J. V. P. Long, C-f, p. 317.

(96) R. Fitzgerald, Adv. X-ray Analysis, 7, 369 (1964).

(97) A. E. Ennos, Brit. J. Appl. Phys., 4, 101 (1953).

(98) J. Hillier, J. Appl. Phys., 15, 663 (1944).

(99) A. J. Campbell and R. Gibbons, C-d, p. 75.

(100) D. E. Day, Nature, 200, 649 (1963).

(101) W. L. Baun and D. W. Fischer, Nature, 204, 642 (1964).

(102) E. W. White, et al., Proceedings, Seventh Annual Conference on Industrial Applications of X-ray Analysis, Univ. of Denver, 1958.

(103) D. M. Koffman and S. H. Moll, Adv. X-ray Analysis, 9, 323 (1966).

(104) D. W. Fischer and W. L. Baun, Adv. X-ray Analysis, 9, 329 (1966).

(105) D. W. Wilbur and J. W. Gofman, Adv. X-ray Analysis, 9, 354 (1966).

(106) B. L. Henke, Adv. X-ray Analysis, 9, 430 (1966).

(107) J. E. Holliday, C-d, p. 3.

(108) L. A. Ferguson, Adv. X-ray Analysis, 9, 265 (1966).

(109) D. F. Kyser and D. B. Wittry, C-d, p. 691.

(110) D. B. Wittry and D. F. Kyser, J. Appl. Phys., 38, 375 (1967).

(111) D. F. Kyser, J. McCoy, and D. B. Wittry, 2nd National Conference on Electron Microprobe Techniques, Boston, Mass., Paper #50 (1967).

(112) R. T. Greer and E. W. White, 2nd National Conference on Electron Microprobe Techniques, Boston, Mass., Paper #51 (1967).

(113) K. F. J. Heinrich, Adv. X-ray Analysis, 7, 325 (1963).

(114) J. N. Ramsey and P. Weinstein, C-d, p. 715.

(115) C. C. Nealy, et al., C-d, p. 748.

Chapter 18

QUANTITATIVE STEREOLOGY

R. T. DeHoff

University of Florida, Gainesville, Florida

I. INTRODUCTION

The problem of the quantitative characterization of structure on a microscopic scale occupies an important role in many diverse areas of applied science. It is generally true that, while information about some three-dimensional structure is required, the method of observation necessarily yields a sample of the structure which is either a two-dimensional section, in the case of opaque materials, or a thin slab of the structure, in the case of transparent materials.* The problem of estimating geometric properties of three-dimensional structures from quantitative measurements made upon a sample which is essentially two dimensional is a purely geometric problem. The body of information which deals with this interdisciplinary problem has been called "quantitative stereology."

This chapter is essentially a catalog of the relationships that exist in the field. There is no attempt to derive or justify these relationships; the reader interested in such detail may consult the references supplied with each relationship.

A large number of geometric properties of three-dimensional microstructures can be estimated, and a proportionately large number of two-dimensional measurements are defined. Accordingly, it is necessary to create a list of symbols which are both descriptive and simple. The International Society for Stereology has recommended such a notation(1). Table I summarizes these recommendations. Efficient use of the information contained in this chapter requires some degree of familiarity with these symbols.

Observation of opaque materials requires reflection microscopy; observation of transparent materials employs transmission microscopy. In reflection microscopy, the structure observed is, at least ideally, a two-dimensional section through the volume of the structure under study. Contrast between various constituents is usually based upon preferential chemical attack on a polished surface of the material. In transmission microscopy, the projection of all of the structure contained within a thin slab of the material is observed; contrast is based upon differences between the absorption

*Many materials which are opaque to the visible light spectrum are transparent to electrons, so that thin film samples of metals, for example, may be the subject of analysis in transmission electron microscopy.

of the various constituents in the system.* Because the sampling procedure reveals a different aspect of the structure, the mathematical relationships involved in each case are somewhat different. Accordingly, reflection and transmission microscopy will be treated separately.

Within these two categories, the relationships are presented in increasing order of the number of limiting geometric assumptions which are necessary to their application. If a relationship requires no geometric assumptions, and into this category fall the most useful of the results of the field, then only relatively simple statistical assumptions must be satisfied. In every case, it is necessary that the sample of the structure which is observed be a representative sample[(3)], i. e., that it contain a uniform representation of structural gradients and anisotropy that may exist. If these conditions are satisfied, then the measurement of the appropriate quantity on the two-dimensional image provides an unbiased estimate of the corresponding three-dimensional geometric property.

II. QUANTITATIVE REFLECTION MICROSCOPY

Microscopic observation of an opaque sample yields information about a two-dimensional section, or cut, through the three-dimensional structure under study. The following geometric properties of the structure may be estimated from measurements made upon such a section:

a. No limiting geometric assumptions:
    1. Volume fraction V_V of each phase;
    2. The surface area per unit volume, S_V, of each kind of interface;
    3. Total length of lineal features in unit volume, L_V;
    4. The mean chord intercept of particles, $\bar{\lambda}$;
    5. Interfacial contiguity in multiphase structures;
    6. Interfacial anisotropy;
    7. Particle size anisotropy;
    8. Average mean surface curvature.

b. Particulate systems in which all particles have the same shape:
    1. All of the properties listed is part (a);

*In the application of transmission electron microscopy to the study of imperfections in crystals, contrast is based upon diffraction effects[(2)] (see Chapters 6 and 7).

2. The particle size distribution;
3. The total number of particles in unit volume, N_V.

c. Particulate systems in which all the particles are the same size and shape:
   1. All of the properties listed in parts (a) and (b).

The introduction of limiting geometric assumptions, as in parts (b) and (c), generally reduces the effort that must be expended to obtain a given amount of information. However, it is emphasized that the results of studies based upon geometric assumptions are necessarily only as valid as the assumptions themselves.

III. GEOMETRIC PROPERTIES THAT CAN BE EVALUATED WITH NO GEOMETRIC ASSUMPTIONS

A. Volume Fraction V_V of Each Phase in a Multiphase Structure

There exist three relationships which permit the estimation of the volume fraction.

a. The fraction of the area of a representative section, A_A, which is occupied by the phase under study, see Fig. 1(a), is an unbiased estimator of the volume fraction of that phase[4]:

$$\overline{A_A} = V_V \tag{1a}$$

b. If a representative section is uniformly sampled with a test line, the fraction of the line length which is occupied by the phase under study, L_L, see Fig. 1(b), is an unbiased estimator of the volume fraction[4, 5];

$$\overline{L_L} = V_V \tag{1b}$$

c. If a representative section is uniformly sampled with a test point (or, alternatively, with a set of test points distributed on some grid superimposed upon the section), the fraction of test points which lie within the phase under study, P_P, see Fig. 1(c), is an unbiased estimator of the volume fraction[4, 6]:

$$\overline{P_P} = V_V \tag{1c}$$

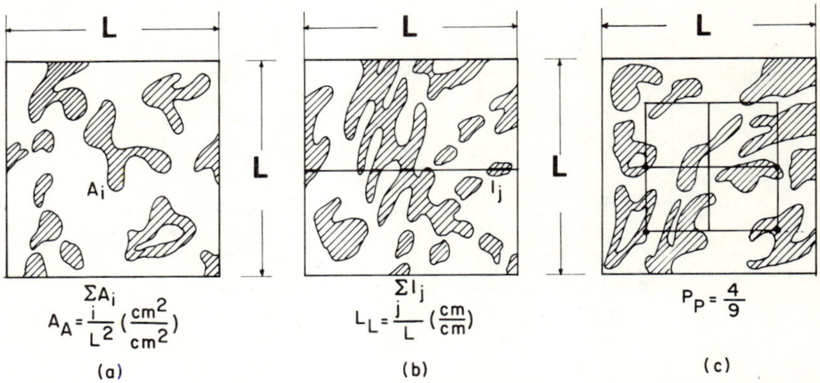

Fig. 1. Illustration of three basic methods for estimating the volume fraction of a phase: (a) the area fraction, (b) the lineal fraction, and (c) the point count.

Equation (1a) is chiefly of historical interest; this method of evaluation of the volume fraction was used in the beginnings of quantitative microscopy. Equation (1b) is used in conjunction with automatic scanning devices[7,8], which traverse the sample area with a linear scan. Hilliard and Cahn[6] have shown that Eq. (1c) provides the most efficient method for estimating the volume fraction, and its use is recommended in all manual measurements of V_V.

B. Total Interfacial Area in Unit Volume, S_V

If a representative metallographic section is sampled with an array of test lines that are uniformly distributed in position and orientation, the ratio N_L of the number of intersections made by test lines in this array with the traces of interface in the system to the total length of test lines constructed, see Fig. 2(a), is an unbiased estimator of the total area of interface in unit volume of structure[4,9,10]:

$$\overline{N_L} = \frac{1}{2} S_V \tag{2a}$$

The test lines in this case need not be straight lines, so long as their total length is known. In a structure possessing anistropy in surface area, the requirement that the test lines be uniformly distributed in orientation may be at least partially satisfied by using a circle for a test line. This effectively averages over orientation, at least in the plane of polish.

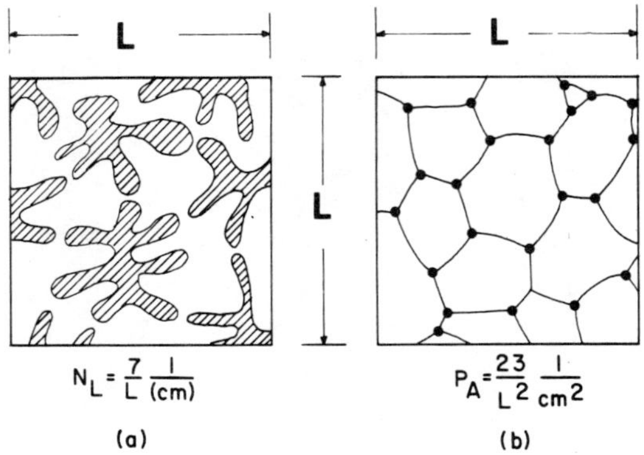

Fig. 2. Illustration of unit measurements required to (a) estimate the total interface area with the line intercept count, and (b) the total length of grain boundary edges in a structure, with the area intercept count.

C. Total Length of Lineal Features in Unit Volume, L_V

Cell structures (polycrystalline metals, plants, and animals) consist of a three-dimensional array of contiguous polyhedra. On a polished section through a cell structure, the edges of these polyhedra appear as points of intersection of traces of the surfaces bounding the cells, see Fig. 2b. The total number of such points of emergence contained in a sample, divided by the area of the sample, P_A, is an unbiased estimator of the total length of cell edges in unit volume[4, 9]:

$$\overline{P}_A = \frac{1}{2} L_V \tag{3}$$

This relationship may be applied to any other lineal feature (e.g., dislocations in crystalline materials), provided the points of emergence of the lineal features on the sectioning plane may be identified.

D. Mean Chord Intercept of Particles, $\overline{\lambda}$

The lineal fraction of a phase, L_L, is the total length of phase traversed by unit length of test line. The $\frac{1}{2} N_L$ is the number of particles intersected in unit length of test line.* The ratio of these two quantities is the mean length of particle intercepts:

$$\overline{\lambda} = \frac{2\overline{L_L}}{\overline{N_L}} \tag{4a}$$

Since the point count P_P gives the same number as the lineal fraction L_L (compare Eq. (1b) and (1c),

$$\overline{\lambda} = \frac{2\overline{P_P}}{\overline{N_L}} \tag{4b}$$

This result may be applied to each phase in a multiphase structure, by inserting the corresponding values of P_P and N_L. In a two-phase structure consisting of particles dispersed in a matrix, the mean chord intercept of the matrix phase is also the mean free path between particles[11].

E. Contiguity in Multiphase Structures

A structure composed of two phase, α and β, may have three species of interfaces: $\alpha\beta$ interphase boundaries, and $\alpha\alpha$ and $\beta\beta$ grain boundaries. The surface area of each of these three kinds of boundaries may be independently determined by applying Eq. (2) to each specie in turn. This requires separate counts of $N_{L\alpha\beta}$, $N_{L\alpha\alpha}$, and $N_{L\beta\beta}$. These three quantities may, in turn, be combined to yield a contiguity parameter for each phase. This quantity has been defined[12,13] as the ratio of the area of cell boundaries which are contiguous to the same phase to the total area of cell boundaries of that phase. This, for the phase,

$$C_\alpha = \frac{2S_{V\alpha\alpha}}{2S_{V\alpha\alpha} + S_{V\alpha\beta}}$$

---

*Since N_L is defined as a count of intersections with particle boundary, in an array of dispersed particles each particle intercept is equivalent to two boundary intersections. If the structure is partially contiguous, the value of N_L appropriate to this measurement is N_L (eff) = N_L (unlike boundaries) + $2N_L$ (contiguous boundaries).

Expressed in terms of the counting measurements involved,

$$C = \frac{2N_{L\alpha\alpha}}{2N_{L\alpha\alpha} + N_{L\alpha\beta}} \tag{5}$$

A similar definition holds for the β phase in a two-phase structure. In a multiphase structure, a contiguity parameter may be defined for each phase.

In principle, Eq. (2) may be applied to any portion of interface in the system, provided it is uniquely identifiable on a section, for example, it is possible to measure independently the total grain boundary area and twin boundary area in a one-phase system(14). This principle has also been applied to the estimation of total surface area of the interparticle welds that form in lightly sintered materials(15).

F. Interfacial Anisotropy

Structures which are the result of directional solidification or deformation are common in the materials sciences. In such systems there may be some tendency for interfaces to be aligned in a more or less complex fashion. Figure 3 is an example of an exaggerated case of interfacial anisotropy. Let such a structure be sampled with an array of test lines that are uniformly distributed in position, but all have the same orientation. A count of the number of intersections per unit length on this array of test lines provides an unbiased estimate of the total projection of all interface in the system upon a plane whose orientation is perpendicular to the direction of the test lines(16):

$$N_L \text{ (directed)} = S_V \text{ (total projection)} \tag{6}$$

A map of anisotropy of interface in a structure may be generated by making an N_L count for a set of different directions of test line arrays, and plotting the result in spherical coordinates. In two dimensions, such a plot is called the "rose-of-the-number-of-intersections"(4). In three dimensions, this plot would resemble a pole figure.

For a complete discussion and classification of the kinds of interfacial anisotropy that a system might possess, see Ref. (16).

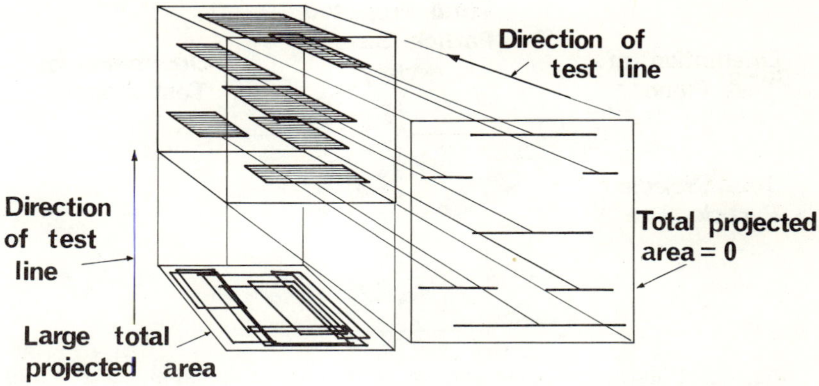

Fig. 3. Illustration of a totally anisotropic array of interface. A test line oriented vertically measures the total projected area, upon a horizontal plane. A horizontally oriented test line intercepts no interface.

G. Particle Size Anisotropy

If a structure consisting of separate particles distributed in a maxtix is directionally deformed, or solidified, the particles will tend to be elongated or flattened in certain directions. A count of the number of particle sections observed upon a plane of polish, divided by the area of the plane included in the observation, defined as the N_A count, is related to the number of particles in unit volume of structure, and to the average particle dimension in a direction which is perpendicular to the polished section:

$$N_A \text{ (directed)} = N_V \overline{D_V} \text{ (directed)} \quad (7)$$

The "directed average particle size," $\overline{D_V}$ (directed), is obtained by measuring the distance between parallel tangent planes in a direction perpendicular to the sectioning plane, and averaging over all of the particles in the structure. An extreme example of particle size anisotropy, and the relation of the oriented N_A count to the anisotropy of particle dimensions is illustrated in Fig. 4.

H. Average Mean Surface Curvature

Surface curvature is defined in terms of the two principal normal curvatures, κ_1 and κ_2, at any point on the surface[17]. The mean surface curvature H is the average of these two quantities and is a geometric

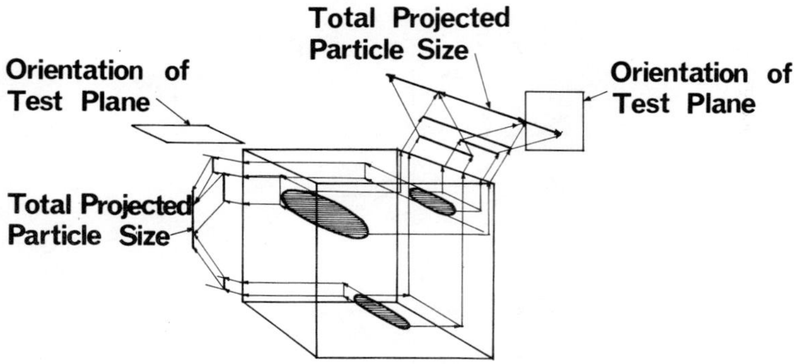

Fig. 4. Illustration of anisotropy in particle dimension. The "total projected particle size," as measured by the N_A count, is shown for two orientations of test lines.

property of fundamental importance in systems in which capillary effects are important:

$$H = \frac{1}{2} (\kappa_1 + \kappa_2)$$

In an arbitrary system, H varies from point to point on the surface of the system. It thus has a distribution of values, and some average value:

$$\overline{H} \equiv \frac{\iint_S H(u, v)\, dS(u, v)}{\iint_S dS(u, v)}$$

where the integration is over the surface area involved. It can be shown that[18,19]:

$$\overline{H} = \pi \frac{N_{Anet}}{N_L} \tag{8}$$

where N_{Anet} is the difference between the number of closed loops of particle outline in unit area that enclose particle sections, and the number that enclose holes in the particle sections. Figure 5 illustrates the counting procedure for a complex structure. N_L has the same meaning as defined for Eq. (2).

It is possible to evaluate $\overline{H}$, the average mean surface curvature, for identifiable parts of interfaces in a structure; e.g., for interparticle welds

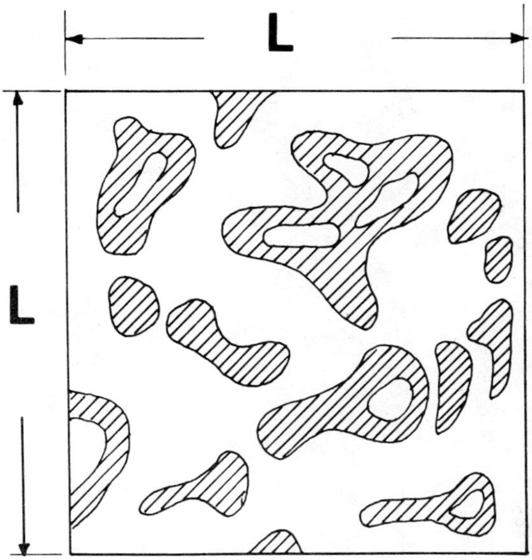

$$N_{A+} = \frac{12.5}{L^2}; \quad N_{A-} = \frac{6.5}{L^2}$$

$$N_{A\,net} = \frac{5}{L^2}$$

Fig. 5. Illustration of the method for determining the net N_A count for a structure containing both positive and negative loops.

in sintering, by substituting for the N_A count a count of the number of tangents formed by a sweeping test line with those segments of particle outline which are under study. For a complete exposition of this method, consult Ref. (18).

The relationships presented in this section may be applied to structures of arbitrary geometry; their validity requires only that a representative sample of the structure be uniformly sampled. All of the information dealt with in this section may be determined from the three simple counting measurements, P_P, N_L, and N_A (or P_A) applied to appropriate parts of the structure, or to the structure as a whole.

I. Evaluation of Geometric Properties in Particulate Systems in Which all Particles Have the Same Shape

There remain three aspects of the geometry of a microstructure which cannot be determined from the counting measurements alone, without additional assumptions: particle shape, particle size distribution, and the number

Fig. 6. A partially sintered aggregate of spherical copper powder, illustrating the wide variety of section shapes that result in this multiply connected structure.

of particles, N_V. Of these, only the latter has rigorous meaning for an arbitrary structure. It is not possible to give mathematical precision to the shape or size of particles unless it is assumed that all of the particles have the same shape, and that this shape is simple enough to be described mathematically. Accordingly, in developing methods for estimating particle size and shape, it has been necessary to introduce the assumption of constant shape.

J. Estimation of Particle Shape

In a qualitative description of particle shape, it is essential to take account of the fact that microscopic observation produces a two-dimensional section through a structure. If the three-dimensional particles are disc-shaped (e. g., flakes), most sections will be elongated. If, in three dimensions, the particles are cigar-shaped (e. g., needles), sections will produce mostly circular, or at least equi-axed, particle traces. More complex three-dimensional shapes will produce a greater degree of variety in the shapes of the sections that result. Examples of this variety are shown for a partially sintered structure, Fig. 6, and for a dendritic structure, Fig. 7.

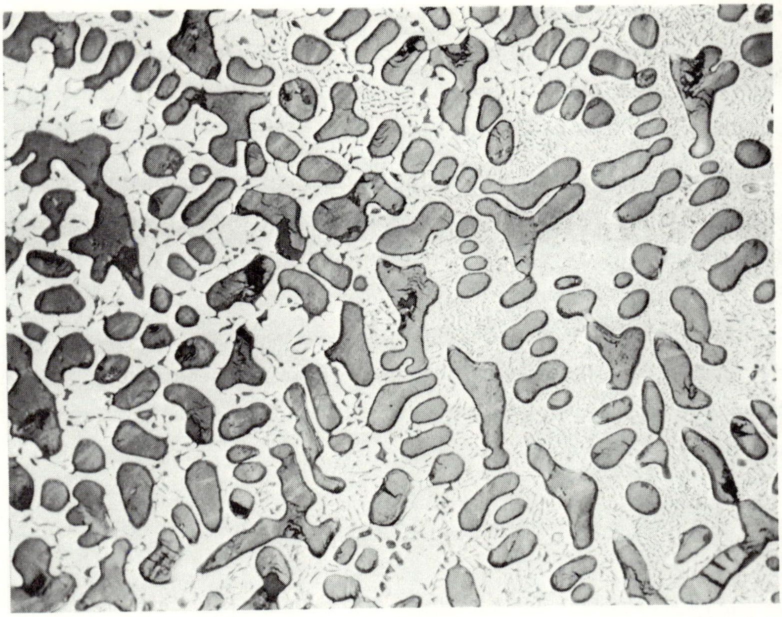

Fig. 7. Illustration of the broad range of section shapes characteristic of a dendritic structure.

In estimating particle size distribution and number of particles with the methods presently available, it is necessary that the particle shape be constant, and also that it be quantitatively determined. For ellipsoidal particles, whether elongated or flattened, this requires the evaluation of the axial ratio of the generating ellipse. As the particle shape becomes more complex, it becomes less likely that the assumption that all of the particles are the same shape is valid. Accordingly, application of the procedures presented in the following sections is restricted to ellipsoids of revolution. Methods for estimating size distribution, or counting N_V, for more complex shapes require a synthesis of the three-dimensional structure itself[20, 21], and will not be dealt with in this chapter.

In order to estimate particle shape in an aggregate of ellipsoids of revolution of arbitrary size distribution, it is first necessary to determine whether the particles are elongated (prolate), or flattened (oblate). This determination requires the comparison of the largest circular (equi-axed) section with the largest of the most elongated sections. If the particles are flattened (i. e., approaching disc-shaped), the diameter of the circular section will compare in size with the major axis of the elongated particle. If they are elongated, the diameter of the circular section will be the same size as the minor axis of the most elongated section.

The quantitative shape q is defined as the value of the axial ratio of the generating ellipse. It can be shown that the axial ratio of the most elongated particle section observed is the same as that of the ellipsoids in three dimensions. Thus, this quantity may be directly measured on the metallographic section. If the measurement of shape requires more statistical weight, it may be estimated, with a much greater expenditure of time and effort, from measurements of the axial ratios of all of the particle sections in the sample[22].

K. Particle Size Distribution

There are probably more proposals for methods for estimating the size distribution of particles than any other property measured in quantitative microscopy. With one exception[23], all methods require some measurement of the distribution of sizes of particle section in the sample. Some are based upon the measurement of section areas[4, 24]; others on the measurement or some dimension of the particle section, e.g., a diameter[25], or maximum dimension[26]; and still others on the distribution of lineal intercepts[27-29]. Because the latter is the simplest method, both mathematically and from the standpoint of the experimental method, it will be described in this section. The measurement of lineal intercepts is also more directly adaptable to automatic scanning devices[7, 8].

The procedure is a relatively complex one and is presented in outline form. For a complete exposition of procedure, consult Refs.[27, 28, or 29].

An array of test lines is uniformly dispersed over the microsection. Measurements are made of the length of each intercept with particles of the phase under study. The measurements are then classified according to length; the number of size classes, n, is arbitrary, but should probably be between 8 and 15, depending upon the resolution desired. A sufficient number of measurements should be taken so that an average of 40 to 50 readings are contained in the individual size classes. This means from about 300 to

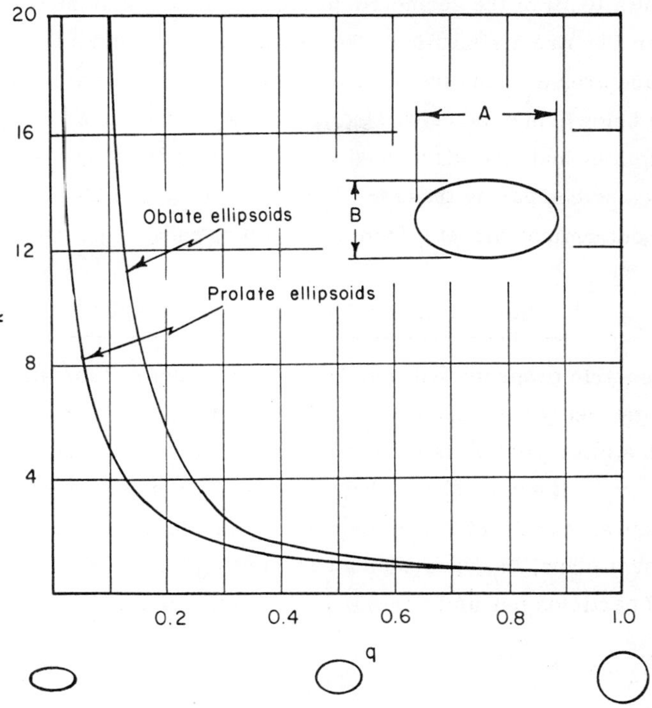

Fig. 8. The shape factor required in Eq. (9) for estimating the size distribution of ellipsoidal particles: q is the axial ratio of the generating ellipse.

1000 individual intercepts must be measured, depending upon the resolution and accuracy required.

Let Δ be the width of the interval between size classes. Let N_{Li} be the number of intercepts, normalized to per unit length of test line, in size class i, i.e., with lengths between $(i-1)\Delta$ and $i\Delta$. The number of particles contained in this size class per unit volume of structure is given by

$$N_{Vi} = \frac{1}{2k\Delta^2}\left[\frac{N_{L(i-1)}}{i-1} - \frac{N_{Li}}{i}\right] \tag{9}$$

where k is a shape factor, which may be evaluated from the measurement of the axial ratio q and Fig. 8.

Essentially all of the geometric properties of the system may be determined from the size distribution. For example, the volume fraction V_V, the total surface area S_V, and the product $N_A\overline{D_V}$, which is related to the N_A count (see below), may be calculated from this information. The validity of the assumptions and procedure used in determining the size distribution may be assessed by comparing these total or extensive properties with the same quantities determined directly from the counting measurements.

L. Number of Particles per Unit Volume, N_V

A geometric property that may be rigorously defined for structures of arbitrary geometry but cannot be rigorously determined from measurements made upon a microsection is the number of particles per unit volume of structure, N_V. It can be shown that, at least for particles of relatively simple shapes, a count of the number of particle sections observed in unit area of a test plane, N_A, provides an unbiased estimate of the product of the number of particles per unit volume, and an average particle size[22]:

$$\overline{N}_A = N_V\overline{D_V} \tag{10}$$

where $\overline{D_V}$ is defined as the distance between parallel tangent planes averaged over orientation for each particle, and subsequently averaged over all of the particles in the structure. This product cannot be separated into its component parts without the introduction of some simplifying geometric assumptions.

Several methods for estimating N_V have been proposed which involve the determination of the harmonic mean of some dimension of particle section[11, 22]. However, this determination requires the same expenditure of effort as does the estimation of the particle size distribution, which yields the number of particles per unit volume in each size class, see Eq. (9). It is recommended that N_V be determined by measuring the size distribution. Then, the total number of particles in unit volume is obtained simply by summing the number in each size class.

M. Evaluation of Geometric Properties in Particulate Systems in Which All of the Particles Have the Same Size and Shape

In the preceding sections, all of the geometric properties of structures which are accessible from measurements made upon a microsection have been exposed. The introduction of the assumption of constant size does not lead to the determination of properties that are not otherwise measurable. However, in cases where this assumption is valid, the effort required to obtain complete geometric information about a structure is greatly reduced.

For example, if all particles are the same size and shape, and the shape is know, the particle size X and number of particles N_V may be estimated from any two counting measurements:

$$N_V = \frac{N_A^2 \; k_2}{2N_L \; k_1^2} \tag{11a}$$

$$X = \frac{2N_L \; k_1}{N_A \, k_2} \tag{11b}$$

where k_1 and k_2 are shape factors[30]. A knowledge of the size and number of particles completely determines the structure in this simple case, so that the other properties, i. e., total and average surface area, and total and average particle volume, may be determined from these two counting measurements.

If the particle size is also known, e. g., by measuring the size of the largest section in the microstructure, then the evaluation of any one of the counting measurements permits the estimation of N_V, and the structure is completely determined. Thus, if X is known[30],

$$N_V = \frac{2}{k_2 X^2} N_L \tag{12}$$

It is evident that there is a correspondence between the effort that it is necessary to expend and the number of geometric assumptions that may be introduced in any given case. In the materials sciences, the assumption of constant size and shape is generally invalid, so that results obtained from such an assumption must be considered to be approximations. Such a procedure might be used to provide an order-of-magnitude estimate of N_V, for example. In the life sciences, on the other hand, the structures involved tend to be more regular, and these assumptions may be applied with greater accuracy. For a complete discussion of the relationships that apply in this case, see Ref. (30).

III. QUANTITATIVE TRANSMISSION MICROSCOPY

Specimen preparation for observation by transmitted light, whether for histology, petrography, or electron microscopy, involves the preparation of a thin slab of the structure. The image observed is a projection of all of the structural features in this slab. From a geometric point of view, this image provides information which is fundamentally different from that obtained in reflection microscopy.

If the thickness of the slab is negligible in comparison with the dimensions of the structural features it contains, the projected image will be negligibly different from a reflected image. In this case, the quantitative analysis of the system simplifies to a problem in reflection microscopy, and all of the methods developed in the previous section are valid.

For a relatively thick slab, the projected image contains information about structural elements that intersect the top and bottom surfaces, and about features wholly contained within the volume of the sample. An additional problem arises from the fact that the projected images of structural features may overlap to a greater or lesser degree. This distortion, called the Holmes effect(31), increases as the section thickness increases. It generally leads to the underestimation of geometric properties. Corrections for this effect have been proposed, but they generally rely upon simplifying assumptions; i. e., that the structure consists of spherical particles, whose centers are randomly distributed in space(32). It is not possible, at the current state of development, to make quantitative assessments of the degree to which any given structure may satisfy these requirements.

Because of these limitations, the relationships between information contained in the observed image and the properties of the structure for which it is obtained do not possess the same degree of independence of geometry as do those derived for reflection microscopy. As a consequence, it is necessary to present the basic relationships for transmission microscopy in terms of what properties the counting measurements are dependent upon, rather than what properties they may be used to determine.

A. Counting Measurements in Transmission Microscopy

The following relationships are valid for structures of arbitrary geometry. It is necessary, however, to assume that the Holmes effect may be neglected. This assumption effectively limits the validity of these relations to very thin sections, or to small volume fractions of the particulate phase. For this case, the three counting measurements are related to the following structural features[32, 33]:

a. The Point Count, P_P:

$$\overline{P_P} = V_V + N_V t \overline{A_{proj}} \tag{13}$$

where t is the thickness of the sample, and $\overline{A_{proj}}$ is the average projected area of all features in the structure.

b. The Line Intercept Count, N_L:

$$\overline{N_L} = \frac{1}{2} S_V + \frac{2}{\pi} N_V t \; \overline{L_{proj}} \tag{14}$$

where $\overline{L_{proj}}$ is the average projected perimeter of features in the structure.

c. The Number Per Unit Area Count, N_A:

$$\overline{N_A} = N_V \; (\overline{D_V} + t) \tag{15}$$

where all symbols have the same definitions as used in previous sections.

In each case, the first term of the right side of the equation is the contribution from the top and bottom faces, and is identical to the result that obtains for reflection microscopy. The second term is the contribution from features contained within the volume of the slab. The three counting measurements are related to the section thickness t and to six geometric properties of the structure: V_V, S_V, D_V, N_V, $\overline{A_{proj}}$, and $\overline{L_{proj}}$. Thus, the experimental determination of the three counting measurements does not uniquely determine the properties of the three-dimensional structure; additional geometric information or assumptions are required.

B. Assumption of Constant, Spherical Shape

The introduction of the assumption that all of the particles are spherical does not provide sufficient simplification to permit the direct application of the counting measurements. Further information about the size distribution is required. For example, if it is assumed that the structure is characterized by a two-parameter distribution function[34] and the total number of particles N_V, then all six three-dimensional properties appearing in Eq. (13) through (15) may be evaluated. Under these conditions, i. e., that the structure depends upon only three parameters, the evaluation of the three counting measurements may be suitably manipulated to provide a unique description of the geometry.

Hilliard[32] has presented a method for the quantitative determination of the particle size distribution, subject to the assumptions that the particles are spherical and that the Holmes effect is negligible. This treatment provides the most detailed information about structures that can currently be obtained from projected images.

C. Assumptions of Constant Size and Shape

Let it be assumed that all of the particles are the same shape, e. g., spherical, and that their size is constant. The sphere diameter D may be measured directly on the projected image. The five metric quantities may evaluated directly; the surface area S is πD^2; the volume V is $\frac{\pi}{6} D^3$; $\overline{D_V} = D$; $A_{proj} = \frac{\pi}{4} D^2$; and $L_{proj} = \pi D$. Any one of the counting measurements may be used in conjunction with this information to estimate N_V. The evaluation of number and size then describes the structure completely. If it is desired, it is possible to make an assessment of the validity of the assumptions of constant shape and size by evaluating the remaining counting measurements from the number and size of particles and comparing the result with experimental measurements of these quantities.

IV. DESIGN OF EXPERIMENTS

In the derivation of all of the relationships presented in this chapter it is assumed that a representative section has been uniformly sampled in obtaining the counting measurements, or the distribution of section sizes. This section clarifies the meaning of a representative sample and discusses some aspects of design of experiments for specified statistical accuracy.

A. Representative Sample

In general, structures may contain property variations with position (structural gradients) and with orientation (structural anisotropy) In order to be representative, a sample obtained from a structure must contain a uniform representation of all gradients and anisotropy[3, 4].

If there is independent information, or experience, which proves that a given structure is homogeneous and isotropic, then a single sample, whether it be a test plane, line, or point, is representative. The problem of experimental design then simplifies to the determination of the statistical accuracy required, and the number of measurements that must be taken to attain that accuracy.

If the inhomogenieties are one dimensional, or planar, then it is possible for a single plane section to contain a uniform representation of structural gradients and anisotropy. It is only necessary that this plane be chosen so that it contains the gradients.

In more complex distributions of geometric properties, or in cases where prior experience is lacking, it is necessary to sample the structure with test planes (or lines) which are uniformly distributed in position and orientation. The number of different planes distributed will depend upon an optimum between the effort that may prudently be expended, and the resolution and accuracy desired in the analysis.

B. Statistical Accuracy and the Number of Measurements

The problem of assessing the accuracy of a set of statistical data is common. The simple, general statistical approach seems to be the most useful one for microstructural analysis. A set of data is collected containing a relatively small number of readings of the property being measured. The standard deviation of the population is estimated from these readings, using the formula[35]

$$\sigma_x = \left(\frac{\sum_{i=1}^{n} (x_i - \bar{x})^2}{n=1} \right)^{1/2} \qquad (16)$$

where the x_i are the individual readings, $\bar{x}$ is their average value, and n is the number of readings taken. The standard deviation of the sample mean, $\sigma_{\bar{x}}$, is then calculated:

$$\sigma_{\bar{x}} = x/\sqrt{n} \ . \tag{17}$$

The range of values of x specified by $\bar{x} \pm 2\sigma_{\bar{x}}$ is called the 95% confidence interval for this experiment. This statement infers that there is a probability of 0.95 that the true mean value of x for the population is within $\pm 2\sigma_{\bar{x}}$ of the sample. Equation (17) may then be used to determine the number of readings that must be contained in the sample to achieve any predetermined accuracy, i.e., confidence interval. It is usually necessary to revise downward the accuracy required in order to obtain a reasonable number of readings in the experimental design.

V. CONCLUSION

The quantitative estimation of the three-dimensional properties of structures is an established discipline. The fundamental relationships which underly this discipline have been presented. It is observed that the three simple counting measurements, P_p, N_L, and N_A (or P_A for linear features), form the core of methods of analysis. Additional structural information requires additional kinds of measurements, e.g., the measurements of section size distributions, which generally require considerable additional effort.

REFERENCES

(1) E. E. Underwood, "A Standardized System of Notation for Stereologists," Stereologia, Vol. 3, 1964, p. 5.

(2) G. Thomas, "Transmission Electron Microscopy of Metals," Wiley, New York, 1962.

(3) R. T. DeHoff, "Sampling of Materials and Statistical Analysis in Quantitative Stereology," Proceedings of the Second International Congress for Stereology, Springer-Verlag, Berlin, 1967.

(4) S. A. Saltykov, "Stereometric Metallography," 2nd Edition, Metallurgizdat, Moscow, 1958.

(5) A. Rosiwal, "On Geometric Rock Analysis. A Simple Surface Measurement to Determine the Quantitative Content of the Mineral Constituents of a Stony Aggregate," K. K. Geologischen Reichenstalt, Vol. 5-6, 1898, p. 143.

(6) J. E. Hilliard and J. W. Cahn, "Evaluation of Procedures in Quantitative Metallography for Volume Fraction Analysis," Trans. Met. Soc. AIME, Vol. 221, 1961, p. 344.

(7) M. Cole, "The Metals Research Quantimet," The Microscope and Crystal Frontier, Vol. 15, 1966, p. 148.

(8) G. A. Moore, "Application of Computers to Quantitative Analysis of Microstructures," Nat. Bur. of Stds., Report #9428, U. S. Dept. of Commerce, 1966.

(9) C. S. Smith and L. Guttman, "Measurement of Internal Boundaries in Three Dimensional Structures by Random Sectioning," Trans. Met. Soc. AIME, Vol. 197, 1953, p. 81-87.

(10) R. J. Duffin, R. A. Meussner, and F. N. Rhines, "Statistics of Particle Measurement and of Particle Growth," Carnegie Inst. of Tech., Tech. Rep. #32, AF-33 (616) - 294, April 1953.

(11) R. L. Fullman, "Measurement of Particle Sizes in Opaque Bodies," Trans. Met. Soc. AIME, Vol. 197, 1953, p. 447.

(12) J. Gurland, "An Estimate of Contact and Continuity of Dispersions in Opaque Samples," Trans. Met. Soc. AIME, Vol. 236, 1966, p. 642.

(13) J. Gurland, "The Measurement of Grain Contiguity in Two Phase Alloys," Trans. Met. Soc. AIME, Vol. 212, 1958, p. 452.

(14) W. J. Babyak and F. N. Rhines, "The Relationship Between the Boundary Area and Hardness of Recrystallized Cartridge Brass, "Trans. Met. Soc. AIME, Vol. 218, 1960, p. 21.

(15) R. A. Gregg, Dept. of Met. and Mats. Eng., Univ. of Florida, Private Communication.

(16) J. E. Hilliard, "Specification and Measurement of Microstructural Anisotropy," Trans. Met. Soc. AIME, Vol. 224, 1962, p. 1201.

(17) D. J. Struik, Lectures on Classical Differential Geometry, Addison-Wesley, Reading, Mass., 1950.

(18) R. T. DeHoff, "The Quantitative Estimation of Mean Surface Curvature," Trans. Met. Soc. AIME, Vol. 239, 1967, p. 617.

(19) J. W. Cahn, "The Significance of Average Mean Curvature and Its Determination by Quantitative Metallography," Trans. Met. Soc. AIME, Vol. 239, 1967, p. 610.

(20) J. Kronsbein, L. J. Buteau, and R. T. DeHoff, "Measurement of Topological Parameters for Description of Two Phase Structures With Special Reference to Sintering," Trans. Met. Soc. AIME, Vol. 233, 1965, p. 1961.

(21) S. N. Postlethwait, "Cinematography of Serial Microscope Sections," First International Congress for Stereology, Congressprint, Vienna, 1963, p. 37a/1.

(22) R. T. DeHoff and F. N. Rhines, "Determination of the Number of Particles Per Unit Volume from Measurements Made on Random Plane Sections: The General Cylinder and the Ellipsoid," Trans. Met. Soc. AIME, Vol. 221, 1961, p. 975.

(23) R. T. DeHoff, "The Estimation of Particle Size Distributions from Simple Counting Measurements Made on Random Plane Sections," Trans. Met. Soc. AIME, Vol. 233, 1965, p. 25.

(24) W. A. Johnson, "Estimation of Spatial Grain Size," Metal Progress, Vol. 49, 1946, p. 87.

(25) E. Scheil, "Statistiche Qefugeuntersuchungen I," Zeit. f. Metallk., Vol. 27, 1935, p. 199.

(26) R. T. DeHoff, "The Determination of the Size Distribution of Ellipsoidal Particles from Measurements Made on Random Plane Sections," Trans. Met. Soc. AIME, Vol. 224, 1962, 474.

(27) G. W. Lord and T. F. Willis, "Calculation of Air Bubble Size Distribution From Results of a Rosiwal Traverse of Aerated Concrete," ASTM Bulletin, No. 177, 1951, p. 56.

(28) J. W. Cahn and R. L. Fullman, "On the Use of Lineal Analysis for Obtaining Particle Size Distribution Functions in Opaque Samples," Trans. Met. Soc. AIME, Vol. 206, 1956, p. 610.

(29) R. T. DeHoff and P. Bousquet, "Estimation of the Size Distribution of Triaxial Ellipsoidal Particles from the Distribution of Linear Intercepts," Jnl. Microscopy, Vol. 92, Pt. 2, 1970, p. 119.

(30) R. T. DeHoff, "The Determination of the Geometric Properties of Aggregates of Constant-Size Particles from Counting Measurements Made on Random Plane Sections," Trans. Met. Soc. AIME, Vol. 230, 1964, p. 764.

(31) A. H. Holmes, Petrographic Methods and Calculations, Murby, London, England, 1927.

(32) J. E. Hilliard, "The Counting and Sizing of Particles in Transmission Microscopy," Trans. Met. Soc. AIME, Vol. 224, 1962, p. 906.

(33) J. W. Cahn and J. Nutting, "Transmission Quantitative Metallography," Trans. Met. Soc. AIME, Vol. 215, 1959, p. 526.

(34) R. T. DeHoff, The Use of Counting Measurements in the Estimation of Particle Size Distributions in Transmission Microscopy, Unpublished manuscript.

(35) G. U. Yule and M. G. Kendall, An Introduction to the Theory of Statistics, 14th Ed., Hafner, New York, 1950.

Chapter 19

APPLIED STEREOLOGY

S. W. Freiman*

IIT Research Institute, Chicago, Illinois

I. INTRODUCTION

The basic concepts of quantitative stereology were presented in the previous chapter. The objective of this chapter will be to show how quantitative stereology can be used for analysis and characterization of ceramics and to describe some of the actual measuring techniques.

There are a number of tools available for the study of microstructures. These include the optical microscope used in both reflection and transmission, the scanning electron microscope, and the transmission electron microscope. Each of these tools has special advantages and disadvantages. Descriptions of all of these microscopes have been presented previously; their application to stereology will be shown here. Scanning techniques, including new automated methods, will also be discussed. In addition, it will be shown how the quantitative measurements of the microstructure are related to some of the physical and mechanical properties of materials.

*Present address: Naval Research Laboratory, Washington, D.C.

II. SCANNING TECHNIQUES

Older methods of quantitative microstructural analysis have largely involved the use of individuals to make the large number of observations necessary for an accurate determination of a stereological parameter. Over the years, various mechanical aids such as cell counting plates and occular grids have been developed. These are purely optical accessories, however. Newer inventions of automatic scanning techniques have both speeded up and increased the accuracy of quantitative analyses.

There are numerous scanning methods available for quantitative analysis of a microstructure. Many of these are discussed by Fischmeister[1] for the operations of counting, size measurement, and measurement of areal fractions.

The oldest method of scanning is by the visual inspection of every feature in the structure and is called feature-to-feature scanning. Other, more systematic methods are known as line and step scanning. Feature-to-feature scanning is used routinely in the measurement of grain size[2, 3], the determination of slag inclusions[4], and in the analysis of porosity[5]. This technique is of particular value for particle size measurements in powders[6, 7]. Instrumentation for sizing is similar to that for simple counting except that a device for size measurement must be added. Tools such as a "photoelectric area determinator" have been used for sizing[8]. It has been suggested that if reliable apparatus became available, photometry may be a quick method for area measurements. Further possibilities have been opened by flying spot scanners and a photoelectric method which has been used for the determination of carbides in steel.

The main application of step-scanning is in the determination of areal and volume fractions by the point-count method, i.e., by determination of P_p. The step-scanning procedure involves movement of a point in regular steps across the microstructure while counting the number of times a point crosses a given feature. Microscope stages have been modified to allow step-scanning [9, 10] to be accomplished in the microscope. Fully automatic step motion has been incorporated into various counting instruments[11].

Line-scanning is of value in the determination of grain-size distributions, in the analysis of the degree of mixing in a two-phase system, and in the measurement of grain contiguity. Instruments have also been developed for this type of counting.

The newest developments in the field of quantitative microscopy are the completely automatic scanning instruments which depend on a photocell to register the microstructural[12, 13] contrast. Potential problems associated with the automatic equipment, however, include lack of resolution, poor quality of image, and background noise. Moore, et al.[14] have recently described a method by which metallographic analyses can be made with the aid of a digital computer.

The amount of analysis that is to be done is likely to determine the choice of scanning equipment. Table I describes the scanning tools which might be employed for occasional, routine, and heavy routine analyses.

TABLE I

Evaluation of Methods of Quantitative Analysis[1]

1. Occasional work	
Counting:	With strip or ruled eyepiece
Sizing:	With eyepiece or screen graticule
Point counting:	With point-counting eyepiece (preferred technique for determination of volume fraction)
Lineal analysis:	On photographs or screen (only for special measurements that cannot be performed by point counting)
2. Routine work	
Counting:	As under (1) but with electrical counter to record results, or with line-scanning stage and reference line perpendicular to scanning direction
Sizing:	With adjustable diaphram instrument or with electrical line-scanning stage equipped for automatic intercept sizing
Point counting:	With point-counting eyepiece and/or point-counting stage. Note that if an electrical line-scanning stage is available volume-fraction determinations combined with other measurements (e. g., mean size of inclusions) are made faster with the stage, and point counting should be used only if nothing but the volume fraction is required
Lineal analysis:	With electrical line-scanning stage
3. Heavy routine work	
Counting and sizing:	(Of isolated structural features) with automatic scanning machine (possible under favorable contrast conditions with existing instruments)
Determination of volume fractions:	By areal analysis with flying spot scanner (under unfavorable contrast conditions, see (2))
Lineal analysis:	(For special measurements only, such as degree of contiguity, etc.): electrical line-scanning stage, until improved automatic scanners with logical facilities or suitable computer programs are available.

III. EFFECTS OF MICROSCOPY

Optical microscopy has been a fundamental tool in the identification and study of materials for over a century. Most of the quantitative analyses of ceramics can be made using optical equipment either in reflection or in transmission.

The electron microscope has been available for analysis of structure for about two decades. It has become an essential tool for the resolution of microstructural features in the 10-500 Å range. Although the microscope has been discussed in detail earlier(5), the features of the electron microscope technique pertinent to quantitative stereology will be mentioned here.

Specimens can be viewed in the electron microscope in two ways. Either a thin section of the material can be prepared, the desired thickness being less than 1000Å, or a replica of the sample surface can be taken. While the former method affords better resolution, the latter is much easier to prepare.

In taking a replica from a sample surface for quantitative analysis, one should be sure that a representative portion of the structure is obtained. The accuracy of the counting measurements will be only as good as the accuracy in obtaining the replica. If there is doubt about the representativeness of a given area of the surface, a number of different replicas should be taken. Because the resolution of any given replica will possibly depend on the etching treatment given the sample, this factor must also be considered.

Because only the surface is represented on the replica, the quantitative relations given in Chapter 18 are correct as derived, and no correction for thickness is required.

Samples to be used in the scanning electron microscope (SEM) are viewed in reflection. One of the great benefits of this tool is the ease of sample preparation, since bulk samples such as fracture surfaces can be examined. The same techniques as used in optical counting measurements can be employed, except that the depth of focus as well as the degree of resolution in the microscope is much improved. Because of the large depth of field in the SEM a planar sample surface is required in order for the quantitative relations of Chapter 18 to be valid.

A quite different situation arises when a thin section of material is examined in transmission in either the optical or electron microscopes. It is well recognized that accurate quantitative analysis requires that correction factors be applied which will take into account errors occurring due to the thickness of the section.

When the sample thickness to particle diameter ratio (t/b) for a two-phase microstructure is greater than about 2, a considerable error can be introduced if the effect of thickness is neglected. This error can be greatly magnified since on many electron or optical microscopy studies it is possible to achieve a t/b ratio of 20 or more.

As discussed in the previous chapter, in reflection microscopy three fundamental stereological relations permit the estimation of the geometric properties of a three-dimensional structure from simple counting measurements on a random plane section[16]:

$$P_p = V_V \quad (1)$$

(I)
$$N_L = 1/2\, S_V \quad (2)$$

$$N_A = N_V \overline{D}_V \quad (3)$$

The quantities on the left of these equations are the counting measurements, which are performed under the microscope or on photomicrographs: P_p is defined as the point fraction, N_L is the number of intersections per unit length of a test line, and N_A is the number of particles observed per area of observation.

The quantities on the right side of Eqs. (1) to (3) are properties of the structure which may be estimated from the counting measurements: V_V is the volume fraction of the phase of interest in the sample, S_V is the total area of interface between the matrix and the phase of interest in a unit volume of sample, N_V is the number of particles in a unit volume, and $\overline{D}_V$ is their average size (caliper diameter). These relations are valid for structures of arbitrary geometry, i. e., for aggregates of particles of arbitrary shape and size distribution.

When a sample is observed in transmission microscopy, the image contains a projection of all of the structure contained within the volume of the thin section under study. Particles that intersect the top and bottom surfaces of the section contribute to the projected image in proportions essentially equal to those that would be observed in reflection microscopy. Two additional kinds of contributions to the projected image must also be taken into account. These alter and complicate the relations between the counting measurements and the properties of the three-dimensional structure in the thin section.

Particles whose centers lie within the volume of the section contribute projected images that must be added to those that result from particles that intersect the surfaces of the section. The equations in Set (I) become, as discussed in the previous chapter:

$$P_p' = V_v + N_v + \overline{A}_{proj} \tag{4}$$

(II)

$$N_L' = 1/2\, S_v + 2/\pi\, N_v + \overline{L}_{proj} \tag{5}$$

$$N_A' = N_v \overline{D}_v + N_v\, t \tag{6}$$

where P_p', N_L', and N_A' are the values of the counting measurements measured in transmission microscopy, $\overline{A}_{proj}$ is the average projected area of particles in the structure, $\overline{L}_{proj}$ is the average perimeter of the projections of all particles in the structure, and t is the section thickness. Thus, if the equations in Set (I) are used to estimate properties of the structure, the values of these properties will tend to be overestimated by amounts corresponding to the second term in each of these equations.

One exception to the use of the above equations in transmission occurs if the second phase was formed in the section after it was prepared rather than before the thinning process. In this instance the contribution to the equations of particles lying outside the section should be ignored[17].

If the section thickness is large compared to the particle size, the projections of some particles will be partly or wholly obscured by the projections of other particles in the system. This effect of particle overlap, called the Holmes effect[18], tends to yield an underestimate of the geometric parameters of the structure. The Holmes effect increases as the section size increases and as the volume fraction of the second phase increases. Hilliard[17] showed that if $t/\overline{D}_v > 2$, where t and $\overline{D}_v$ are defined as above, then the effect of overlap could be corrected. If $V_v(t/\overline{D}_v) > 0.15$, however, the effect can not be accurately eliminated by mathematical corrections. When $t/\overline{D}_v \sim 1$ or $V_v(t/\overline{D}_v) < 0.04$, the Holmes effect can be neglected.

If it is assumed that all of the particles in the structure are the same size and shape, the geometric properties of the structure may be written in terms of the particle size, b, and shape factors:

$$\overline{D}_V = k_1 b \tag{7}$$

$$S_V = N_V S = N_V k_2 b^2 \tag{8}$$

(III) $$V_V = N_V V = N_V k_3 b^3 \tag{9}$$

$$\overline{L}_{proj} = k_4 b \tag{10}$$

$$\overline{A}_{proj} = k_5 b^2 \tag{11}$$

where the constants k_1, k_2, k_3, k_4, and k_5 are shape factors which must be evaluated for a given particle shape and choice of size parameter. Further, it can be shown that, for convex particle shapes,

$$k_4 = \pi k_1 \tag{12}$$

$$k_5 = 1/4\ k_2 \tag{13}$$

Substitution of these results into the equations of Set (II) gives:

$$P_p' = N_V b^2 \left[k_3 b + 1/4\ k_2 t\right] \tag{14}$$

(IV) $$N_L' = N_V b \left[1/2\ k_2 b + 2k_1 t\right] \tag{15}$$

$$N_A' = N_V \left[k_1 b + t\right] \tag{16}$$

The experimental determination of the three counting measurements on the projected image, coupled with the evaluation of the three shape factors, thus yields three simultaneous equations which may be solved for the three unknowns N_V, b, and t. The insertion of the number of particles and their size into Eqs. (8) and (9) then yields a complete description of the structure in terms of its extensive geometric properties: volume, surface area, number, and size of particles. In practice, it is necessary to obtain an independent measurement of sample thickness, t, to solve the equations in Set (IV) with accuracy.

An experimental study was undertaken[19] to quantitatively determine the errors which can occur if the corrections for thickness are ignored and to demonstrate that complete stereological treatment of the transmission microscopic analysis yields calculated values comparable to the experimental values obtained from reflection measurements. A fairly detailed review of this work will be presented because it represents one of the few applications of complete stereological analysis to ceramic systems. Thus, it illustrates not only the details of the methods required for complete structural characterization but also the potential power of structural analysis.

The system used in the study was a partially devitrified Li_2O-SiO_2 glass which could be examined in both reflection and transmission over a range of volume fractions of crystals. The partially crystalline glass was prepared according to the procedure outlined by Freiman and Hench[20]. A petrographic thin section, 20± 5 μm thick, was prepared from each sample. The remainder of each sample was retained for use in the reflection studies.

The thin sections were studied in transmission in a polarizing microscope. Because of the birefringence of the devitrified phase, the $Li_2Si_2O_5$ crystals appeared light against a dark background when the section was viewed under crossed Nicols. A typical microstructure is shown in Fig. 1.

The quantitative stereological parameters P_p', N_L', and N_A' were measured in transmission by placing an 8 by 8 grid in the eyepiece of the microscope. Twenty-five measurements of each parameter were made on each section, and the average was calculated. At any given position of the microscope stage, a constant fraction of $Li_2Si_2O_5$ crystals was invisible under crossed Nicols because of partial extinction. The quantitative measurements were corrected for the extinction effect by multiplying the calculated average values of P_p', N_L', and N_A' by a factor which accounted for the invisible fraction of the particles. By measuring the angular extinction of 100 particles this factor was determined to be 1.2.

A typical microstructure of a specimen prepared for reflection measurements is shown in Fig. 2. An average of the twenty-five measurements of P_p, N_L, and N_A made in a metallograph was calculated. The thickness of each thin section was also measured in the metallograph.

The evaluation of the average particle size and number of particles in the system requires a determination of the three shape factors k_1, k_2, and k_3 as shown in Eqs. (14) - (16). Consequently, an assumption or determination of the shape of the particles is involved. Examination of the microstructure of

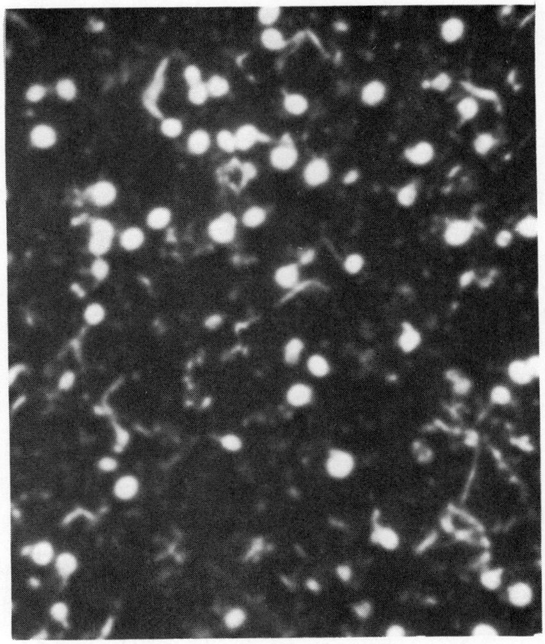

Fig. 1. Transmission microstructure of a partially devitrified Li_2O-$2SiO_2$ glass, $V_V = 0.045$; crossed Nicols X 125.

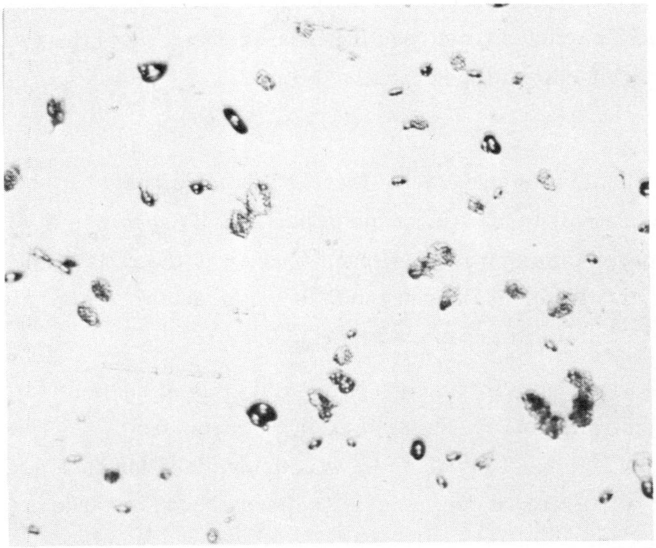

Fig. 2. Reflected light micrograph of a partially devitrified Li_2O-$2SiO_2$ glass, $V_V = 0.045$ (X250).

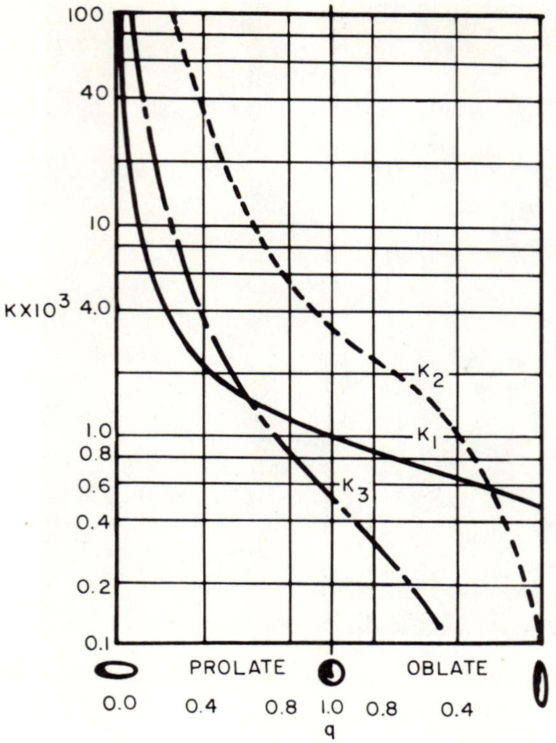

Fig. 3. Various shape factors used for estimating the number and size of ellipsoidal particles from counting measurements; q is the ratio of the minor to major axis of the generating ellipse.

the system under study suggested that the particles could be modeled as ellipsoids of revolution. The shape of particles is specified in terms of the axial ratio q of the generating ellipse, where q is the ratio of the minor to the major axis (b/a). Values of the three shape factors were calculated as a function[21] of q and are plotted in Fig. 3.

It has been shown that the average axial ratio of sections through ellipsoids, $\overline{F}$, is related to the value of q for the structure [22]. This relation is plotted in Fig. 4. The value of $\overline{F}$ was determined for each sample by measuring the axial ratio of 100 particle sections chosen at random. The axial ratio of the three-dimensional particles, q, was then determined from Fig. 4, and the corresponding shape factors were read from Fig. 3.

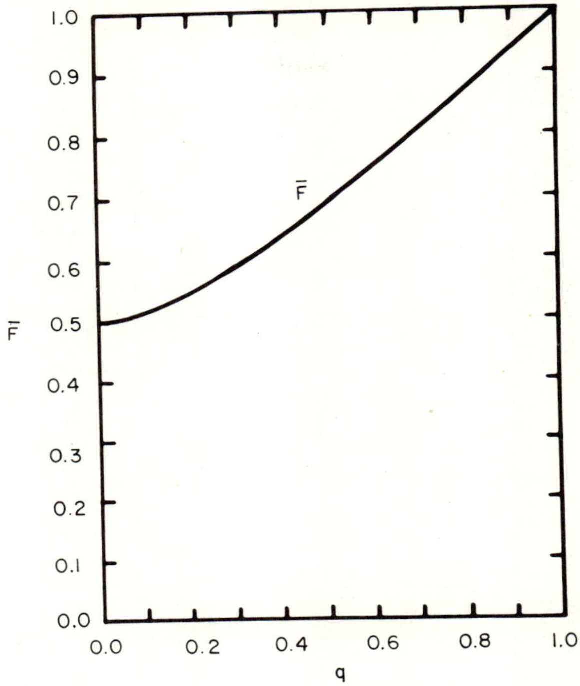

Fig. 4. The average axial ratio of cuts through prolate ellipsoids as a function of the axial ratio of the generating ellipse.

The results of this investigation are summarized in Tables II and III. In Table II the values of the parameters measured in transmission are given in the first column for each sample, the calculated reflection values in the second column, and the values measured in reflection in the third column. The fourth column lists the "thickness error," the error which would be encountered if the values measured in transmission were not corrected for the thickness of the thin section. The last column gives the error after correction, the difference between the calculated values, and the values measured in reflection.

The correction error could be attributed to a number of causes discussed by the authors(19).

Table III summarizes the complete microscopic analysis of the devitrified glass samples measured in transmission and corrected for thickness effects.

The volume fraction of $Li_2Si_2O_5$ spherulites, the number of spherulites per unit volume, the spherulite size, the surface area of the spherulites, and

TABLE II

Summary of Thickness Error Analysis

Sample No.	Transmission	Reflection, calc.	Reflection, meas.	Thickness error (%)[a]	Correction error (%)[b]
		P_p			
1	0.090	0.023	0.020	291	15.0
2	0.106	0.025	0.029	324	14.0
3	0.213	0.045	0.053	373	15.1
		N_L			
1	177	103	107	71	3
2	154	86	85	79	1
3	232	135	106	72	27
		N_A			
1	9.3×10^4	5.0×10^4	4.8×10^4	66	4
2	7.0×10^4	4.1×10^4	4.1×10^4	71	0
3	8.2×10^4	3.4×10^4	4.6×10^4	141	26

[a] Thickness error $= \dfrac{\text{transmission} - \text{(reflection) calc.}}{\text{(reflection) calc.}}$

[b] Correction error $= \dfrac{\text{(reflection) calc.} - \text{(reflection) meas.}}{\text{(reflection) meas}}$

TABLE III

Summary of Particle Parameter Analysis

Sample No.	Section thickness (μ)	Volume fraction, V_V	No. particles/ unit vol, $N_V (X10^{-7}/cm^3)$	Ave particle size $\overline{b}$ (μ)	Surface area of particles/ unit vol, $S_V (cm^2/cm^3)$	Ave distance between particles, λ (μ)
1	9.9	0.023	5.3	7.5	207	188
2	12.2	0.025	3.2	9.2	172	227
3	20.4	0.045	2.5	12.8	270	141

the average distance between the particles is presented for three samples. The average distance between particles is obtained from the relation $\lambda = 2\left[1 - \overline{P}_p\right]/\overline{N}_L$. It can be seen that, although the same volume fraction of crystals can always be obtained from different heat treatment schedules, the other structural parameters may vary depending on nucleation treatments, differences in glass composition, experimental scatter, or other factors.

It is clear from Table II that the quantitative stereological parameters measured from transmission specimens differ considerably from the true parameters. This is demonstrated by the magnitude of the thickness error. The measured and calculated reflection values agree very closely, however, justifying the assumptions made in the analysis regarding particle size and shape. Comparison of the values of V_V determined in this study with those expected from the crystallization kinetics determined from x-ray diffraction data[20] also confirms the accuracy of the present analysis. This investigation showed that quantitative stereological parameters can be overestimated by as much as 300% when measured in transmission if corrections for thickness effects are not made. It was also shown that structural parameters corrected for thickness effects correspond with values obtained from reflection measurements and x-ray analysis within about 10%. Consequently, it is possible to obtain quantitative correlations of physical properties and process variables with ceramic microstructures from either reflection or transmission microscopy if the above analysis is followed.

While it is obvious from the preceding discussion that the analysis in transmission is considerably more complicated than that in reflection, there are situations when the measurements must be made in this way. The greater resolution possible through the use of thin sections in the transmission electron microscope at times makes it necessary to make measurements in transmission. While the study just described was performed in an optical microscope, it is equally applicable to the electron microscope.

Another possible source of error which must be considered occurs due to differential polishing or etching of the microstructure, causing the volume fraction to be overestimated. Equations[23] have been developed, however, which sufficiently correct for such errors.

Other errors can arise in the quantitative analysis of a microstructure. These errors have been described by Hilliard[24] as bias of the results, and are listed below:

1. Errors introduced due to transmission measurements have already been discussed.

2. The presence of anisotropy in a structure will yield incorrect results, if corrections are not applied.

3. The termination of an analysis after too few counts will result in error.

4. Lack of resolution of the microstructure can introduce bias.

5. A subjective bias can arise if one searches for a "typical area."

Because quantitative analyses are of a statistical nature, the accuracy of any analysis is dependent on the number of measurements which are taken (see Chapter 18). In regard to the statistics of counting, Hilliard and Cahn[25] found that a point count with a two-dimensional grid was the most accurate method for the determination of volume fraction. They showed that the magnification of a microstructure relative to a grid opening should be such that each structural feature should not occupy more than one grid corner. The standard deviation in the determination of volume fraction was found to be:

$$\sigma_{V_V} = \frac{V_V}{(N)^{1/2}} \tag{17}$$

where V_V is the volume fraction of the phase of interest and N is the number of points of the grid falling on this phase. This equation can be used to determine the number of counts which must be taken to achieve a desired degree of accuracy.

IV. APPLICATIONS OF QUANTITATIVE STEREOLOGY

The quantitative stereological techniques which have been described can be applied to a variety of processes and phenomena in metals, ceramics, and composites. While many of these applications which appear in the following sections have been discussed in the literature, new ones are also suggested.

A. Sintering and Grain Growth

Although sintering and grain growth are often treated simultaneously, they are fundamentally different. One associates with grain growth a change in size and morphology of grains with time. Sintering in the broadest sense implies the coalescence of particulates, with particular emphasis on the volume fraction, shape, size, and distribution of porosity.

The use of quantitative microscopy in the study of grain growth is well established in metallic and nonmetallic systems. Routine analysis is often performed by a comparative technique either directly using a grain size reticle in the viewing instrument or with transparent overlays on micrographs. This method is covered in ASTM E-112. More precise measurements are made using a lineal intercept technique (N_L count). If one assumes all grains have the same shape, a size distribution may be obtained (see preceding chapter). Applications of this technique are the work of Burke on alpha brass[26], MacEwan on UO_2[27, 28], Haertling on lead zirconate-lead titanate ceramics[29], and Coble on Al_2O_3[30]. In each case it was found that data could be fitted to an expression of the form

$$D^2 - D_o^2 = K_o t^n \exp(-Q/RT) \tag{18}$$

where D_o = initial grain size
D = grain size after terminal treatment of time t
K_o = a constant
n = the time exponent
Q = activation energy for grain growth

Although the theoretical value of n is unity, it is frequently found experimentally that n is somewhat less than 1.

Underwood and Manning[31] in their study of grain growth pointed out that the presence of inclusions had a restraining effect on grain growth. They used an expression which showed the grain size (R) to be approximately inversely proportional to N_L and directly proportional to the mean spacing between particles.

The bulk of sintering studies in both metallic and nonmetallic systems have been based on microscopic models such as a sphere on plate[32, 33] or on measurement of bulk density[33] as a function of variables such as initial particle size, time, temperature, green density, etc. Observations of microstructure have been qualitative or descriptive in nature.

The application of quantitative microscopy to sintering has not received a great deal of attention in the ceramic literature. Francois and Kingery[34] measured pore and grain size in an attempt to correlate pore distribution with heating rate and impurity content in UO_2. Gupta and Coble[35] have studied the sintering of ZnO and found a density decrease during final stage sintering. Using a lineal analysis, the authors found that pore size increased and that the rate of growth correlated well with the lattice diffusion of zinc.

Rhines[36] has shown that it is possible to quantitatively and completely describe the evolution of microstructure by the introduction of topological concepts. For example, it is well known that at intermediate stages of the sintering process, porosity ceases to be continuous and begins to isolate itself. This change in the character (not necessarily size or volume fraction) of the pore phase reflects itself in a quantitative way in the measurement of the genus of the body. The genus, or more correctly, the genus per unit volume is a measure of the "connectivity" of the pore phase. It may be obtained directly from measurements made on a series of plane sections through the specimen. Details of the counting procedure are given in Ref. (37). This complete characterization of the three-dimensional structural changes occurring in sintering has recently been applied for the first time to a copper sinter body[38].

Desse[39] reported on the investigations of the reaction kinetics of the transformation $UAl_3 + Al \rightarrow UAl_4$ through the use of quantitative microscopy. The volume fraction of UAl_3 in the microstructure was measured under polarized light. The total amount of the two aluminides was determined using dark-field illumination. The dependence of both of these quantities on time and temperature yielded the desired information on the reaction kinetics. In another study, the growth kinetics of the superconducting compound Nb_3Sn have also been determined by quantitative microscopy[40].

B. Phase Transformations

The processes of nucleation and growth of second phases are important to the production of materials such as glass-ceramics as well as to crystallization in ceramic glazes. One of the major problems associated with these phenomena is the difficulty in separating the processes of nucleation and growth from the overall reaction by observation of the microstructure after various times.

DeHoff[41] has shown through a geometrical derivation that, in fact, both the nucleation rate and the growth rate of a second phase can be determined from quantitative stereological parameters. Basically, it was shown that the particle size distribution expressed in terms of the number of particles per unit volume in the system that are larger than size R at time t, $N_{V>}(R, t)$, is related to the nucleation and growth rates of the phase by

$$N_{V>}(R, t) = \frac{N[\tau(R, t)]}{G(R)} \tag{19}$$

where $N[\tau(R, t)]$ is the nucleation rate as a function of the time of nucleation, and G(R) is the growth rate, which is assumed to be a function only of the particle radius, R. It should be recalled from the previous chapter that the particle size distribution can be determined from measurements which can be made on the microstructure.

C. Mechanical Properties

It is an established fact that the presence of a second phase in a microstructure can influence the elastic modulus, strength, and creep behavior of the material. The second phase can be either pores or particles having either a higher or lower modulus than the matrix. The effect of these second phases should be known quantitatively in order to predict their effect on the properties mentioned above.

Expressions for Young's modulus as a function of volume fraction of second phase have been developed by Paul[42] and by Hashin and Shtrikman[43]. The expressions show that the value of the elastic modulus for a composite containing higher modulus particles in a lower modulus matrix is directly dependent on the volume fraction of particles. This volume fraction can usually be determined most accurately by quantitative microscopy.

Knudsen[44] showed experimentally that the effect of porosity on the elastic modulus could be expressed in terms of Sprigg's equation[45] given by:

$$E = E_o e^{-kP} \tag{20}$$

where E and E_o are the moduli of the porous and nonporous material: K is a material constant and P is the volume fraction of pores.

The volume fraction of pores can be determined by both density and microscopic measurements, but the shape of the pores, also an important factor, can only be determined microscopically.

Increases in strength for two-phase composites can also be explained in terms of the interaction of flaws with the microstructure. Hasselman and Fulrath[46] showed that when special alumina particles in sufficient numbers were added to a glass matrix, the strength of the composite is given by:

$$\sigma = \left(\frac{4\gamma E}{\pi \lambda}\right)^{1/2} \tag{21}$$

where γ is the fracture energy of the matrix, E is the elastic modulus, and λ is the mean free path between particles.

It was shown by Fullman[47] that, for spheres of uniform size, the mean free path between particles is given by:

$$\lambda = \frac{4R(1 - V_V)}{3V_V} \tag{22}$$

where V_V is the volume fraction of particles and R is the particle radius. Fullman also showed that λ could be determined from the expression

$$\lambda = \frac{1 - P_P}{N_L} \tag{23}$$

regardless of the shape or size distribution of the particles. The two parameters on the right-hand side of the expression are easily obtained in any quantitative microstructural measurements, making the determination of λ a relatively easy task.

Substituting Eq. (22) into (21), one obtains

$$\sigma = \left[\frac{3\gamma E V_V}{\pi R(1 - V_V)}\right]^{1/2} \tag{24}$$

By showing that a plot of strength versus $\lambda^{-1/2}$ gave a straight line, Hasseiman and Fulrath[46] concluded that Eq. (21) holds for mean free paths less that about 50 μ, which is the value calculated for the average flaw size in the original glass. They took these results to mean that the introduction of the particles into the glass reduced the flaw size, thereby increasing the strength. Freiman and Hench[48] also reported data showing the importance of the mean free path between crystals in a glass-ceramic in determining the strength of the material.

Lange[49] has recently shown that the fracture energy of a two-phase body, i. e., the energy necessary to propagate a crack in the body, is related to the mean free path between second-phase particles through the expression

$$G = \gamma_o \left[1 + \frac{C}{3\lambda}\right] \tag{25}$$

where γ_o is the fracture energy of the matrix

C is the flaw size, which is assumed to remain constant even though the fracture energy varies, and

λ is the mean free path between particles.

The mean free path between second-phase particles has also been shown to be related to yield strengths of alloys[50, 51]. Gurland and Bardzil[52] showed that the rupture strength of Co dispersion strengthened with WC was dependent on the mean free path between WC particles.

Another possible use of quantitative stereology in the prediction of the strength of brittle materials is in the analysis of the flaw distribution in glass surfaces. Ernsberger[53] has shown that the flaws present in any glass surface can be revealed by an ion exchange treatment in which ions smaller than those in the glass are introduced into the surface from a molten salt bath. Tensile stresses are then produced which open up already existing flaws. By using this treatment on small selected areas, the number, size distribution, and possible orientation of the flaws can be determined. This type of analysis might be of benefit to the glass container industry where the flaw distribution could be correlated with parameters such as breakage rates.

Underwood[54] discussed the analysis of cracking in polycrystalline ceramics by counting prodecures. He proposed that the total surface of cracks in a two-phase ceramic microstructure such as that represented by quartz particles in a whiteware body was given by

$$Sv = (S_V)_i + (S_V)_{per.} + (S_V)_m \tag{26}$$

where the subscripts i, per, and m refer to internal, peripheral, and matrix cracks, respectively, and S_V is the total crack area per unit volume. It is known that $S_V = 2P_L$, where P_L is the number of crack traces intersected by a test line.

D. Composites

Because of the increased use in recent years of both ceramic-ceramic and ceramic-metal composites for numerous applications, the need to understand the relationships between their structure and properties has become greater.

It is known, for instance, that the degree of anisotropy of the fibers in a fiber-reinforced composite is one of the most important parameters in determining both the strength and elastic modulus of the material. This parameter as well as information regarding fiber lengths, particle size distributions, volume fractions of phases, and the amount of interfacial area can be easily obtained by quantitative microscopy. One gets the feeling, however, that not enough use is made of this tool in obtaining this information regarding composite structure.

In other aspects of composite behavior, Jesse and Ondracek[55] investigated the microstructure of several composites by quantitative stereological procedures. They reported on the determination of UAl_3 concentrations in UAl_3-Al reactor fuel samples, and showed that the strength of the composite could be correlated with the volume fraction of UAl_3. They showed that the degree of porosity as measured by quantitative microscopy in UO_2-Cu composites agreed very well with that determined by density measurements. These authors also determined surface quality in UO_2-Mo and Al_2O_3-Mo composites as a function of composition and polishing conditions.

If we can consider ferrous alloys to be in the general area of metal matrix composites, then they represent a composite system in which quite a bit of analysis has been performed by quantitative stereology. Cole[56] discussed the successful application of this technique to the measurement of the volume fraction of pearlite, graphite, delta ferrite, or carbides in steels as well as to the determination of the mean size and spacing of particles. The degree of banding and segregation of carbides in high speed steels has been determined by measuring the volume fraction of banded constituent. The carbon gradient in case-hardened steel was found by measuring the volume fraction of pearlite across the case. Grain size and shape distributions have been determined. Widdowson[57] showed that the volume of nonmetallic inclusions in steel could be measured by automatic counting.

E. Other Applications

Other possible applications for quantitative stereology include the determination of defects in fiber glass and other glass products. These defects can come from devitrification, unreacted batch, degradation of the furnace refractory, gas inclusions, or compositional variations.

The use of stereology in the measurement of particle size distributions has been discussed earlier. Knowledge of this kind is of great importance in the pressing of ceramic products. The fact that grain size distributions can also be determined using these methods has also been discussed. The previous chapter has described the methods by which anisotropy of grain size can be established.

This chapter has attempted to discuss some of the actual techniques of quantitative stereological measurements. Some of the ways in which this analytical technique can be applied toward the characterization of ceramics have also been discussed. There are, of course, many applications for quantitative stereology which have not even been mentioned here. It is hoped that the reader will find numerous applications of his own.

ACKNOWLEDGMENT

The author would like to acknowledge the assistance of W. D. Tuohig in the writing of the sintering and grain growth section.

REFERENCES

(1) H. F. Fischmeister, Quantitative Microscopy, pp. 336-379 (R. T. DeHoff and R. N. Rhines, eds.), McGraw Hill, New York (1968).

(2) Tentative Methods for Determining the Average Grain Size of Metals, ASTM E112-55T.

(3) L. L. Wyman, The New ASTM Grain Size Methods," ASTM Bull., 59-61 (July 1956).

(4) B. Rinman, H. Kjerrman, and B. Kjerrman, En Skala; för bedomning av Slag gruneslutningar i järn och Stal, Jernkoutorets Ann., 91, 199-226 (1936).

(5) Recommended Practice for Evaluating Cemented Carbides for Apparent Porosity., ASTM B 276-54.

(6) C. Orr and J. M. Dalla Valle, Fine Particle Measurement, Macmillan, New York (1959).

(7) G. Herdan, Small Particle Statistics, Butterworth and Co., London (1960).

(8) Photometric Area Determinator, American Instrument Co., Silver Springs, Md.

(9) F. Chayes, A Simple Point Counter for Thin-Section Analysis, Am. Mineralogist, 34 [1] (1949).

(10) P. S. Keeling and C. Stone, "A Method of Quantitative Mineralogical Analysis," Trans. Brit, Ceram. Soc., 53, 67, (1954).

(11) Swift Automatic Point Counter, S. E. 5, J. Swift and Sons, London.

(12) S. Larsson, "A Mechanical Microscanner," Exptl. Cell. Resch., 12, 666 (1957).

(13) C. R. Adler, A. M. Mark, W. R. Marshall, and R. J. Parent, "Scanning Device for Determining the Size Distribution of Spray Droplet Images," Chem. Eng. Progr., 50, 14-23 (1954).

(14) G. A. Moore, L. L. Wyman, and H. M. Joseph, "Comments on the Possibility of Performing Quantitative Metallographic Analyses with a Digital Computer," pp. 380-402, Quantitative Microscopy (R. T. DeHoff and R. N. Rhines, eds.), McGraw Hill, New York (1968).

(15) R. W. Newman, Chapter 7, this volume.

(16) R. T. DeHoff, "A Synopsis of Fundamentals of Quantitative Stereology," Chapter 18, this volume.

(17) J. E. Hilliard, Trans. AIME, 224, 906-917 (1962).

(18) A. N. Holmes, Petrographic Methods and Calculations, Thomas Marby, London (1927).

(19) C. V. Gokularathnam, S. W. Freiman, R. T. DeHoff, and L. L. Hench, "Thickness Error in Quantitative Transmission Microscopy of Ceramics," J. Am. Ceram. Soc., 52, 327-331 (1969).

(20) S. W. Freiman and L. L. Hench, "Kinetics of Crystallization in Li_2O-SiO_2 Glasses," J. Am. Ceram. Soc., 51, 382-387 (1968).

(21) R. T. DeHoff, "Determination of the Geometric Properties of Aggregates of Constant-Size Particles from Counting Measurements Made on Random Plane Sections," Trans. AIME, 230, 764-769 (1964).

(22) R. T. DeHoff and F. N. Rhines, "Determination of Number of Particles per Unit Volume From Measurements Made On Random Plane Sections: General Cylinder and the Ellipsoid," ibid, 221, 975-982 (1961).

(23) J. E. Hilliard, "Measurement of Volume in Volume," 57, Quantitative Microscopy (R. T. DeHoff and F. N. Rhines, eds.), McGraw Hill, New York (1968).

(24) J. E. Hilliard, "Estimation of Microstructural Properties in the Presence of Anisotropy," Ceramic Microstructures, (R. M. Fulrath and J. A. Pask, eds.), John Wiley, New York (1966).

(25) J. E. Hilliard and J. W. Cahn, Trans AIME, 221, 344 (1961).

(26) J. E. Burke, Trans. AIME, 180, 73 (1949).

(27) J. R. MacEwan, J. Am. Ceram. Soc., 45, 37 (1962).

(28) J. R. MacEwan, J. Hayashi, Proc. of the British Ceramic Society, 7, 245 (1967).

(29) G. Haertling, J. Am. Ceram. Soc., 49, 113 (1966).

(30) R. L. Coble, J. App. Physics, 32, 787 (1961).

(31) E. E. Underwood and G. K. Manning, Rev. Metallurg., 60(9), 637 (1963).

(32) G. C. Kuczynski, Trans. AIME 185, 169 (1949).

(33) R. L. Coble, J. Am. Ceram. Soc., 41, 65 (1958).

(34) B. F. Francois and W. D. Kingery, Sintering and Related Phenomena, (G. Kuczynski, N. A. Hooten, and C. F. Gibbon, eds.), Gordan and Breach, New York (1967).

(35) T. K. Gupta and R. L. Coble, J. Am. Ceram. Soc., 51, 521 (1968).

(36) F. N. Rhines, Plansee Proceedings, 3rd Seminar, Rentte, Tyrol, 33 (1958).

(37) R. T. DeHoff, F. N. Rhines, et al., Trans. AIME, 233, 1961 (1966).

(38) E. H. Aigeltinger and R. T. DeHoff, to be published.

(39) A Jesse, "Quantitative Image Analysis in the Field of Nuclear Materials," The Microscope, 19 (1), 65-75 (1971).

(40) C. F. Old and I. MacPhail, "The Mechanism and Kinetics of Growth of the Superconducting Compound Nb_3Sn," J. Materials Science, 4, 202-207 (1969).

(41) R. T. DeHoff, Met. Trans., in press.

(42) B. Paul, Trans. AIME, 218, 36 (1960).

(43) Z. Hashin and S. Shtrikman, J. Mech. Phys. Solids, 11, 127 (1963), 10, 335 (1962).

(44) F P. Knudsen, J. Am. Ceram. Soc., 45, 94 (1962).

(45) R. M. Spriggs, J. Am. Ceram. Soc., 44, 628 (1961).

(46) D. P. H. Hasselman and R. M. Fulrath, "Proposed Fracture Theory of a Dispersion Strengthened Glass Matrix," J. Am. Ceram. Soc., 49, 68-72 (1966).

(47) R. L. Fullman, "Measurement of Particle Sizes in Opaque Bodies, Trans. AIME, 197, 447 (1953).

(48) S. W. Freiman and L. L. Hench, Presented at the International Commission on Glass, London (1968).

(49) F. F. Lange, "Theory on Dispersion Toughening of Brittle Materials," Tech. Report No. 2, N00014-68-C-0323, Westinghouse Electric Corporation.

(50) A. M. Turkalo and J. R. Low, Jr., Trans. AIME, 212, 757 (1958).

(51) W. H. Meiklejohn and R. E. Skoda, Acta Met., 7, 675 (1959).

(52) J. Gurland and P. Bardzil, Trans. AIME (J. of Metals), 313 (Feb., 1955).

(53) F. M. Ernsberger, Phys. and Chem. of Glasses, 1, 37 (1960).

(54) E. E. Underwood, in Ceramic Microstructures (R. M. Fulrath and J. A. Pask, eds.), Wiley, New York (1966), p. 699.

(55) A. Jesse and G. Ondracek, "Application of Quantitative Area Analysis to Composite Materials," The Microscope, 16 (2), 115-121 (1968).

(56) M. Cole, "The Metals Research Quantimet (Q. T. M.)," The Microscope, 15 (4), 148-160 (1966).

(57) R. Widdowson, "Automatic Counting of Non-Metallic Inclusions in Steel," The Microscope, 15 (3), 116-125 (1966).

PART 5

Surfaces

Chapter 20

CHARACTERIZATION OF CERAMIC SURFACES

L. Berrin and R. C. Sundahl
Bell Telephone Laboratories, Inc.
Allentown, Pennsylvania

I. INTRODUCTION

There are numerous ways to describe a ceramic surface. Quite naturally, the properties used for such a description are usually limited and related to the manner in which the surface is utilized. Since the characterization of powder surfaces has been treated in Chapter 16, the discussion in this paper will be exclusively centered on the characterization of bulk ceramic surfaces, polycrystalline and single crystal. Other restrictions result from the authors' familiarity with the use of ceramic surfaces in the electronics industry, namely, as substrates or carriers for metal films. Substrate materials in use, or proposed for use, include Al_2O_3, BeO, porcelains, glazed ceramics, glass, and single-crystal insulators. The emphasis will be on Al_2O_3 which is one of the oxide systems most exhaustively studied and has found general favor in the electronics industry because of its excellent chemical and thermal stability, high electrical resistance, and good mechanical strength. For convenience in handling the subject, it is divided into two sections as follows:

The first section, "Criteria for Substrate Surface Evaluation," discusses the topographical, chemical, electrical, and thermal properties and the role of the ceramic substrate surface as part of an electronic device. It will be seen that the criteria presented here are at least as stringent as those applied to ceramic surfaces used for other applications.

The second section, "Methods for Surface Characterization," deals with standard methods as well as some newer, more sophisticated methods of studying the surface properties outlined in the preceding section. Direct and indirect observations are discussed. In this latter category conclusions are drawn about surface characteristics after the surface is completely coated with a metal film. Although the techniques described in this section are again directed toward the substrate surface, many of the techniques have general application.

II. CRITERIA FOR SUBSTRATE SURFACE EVALUATION[1-4]

A. System as a Whole

The analysis and characterization of surfaces resulting from ceramic forming, machining, and finishing methods is necessary to obtain a fundamental understanding of the effects of such operations. It is through this understanding that ceramic surfaces can be improved so as to meet the rigid requirements of the new technologies. In the electronics industry the ceramic substrate may be called upon to support many integrated circuit chips, each of which may require a large number of interconnections. An example of

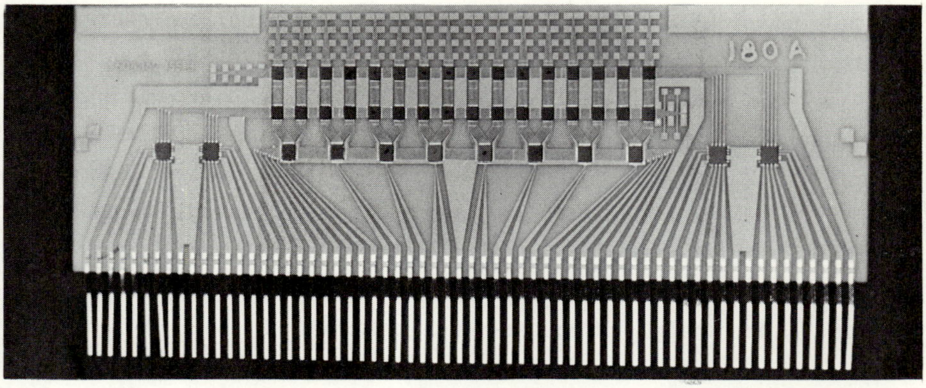

Fig. 1. An alumina substrate, approximately 3 cm x 8 cm, supports the thin film interconnect circuit for a 1024 bit memory. The 32 chips in the top center of the circuit are 32 bit memory chips, each with 26 leads, 50 cm wide and on 100 cm centers.

such an IC interconnect pattern is shown in Fig. 1. Ideally the substrate surface would only interact with the deposited thin form metallization sufficiently to provide adhesion. Other factors important in determining compatibility of the ceramic with the integrated circuit assembly requirements are resistance to chemical and thermal degradation during processing, high electrical resistance, appropriate dielectric constant, and sufficient mechanical strength to withstand bonding operations and handling.

B. Topographical Properties

For the purpose of providing fine metal line definition, film uniformity, and hence reproducible electrical properties the ceramic surface must be smooth. The degree of smoothness required depends upon the circuit application. For resistors, the sheet resistivity and effective path length may depend upon the degree of surface smoothness or grain shape but the dependency is undoubtedly complex and related to the mode of metal deposition. In very thin films local hot spots and failures may result from variable sheet resistivity. For conductor lines the surface smoothness does not appear to be a critical requirement, at least for line widths greater than 2 μ m or spaced no closer than about 2 μ m. When line widths of less than 1 μ m become desired the surface smoothness as regards conductor lines will gain renewed importance. Great demands on surface smoothness are made when circuits include capacitors. In order to obtain tolerable yields, sharp discontinuities in the ceramic surface must be eliminated, otherwise these regions could lead directly to short circuits across the dielectric film. Alternatively replication of surface discontinuities by the dielectric film could result in dielectric failure due to the enhancement of the electric field at a point. Discontinuities greater than about 250 Å are considered harmful. While the average microsmoothness as described above may be important, it is the large discrete surface flaws that have a pronounced deleterious effect upon the integrity of fine-line metal patterns and should, therefore, be eliminated.

One method of placing a circuit pattern onto a ceramic substrate is to deposit the metal and etch this film using a photolithographic process. Fine-line definition is adversely affected by surface waviness or surface camber which may be caused by warping during firing. The deviation from flatness that can be tolerated depends upon whether the substrate or the photographic mask can bend enough to permit intimate contact. Surface deviations of 50 μm/cm are common to as-fired alumina but they are presently acceptable since they are associated with a large radius of curvature. However, a better surface flatness would facilitate the fitting of standard fixtures in processing.

C. Chemical Properties

There are three aspects of the ceramic surface related to the chemistry of that surface which are important for the performance during processing of a thin film device. The three aspects are good adhesion of the surface to metal films, resistance to chemical attack, and chemical compatibility with given metal systems.

In thin film circuitry interconnections are made among components on the substrate surface. Connections must also be made to external components. Gold leads bonded to the ceramic surface must be sufficiently strong to resist tearing loose during subsequent process steps, handling, and extended use under a variety of conditions. It is generally found that the critical link in the bond between the lead and substrate is the metal-ceramic interface. The strength of the bond at this interface is dependent upon both the substrate and the thin film deposition process.

One reason for the acceptance of Al_2O_3 as a substrate material is its resistance to chemical attack by anodizing solutions and by etchants used to delineate the metallization patterns(4). Surfaces containing a high silica content may be susceptible to attack by hydrofluoric acid, a common etchant for refractory metals. Surface degradation during long-term exposure to moisture must also be minimized. Surfaces with a high soda content are particularly affected by high humidity conditions.

Aging effects in tantalum nitride resistors has been directly correlated to the chemistry of the substrate surface, specifically to alkali metal oxides(4). It is believed that alkali ions migrate and electrochemically react with the film. The stability of resistors tested on glass substrates, from soda lime to alumina borosilicate, points out the possible problems that may result from alkali contaminants on ceramic surfaces.

In the above discussion of surface requirements it was assumed that the chemical composition is identical over the entire surface; however, this may not always be the case. A nonhomogeneous distribution of impurities or additives in the raw material, foreign contaminants, and the substrate forming process can contribute to nonuniform surface chemistry.

D. Electrical Properties

The surface conductivity of ceramics, rather than bulk conductivity, exerts the most influence on the characteristics of the circuit components. This surface conduction is a function of the surface condition. The volume resistivity of ceramic substrate materials are $> 10^{10}$ ohm cm at room temperature. However, the surface resistivity is influenced by mobile impurity cations in the raw material segregating at surfaces, surface contamination during processing, and adsorbed gases and water. In fact, in some materials, increasing the relative humidity can cause a drop in surface resistance of 4 to 5 orders of magnitude.

E. Thermal Properties

Along with its resistance to chemical attack, it is the thermal properties of ceramics that make them superior to that of glasses, fused silica, or procelain for many substrate applications. Essentially these are bulk properties; e. g., stability to decomposition at high temperatures, thermal coefficients of expansion, thermal conductivity, and resistance to thermal shock under conditions of rapid temperature changes. The overall stability of the electrical circuit is dependent upon the rate of heat dissipation from the film, especially for resistors which dissipate large amounts of power. Although heat may be dissipated by radiation or convection cooling to the ambient, the major means for decreasing the surface temperature is by conduction through the substrate to heat sinks. Thus the surface temperature is inversely proportional to the thermal conductivity of the substrate.

II. METHODS FOR SURFACE CHARACTERIZATION

A. Surface Topography

A profilometer, Talysurf[5], or other stylus instrument is used to obtain surface traces and a surface "roughness value" in terms of one or more defining parameters. Although this instrument provides a rapid method of evaluating surface roughness for routine applications, a more quantitative assessment requires a statistical approach with macro and micro visual examination of surface details. Besides an average microroughness as determined by a stylus instrument a topographic description of a ceramic surface should include the size and type of significant surface defects and their frequency per unit of surface area. These latter features can be determined by optical microscopy, the scanning electron microscope, or replica transmission electron microscopy. These tools can be used, for example, in correlating a "ceramic surface quality" to manufacturing yield in ceramic substrate microcircuitry applications.

1. The Stylus Instrument[2, 4, 6]

The profilometer consists of a diamond stylus, with a load of less than 500 kg/cm^2, that traverses about 1 cm of the ceramic surface. The stylus tip radius may be as small as 1 μm but is usually about 2.5 μm. The vertical mechanical movements of the stylus are converted and magnified into an electrical signal that is recorded on rectilinear paper. Magnifications range from X1000 to X50,000 and at the highest magnification, corresponding to 0.2 μm/cm of chart paper, vertical displacements can be resolved to about 50 Å. Horizontal displacements of the stylus are not as sensitive as the normal surface displacements. The highest magnification along the surface is x100 corresponding to 100 μ m/cm of chart paper, or 500 times less sensitive than the vertical displacements. This pictorial distortion, or distortion ratio, should be kept in mind when examining actual recorder traces. Depressions in the surface with only a few degrees slope appear as long sharp peaks.

Some typical recorder traces are shown in Fig. 2. Characteristics of these curves which represent prominent features are (1) an average peak-to-valley distance; (2) the maximum peak-to-valley distance; (3) an average half-width of peaks; and (4) an average number of peaks per unit length. The information obtained from these features are of more significance than when one employs the instrument with a cutoff meter resulting in a center-line average (CLA) over a distance of 0.08 cm. However, the CLA is extensively employed for quality control for the sake of simplicity.

2. Optical Microscopy

Defects of 10 μ m diameter and greater have been found to occur on substrate surfaces at a frequency of $1/cm^2$. It is easily calculable that a 1-cm stylus trace would have only one chance in 400 of intersecting such a defect.

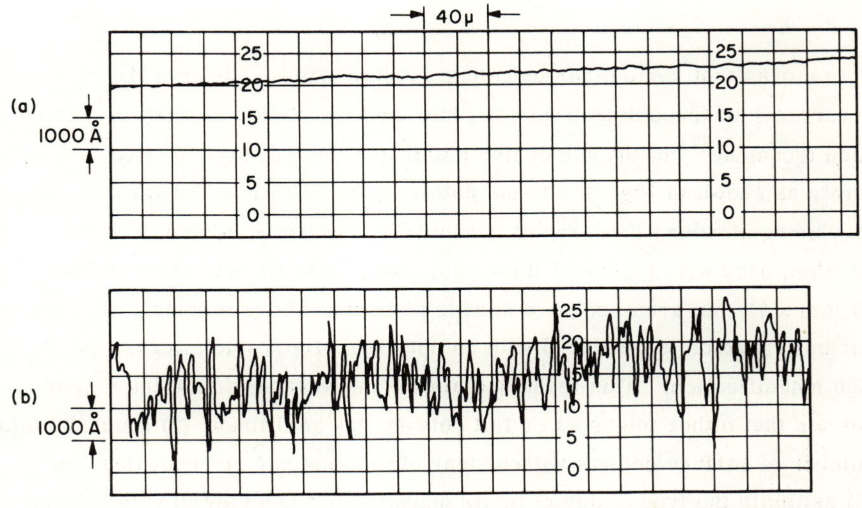

Fig. 2. Profilometer stylus traces: (a) polished surface of sapphire and (b) as-fired surface of 96% dense, tape-cast aluminum oxide.

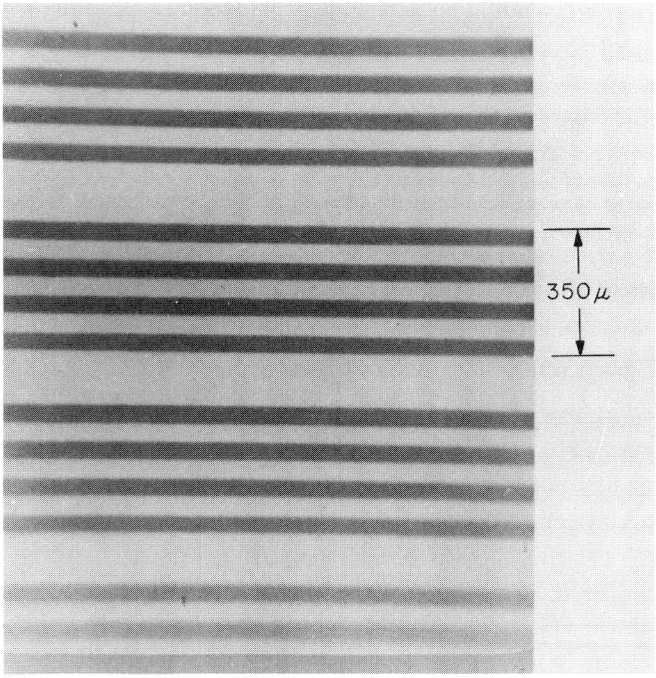

Fig. 3. Test pattern for optical stereoscopic examination of surface flaws viewed by transmitted light.

This means that some other technique must be used to observe the size and occurrence of these defects to complete the topographic evaluation. One such technique[7] is the use of fine metal line test patterns applied to a substrate as shown in Fig. 3. These patterns, generated on the ceramic surface using standard lithographic techniques, consist of a 2000-Å thick titanium base with 15,000-Å thick gold lines. The lines of the resulting 25 mm x 75 mm area are 50 μ m wide and spaced 50 μ m apart. The coated ceramic surface is examined with an optical stereoscopic microscope at X60 magnification. This technique allows one to (1) estimate the size of defects that either intersect or fall between the gold lines, (2) count the total number of surface defects without fear of missing any or repetition, and (3) estimate the type of defect by its appearance when viewed with both reflected and transmitted light. A surface void and a burr-type defect are seen in Fig. 4. Note the unmistakable shadow cast by the burr. This defect

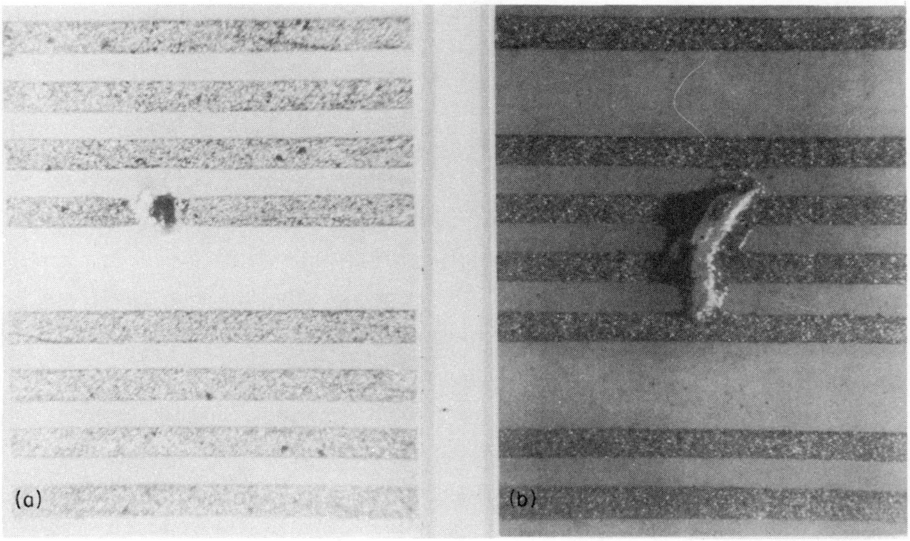

Fig. 4. Optical micrographs taken at a low reflecting angle: (a) surface void; (b) a burr protruding from the surface.

can often be seen without the aid of magnification by holding a substrate at a glancing angle to a high intensity light source.

3. Scanning Electron Microscope[8-11]

Some of the defects observed on substrate surfaces are best viewed with the scanning electron microscope (SEM). The great depth of focus (nearly 300 times that of the optical microscope) and high resolution (~200 Å) of this instrument make it ideal for the direct study of surfaces (e. g., fractures, pits, precipitates, crystal morphology, presence and nature of inhomogenieties, etc.). Details of this technique are described in Chapter 14 of this text.

4. Transmission Electron Microscopy[12-15]

Voluminous papers and monographs written about the transmission electron microscope (TEM) reflect the highly significant work that has resulted from its application. No attempt will be made here to detail the theory of the electron optics or data interpretation since they are described in Chapters 6 and 7.

B. Chemical Properties

In the past it has been the general practice to determine the effect of the chemical properties of the substrate surface through some operational measurement of substrate performance, such as adhesion measurements, aging experiments, or environmental tests. The role of chemistry has then been inferred indirectly from the results of the experiment and, in some cases, knowledge of the bulk chemical composition of the substrate. Recently, however, there has been a proliferation of new analytical techniques which provide direct information about the chemistry of the surface. These include the electron and ion beam methods, such as electron microprobe analysis, Auger electron spectroscopy, photoelectron spectroscopy, and ion microprobe analysis. There are also available to the investigator a series of electromagnetic spectroscopy techniques including infrared spectroscopy, magnetic resonance methods, and ellipsometry. Finally, there are a series of classical techniques available, such as contact angle measurements and gas adsorption studies. The beam techniques have found some application in the characterization of ceramics. However, there has been little work up to this point using the electromagnetic spectroscopy methods to evaluate substrate surfaces. The contact angle measurement is a standard tool for the evaluation of substrate surface cleanliness. Gas adsorption-desorption studies have not been performed as, up to recently, the sensitivity of the technique was limited by the small surface area available.

1. Electron Microprobe[15-18]

The electron beam microanalyzer or microprobe is essentially an "electrons in, photons out" analytical instrument. That is, the sample to be characterized is excited by an incident beam of electrons, which results in the emission of x rays. The x-ray emission spectrum is then analyzed with an x-ray spectrometer. The positions and intensities of peaks in the spectrum are associated with the identities and quantities of the elements making up the sample. As with most instrumentation of this nature, the procedure is simple in concept, but requires a complex and expensive instrument, backed up by even more complex peripheral hardware and software and a competent operator if reliable quantitative results are to be obtained. This technique is described in some detail in Chapter 17 of this text.

2. Auger Electron Spectroscopy[19-22]

The technique of Auger electron spectroscopy (AES) is a new addition to the catalog of analytical tools available for the analysis of surfaces. Because of its high sensitivity (~10^{12} atoms/cm^2), especially to light elements, and its specificity to the surface, the technique has found many applications to problems in surface chemistry. These applications have generally been in the fields of metallurgy (grain boundary segregation, surface diffusion) solid state technology (photoemission from semiconductors, contamination of Si wafers, segregation of impurities at surfaces), catalysis (chemisorption on Pt), and solid state and molecular physics (band structure, molecular spectra). As might be suspected from this brief list, there has been very little done on insulating surfaces (some work has been reported on MgO[23], Al_2O_3[20], and KCl[24]) and almost nothing done on the characterization of complex ceramic surfaces with AES. This is in part due to the newness of the technique and the associated lack of awareness on the part of ceramists of AES, and of those active in the field of problems in ceramics. There are more fundamental limitations which have impeded application, however. After a brief description of the Auger process itself and some of the instrumentation used in AES, we will discuss some of these considerations and how one can, if not eliminate the difficulties, at least live with them.

The Auger process of electron ejection is a relaxation process complementary to x-ray emission. In many ways, as we shall see, AES supplements information gained from the x-ray emission spectrum obtained from the electron microprobe. Whereas the microprobe was described as an "electron in, photons out" device, AES is an "electrons in, electrons out" technique. In both cases the sample surface is excited with an incident beam of electrons [see Fig. 5(a)]. Thus the analyses are performed in a vacuum. The beam characteristics in a typical AES spectrometer might be 3 kV and 100 μA. X-ray emission [Fig. 5(b)] involves the relaxation of the excited atom, in this case Mg, through an electron transition and simultaneous photon emission. The frequency of a particular transition depends on (1) the probability of excitation and (2) the relaxation time for photon emission. The latter is found to increase rapidly as the energy of the transition [$K - L_2$ in Fig. 5(b)] decreases. This is one of the reasons for the difficulties with sensitivity in light element x-ray emission spectroscopy.

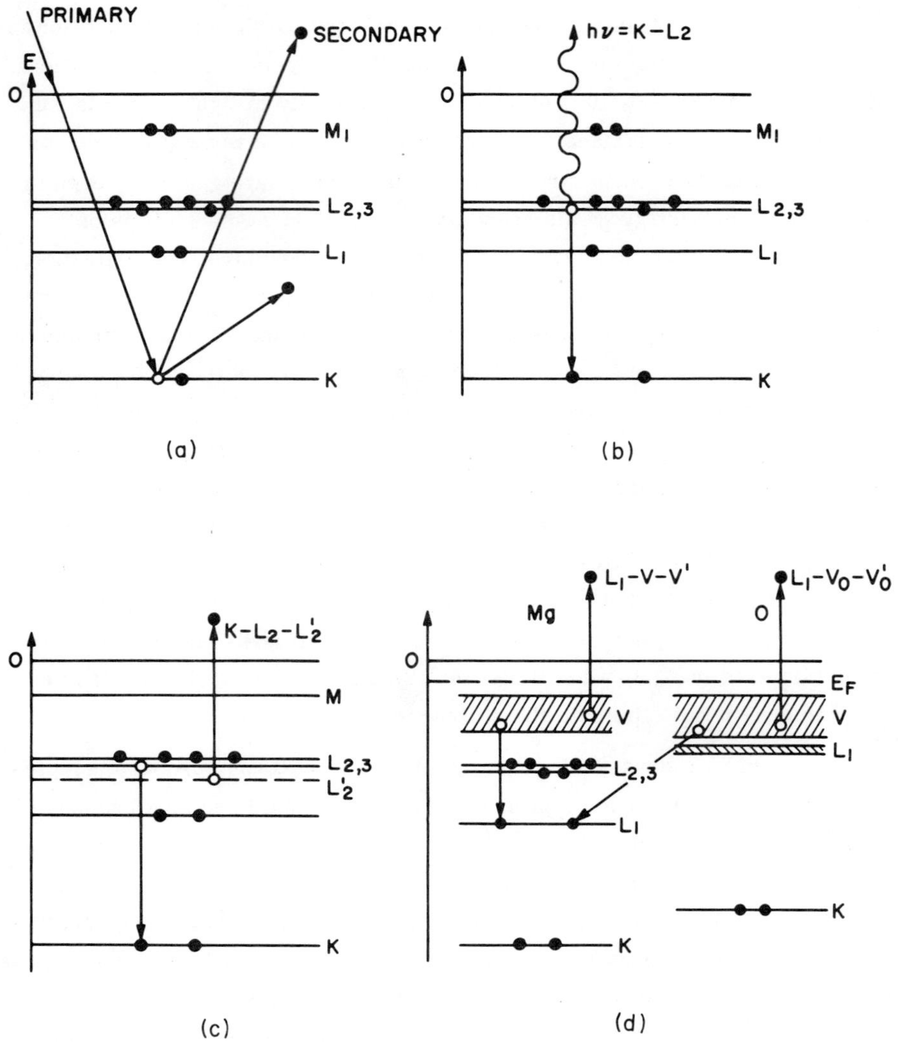

Fig. 5. A comparison of Auger and x-ray emission processes for Mg, MgO. (a) The excitation of the atom with an incident electron is common to both techniques. (b) Relaxation of excited atom via x-ray emission. (c) Relaxation accompanied with Auger ejection of an electron — a KL_2L_2' transition. (d) Auger transitions in MgO.

The long relaxation time for x-ray emission for low energy transitions would not result in a decrease in photon yield and a corresponding decrease in x-ray emission sensitivity if there were no alternative relaxation process available — the excited atoms would have to relax sometime and emit a photon. However, as is true in most processes of this type, nature eventually does find an alternative. In this case it is Auger ejection. In Fig. 5(c) we see the same excitation and decay process, $L_2 \rightarrow K$ but accompanied this time by the ejection of an electron. Conservation of energy dictates that the kinetic energy of the ejected electron be $K - L_2 - L_2'$, where L_2' is the level perturbed by the double ionization of the Mg atom. Selection rules for Auger emission are relaxed from those for photon emission and many "forbidden" transitions are observed. The KLL spectrum of Mg exhibits six transitions; however, the resolution of most spectrometers is such that only three, the KL_1L_1, $KL_1L_{2,3}$, and $KL_{2,3}L_{2,3}$, are detected. An Auger spectrum of MgO is seen in Fig. 6.

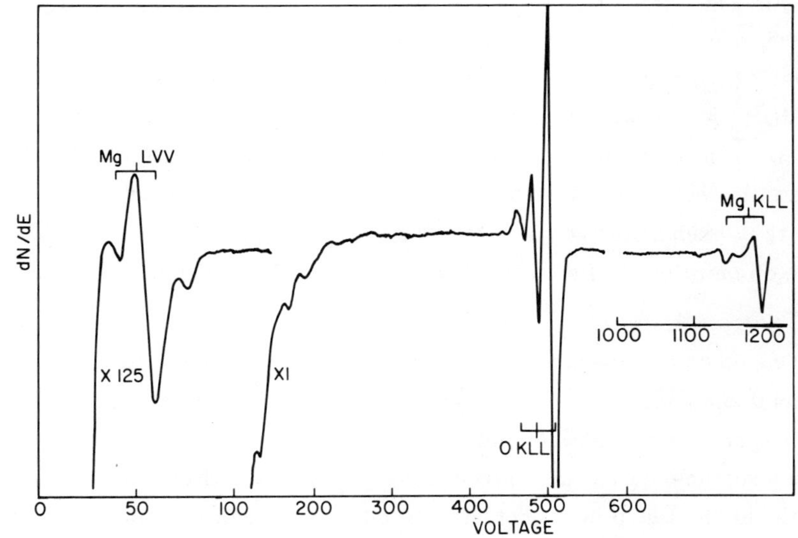

Fig. 6. An Auger spectrum of MgO. The low energy Mg LVV transitions are by far the most intense. A series of weak transitions are observed between 100 V and 300 V which have not been identified. The oxygen and magnesium are KLL transitions but differ in intensity by an order of magnitude as Auger yield drops off above 1000 V.

The relaxation time for Auger emission is approximately independent of the energies involved and, as a result, there is a crossover point on the energy scale for electronic transitions, below which Auger emission is more probable than x-ray emission. Below 1000 eV virtually all relaxations are associated with Auger emission, thus the relatively high sensitivity of AES to the presence of light elements.

The above discussion applied only to gaseous atoms with their well-defined energy levels. The case of a solid is complicated by the broadening of energy levels into bands. In the case of insulators, such as MgO depicted in Fig. 5(d), the problem is made even more complex. Here there is a good deal of uncertainty as to the origin of the electrons which decay into the lower level and of the origin of the Auger electrons. If MgO were purely ionic, the L_1VV' transition would be missing from the Auger spectrum, as would the corresponding L_β x-ray emission spectrum. These transitions are observed and would have been explained alternatively in terms of a deviation from ionicity on the part of MgO, crossover transitions from $O^=$ band, as depicted in Fig. 5(d), or some combination of the two alternatives. There are several inconsistencies in these models when used to explain soft (K_β) and ultrasoft ($L\beta$) x-ray emission spectra of compounds such as MgO, Al_2O_3, and SiO_2. A careful examination of the cation LVV and KLL AES spectra, as well as the oxygen KLL spectrum may resolve some of these difficulties. Meanwhile, however, these complications make the interpretation of Auger spectra of oxides ambiguous in the range of energies below 100 eV. This range of energies should be avoided, if possible, in any analytical investigation.

One other consideration relative to the study of solids is the volume of material analyzed in an AES measurement. It was stated earlier that the Auger spectrum is characteristic only of atoms at, or very close to, the sample surface. That this is indeed the case is seen when one considers that the critical step in the detection of an Auger electron is the diffusion of that electron to the surface. Unfortunately there is little information, either experimental or theoretical, on escape depths of electrons in the energy range 100 to 1000 eV. The experimental evidence that is available[22, 23, 25] indicates that for electrons with an energy of about 500 eV one would expect a mean escape depth of 5 to 15 Å or 1 to 5 monolayers.

Finally we should define the limit of sensitivity of the technique. Chang[22] has estimated that the one method of analysis, the retarding field method, is ultimately limited by shot noise to a signal-to-noise ratio (S/N) of 5000, which corresponds to a sensitivity, at S/N = 1, of 10^{12} atoms/cm^2. This would be equivalent to a bulk concentration of $10^{18}/cm^2$, or 100 ppm. A check of this estimate with $4 \times 10^{14}/cm^2$ chemisorbed oxygen atoms on Si yielded a S/N of 200, in agreement with his calculation. Recent improvements in techniques, described below, have provided several orders of magnitude improvement in S/N. However, it was found that sensitivity did not show a corresponding increase, as it appears that ultimate sensitivity of AES is limited to $\sim 10^{11}$ atoms/cm^2 by the presence of a complex fine structure of undetermined origin in the background spectrum.

The detection of the Auger spectrum from a surface is complicated by the presence of large number of electrons in the secondary electron energy distribution N(E) which are emitted from the surfaces by processes other than Auger emission [see Fig. 7(b)]. The N(E) curve is characterized by a large low energy peak at about 10 eV, a long tail and a peak due to elastic reflection of incident electrons at E_o. There are several features in the spectrum associated with the elastic peak and are due to reflected electrons which have suffered inelastic losses. Superimposed on this high background [exaggerated greatly in Fig. 7(b)] is the characteristic Auger spectrum. The problem to be solved then in an AES measurement is the detection of the weak Auger signals in the presence of a high background level.

There are two types of instruments currently in common use which have solved this signal detection problem. The retarding field analyzer[26, 27], seen in Fig. 8, uses the low energy electron diffraction (LEED) electron optics for electron energy analysis. Many experiments have been performed in which complex crystallographic structures of surfaces have been correlated with the presence (or lack) of stabilizing surface impurities. The instrument is essentially a high pass filter, accepting all electrons at the collector with an energy greater than the bias voltage V_B on the spherical grid structure. The collector current i_c is then the integral of N(E) from eV_B to eV_O [see Fig. 7(a)]. One must then differentiate i_c electronically to obtain N(E). This is accomplished electronically by modulating V_B with an ac signal of frequency ω_o and measuring the first harmonic of i_c. A large gain

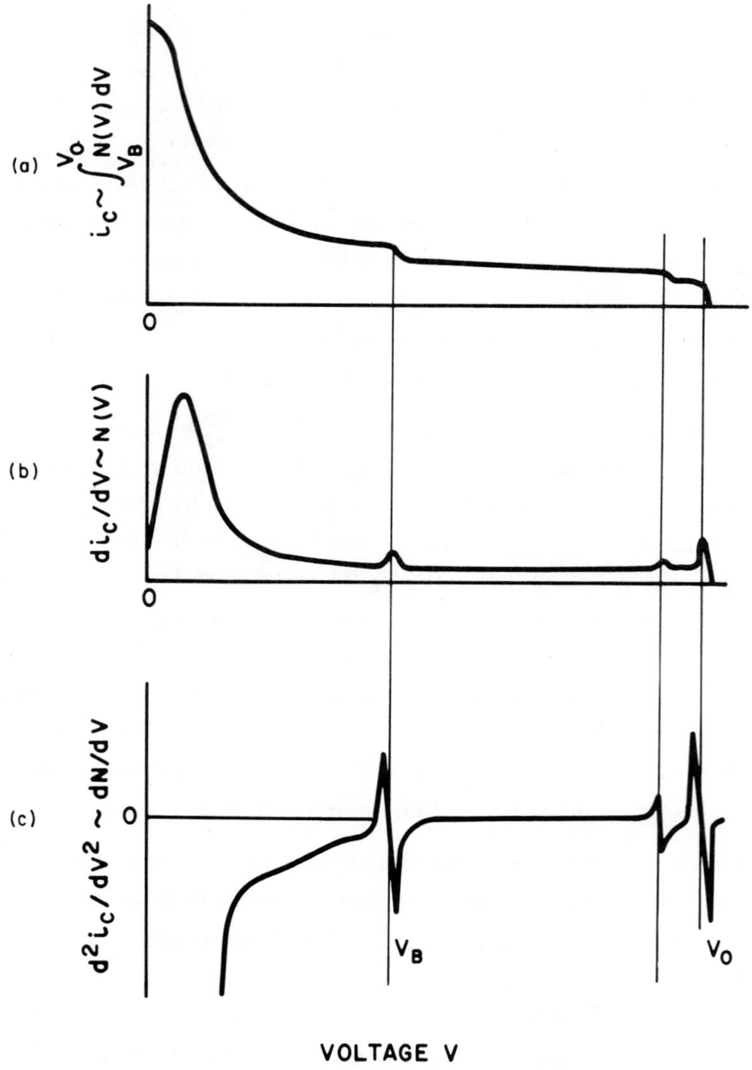

Fig. 7. Schematic diagram showing the relationships between the various harmonics of the collector current i_c and the energy distribution N(E). (a) The dc current is equal to the integral of N(E) from eV_B to eV_o. (b) The first harmonic $i_c(\omega_o)$ is proportional to di_c/dV_B or to $N(eV_B)$. (c) The second harmonic $i_c(2\omega_o)$ is proportional to d^2i_c/dV_B^2 or to dN/dE.

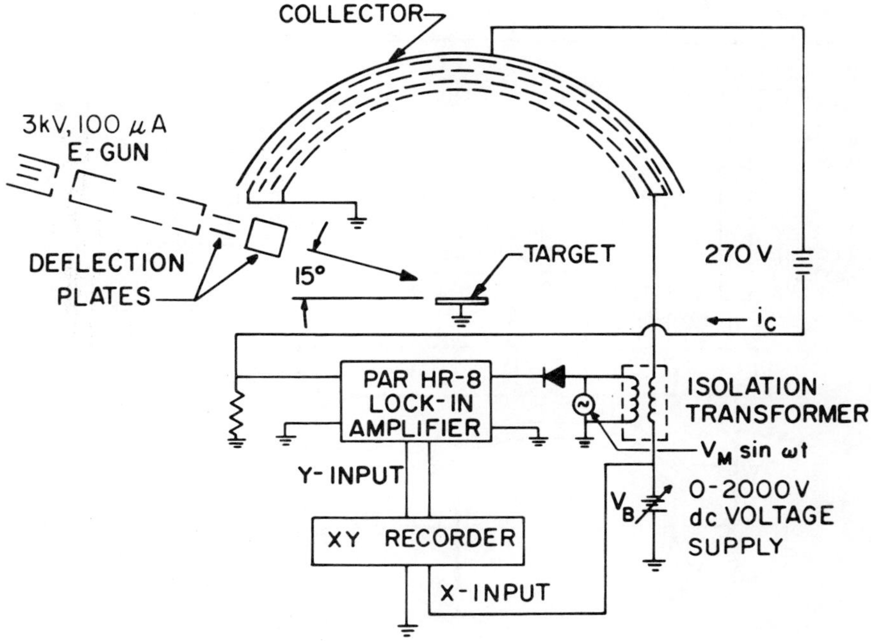

Fig. 8. Schematic diagram of the retarding field analyzer.

in signal-to-background ratio can be obtained by measuring the differential dN/dV_B [Fig. 7(c)]. This is accomplished by measuring the second harmonic (frequency $2\,\omega_o$) component of i_c. It is then the combination of the LEED optics, with its wide acceptance angle, the concept of measuring dN/dV_B rather than $N(V_B)$ and the use of sophisticated signal processing techniques which has made the measurement of Auger spectra a practical tool.

The second method of analysis, first described by Palmberg et al. [28] in 1969, uses a coaxial cylindrical electrostatic energy analyzer for measurement of Auger spectra (see Fig. 9). The cylindrical analyzer is a bandpass device — it admits only those electrons within a narrow energy range ΔE. This reduction in the total number of collected electrons results in a corresponding reduction noise level N as the shot noise contribution to N is proportional to $(i_c)^{1/2}$. The reduction in noise level by several orders of magnitude has been accomplished without loss in signal S as (1) the acceptance angle of this spectrometer is roughly to that of the retarding field

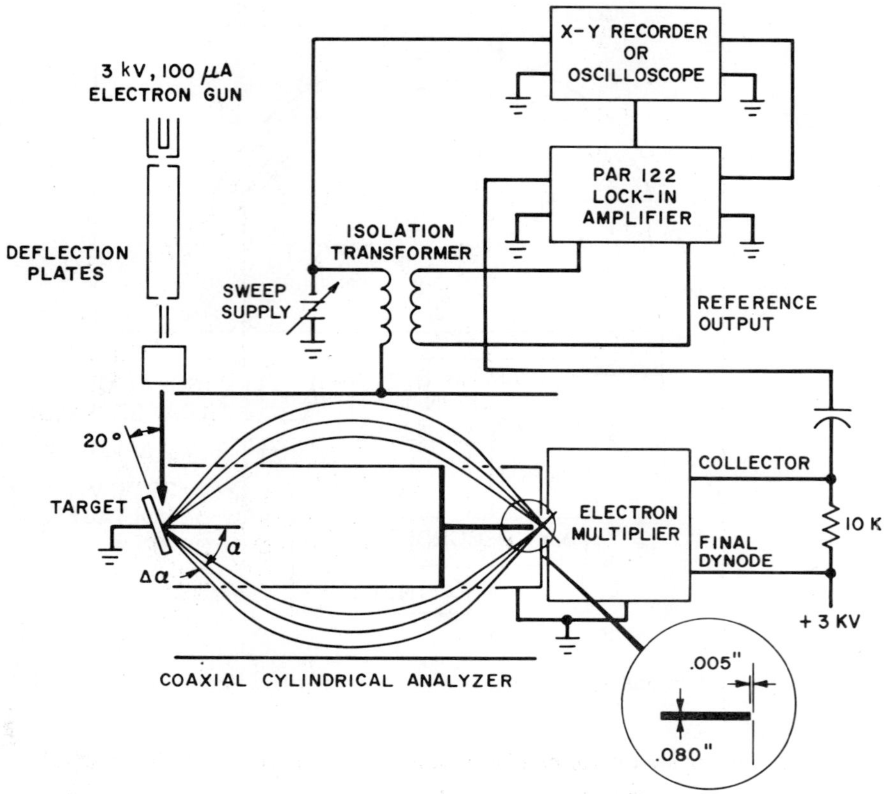

Fig. 9. Schematic diagram of the cylindrical mirror analyzer.

structure and (2) one need only perform a single differentiation on the signal, which is now proportional to N(E). The gain in S/N has made it possible to perform real time in situ experiments, as now sensitivity can be traded off for speed. Voltage sweeps of several hundred volts can be performed in milliseconds and spectra displayed on an oscilloscope, rather than in minutes using a recorder output, as with LEED optics. Alternatively, one can operate the analyzer using a very low beam current such as those available in the scanning electron microscope or electron microprobe. MacDonald et al[29] have described the use of a specially designed cylindrical analyzer in a SEM to study the surfaces of fractured steel.

We now have seen that an Auger spectrum is similar to an x-ray emission spectrum, but with certain important differences: (1) the complex emission process results in a more complex spectrum, (2) a higher sensitivity to light elements, and (3) higher selectivity to the surface. We have also seen that this spectrum can be measured with existing spectrometer designs and impurity levels of 0.1% of the surface can be detected if one is quite careful. Let us now discuss how this technique might be applied to problems peculiar to ceramic technology.

The type of problems which have been studied with AES in other areas of technology can be classified according to the atomistic phenomena examined. These include investigations of the segregation of bulk impurities to surfaces and grain boundaries, the contamination of surfaces during certain process steps, chemisorption of gases on surfaces, depletion of one component of a compound at the surface, and the electronic structure of elements and compounds. These same atomistic phenomena play important roles in problems of interest in ceramic technology, such as sintering, electrical properties of ferrites, leakage across insulator surfaces, epitaxial growth of single-crystal films on insulating substrates, adhesion of various glazes, brazing alloys and films to ceramic bodies, and the effect of surface treatments on the strength of ceramics.

Application of AES to a problem of this nature requires that, just as with the electron microprobe, the ceramist keep in mind those aspects of the technique which will limit its application to his problem. The list of these considerations is quite similar to those for the microprobe (both are "electrons in" methods) and can be categorized under the headings (1) surface charging of insulators, (2) contamination of the surface prior to and during analysis, (3) effects of the e-beam on the sample chemistry, and (4) quantitative analyses.

a. Surface Charging of Insulators. The problem is the same as described briefly in the case of the electron microprobe and the SEM. In this case, however, the surface charge cannot be dissipated with a thin conductive coating, as all that would be observed in the AES pattern would be a spectrum characteristic of the coating. Several observations by this author and others[21, 22] seem to support the conclusion that surface charging effects, although not well understood as yet, can be controlled or avoided in most cases. In general the only effect of charging on the high energy spectrum (>200 eV) is to displace the spectrum uniformly toward a lower voltage (for a positive charge buildup). Since in most cases there is a carbon and/or an oxygen peak present, these peaks can be used as fiducial marks to estimate the extent of the voltage shift. The choice of incident beam voltage should be such that the charging is positive, otherwise the problems are magnified. The spectrum at low voltage (< 100 eV) is more strongly affected by sample charging. The use of a conductive mask over all but the area to be examined tends to reduce charging effects, but it is best whenever possible to avoid the low voltage region as suggested above for different reasons.

b. Surface Contamination. The word "contamination" has a rather negative connotation. It may very well be that it is exactly the contamination that is of interest. Impurities picked up during machining, firing steps, or cleaning operations may play an important role in determining the performance of a ceramic body in certain applications. In other investigations the contamination due to the ambient is only a nuisance. In these cases it is best to prepare the surface to be examined in situ. As an example, the ideal method to study grain boundary segregation on a fractured ceramic surface is to fracture the sample in an ultra high vacuum station and perform the analysis without breaking vacuum. Such experiments have been carried out on steels[(29)]. Alternatively, the sample can be cleaned in the vacuum station by ion beam etching or by the electron beam itself (electron desorption). One must carry out such a procedure with a great deal of care as one doesn't want to remove the "interesting" contaminants in his zeal to rid the surface of the "nuisance" contaminants.

The vacuum station itself can also be a source of contamination. It is well known that organic contaminants and CO present in the vacuum station will decompose under the electron beam leaving a residue of amorphous carbon. It should be remembered that it is the partial pressure of the carbon-bearing gases that is important rather than the total pressure. For this reason it is useful to have a residual gas analyzer available to monitor partial pressures.

c. Electron Beam Effects. In addition to e-beam decomposition of contaminants on the surface, the beam may serve to decompose the surface itself. The effects can be similar to those observed in soft x-ray spectroscopy — the decomposition of SiO_2 to Si has been reported in AES studies[(22)]. The beam may also preferentially sputter away one component of a compound, as in the case of KCl[24]. Finally, if one is interested in impurities on the surface, one has to be especially careful as these are easily desorbed by the electron beam. One of the major advantages of the cylindrical analyzer with its high sensitivity is the ability to operate at low currents (1 μa), reducing electron desorption rates and damage.

d. Quantitative Analysis. The ability to obtain quantitative numbers on surface levels of impurities is still quite limited in the case of AES. There has been no attempt as yet to put the technique on the same quantitative basis as electron microprobe analysis although there have been semiquantitative descriptions of some of the factors which influence signal level[21,22,27,30]. The few quantitative studies carried out have involved the study of a well-characterized clean surface onto which was introduced a known amount of a single impurity[21-23,25]. It was found that signal level was a linear function of impurity level below one monolayer. Such experiments are not feasible for multicomponent ceramic systems, and one is generally limited to semiquantitative correlations between signal level and some other property, such as ceramic performance. It is useful, however, to gain some knowledge of the signal levels to be attained from some standard sample of the impurity under investigation. A measure of the Mg signal from MgO, as an example, will provide a rough estimate of the sensitivity of the instrument to Mg segregation at Al_2O_3 grain boundaries.

3. Photoelectron Spectroscopy[31-34]

The technique of photoelectron spectroscopy (or ESCA, electron spectroscopy for chemical analysis) is closely related to, and sometimes indistinguishable from Auger spectroscopy. ESCA differs in that the usual excitation source is an intense beam of x rays and that the emitted photoelectrons (or Auger electrons) are analyzed with an elaborate, high resolution electron spectrometer. A schematic of the "photons in, electrons out" process of photoelectron spectroscopy is seen in Fig. 10. The attractive feature of the ESCA technique is the inherently narrow linewidth associated with the process. Also, the photoelectron process is a simpler mechanism than either x-ray emission or Auger ejection, and, therefore, easier to interpret. Finally, the elimination of the electron beam source results in a considerable reduction in secondary and back-scattered electrons as source of background noise.

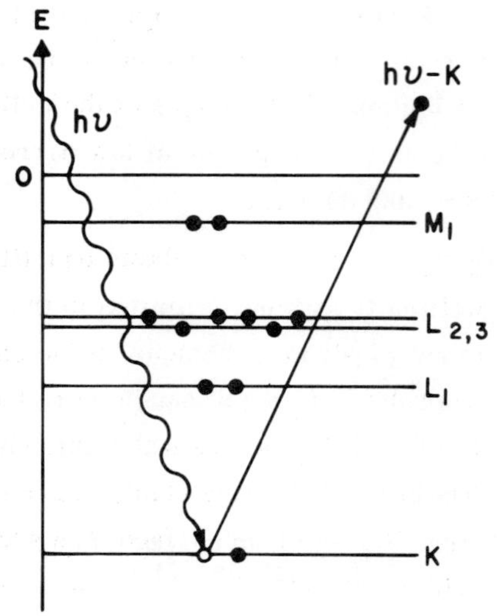

Fig. 10. Emission of a photoelectron from an Mg atom.

The fundamentals, applications, and instrumentation associated with the various forms of ESCA have been described in detail by Seigbahn and his co-workers(31, 32), who pioneered in this technique, in two recent monographs. The instrumentation has been developed to exploit the inherently high resolution of this technique and a number of sophisticated electrostatic and magnetic electron spectrometers are now on the market. Resolutions on the order of $\Delta E/E \sim 10^{-4}$ have been reported.

The sensitivity of the technique is not as high as that of Auger spectroscopy, being limited by the relatively low fluxes available from x-ray sources and the low cross sections for photoelectron ejection. The detection problem has been sloved through the use of multichannel analyzer techniques and, in some cases, small digital computers. A schematic of such an arrangement is seen in Fig. 11(33).

The spatial resolution of ESCA does not compare favorably either with Auger spectroscopy or the electron microprobe. The sample area exposed to the x-ray beam is on the order of 1 cm^2. Surface selectivity is determined by the escape depth of emitted electrons, as in the case of AES. However, photoelectron energies normally are on the order of thousands, rather than hundreds, of electron volts and correspondingly, average escape depths

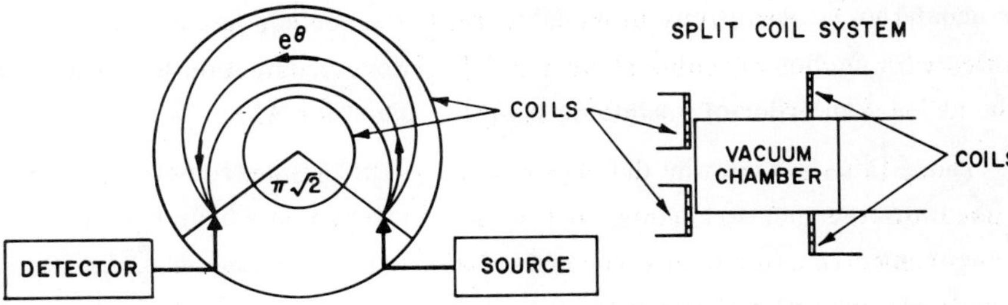

(a)

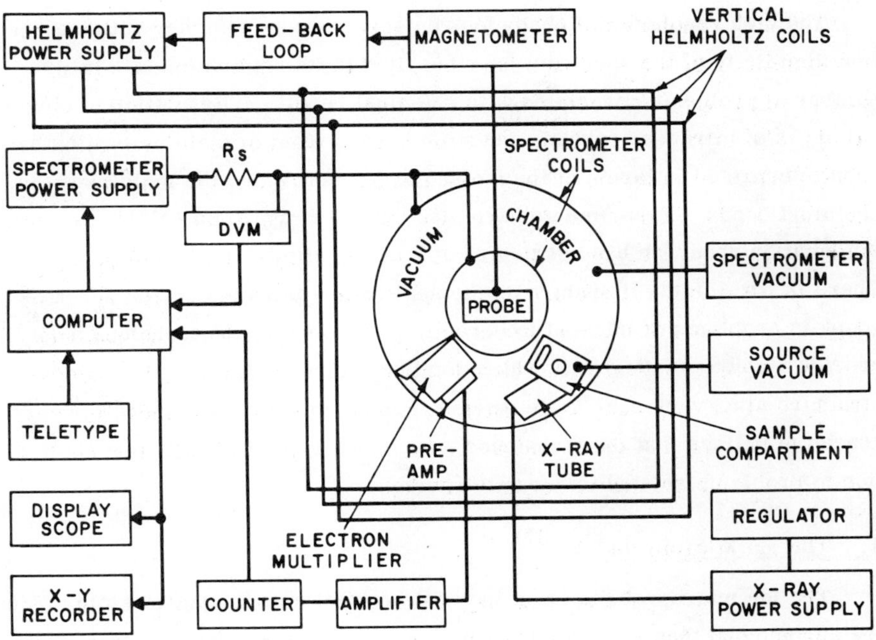

(b)

Fig. 11. Schematic diagram of an electron spectrometer with the signal processing electronics. (a) A double-focusing magnetic spectrometer. Electrons emitted from sample at source diverge vertically and horizontally and are focused back to point at detector. (b) Spectrometer with associated electronics: (1) x-ray power supply and source, (2) power supply for spectrometer electromagnetic coils, (3) detector and signal processing equipment, and (4) residual magnetic field compensation system.

are about 100 Å. Sensitivity to monolayers on the surface has been demonstrated with studies of iodine stearates[31]; however, detection levels appear to be at least an order of magnitude less than found for AES.

There is no reason why the high resolution ESCA spectrometer cannot be used for AES measurements, and in fact measurements have been reported of Auger spectra using both x-ray and e-beam excitation modes. An example of an Auger spectrum obtained for the Mg KLL transition on a high resolution ESCA spectrometer[31] is seen in Fig. 12. It is to be noted that five transitions are now resolved.

The high resolution of photoelectron spectroscopy together with the relative simplicity of the spectrum have resulted in its application to a large number of problems associated with chemical bonding. Resolution of chemical shifts of nitrogen, carbon, and sulfur peaks from organic molecules have been interpreted in terms of the molecular structure and the ionicity of the chemical bonds. Experiments have also been performed on solids, providing information about the band structure of metals, alloys, and compounds. There is little in the literature on the application of photoelectron spectroscopy to problems of interest to ceramists. However, the technique does provide a unique method with which detailed information about the chemical structure at or very near to the surface can be obtained, and there is every reason to believe that the techniques associated with ESCA will find application to problems related to ceramic surfaces.

4. The Ion Microprobe[35-37]

The ion microprobe uses an incident beam of ions to sputter atoms from the surface and then analyzes the sputtered atoms in a mass spectrometer. There are several versions of ion microprobes now on the market. In many ways the ion microprobe appears to be the ideal analytical instrument for characterization of surface chemistry, as well as bulk chemistry. One version provides an intense beam of ions, either positive or negative, which can be focused to a spot from 2 μ to 500 μ diameter. Using the fine focus, the beam can be rastered across the surface, just as with the electron microprobe or SEM, and a two-dimensional visual display of the output can be obtained on an oscilloscope (see Fig. 13). Sensitivity is determined by beam current and mass spectrometer design and sensitivities in the 10^{-9} range are claimed.

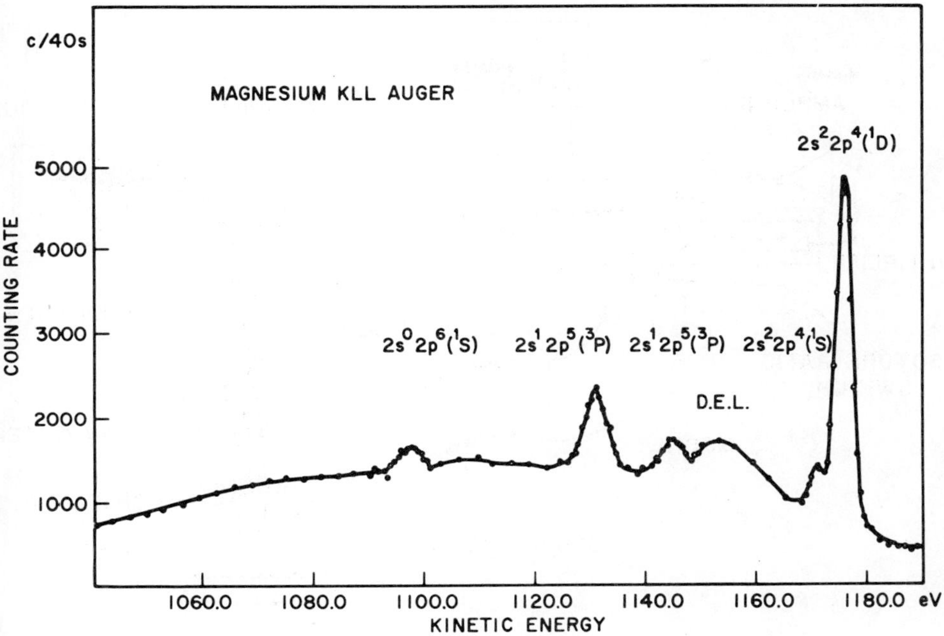

Fig. 12. The Mg KLL spectrum measured on a high resolution electron spectrometer. Five of the predicted six peaks are detected. The sixth peak, at 1183 V, is too weak to be detected.

The surface selectivity is, by definition, confined to the first layer of atoms. In addition to high selectivity, the ion microprobe provides the unique facility for obtaining a composition profile of the surface as the incident beam sputter etches away the surface atoms. There are certain problems associated with this sputter etch method of analysis. As the beam forms an etch pit on the surface, the sides as well as the bottom of the pit contribute atoms to the total yield. This, of course, tends to smear out any composition profile. One solution to this problem is to use a defocused beam and aperture the detector optics such that only atoms from the bottom of the pit are accepted[(35)]. An alternative solution is a finely focused beam which is rastered over an area of the surface. The latter approach is, in principle at least, a better approach but results in very slow etch rates[(36)]. In addition, errors are also introduced into composition profiles by the inherent statistical nature of the sputtering technique (even a delta function profile would exhibit an exponential decay) and by differences in sputtering yields for different components.

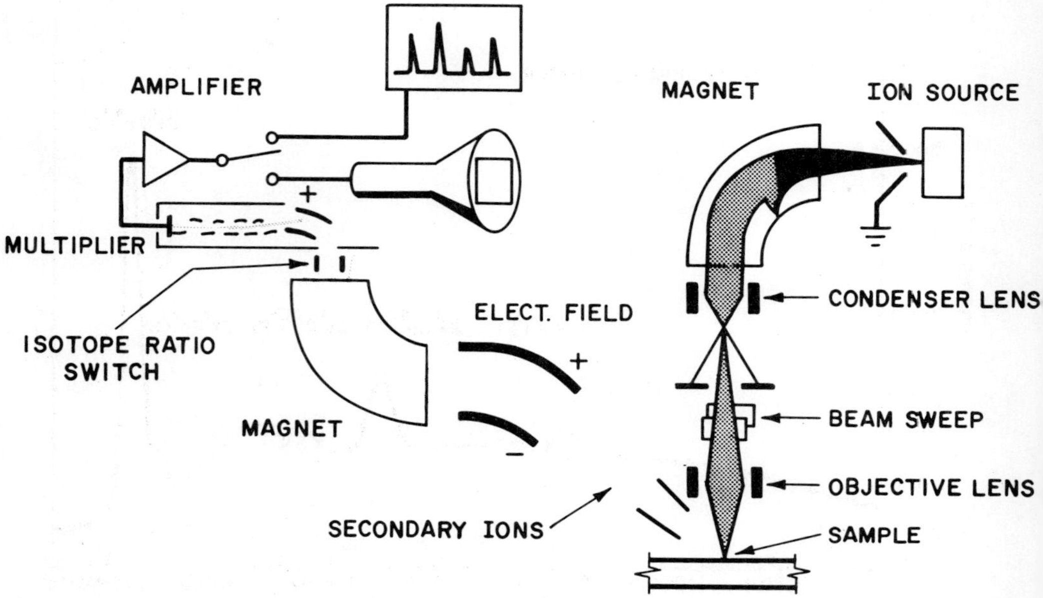

Fig. 13. Schematic diagram of the ion microprobe.

A great deal of effort has gone into the analysis of the various processes associated with the sputtering and detection steps in the ion probe and the technique can provide accurate quantitative information under carefully controlled conditions. An additional feature of the ion probe technique is the ability to perform isotopic analyses, enabling one to quickly and accurately determine diffusion profiles.

An instrument with all these features which have obvious potentialities for the study of ceramic systems must have at least one negative feature. This feature is, of course, the complexity and associated cost of the instrument. The pricing of such instruments is such that only the largest research and development organizations could contemplate the purchase of an ion microprobe.

The techniques described previously provide direct information about the chemical composition of the surface. Next we will briefly examine a series of methods which provide indirect information about the surface chemistry. These methods include infrared spectroscopy, ellipsometry, contact angle measurements, and gas adsorption and desorption studies. Although the data obtained cannot be directly associated with specific surface components, it often is more convenient to make such measurements and, in many cases, these techniques are more sensitive to subtle changes in surface chemistry than are the beam techniques. Also, the information, which in many cases is related to some electrical property of the surface, might be useful in evaluation of the performances of the ceramic surface as part of a device.

5. Infrared Spectroscopy[38-40]

The absorption spectrum of a material in the infrared spectrum (0.001 to 1 eV) is determined by the excitation of electronic, vibrational, and rotational states in the material. The adsorption of a molecular species on a surface results in a change in the IR spectrum as (1) the vibrational modes associated with the surface will be modified and (2) the rotational and vibrational modes of the molecule will be changed drastically. Adsorption bands thus not only provide information about the molecular species present, but also about how the environment of the surface modifies these species. Unfortunately, the sensitivity of the technique is such that in most cases an extremely large surface area is required, and the application of IR spectroscopy has been limited mainly to the characterization of powders of such materials as SiO_2, Al_2O_3, and MgO. A significant amount of information is available in the literature on topics such as the hydration of these surfaces, but to what extent this information is applicable to solid ceramic surfaces is not immediately obvious.

Alternative methods of IR spectroscopy have been developed to cope with this problem, and the technique has been used to evaluate thin metal films and thin glass films, such as glazes, on Si substrates in transmission. A multiple reflection method has been used to study the surface of thin plates of materials such as Al_2O_3, MgO, and quartz. Thus IR spectroscopy can be applied to the characterization of "idealized" forms of ceramics, that is, single crystals which are transparent in the infrared. The application to polycrystalline ceramic surfaces, however, is not feasible.

6. Ellipsometry[39, 41, 42]

The ellipsometer is used to measure changes in the state of polarization of light refected from a surface. In the case of a clean surface, the degree of polarization and phase change can be directly related to the real and imaginary components of the index of refraction n_s at the surface. If the surface is coated with a film, then the optical behavior is related in a very complex way to the angle of incidence, the optical constants of the substrate and film, and the film thickness. In this case it may be necessary to make additional assumptions to define the problem, such as assuming the imaginary component n_2^* of the film index of refraction to be zero. Under the best conditions (large difference between optical constants of substrate and film) the technique is sensitive to small fractions of a monolayer. However, losses of sensitivity occur in the case of low reflectivity surfaces such as polycrystalline ceramics. Still, the technique exhibits a high sensitivity to subtle changes in surface properties and is found to be useful as a measure of surface cleanliness[42].

7. Contact Angle Measurements[45-47]

Contact angle and related measurements fall into two classifications. The one type of measurement is used as a routine method for determination of the presence of hydrophobic contaminants (mainly organic residues) on ceramic substrate surfaces prior to metallization[4, 43]. This technique involves the observation of water droplets on the ceramic surface. In its quantitative form, an optical goniometer is used to measure the contact angle between the water droplet and the ceramic. Contact angle measurements are currently used as a standard inspection procedure for ceramic and glass substrates following the cleaning step in the thin film metallization process.

There is a large body of literature which describes more elaborate experiments in the same vein which have been performed on ceramic surfaces. Measurements of contact angles of ceramic-liquid interfaces, combined with surface tension data, provide a wide variety of information about parameters such as interfacial energies, work of adhesion, and tendencies toward surface and interfacial segregation of impurities[44]. A typical experiment of this type involved the study of liquid Ni sessile drops, with various impurities present, on sapphire surfaces[45]. The measurements of the drop shape and the contact angle provide the information necessary to calculate the liquid-vapor and liquid-solid surface free energies γ_{LV} and γ_{LS}. Changes in γ_{LS} and γ_{LV} with impurity content can be directly related to the excess interface concentration of the impurity[46]. Studies of contact angle variations as a function of time[47], together with metallographic examinations of frozen interfaces, provide information about the nature of the reaction between the liquid metal and solid oxide. This type of information is of special interest to those concerned with adherence of metal films or brazing alloys to ceramics. In one case the adherence of the frozen sessile droplets has been directly related to the work of adhesion and the reaction at the interface[45].

Most data on contact angle measurements are performed on single-crystal oxides, such as sapphire, MgO, NiO, and CaO. One has to have certain misgivings about many of the investigations as they are usually carried out in vacuum systems with relatively high ($\sim 10^{-6}$ Torr) pressures. The nominally high purity substrates were not cleaned in situ, and results will be characteristic of the contaminated surface. It might be hoped that reactions at the interface would alleviate this problem.

In the case of ceramic systems it is of vital interest how surface contaminants influence interfacial energies. Experiments in which measurements are made as a function of ceramic history should provide useful information about the interrelation of contamination and adhesion.

8. Adsorption and Desorption of Gases[44, 48-50]

The last in this particular catalog of techniques is a method, actually a group of methods, used to determine the extent to which a gas phase adsorbs on a surface. As discussed in Chapter 16, the usual approach[44, 48] to this problem is to expose a powder sample to the gas in question and measure change in volume or pressure change of the gas, or the weight change of the sample[50]. Alternatively, the sample, with the gas adsorbed, can be heated and the thermal desorption of the adsorbate studied. These techniques, like many others described earlier, are easily applicable to powders with large surface areas but are not useful in the study of bulk materials.

Sensitive gas pressure measurement devices and mass spectrometers, together with ultrahigh vacuum techniques, have made it possible to monitor adsorption and desorption on low surface area solids[49]. After exposure of a clean surface to a gas at a given pressure and temperature and for a given time, the sample is rapidly heated, flash desorbing the gas. The pressure rise in the vacuum system is directly related to the surface coverage. A careful examination of the pressure-temperature dependence many times reveals peaks in the P(T) plot. These peaks can be associated with the adsorption of the gas at a particular surface site. A detailed study of the time-temperature-pressure relationships for thermal desorption provides information about the extent of gas coverage and the kinetics of adsorption and desorption.

Surface physicists often combine this type of experiment with measurements of such parameters as the work function, photoemissive properties, crystallographic structure, and Auger spectrum. It is this combination of several techniques in one investigation which provides a detailed description of the surface. This type of information would be of use to those interested in the chemical activity of a ceramic surface, as well as to anyone concerned with the effect of adsorbed species such as H_2O and CO_2 on some technological process, e.g., adhesion of metallization systems or electrical leakage at the surface.

C. Crystallography of Ceramic Surfaces

In most cases the surface has crystallographic structure identical to that of the bulk material. Exceptions to the rule, however, can be of some importance, particularly in the field of electronic ceramics. Measurements on single-crystal surfaces suggest that reconstruction of the crystallographic structure of the first few monolayers is a common occurrence[51-53]. The reconstructed surface might be stabilized by the presence of impurity atoms, such as oxygen on surfaces such as Ni or W, or reconstruction may occur without impurities present. The latter phenomenon is still a somewhat controversial subject however. The significance of surface reconstruction, either impurity stabilized or not, to the behavior of the complex polycrystalline ceramic surface is not apparent. There has been some work on single-crystal insulating surfaces which dealt with the effect of surface structure on the epitaxial growth of thin films of silicon[54, 55].

On a macroscopic level an important crystallographic characteristic of the polycrystalline ceramic surface is the texture, or orientation distribution of the individual crystallites. One might expect that there would be no preferential orientation of crystallites in a ceramic and in most situations none is observed; the material exhibits a random texture. However whenever the fabrication process includes some nonisotropic generalized stress in a process step, the ceramic body will tend to "remember" the bias introduced into the orientation and, in some cases, the effect will be exaggerated in later steps. Such a "stress" might be the nonisotropic mechanical stresses introduced in some hot pressing operations, a magnetic field during wet pressing of a magnetic ceramic or the effect of gravity and vibration during packing of an acicular powder into a mold prior to pressing. These effects would be reflected in the behavior of the bulk properties rather than the surface alone.

The surface itself might be thought of as a uniaxial "stress" which would affect texture at the surface alone. The mechanism by which this stress affects crystallite orientation is the orientation dependence of surface energy. Some crystallites at the surface will happen to be oriented so as to expose a low surface energy plane (usually a low index plane). These will grow at the expense of less favorably oriented grains during the grain growth stage in sintering. One would expect then that as-fired surfaces of polycrystalline ceramics would exhibit a uniaxial or fiber texture. Such behavior has been observed on tape cast alumina substrates, where a strong (0001) fiber texture was measured in a thin surface layer[56, 57]. Since the texture is related to surface energy one might expect that changes in ceramic composition which result in changes in surface chemistry might also result in a change in the degree of texture or, in some cases, a change in the basic texture orientation.

The surface texture might affect the performance of an electronic ceramic component in several possible ways. The growth of thin films deposited on a surface could be influenced by the texture, changing the usual thin film texture seen for a particular system. The dielectric constant, which is a second-order tensor, is known to be strongly dependent on orientation for anisotropic materials, such as corundum[57]. High frequency thin film circuits fabricated on textured surfaces will exhibit characteristic stripline impedances which differ from those calculated for an isotropic material. The same considerations apply for other tensor properties such as thermal conductivity and thermal expansion coefficient.

1. Electron Diffraction Methods — Low Energy Electron Diffraction[51-53]

The techniques available for analysis of the crystal structure of ceramic surfaces are limited. The two most powerful tools for surface structure analysis are low energy electron diffraction (LEED) and reflection high energy electron diffraction (variously, RHEED, RED, or HEED). The low energy mode utilizes a beam of electrons of an energy of about 100 eV to 1000 eV at normal incidence to the surface. The LEED optical arrangement is similar to the AES electron optics in Fig. 8, but with a small electron gun directed at the sample through the collector and grid structure, along the central axis of the optics. The electrons elastically reflected from the sample pass through the grid structure, biased to reflect all inelastic electrons, and are accelerated by a high positive potential toward the phosphor-coated collector. The diffraction pattern, which is similar in appearance to a Laue pattern, is then observed on the collector. Alternative methods have been devised which use a small Faraday cup on a diffractometer as a collector, enabling the investigator to make quantitative measurements of spot intensities directly. As will be seen, however, the state of the art in LEED is such that for most purposes the semiquantitative data available from diffraction photographs are more than adequate.

Information available from LEED studies is quite difficult to interpret, and applications of interest to the ceramist are restricted. However there are aspects of this technique which may be useful in specific areas. Several characteristics of LEED phenomena relevant to its application to ceramic problems are listed below.

1. The low energy electrons interact very strongly with the sample. Because of the high probability of elastic collisions, as well as inelastic collisions, the diffraction pattern is representative of only the first few monolayers of the surface.

2. Since diffraction takes place in a virtual two-dimensional crystal, rather than three-dimensional crystal, the conditions for diffraction are relaxed. As a result, diffraction will occur from a single-crystal surface irrespective of its alignment with the beam in directions which satisfy Bragg's law. A polycrystalline surface then reflects electrons in all directions, and no diffraction pattern is observed. Again, one might imagine an analogy to the Laue pattern. As a consequence, LEED studies are limited to single crystals.

3. Another consequence of the strong interaction is the difficulty in interpretation of LEED data. The complexity of the phenomena (multiple scattering and inelastic scattering effects dominate the behavior) make it virtually impossible to develop a model of the crystal surface from the diffraction data in a rigorous manner. Some progress has been made in the inverse problem, that of calculating diffraction intensities from the model. Until a breakthrough is accomplished in LEED theory, analysis is limited to a combination of symmetry arguments based on the symmetry of the LEED pattern, a simple kinematic interpretation of LEED intensities and information gained from other techniques such as gas adsorption studies, work function, and optical measurements, and AES and ion scattering studies. This approach has been successful in constructing physically plausible models of surface structures for materials such as cleaved Si and Ge and metals with adsorbed gases present.

4. As with other electron beam techniques surface charging of insulators presents a problem. Investigations have demonstrated that operation above the "crossover point" ($\delta > 1$ and positive charging) leads to undistorted diffraction patterns. An alternative approach has been to heat the sample to a temperature sufficient to reduce the spreading resistance to a reasonable value. Additional problems are encountered if the insulating surface is heterogeneous. The presence of electron trap regions may lead to unusual electric field configurations at the surface and distort the diffraction pattern.

5. In addition to information about the symmetry of the surface structure, LEED experiments also provide knowledge about such phenomena as structural disorder, thermal vibrations, surface steps, and faceting and magnetic order at the surface.

6. The sensitivity of the technique is such that if less than 5 - 10% of the surface has a particular crystallographic structure, that structure will go undetected. As a result the LEED method is not suitable for the study of small nuclei or crystallites present on a surface.

Despite some of the limitations described in this list, the LEED method is capable of providing useful information about insulator surface structures. Studies have been made of the nucleation and growth of films on such substrate materials as mica, KCl, MgO, and Al_2O_3. Of particular interest are a series of experiments reported on the structure of Al_2O_3 surfaces and the effect on epitaxial growth of Si films[54, 55]. It was determined that the crystal structure of an (0001) sapphire surface is highly dependent on sample history. Deposition of Si at high temperature however tends to etch any reconstructed layer away as some Si forms SiO and evaporates. The result is an (0001) surface identical to the bulk structure under the epitaxial Si.

2. Reflection High Energy Electron Diffraction[53, 56-59]

The reflection high energy electron diffraction technique differs from LEED both in the mode of the experiment and in the phenomena observed. The experimental arrangement, shown in Fig. 14 is determined by the short wavelength ($\lambda = 0.1$ to 0.05 Å) of the high energy electrons (10 keV to 100 keV). Because a significant amount of scattering of such a beam occurs in a spread of a few degrees, an incident beam angle ϕ of about 1° is necessary to obtain a measurable number of scattered electrons exiting the surface. As might be imagined, alignment of the electron beam in this configuration is more critical than for the LEED method. For this reason the electron optics of the e-gun in Fig. 14 are more complex than encountered in LEED. Experimental difficulties with contamination are similar to those described for other surface characterization techniques, and if one is interested in the surface it is best to conduct RHEED measurements at pressures lower than 10^{-9} Torr. This technique is not as sensitive to surface contamination as LEED however, and investigations of thin film structures have been conducted in commercial systems operating at pressures of 10^{-7} Torr and higher. Consistent with importance of alignment, it is necessary to provide some means by which the sample can be moved precisely with several degrees of freedom. In many instances the manipulator is also equipped with a facility for heating, cooling, or even cleaving.

The preparation of the sample surface itself is quite critical in RHEED studies. Again, as a consequence of the low angle of incidence, the diffraction pattern is quite sensitive to surface topography. In fact, the whole nature of the diffraction pattern changes with subtle changes in topography. The two extreme situations for diffraction from a single-crystal surface are illustrated in Fig. 15. The diffraction pattern from an atomically flat surface, in Fig. 15 (a), consists of a series of vertical streaks, a consequence of the low penetration of the beam into the surface (about 10 Å) and the effects of refraction. This pattern is usually displayed on a phosphor screen. One

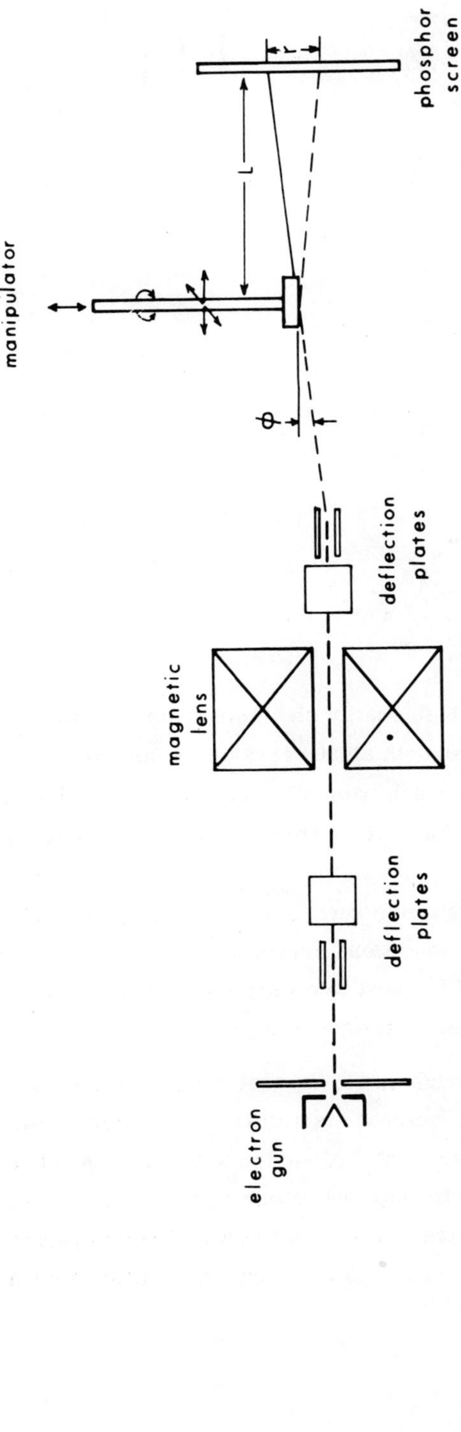

Fig. 14. Schematic diagram of reflection high energy electron diffraction system.

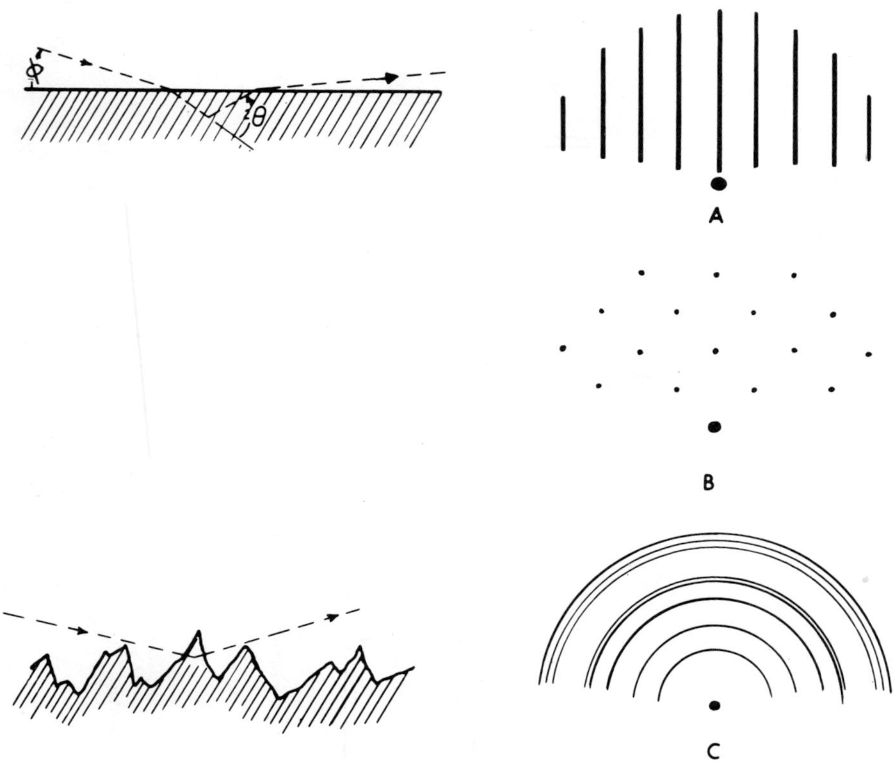

Fig. 15. Diffraction of high energy electrons from a surface. (a) Diffraction from an atomically smooth single-crystal surface produces a rod diffraction pattern. (b) Diffraction from a rough single-crystal surface produces a spot pattern. (c) A ring pattern from a rough polycrystalline surface.

might imagine that a polycrystalline surface would produce no diffraction pattern at all if such surface conditions prevailed. However, it is more likely, especially in cases of interest to ceramists, that polycrystalline surfaces will exhibit some degree of surface roughness.

The diffraction from a rough single-crystal surface is seen in Fig. 15(b). We see here that the angle of incidence is actually much larger than ϕ, so refraction is no longer as important. Also, the diffraction is actually a transmission rather than a reflection phenomenon, and there is no loss in resolution normal to the surface. The result is a diffraction pattern which is equivalent to a plane section through the reciprocal lattice normal to the sample surface. A polycrystalline surface will, under the same circumstances, result in the formation of Debye-Scherrer rings, illustrated in Fig. 15 (c) for the case of randomly oriented crystallites. One of the major advantages of the RHEED technique over LEED then is the ability to study the structure of polycrystalline surfaces. Furthermore, the diffraction pattern not only provides information about the crystal structure but also the topography and orientation of the surface. It must be remembered however that statistically the "sample" examined by RHEED is heavily weighted toward the protrubances in Fig. 15 (b). If the surface is inhomogeneous the pattern will exaggerate those components present which tend to protrude from the surface.

Interpretation of reflection diffraction patterns is also complicated by the geometry of the experiment. There are situations however which allow for quantitative analysis of intensity data in terms of a crystallographic model. This feature of RHEED experiments provides another advantage over the LEED technique. Three characteristics which limit the quantitative interpretation of reflection patterns are (1) effects of geometry and topography, (2) low resolution, and (3) dynamical scattering phenomena. As was implied above, the diffraction pattern is strongly dependent on surface topography. Most observed single-crystal patterns will consist of some combination of Figs. 15(a) and (b). In the case of the polycrystalline surface the shape of the individual crystallites can affect the intensity and width of the Debye rings. This phenomenon can be used as a method for characterizing particle sizes, in analogy to x-ray line broadening studies. However, the interpretation would be at least as complicated as the x-ray case.

The geometric aspects of the reflection mode also lead to difficulties in interpretation. Determination of d spacings from powder pattern measurements depend on knowledge of the "camera constant" $\lambda L \simeq dr$ (see Fig. 14). This constant can be determined empirically with some standard, such as a MgO aerosol. However the value of λL is quite sensitive to systematic errors introduced by refraction effects, surface roughness, and alignment.

In addition to systematic errors, resolution ($\Delta r/r$) is also limited by several line broadening effects and measurements of d-spacing of better than 1% are difficult to obtain. Particle size broadening, surface strain and texture, as well as beam spot size, magnitude of the camera constant λL, resolution of the phosphor screen or photographic plate, and ripple in the e-gun supply all have an influence on the size of diffraction spot or width of rings. As a result it is difficult to make an identification of an individual structure if it is complex, and it is nearly impossible to separate out the structures of several phases without corroborating information from techniques such as microprobe analysis or AES.

The above considerations tend to limit the application of RHEED to the identification of surface phases. Application of RHEED to quantitative studies of crystal structure analysis, texture, topology, and crystal perfection are also limited by the complexity of phenomena which determine diffraction intensities. The analysis of x-ray data is based on the kinematical theory, the basic assumption of which is that only a small fraction of the incident beam is scattered by the sample (Born approximation). Because of the strong interaction of the charged electrons with matter, unfortunately, this simplifying assumption is invalid for electron diffraction and the more complex dynamical theory must be used. This theory has only been developed in a quantitative form in a few idealized situations and cannot be applied to the quantitative analysis of reflection data. As a result, intensity analyses have been limited in general to semiquantitative treatments. As an example, changes in the intensity of diffraction spots associated with a corrosion product can be correlated with the progress of a corrosion reaction. If the material studied is made up of very small particles or thin films (e. g. 200 Å for Al) the simple kinematic treatment can be applied to the intensity analysis.

Inevitably, as with other techniques employing an ionized beam, charging of insulating surfaces has to be considered in a RHEED experiment. The problem is particularly critical for this technique:

1. The high energy of the incident beam usually results in a secondary yield coefficient $\delta < 1$ (negative charging).

2. Because of the geometry of the reflection experiment, alignment is especially difficult in the presence of a negative charge.

3. The high degree of sensitivity of the surface precludes the use of a conductive coating. One common solution to this problem is the use of a low voltage (200 - 1000 eV) electron gun to discharge the sample. The production of excess secondary electrons ($\delta > 1$) in this energy range can neutralize the surface charge.

Application of RHEED to the characterization of ceramic surfaces has been limited. There are however a number of examples of applications in the literature which are of direct interest to the ceramic engineer. Two such applications are the study of surface damage introduced by fabrication processes and the characterization of amorphous insulating films.

One of the first practical applications of RHEED was the examination of the "Beilby layer" on polished metal surfaces[59, 60]. Diffraction patterns indicated the presence of a thin amorphous layer which overlay a deformed crystalline layer. There is still some controversy as to the nature of this amorphous layer and to what extent oxide particles stabilize the layer. One might presume that damage introduced by grinding or polishing a ceramic would differ from a metal surface. The RHEED technique should be capable of supplying a great deal of information about the nature of this damage[59]. Line broadening of the Debye-Scherrer rings can be directly related to distortion and fragmentation of the polycrystalline structure, due either to a surface treatment or bulk mechanical working.

The extreme case of line broadening is, of course, the diffuse pattern observed for metal surfaces. The circumferential intensity distribution in these same rings can be quantitatively related to surface texture, in a manner similar to the x-ray analysis of texture. If the instrument manipulator is equipped with an adequate heating capability one can perform an in situ study of the effect of annealing temperature and time on these parameters. Finally, if the instrument is equipped with an ion bombardment gun, successive surface layers can be removed by ion etching, providing a unique method of studying the nature of surface damage as a function of sample depth, again in situ.

X-ray diffraction studies of bulk amorphous materials such as silica and other types of glasses have demonstrated that the diffuse pattern obtained from these materials contains considerable information about near-neighbor distances and distributions in these structures. Diffuse electron diffraction patterns contain the same information about thin films on surfaces of amorphous materials[57]. The intensity data can be analyzed quantitatively using the simple kinematic model in this case as dynamical effects are minimized. An example of such an investigation is a comparative study of thin SiO_2 films grown of Si by pyrolytic decomposition and by oxidation of Si in steam, dry

O_2, and an oxygen plasma[61]. It was found that all films exhibited a high degree of short range order and that upon aging at room temperature certain films devitrified.

REFERENCES

(1) B. Schwartz, in Electronic Ceramics, p. 12, American Ceramic Soc., Special Publ. No. 3, Columbus, Ohio, 1969.

(2) M. D. Rigterink, in Electronic Ceramics, p. 27, American Ceramic Soc., Special Publ. No. 3, Columbus, Ohio, 1969.

(3) D. G. Thomas, Physics of Electronic Ceramics, (L. L. Hench and D. B. Dove, eds.), M. Dekker, New York (1971).

(4) R. W. Berry, P. M. Hall, and M. T. Harris, Thin Film Technology, Chap. 9, Van Nostrand, Princeton, 1968.

(5) Product of Taylor Hobson Co. Ltd., England.

(6) N. Schwartz and R. Brown, Trans. 8th Natl. Vac. Symposium, New York, 1961, p. 836, Pergamon, New York, 1962.

(7) R. C. Sundahl, L. Berrin, and E. J. Sedora, Effects of Alumina Substrate Defects on Thin Film Interconnect Patterns, presented at the 72nd Annual Meeting of the Amer. Ceram. Soc., Phila., May, 1970.

(8) P. R. Thorton, Scanning Electron Microscopy, Chapman and Hall, London, 1968.

(9) R. F. M. Thornley and L. Cartz, J. Amer. Ceram. Soc., 45, 425 (1962).

(10) G. Remond, S. Kimoto, and H. Okuzumi, in Scanning Electron Microscopy/1970 (O. Johari, ed.), p. 33, IIT Research Inst., Chicago, 1968.

(11) C. W. Oatley, W. C. Nixon, and R. F. Pease, in Advances in Electronics and Electron Physics (L. Marton, ed.) Vol. 21, p. 181, Academic, New York, 1965.

(12) C. E. Hall, Introduction to Electron Microscopy, McGraw-Hill, New York, 1966.

(13) R. D. Heidenreich, Fundamentals of Transmission Electron Microscopy, Interscience, New York, 1964.

(14) G. Thomas and W. L. Bell, in Techniques of Metals Research (R. F. Bunshah, ed.), Vol. 2B, Interscience, New York, 1971.

(15) W. Reuter, in Modern Methods of Surface Analysis (P. Mark and J. D. Levine, eds.), Surf. Science, Dec. 1970.

(16) The Electron Microprobe (T. D. McKinley, K. F. J. Heinrich, and D. B. Wittry, eds.), Wiley, New York, 1966.

(17) T. J. Gray and J. K. Zope, in Materials Science Research (W. W. Kriegel and H. Palmour III, eds.) Vol. 3, p. 301, Plenum, New York.

(18) S. N. Ruddleson, Trans. Brit. Ceram. Soc., 66, 587 (1967).

(19) E. H. S. Burhop, The Auger Effect, University Press, Cambridge, 1952.

(20) L. A. Harris, J. Appl. Phys. 39, 1419 (1968).

(21) N. J. Taylor, in Techniques in Metals Research (R. F. Bunshah, ed.) Vol. 7, Interscience, New York, 1971.

(22) C. C. Chang, in Modern Methods of Surface Analysis (P. Mařk and J. D. Levine, eds.) To be publ. in Surface Science, Dec. 1970.

(23) P. W. Palmberg and T. N. Rhodin, J. Appl. Phys., 39, 2425 (1968).

(24) P. W. Palmberg and T. N. Rhodin, J. Phys. Chem. Solids, 29, 1917 (1968).

(25) R. E. Weber and A. L. Johnson, J. Appl. Phys., 40, 314 (1969).

(26) R. E. Weber and W. T. Peria, J. Appl. Phys., 38, 4355 (1967).

(27) P. W. Palmberg, in The Structure and Chemistry of Solid Surfaces (G. A. Somerjai, ed.), pp. 29-31, Wiley, New York, 1969.

(28) P. W. Palmberg, G. K. Bohn, and J. C. Tracy, Appl. Phys. Letters, 15, 254 (1969).

(29) N. C. MacDonald, H. L. Marcus, and P. W. Palmberg, in Scanning Electron Microscopy/1970 (O. Johari, ed.) IIT Research Inst., Chicago, 1970, p. 70.

(30) L. A. Harris, Surface Science, 15, 77 (1969).

(31) K. Sieghahn, C. Nordling, et al., ESCA, Almquist and Wiksalls AB, Upsalla, Sweden, 1967.

(32) K. Siegbahn, C. Nordling, et. al. ESCA Applied to Free Molecules, North-Holland, Amsterdam, 1970.

(33) D. M. Hercules, Anal. Chem., 42, 20 (1970).

(34) C. R. Brundle and M. B. Robin, in Determination of Organic Structures by Physical Methods (F. Nachod and G. Zuckerman, eds.) Academic, New York, 1971.

(35) A. J. Soucha, in Modern Methods of Surface Analysis (P. Mark and J. D. Levine, eds.)) to be publ. in Surface Science, Dec. 1970.

(36) C. A. Anderson, J. Mass Spectr. and Ion Phys. 2, 61 (1969).

(37) C. A. Evans, Jr. and J. P. Pemsler, Anal. Chem., 47, 1061 (1970).

(38) M. L. Hair, Infrared Spectroscopy in Surface Chemistry, Dekker, New York, 1967.

(39) W. A. Pliskin, in Physical Measurements and Analyses of Thin Films (E. M. Murt and W. G. Guldner, eds.) p. 168, Plenum, New York, 1969, p. 1.

(40) R. K. Webster, T. G. Jones, and P. J. Anderson, in The Physics and Chemistry of Ceramic Surfaces, Brit. Ceramic Soc., Stoke-on-Trent, 1965, p. 143.

(41) Ellipsometry in the Measurement of Surfaces and Thin Films (E. Passaglia, R. R. Stromberg, and J. Krugers, eds.) NBS Misc. Publ. 256, U. S. Govt. Printing Off., Washington, D. C., 1964.

(42) R. R. Stromberg and F. L. McCrackin, in Clean Surfaces (G. Goldfinger, ed.), p. 65, Dekker, New York, 1970.

(43) M. L. White, in Clean Surfaces (G. Goldfinger, ed.), p. 361, Dekker, New York, 1970.

(44) A. W. Adamson, Physical Chemistry of Surfaces, 2nd Ed., pp. 345, 414, Interscience, New York, 1967.

(45) W. H. Sutton and E. Feingold, in Materials Science Research (W. W. Kriegel and H. Palmour III, eds.), p. 577, Vol. 3, Plenum, New York, 1966.

(46) C. R. Kurkjian and W. D. Kinergy, J. Phys. Chem. 60, 1961 (1956).

(47) R. D. Carnahan, T. L. Johnston, and C. H. Li, J. Amer. Ceram. Soc., 41, 343 (1958).

(48) The Solid-Gas Interface, (E. A. Flood, ed.) Dekker, New York, 1967.

(49) Molecular Processes on Solid Surfaces (E. Drauglis, R. D. Gretz and R. I. Jaffee, eds.), McGraw-Hill, New York, 1969.

(50) A. W. Czanderna, in Clean Surfaces (G. Goldfiner, ed.), p. 133, Dekker, New York, 1970.

(51) E. Bauer, in Techniques in Metals Research (R. F. Bunshah, ed.), Vol. 2, Part 2, p. 559, Interscience, 1969.

(52) J. J. Lander, in Progress in Solid State Chemistry (H. Reiss, ed.), Vol. 2, Pergamon, New York, 1965, p. 26.

(53) P. Estrup, in Modern Methods of Surface Analysis (P. Mark and J. D. Levine, eds.), to be published in Surface Science, Dec. 1970.

(54) C. C. Chang, in The Structure and Chemistry of Solid Surfaces (G. Somorjai, ed.), p. 77-1, Wiley, New York, 1969.

(55) J. M. Charig and D. K. Skinner, in The Structure and Chemistry of Solid Surfaces (G. Somorjai, ed.), p. 34-1, Wiley, New York, 1969.

(56) Z. G. Pinsker, Electron Diffraction, Butterworths, London, 1953.

(57) B. K. Vainshtein, Structure Analysis by Electron Diffraction, Pergamon, N. Y., 1964.

(58) R. B. Marcus, in Measurement Techniques for Thin Films (B. Schwartz and N. Schwartz, eds.), The Electrochemical Society, N. Y., 1962.

(59) E. Bauer, in Techniques of Metals Research, v. II. 2, p. 501, Wiley, New York, 1969.

(60) F. P. Bowden and D. Tabor, Symposium on Properties of Metallic Surfaces, Inst. of Metals, London, 1952, p. 145.

(61) J. Drobek, Electron Diffraction Study of Phase Transformations in Grown and Deposited SiO_2 Films, to be published.

AUTHOR INDEX

Underlined numbers in parentheses are reference numbers and indicate that an author's work is referred to although his name is not cited in the text. Underlined numbers without parentheses give the page on which the complete reference is listed.

H

T

U

SUBJECT INDEX

T